The Probabilistic Revolution

The Probabilistic Revolution
Volume 2: Ideas in the Sciences

edited by Lorenz Krüger, Gerd Gigerenzer, and Mary S. Morgan

A Bradford Book
The MIT Press
Cambridge, Massachusetts
London, England

A substantial part of the research work for this publication was supported by the Stiftung Volkswagenwerk, Hannover, Germany, and by the Zentrum für interdisziplinäre Forschung, Universität Bielefeld, Federal Republic of Germany.
Gedruckt mit Unterstützung der Universität Bielefeld, Bundesrepublik Deutschland.

Second Printing, 1989

This book was set in Times New Roman by Asco Trade Typesetting Ltd., Hong Kong, and printed and bound by Halliday Lithograph in the United States of America.

Library of Congress Cataloging-in-Publication Data

The probabilistic revolution.

"A Bradford book."
Includes bibliographies and indexes.
 Contents: v. 1. Ideas in history/edited by Lorenz Krüger, Lorraine J. Daston, and Michael Heidelberger—v. 2. Ideas in the sciences/edited by Lorenz Krüger, Gerd Gigerenzer, and Mary S. Morgan.
 1. Probabilities—History. 2. Science—History. 3. Social sciences—History.
I. Krüger, Lorenz.
QA273.A4P76 1987   509   86-17972
ISBN 0-262-11118-7 (v. 1)
ISBN 0-262-11119-5 (v. 2)

Contents of Volume 2

    Probabilistic ideas were used in experimental psychology to serve
    the traditional ideals of classical natural science: determinism and
    objectivity. In particular, inferential statistics provided the illusion
    of an objective, mechanized form of inductive inference. This
    illusion was created by the neglect of controversies and alternative
    theories and by institutionalization.

    Rapid adoption of analysis of variance techniques by psychologists
    depended on a prior shift of research practice promoted by the
    requirements of applied psychology, from individuals to group
    data. The treatment group concept was introduced in order to
    permit causal inferences from group data.

    Egon Brunswik made the uncertainty of perceptual cues the basis
    of a Darwinian view of perception, and considered the brain as an
    intuitive statistician. However, his ideas deviated so far from those
    of his contemporaries with respect to the questions "What to look
    for?" and "How to proceed?" that they were misunderstood and
    rejected.

Contents of Volume 1

List of Contributors to Volumes 1 and 2

Contributors' names are marked with asterisks as follows: one asterisk, Fellows of the Zentrum für interdisziplinäre Forschung, Universität Bielefeld, 1982–1983; two asterisks, Fellows of the Zentrum für interdisziplinäre Forschung, supported by the Stiftung Volkswagenwerk; three asterisks, Fellows of the Zentrum für interdisziplinäre Forschung, 1982–1983, supported by the Alexander von Humboldt-Stiftung.

John Beatty*
Department of Ecology and
Behavioral Biology
University of Minnesota
Minneapolis, Minnesota

Marie-Noëlle Bourguet
Département d'Histoire
Université de Reims
Reims, France

Nancy Cartwright*
Department of Philosophy
Stanford University
Stanford, California

I. Bernard Cohen*
Department of the History of Science
Harvard University
Cambridge, Massachusetts

William Coleman**
Department of the History of Science
University of Wisconsin
Madison, Wisconsin

Kurt Danziger
Department of Psychology
York University
Downsview, Ontario, Canada

Lorraine J. Daston*
Department of History
Brandeis University
Waltham, Massachusetts

Gerd Gigerenzer**
Fachgruppe Psychologie
Universität Konstanz
Constance, Federal Republic of
Germany

Ian Hacking*
Institute for History and Philosophy
of Science and Technology
University of Toronto
Toronto, Ontario, Canada

Michael Heidelberger*
Philosophisches Seminar
Universität Göttingen
Göttingen, Federal Republic of
Germany

M. J. S. Hodge
Division of History and Philosophy
of Science
The University of Leeds
Leeds, England

Robert A. Horváth*
Josef Attila Tudományegyetem
Allam-es Jogtudomńyi Karának
Statisztikai Tanszéke
Szeged, Hungary

Gérard Jorland**
Centre de Recherches Historiques
Ecole des Hautes Etudes en Sciences
Sociales
Paris, France

Andreas Kamlah**
Fachbereich für Kultur- und
Geowissenschaften
Universität Osnabrück
Osnabrück, Federal Republic of
Germany

Eberhard Knobloch
Fachbereich Mathematik
Technische Universität Berlin
Berlin (West), Germany

Lorenz Krüger*
Philosophisches Seminar
Universität Göttingen
Göttingen, Federal Republic of
Germany

Thomas S. Kuhn
Department of Linguistics and
Philosophy
Massachusetts Institute of Technology
Cambridge, Massachusetts

Bernd-Olaf Küppers
Max-Planck-Institut für
biophysikalische Chemie
Göttingen, Federal Republic of
Germany

Bernard-Pierre Lécuyer
Institut d'Histoire du Temps Présent
Paris, France

Claude Ménard
Sciences Economiques
Université de Paris I, Panthéon-
Sorbonne
Paris, France

Karl H. Metz*
Historisches Institut
Ludwig-Maximilians-Universität
Munich, Federal Republic of
Germany

Mary S. Morgan**
Department of Economics
University of York
York, England

David J. Murray**
Department of Psychology
Queen's University
Kingston, Ontario, Canada

Anthony Oberschall**
Department of Sociology
University of North Carolina
Chapel Hill, North Carolina

Jan von Plato*
Department of Philosophy
University of Helsinki
Helsinki, Finland

Theodore M. Porter*
Corcoran Department of History
University of Virginia
Charlottesville, Virginia

Ivo Schneider*
Institut für Geschichte der
Naturwissenschaften
Ludwig-Maximilians-Universität
Munich, Federal Republic of
Germany

Stephen M. Stigler
Department of Statistics
The University of Chicago
Chicago, Illinois

Zeno G. Swijtink***
Department of Philosophy
State University of New York
at Buffalo
Buffalo, New York

John R. G. Turner
Department of Genetics
University of Leeds
Leeds, England

M, Norton Wise*
Department of History
University of California
Los Angeles, California

Preface to Volumes 1 and 2

Lorenz Krüger

During the academic year 1982–1983 an international, interdisciplinary group of twenty-one scholars gathered in the Federal Republic of Germany under the auspices of the Zentrum für interdisziplinäre Forschung (ZiF) of the University of Bielefeld and the Stiftung Volkswagenwerk to study the "Probabilistic Revolution." We used that somewhat tendentious shorthand to encompass the web of changes that made probability a part of philosophy, scientific theories and practice, social policy, and daily life between circa 1800 and 1950. We worked and lived together for almost a full year, and from time to time invited other colleagues to join us for conferences and seminars on related themes. In all, the project spanned some four years, being over two years in the planning, with a preliminary conference in September 1981, and followed up by an intensive week-long reunion of the group members in August 1984 to discuss final drafts and to organize their publication.

The prehistory of the project began quite a few years earlier when, in 1974, I had the privilege of participating in a research seminar given by Thomas Kuhn on the development of statistical physics. The idea of assembling an interdisciplinary group to study the rise of probability in the sciences occurred to me as a result of a conference of the International Union of the History and Philosophy of Science, held at Pisa in 1978, to which I was invited by Jaakko Hintikka. My understanding of the depth and the breadth of the problem grew in conversations with colleagues whom I met there, among them Ian Hacking and Nancy Cartwright. In 1980, when the project had been approved by the ZiF, Michael Heidelberger joined it. His active part in conceiving and guiding it throughout, as well as the assistance of Rosemarie Rheinwald during 1982–1983 were essential for its execution.

These two volumes represent the fruit of the year the research group spent together, not only of the individual essays conceived then, but also of the innumerable—indeed, well nigh constant—discussions, over the seminar table and in the laundry room, gathered in plenary session and in twos and threes. Thus, the present work represents more than an anthology of far-flung authors and topics, or the proceedings of a conference. Rather, it records a sort of scholarly experiment in loose-knit but sustained collaboration over disciplinary and national boundaries.

We are grateful to the institutions, and especially to the individuals who are the human face of such institutions, that made this somewhat utopian project both possible and pleasant. The Stiftung Volkswagenwerk generously supported a subgroup of seven members particularly concerned with the impact of the Probabilistic Revolution on man and society. The Zentrum für interdisziplinäre Forschung provided occasion, funds, research facilities, a friendly, helpful staff, and a sylvan setting in the Teutoburger Wald. Families of foreign group members abandoned home and hearth for the year and applied themselves to learning German; families of native group members and their colleagues were hospitable beyond reckoning.

We thank all most heartily.

The Probabilistic Revolution

Introduction to Volume 2

John Beatty, Nancy Cartwright, William Coleman, Gerd Gigerenzer, and Mary S. Morgan

The probabilistic laws of classical statistical mechanics were supposed to be a function of human ignorance; those of quantum mechanics, to reflect the structure of nature. It is tempting to take this celebrated change as the typical expression of the probabilistic revolution. Yet this epoch in the history of physics does not provide an adequate model for the actual transformation in scientific thought; rather it is a single, comparatively recent transition, among a variety of innovations in different sciences, each of which constituted a probabilistic revolution where it occurred.

The emphasis on the shift from statistical to quantum mechanics seems essential to one common view of science but this is not the view underlying these chapters. The conventional view downplays measurement, application, data gathering, and experiment, and focuses instead on theory and on law; it supposes that all science knows of reality is what can be projected from its laws and theories. But nature is recorded equally in the methods of science. Methods, like laws, are not a free choice of the human intellect; and we learn about reality not only from the laws we transcribe but also from the constraints it imposes on our methods. Here nature has shown itself to be recalcitrantly probabilistic. The lessons may be harder to unify into a tidy picture; nevertheless, we have learned a significant and stubborn fact about reality, that probabilities must be used to measure it properly or, in many disciplines, to reason about it. This volume reveals how, over the last 100 years or so, probability has become necessary in various ways to the working business of one science after another. These chapters share in common the assumption that a probabilistic revolution occurs whenever probability becomes indispensable to a science.

The companion volume of this study (*The Probabilistic Revolution, volume 1: Ideas in History*) concentrates on general conceptual, philosophical, and historical issues. In this volume, we focus on a variety of specific modern (until circa 1950) contexts in which probability has proved essential. Reading through the various chapters, we see that probability and statistics work to solve four distinct problems in science: (1) measurement, (2) description of data, (3) inference, and (4) theory construction.

1 Measurement

Nature does not speak with one voice. In repeated measurements of even a single quantity, the observer faces a multitude of differing values. Science seeks the true value, and, as we see, statistical and probabilistic methods provide a way of mechanizing and "objectifying" this search. Variations were initially construed as errors grouped around a single true value. Later such variations were perceived as an inherent feature of the true material rather than as observational error. But under either interpretation, the need for statistical methods seems ineliminable. The case of psychology provides an example of the role of statistical methods in measurement and how the change in perceptions affected scientific practice.

2   Data Description

Even if it is no surprise that the values of quantities are distributed in nature, the fact that there are certain complicated regularities about their distribution may still be unexpected. The companion volume discusses various historical attempts to describe these stabilities and to understand what they can mean; and it is important to note here that statistical science itself was born out of the obsession with such regularities and their explanation. In the social and biological sciences, descriptive statistics and probability distributions were used to describe data derived from both experimental and nonexperimental sources. Yet as we see in this volume, in at least one modern science, physiology, statistical description is closely tied to the development of experimental procedures: without the latter the existence of stable frequencies would not have been recognized. Indeed the failure of early statistical methods to reveal the required information might well have encouraged the development of controlled observation. In many disciplines (including physiology), there was a good deal of resistance to the assumption that the stability of the statistics arose from the fact that fundamental laws of nature were themselves statistical.

3   Inference

The prestigious position of classical mechanics during the nineteenth century assured widespread acceptance of the ideal of certainty in inference: the true representation of reality would emerge from the data. This form of determinism dominated scientific research, and the occasional efforts in the nineteenth century to introduce statistical or probabilistic methods of inference were countered by the prevailing deterministic ethos.

The changing perception of the material under study—that reality presented variability rather than unambiguous true values—required more complex methods of inference. From about 1900, the informal nonnumerical procedures of inference gave way to formal statistical procedures based on probability theory. These methods were exemplified by new schools of experimental design, and experimental investigation has increasingly demanded that the implementation of research procedures serve the needs of these new methods of inference. This occurred in biology and in the social sciences, but never in physics. Further, we find that in psychology at least the introduction and acceptance of these new methods of inference have led to a mechanical evaluation of applied work and thus influenced theory choice. The use of statistical methods and inference based on probability ideas also found a central role in sciences where experiment is the exception rather than the rule. In economics, for instance, where data are rarely the result of experimental procedures, statistical techniques were introduced specifically as a substitute for the conventional experimental tasks of control and manipulation of influencing factors. Probability theory appeared to provide an answer to the problems of drawing inferences from data subject to a variety of uncontrolled influences and the need to find rules for theory evaluation in these circumstances.

4 Theory Construction

Probability has not been a basic building block of scientific theorizing except in physics and in evolutionary biology. In theory construction, the locus of the probabilistic revolution might seem to lie in the distinction between epistemic and ontic interpretations of probability. But this is not the only issue, nor even the main one. We can see from the chapters in this volume that not all scientists are concerned with representational pictures of reality—some are concerned, for instance, first and foremost with empirical adequacy and conceptual integrity. For these reasons, scientists have found it necessary to construct theories on probabilistic foundations. In physics, for example, probability has played a notorious role in both classical statistical mechanics and in quantum mechanics. It is commonly said that quantum physics brought about a revolutionary change in two respects: the world picture, previously deterministic, became indeterministic, and epistemic probabilities became ontological. Yet the ergodic tradition in statistical mechanics supposed a physical basis for its statistical laws, and equally, the fact that probabilities may be supposed ineliminable in a quantum picture does not guarantee that they are to be a part of nature itself rather than a consequence of the scientists' interactions with nature or of the structure of the theory. Evolutionary biology was reconstructed in the 1930s on probabilistic foundations; since then no evolutionist has disputed the stochastic formulation of the theory or pursued the Laplacian dream of ultimately eliminating the probabilities. This does not mean that they all support a fundamentally probabilistic representation of reality—rather, empirical adequacy and conceptual coherence dictate that the theory be stochastic.

Although randomness, probability, and uncertainty have played their part in the treatment of specific theoretical issues in other sciences, theoretical structures have largely remained true to the ideal of Laplacian determinism. Economists, for example, construct their theories as if they were mechanical systems; yet in application these theories are given a stochastic representation for the purposes of empirical adequacy, and on the whole economists are content that there should be a large amount of uncertainty in their results. Psychology in its turn has, until recently, largely suppressed the notion of uncertainty except as the measure of scientific ignorance.

In concluding, it is important to qualify our general thesis in two respects. First, to say that modern scientists have increasingly relied upon probability and statistics to solve the four sorts of problems discussed here is not to say that statistical reasoning has gone hand in hand with an appreciation of probability notions. Modern scientists have, frequently enough, employed statistical methods of considerable complexity without bothering themselves with probabilistic justifications of those methods. Second, to say that modern scientists have increasingly relied upon probability and statistics is not to say that there is any unexceptionable trend in their assimilation. If there have been probabilistic and statistical revolutions, there have been counterrevolutions as well.

The chapters in this volume are arranged in parts according to scientific disciplines. Each science has experienced a probabilistic revolution at one or more levels of its scientific practice and, for the benefit of the interdisciplinary reader, each part (apart from sociology and physiology) is introduced by a brief overview of the role of probability in that science. The structure of this volume reflects the twentieth-century fragmentation of knowledge into disciplines, in contrast to the more fluid nineteenth-century situation treated in the companion volume. Nonetheless, the modern developments analyzed here also reveal patterns and themes that span more than one discipline. Therefore, we hope that this interdisciplinary research will encourage interdisciplinary reading.

I PSYCHOLOGY

The Probabilistic Revolution in Psychology—an Overview

Gerd Gigerenzer

Psychology won its independence from philosophy by introducing measurement, and with measurement, variability entered. There were two major programs, which originated in the second half of the nineteenth century in Germany and England, and they differed sharply in the interpretation of what this variability signifies. In the Fechner-Wundt tradition, or "experimental psychology," both interindividual variability and intraindividual variability came to be considered as *error* around a *true value* that characterizes a real natural process. In the Galton-Pearson tradition, however, interindividual variability was understood as a *sign of progress* rather than as error. Charles Darwin's theory of variability as the motor of evolution was made the cornerstone of the new British psychology, later called "differential" or "correlational" psychology. Interindividual differences were investigated via concepts like the normal distribution, correlations, and other statistical techniques developed by Karl Pearson. Experimental psychology, on the contrary, was modeled on classical natural sciences like physics; the interpretation of variability was analogous to observational errors in astronomy.

1 Measurement and Data Description

In the experimental program, the true values to be measured (e.g., for sensory thresholds) were defined from the very start, by Fechner, in terms of (normal) distributions of intraindividual repeated measures. In the measurement theories of L. L. Thurstone and S. S. Stevens, a shift was made to distributions of *inter*individual repeated measurements (Thurstone), or, simply, to *inter*individual averages, (Stevens). Since both sources of variability were considered as *error*, measurement theories in this tradition failed to distinguish between them, both conceptually and mathematically.

In the Galton-Pearson program, the primary focus was on data description rather than on measurement theory. Pearson's statistical positivism forebade going beyond phenomena to seek hidden causes or true values, as the early experimenters had. However, psychologists like Spearman soon distorted this view by their reification of hidden causes underlying correlations, such as "general intelligence." To summarize, on the level of measurement and data description, probabilistic concepts like the normal distribution were used from the outset to tame variability, although different interpretations were (and still are) given to the nature of the variability.

2 Inference

In the mid-1930s, R. A. Fisher's analysis of variance and Student's *t*-test were applied for the first time, and after World War II, a mixture of Fisher's and J. Neyman and E. S. Pearson's theories of inference became a methodological imperative. Inferential statistics not only replaced the experimenter's judgment (and the critical ratio, an earlier index), but this mechanization of inference also promoted inductive inference to a methodological issue of the first importance. The first great success of statistics in experimental psychology came with the introduction of inference techniques.

3 Theory Construction

Fechner's indeterminism stood at the beginning of experimental psychology, postulating that in both the physical and psychical world there emerge novel elements that have *no* antecedent causes, although cause-effect relationships per se were considered deterministic. Possibly because Fechner's indeterminism and panpsychism proved embarrassing to later psychologists, Wilhelm Wundt was awarded the honor of founding experimental psychology, in 1879. Wundt's disciple, Titchener, and his "structuralist" school set out to fulfill the associationists' promise, to measure the elements of mind (e.g., just noticeable differences in brightness, color, etc.—Titchener counted circa. 44,000 of them), and to prove that mental structures are nothing but sums of these elements. No probabilistic ideas were needed in such a theory of the mind as a puzzle.

Structuralism died with Titchener in 1927, the victim of new programs like Gestalt theory and behaviorism. Although both were very strongly opposed to one another on the questions of whether theory should deal with mental events or with behavior, with elements or with wholes, they shared with structuralism a Laplacian, deterministic account of psychological theory.

From the 1950s and '60s on, behaviorism lost its control over theory and new mentalist programs emerged. This "cognitive revolution" coincided with an avalanche of probabilistic formulations of psychological theory and with a new cognitive metaphor of man as an "intuitive statistician." Theories of perception, memory, decision making, and learning were written in probabilistic language. Studies of the "intuitive statistician" were influenced by different interpretations of the probability concept; for instance, the interpretation implicit in Piaget and Inhelder's experiments on the idea of chance in children is the classical Laplacian definition of probability as the ratio of favorable cases to the total number of equally likely cases. Studies on probability learning can be related to von Mises's frequency interpretation; children or adults had to learn relative frequencies empirically during the experiment, instead of being informed about the equally likely cases before the experiment. Finally, inductive thinking and decision making has been studied by using Bayes's theorem, either as a descriptive or as a normative model.

This coincidence of the cognitive revolution and probabilistic ideas at the level of theory must be qualified in two respects. First, probabilistic formulations were neither implied by nor restricted to cognitive approaches. For instance, as early as 1950, Estes suggested a probabilistic formulation of behaviorist learning theory. Second, probabilistic ideas were put forward before the cognitive revolution, although such attempts were not accepted at the time. Most of these applications are too recent to be included into this historical treatment of the subject, and the first three in the following series of four chapters are addressed to the time interval before that recent avalanche of probabilistic formulations.

The series opens with a chapter on experimental psychology by Gerd Gigerenzer, concerning the symbiosis between probabilistic ideas and the ideals of determinism and objectivity. Kurt Danziger asks a closely related question in the second chapter: Why did Fisher's analysis of variance become a hallmark of psychological experimentation? Like Danziger, Gigerenzer deals in the following chapter with the contrast between the Fechner-Wundt and the Galton-Pearson programs, in terms of ideas of causality, experiment, and the role of practical application. The topic is Egon Brunswik's Darwinian approach to perception, postulating uncertain perceptual cues and, by implication, experiments using uncertain environments rather than the principles of isolation and control. The concluding chapter by David J. Murray, covers a broader time interval than the preceding ones. He identifies sources of dissatisfaction with deterministic theories and follows them from the seventeenth century up to the present day.

1 Probabilistic Thinking and the Fight against Subjectivity

Gerd Gigerenzer

The role of probabilistic thinking in experimental psychology between the late 1920s and circa 1950 is discussed. My thesis is that probabilistic thinking was used in the service of traditional ideals of classical natural science, in particular those of determinism *and* objectivity. *(1)* Determinism: *At the level of theory construction, probabilistic models were welcomed with the caveat that the uncertainty involved reflected the experimenter's ignorance rather than any uncertainty in the subject matter itself: Man was viewed as subject to deterministic laws. (2a)* Objectivity *as independence from the experimenter: At the level of inference, probabilistic thinking was used as a means toward achieving objectivity in the classical sense of considering knowledge as independent of the experimenter. "Inferential statistics" provided the illusion of a* mechanization *of that part of the knowledge process known as inductive inference. This illusion was created by (i) silence concerning extant controversies, (ii) the neglect of alternative theories of inference, (iii) anonymous presentation, and (iv) institutionalization. Some consequences for the direction that experimental research took are discussed. (2b)* Objectivity *as independence from the individual: Besides experimenters, psychology must deal with a second class of subjects, the individuals investigated. Therefore, there seemed to exist a second source of "subjectivity" that had to be eliminated. At the level of measurement, the elimination of the individual by identifying individual differences with "error" around a single "true" value was a second service that probabilistic thinking rendered to the ideal of objectivity.*

From the very beginning, "science" has meant *certain* knowledge, to be found in mathematics and in the natural sciences. The alternative, in Aristotle's view, was mere *opinion*, to be found in ethics and politics. The certain knowledge of physics rather than philosophy became the ideal for an experimental psychology that emerged in the late nineteenth century and also for behavioristic and reductionistic theories in the first half of our century. In Clark Hull's words, experimental psychology was "a fullblown natural science." [1]

Given this early fascination with classical physics, I would like to pose the following question: Why did experimental psychology not follow the lead of quantum theory? Could it be that psychologists simply did not understand quantum physics? I cannot agree with this for two reasons. First, one need not understand something fully to imitate it; it is sufficient to imitate one's conception of that something. Clark Hull is not the only psychologist who could be mentioned in this instance. Second, quantum theory was indeed discussed in the 1940s and '50s within psychology. However, it was unequivocally rejected as a new ideal of science. [2] Once again we must pose the question: Why did psychology not follow the new physics? My answer is that quantum theory seemed to violate two ideals connected with the struggle for *certain* knowledge: *determinism* and *objectivity*. The ideal of *determinism* was given up for an indeterminism in the sense of a microscopic chaos that results in macroscopic order. The second ideal, *objectivity*, seemed to be violated by physicists like Werner Heisenberg, who rejected the assumption that the scientists (and their measurement instruments) are mere "tools" for passively registering reality. Instead, he argued, both *interact* to produce the data.[3] More generally put, knowledge is not about reality; it is about reality *and* the knower.

The idea that probability is about *the subject matter itself* and that knowledge is *not certain* in the sense of being independent from a knower—that seemed too much even for psychologists otherwise willing to follow physics anywhere. However, while psychologists roundly rejected probability at the level of theory construction, they enthusiastically welcomed it at the level of inference. For their part, physicists of neither classical nor quantum persuasion ever accepted probability at the inference level. Why did psychology thus part company with both the old and the new physics, although physics remained its ideal of a science?

The refusal to follow contemporary physics seems to open a door for understanding more about the ideals governing the historical development of psychology. The time interval I want to consider is the period following the advent of quantum theory in the 1920s, roughly between 1927, when Louis Leon Thurstone had the first great success with his probabilistic "law," and the years immediately following World War II, when R. A. Fisher's analysis of variance became the dominant research instrument of experimental psychology. The symbiosis of the three facets, *determinism, objectivity,* and *probabilistic thinking,* during that time suggests the following theses: Probabilistic thinking was used in the service of traditional ideals of "certain knowledge," of determinism and objectivity. In particular:

1. *Determinism*: With very few exceptions, probabilistic thinking could not be considered as a source of models for how man functions. That is, probability could not enter at the level of theory construction.

2. *Objectivity as independence from the experimenter*: Probabilistic thinking was used as a means toward objectivity in the classical sense of separating the experimenter from his knowledge. Such was the role of inferential statistics as a mechanization of the experimenter's inference from data to hypothesis. The connection I shall draw between the kind of inferential statistics established in psychology and objectivity is based on the following observations: (1) There was *a single dominant theory* of inferential statistics that (2) was taught *anonymously* (i.e., without indicating its multiple and sometimes contradictory sources) as "truth" per se. (3) Problems stemming in part from the fact that the theory was spliced together from these different sources went *unacknowledged,* (4) alternative theories were *neglected,* and (5) this dominant hybrid theory was *institutionalized* by editors of journals and internalized by authors as the one "true" path to experimental knowledge (and therefore toward publication).

3. *Objectivity as independence from the individual*: Probabilistic thinking was used as a means toward objectivity in a second sense that is peculiar to psychology. Unlike the natural sciences, psychology has to deal with *two* classes of subjects, the *experimenters* and the *individuals* investigated. Therefore, there seemed to exist two sources of "subjectivity" that had to be eliminated. The elimination of the individ-

ual by matching individual differences with *error* around a single "true" value was another use of probabilistic thinking as a means toward this second kind of "objectivity." This use concerns the level of measurement and data description.

I consider these three functions of probabilistic thinking as important, but not as exhaustive. In the following three sections I shall discuss these functions for the time interval circa 1920–1950 with reference to American psychology.[4] Each section follows the same format: first, I state the ideal; then I provide illustrations for how probabilistic thinking has been used in the service of that ideal.

1   Determinism Reigns

1.1   The Ideal

Laplace could have been a representative member of the American Psychological Association around the middle of this century. To quote him from 1814: "Given ... an intelligence which could comprehend all the forces of which nature is animated and the respective situation of the beings who compose it—an intelligence sufficiently vast to submit these data to analysis ... nothing could be uncertain and the future, as the past, would be present to its eyes." [5]

Edwin Boring summarized the situation for psychology as late as 1963 in only two words: "Determinism reigns." [6] How could one be so confident that man is completely controllable and predictable via deterministic causal laws despite the contradictory experimental experience of unwanted variability, unpredictability, and lack of control? Somehow one could, and this confidence may be best illustrated by David Krech's determinist's confession of faith of 1955: "I have always made it a cardinal principle to live beyond my income. And although I have yet to find a one-to-one correlation in psychology ... I am always ready to make another promissory note and promise that if you bear with us we *will* find uniform laws.... And if I can't pay off on my first promissory note I will come seeking refinancing.... I have faith that despite our repeated and inglorious failures we will someday come to a theory which is able to give a consistent and complete description of reality. But in the meantime, I repeat, you must bear with us." [7]

Inglorious failures and deficit financing, however, went on. Learning from experience seems not to take place with regard to basic ideals of science, be they determinstic or otherwise. The crucial point seems to lie in *styles of explanation* rather than in experience. Inglorious failures can always be attributed either to external causes, like the scientist's ignorance, or to internal causes, like the probabilistic nature of the subject matter, that is, to the "very nature" of man himself. The explanatory style adopted by the experimental community was almost universally that of external attribution. The ideals protected by this style formed the core of a deterministic view similar to that introduced into physiology by Claude Bernard in the midnineteenth century: *causal laws*, like Newton's law of falling bodies, instead of principles like multiple mediation or substitutability of means to an end;[8] *prediction* of the behavior of man similar to that of physical objects instead of, for instance, a Darwinian interest in nonrepeating and unpredictable patterns of evolution;[9] *control* and *isolation* of man's behavior instead of, for instance, an interest in pure description and nonintervention; and finally, *nomothetic laws*, that is, laws that are designed to hold for every individual at all times and places rather than laws of the individual. These ideas about what psychology should be are prototypical, but it does not follow that every experimentalist at the time gave them the same weight. Let us consider some examples to illustrate the breadth of the deterministic ideal.

Concerning causal laws, initially there was Wilhelm Wundt's idea that human responses were the manifestations of an underlying "psychic causality." This causality was to be understood analogously to the responses of physical objects seen as the manifestations of an underlying physical causality.[10] Outside of the experimental tradition, Sigmund Freud's man was determined in every bit of his behavior—right up to slips of the tongue—by a causal mechanism. For this reason Freud has been called the "Sherlock Holmes of the unconscious."[11] In the 1930s, Kurt Lewin argued that there is no place for statistics in a strictly nomothetic, or, as he called it, systematic discipline. He claimed that a single observation would be sufficient to ascertain a deterministic law once and forever. Lewin called this the "pure case" and referred to Galileo's study of falling bodies as a prototype.[12] In the 1940s, Clark Hull argued that uniform nomothetic laws of behavior exist with the same degree of certainty as the law of falling bodies. Present failures to detect those laws were attributed to measurement difficulties rather than to the probabilistic nature of the laws.[13] In the 1950s, Ernest Hilgard, Leo Postman, David Krech, and other leading experimentalists made common cause against a probabilistic conception of man, against the replacement of causality by "vicarious functioning" or "cue substitutability," against a descriptive analysis of behavior of single individuals in their natural environments, and for isolation, control, and nomothetic laws.[14]

An interesting case where a deterministic starting point turned into an obstacle for a subsequent theoretical development is B. F. Skinner's tardy concept of the operant.[15] In 1931, the 27-year-old graduate student set forth his plan for a behavioral science studying the reflex. On the one hand, his program was to "reduce" explanation to description and to "substitute" function for causation; on the other, he continued to speak of the reflex as a *necessary* relation between stimulus and response: "The reflex is important in the description of behavior because it is by definition a statement of the *necessity* of this relationship."[16] In the same year Skinner began having rats press levers in his experiments. The question was, how to account for the rat's first level press. In 1935 he claimed that the initial lever press belongs to a class of unconditioned reflexes he labeled "investigatory" or "food seeking" reflexes.[17] In 1937, he tried to solve the problem by distinguishing between operant and respondent reflexes and had to admit that operant responses such as lever presses are *not* elicited.[18] However, it took him over 15 more years until in 1953 he redefined the operant as nonreflexive, as a response that is initially "randomly emitted."[19] Why did it take Skinner so long to give up the principle of the reflex and make adequate revisions? The answer seems to be that he clung to the deterministic principle of the reflex until he finally found himself forced to retreat from a deterministic to a probabilistic description. Even then, Skinner maintained that the probability belonged to the scientist, not to the subject matter: the operant is "random" only because we are presently ignorant of its causes.[20] Skinner finally returned to Laplace, celebrating determinism in probabilistic language.

1.2   Probabilistic Thinking in the Service of the Deterministic Ideal

How could such widespread determinism coexist with probabilistic thinking during the first half of this century? My answer is that probabilistic thinking was enlisted in the service of determinism. I shall give two typical examples, Louis Leon Thurstone's *law of comparative judgment* and R. A. Fisher's *analysis of variance*. Together, they can be considered as the most "successful" applications of probabilistic ideas in the first half of this century.

In 1900, Leon Solomons provided a new explanation of Weber's law. He argued that the threshold in Weber's law is caused by the variability (irritability) of brain activity.[21] The size of the threshold measures simply the range of this variability; the smaller the variability, the smaller the threshold. In his analysis of the nature of this irritability, he presented a completely Laplacian account, suggesting that should we find all the individual factors causing this variability, we would then have complete control over the variability of brain activity. In 1927 the same idea reappeared (without reference) in Thurstone's writings—as Thurstone himself admitted, he did not read widely in psychological literature and was therefore prone to "independent" discoveries.[22] However, it was Thurstone who gave an explicit mathematical formulation of the connection between variability and threshold (more generally, between variability and the subjective difference between two stimuli). This probabilistic model is known as the "law of comparative judgment." The aim of the model was, in short, to determine the subjective scale values of physical stimuli (e.g., perceived brightness) on the basis of paired comparisons. As with Solomons, no attempt was made to identify the probabilistic concepts with something probabilistic in the way man functions. Thurstone explicitly refrained from giving any interpretation of his "discriminal dispersions" (i.e., the variability of responses to the same stimulus), or "discriminal distributions" (the distribution of these responses) in psychological terms: "This analysis has nothing really to do with any psychological system."[23] He used "discriminal dispersion" synonymously with "error." Only the mean values—that is, neither the distributions nor the dispersions—were considered as psychologically meaningful entities, i.e., as the subjective scale values. Thurstone's probabilistic mathematics was only for the *measurement* of those values, not for theory construction.

In the 1940s, Thurstone's model came to be used as a means for the quantification of Hull's "habit strength." (In Hull's theory, the counterpart to Thurstone's discriminal process was the "momentary effective reaction potential."[24]) Again, the probabilistic model was used for quantification within a deterministic framework, not as a model for probabilistic processes in the subject matter itself:

According to Hull the variability and uncertainty in this measure arises "because we always lack absolutely exact knowledge concerning *conditions*."[25] This interpretation is exactly analogous to Laplace's theory of observational errors in astronomy.

The second example is analysis of variance, which entered psychology in the mid-1930s and, after World War II, became the most used research "tool" in experimental psychology. This statistical model has been used in psychology as if it were the battlehorse of Claude Bernard's determinism: (1) Analysis of variance, it was assumed, can answer the question whether a *causal* relationship exists between variables. However, influenced by David Hume, psychologists of this century were somewhat reluctant to speak of causes in print. Instead one spoke of "effects" only, of the effect of X on Y, or of "conditions," an expression well chosen for its ambiguity without damaging the idea of necessity. In the terminology of analysis of variance, causes and effects were replaced by "independent" and "dependent" variables. (2) Analysis of variance was based on the principles of *isolation* (of a few "independent" variables and the "dependent" variable) and *control* (of other relevant variables), and (3) it was used in accordance with the *nomothetic* ideal, that is, treating individual differences simply as the "error term." What was considered as psychologically important were mean values only; distributions and dispersions were viewed as reflecting some technical "assumptions" concerning the applicability of the instrument. Psychological theories were not about those "assumptions."

To summarize, probabilistic thinking was not tolerated as a model for how man functions; it was tolerated and used in the spirit of Laplace, as an expression of the experimenter's ignorance. Probabilistic thinking seldom threatened psychological determinism. If it did, as with Egon Brunswik, who postulated irreduceable uncertainty in the subject matter itself, such attempts were suppressed or forgotten with the understanding that this was not the psychologist's way toward an acceptable science.[26]

2 Mechanization of Knowledge

2.1 The Ideal

The ideal of the classical natural sciences was to consider knowledge as independent of the scientist and his measuring instruments.[27] "Objective knowledge" in this sense implies that both the scientist and his instrument function as mere *tools* that mirror reality. Consequently both could be "eliminated" for epistemological purposes. Opposed to this ideal was the view that knowledge emerges from an *interaction* between the scientist and reality, as exemplified in Niels Bohr's famous dictum that man is an actor rather than an observer in the spectacle of life. In psychology there was—at least since Max Wertheimer's famous demonstration of the Phi phenomenon in 1912[28]—experimental evidence available for such an interactive view: Even elementary perceptions like perceived motions did not mirror the reality of physical measurements, but seemed to emerge from an interaction of an active perceptual apparatus with that reality. Such results, however, were not used to threaten the ideal of objective knowledge.

For instance, much of the work concerning measurement can be understood on the assumption that psychological entities like perceived loudness, intelligence, or neuroticism can be considered independent of the particular measurement instrument. I take as an example Stanley S. Stevens, who has been called "without question, the strongest voice in psychophysics since G. T. Fechner."[29] With the assumption of independence in mind, Stevens was surprised to find that two standard tasks given to his subjects consistently resulted in values that were *not* linearly related. The two tasks were magnitude estimation (e.g., judging the ratio of the loudness of two tones) and category rating (judging the interval). This non-linearity was quite disconcerting to him, since he wanted to measure *the* perceived loudness, brightness, etc., and considered the measurement instrument, the task, of no theoretical relevance. For, if one assumes that there exist subjective loudnesses independent of the particular task, both measures should be linearly related. If not, only one could be "correct." Such was Stevens's answer. He found himself forced to favor one of the tasks (magnitude estimation) and to ban the other. The alternative to such an understanding of "objectivity" would have been to consider the tasks as theoretically relevant rather than as mere so-called data generation techniques. This would change both the questions posed and the answers chosen. Subjective values, subjective strategies, etc., would be considered as processes that were elicited by or dependent upon certain tasks rather than independent of them.[30]

Perhaps the temptation to consider that which can be quantified as real in the sense of independent of the particular measurement instrument is part of the initial enthusiasm needed to set a science in motion. In Stevens's case there is the astonishing line of cleavage between his theoretical writings and his experimental work. He was one of those psychologists who promoted the use of words like "construct," "operation," and "criterion." In his experimental work, however, he returned to the world of "real" loudnesses. Stevens is an illustrative example of the general ideal that subjective processes can be considered independently of the methods used for measurement. And the example shows that such a metaphysical assumption has an influence on both the questions asked and the answers looked for. It is *not*, as is often assumed, of mere philosophical interest.

In the following I shall deal exclusively with probabilistic thinking introduced to establish the independence of knowledge from the knower at a certain point, inductive inference, i.e., the inference from the given data to the validity of the hypothesis in question. Here, the ideal is to get rid of the experimenter's subjective judgment, and therefore demands a complete mechanization of the inference process.

One of the first attempts at a mechanical form of inference stems from Laplace. Basing his attempt on a set of assumptions—exemplified by random sampling from an urn—he derived his "rule of succession" as a formula of inference. Although those assumptions virtually never hold in actual scientific work, the dream of an automatic form of inductive inference has reappeared. Rudolf Carnap's inductive logic is an example of a modern attempt that seems to have been finally given up.

2.2   Probabilistic Thinking in the Service of the Ideal

My thesis is that the enthusiastic reception of probabilistic thinking at the level of inference is connected with the traditional ideal of objectivity: to produce the illusion of a mechanized inference process. I shall distinguish four components that created that illusion: (1) the neglect of extant controversies, (2) the neglect of alternative theories of statistical inference other than those ideas presented as "inferential statistics" per se, (3) the anonymous presentation of the ideas, and (4) the institutionalization of the ideas. First, however, it is of interest to note what experimental psychologists did before significance testing was introduced into psychology. Before circa 1935, when the t-test and analysis of variance were first applied, there was not much discussion of inference.[31] For instance, the typical article in the *Journal of Experimental Psychology* around 1925 used extensive descriptive statistics for individual data. No standard method of drawing inferences from the data existed; indeed, some authors drew no inferences at all, while others used critical ratios or simple judgment. Typically, inductive inference was based on a kind of consensual bargaining between the author and the reader: on nonstatistical judgment, on the eyeballing of curves, and expressed by statement like "It will be noticed that they [the curves] all have the same general form."[32] Most important, the issue of inductive inference was incidental rather than essential, and the processes of data description and inference were often not even distinguished. Only when the dream of a mechanized inference process seemed to have become a reality did the inference process assume a dominant role in experimental research. Let us consider the four components that finally created that illusion one by one.

Neglect of Extant Controversies   What became known in experimental psychology as "inferential statistics" was a mixture of the positions of R. A. Fisher on the one hand and of Jerzy Neyman and Egon S. Pearson on the other. What was not widely known was that a heated controversy between the two existed, which even became acrimoniously personal. Fisher seems to have contributed most to this bitter tone. He has been described by friends as relentlessly arrogant and polemical, whereas the Neyman-Pearson camp prefered a modest and conciliatory tone. Fisher's death in 1962 changed the tone of the debate, but it remained far from being resolved. What were the controversial issues?

First, Fisher never granted the necessity of specifying alternatives to the "null hypothesis" in hypothesis testing, whereas Neyman and Pearson argued for the symmetry of the hypotheses tested, allowing for the determination of Type I *and* Type II errors and therefore of the power of the test.[33] Null hypothesis testing was introduced by Fisher[34] and rejected by Neyman and Pearson. Second, the interpretation of what a significance level means was quite different in the two cases. For Neyman and Pearson, it was understood as the proportion of Type I errors in a long series of similar tests. This revealed that the Neyman-Pearson theory was relevant mainly to applications, such as quality control in manufacturing, where repetitive sampling was a reality. Fisher's interpretation was ambivalent, and oriented to individual tests rather than to long sequences; he even allowed the specification of the rejection region *after* the data had been analyzed.[35] Third, Fisher as well as

Neyman and Pearson initially hoped to fulfill the Laplacian promise of statistical induction. Neyman and Pearson, however, gave up the pretense of creating a theory of inference in order to preserve the integrity of their theory. This restricted their theory to "direct" probability statements, to probabilities of observations given some hypothesis, which was clearly *not* what psychologists wanted. In addition, they emphasized that the meaning of statistical inference is a "decision" based on balancing of the weighted costs of the two types of errors. Fisher, on the contrary, continued to present his position as if it were *the* perfectly rigorous method of scientific inference.

My point here is that these controversial issues and others were more or less completely neglected in psychology although the "inferential statistics" that spread through psychology was based on the ideas of both fighting camps. How could this work? Michael Acree used a Freudian analogy to describe the assimilation of the two. "The Neyman-Pearson theory serves very much as the superego of psychological research. It is the rationale given, if still anonymously, by the most sophisticated and respected textbooks in the field, which teach the doctrine of random sampling, specifying significance levels and sample sizes in advance to achieve the desired power, and avoiding probability statements about particular intervals or outcomes. But it is the Fisherian ego which gets things done in the real world of the laboratory, and then is left with vague anxiety and guilt about having violated the rules." [36]

How could it happen that the controversial issues of the two camps were presented as a single, apparently noncontroversial theory of inferential statistics? Let history speak. The first practical application of analysis of variance was in agricultural research in 1923, when Fisher and Mackenzie investigated the effects of manure on the rotation of potato crops. [37] Before World War II, small-sample theory like Fisher's work was usually regarded as a specialized topic. The focus was on the large-sample statistics of Karl Pearson, who together with Francis Galton founded the "correlational discipline" within psychology. For instance, in Yule's famous *An Introduction to the Theory of Statistics*, first published in 1911, the *t*-test was not discussed until the 11th edition of 1937. Under the influence of the correlational discipline, the first significance tests were categorized in psychological texts (around 1936) under the heading of "reliability." [38] The main difficulty for psychologists in understanding Fisher's own writings seemed to be his agricultural language and his mathematics. In 1943, Garrett and Zubin complained that psychologists would have difficulty with a language that deals with "soil fertility, weights of pigs, effectiveness of manurial treatments and the like." [39] Already in 1940, Lindquist had noted as a further obstacle that the mathematical argument in Fisher's two books was too incomplete to be understood by a psychologist without considerable mathematical training.

Therefore proselytizers were needed to translate Fisher's message into the language that psychologists understood. The first Fisherian prophet came in 1937 with Snedecors's *Statistical Methods*. [40] Here is the reaction of a contemporary experimentalist: "Thank God for Snedecor, *now* we can understand Fisher." [41] The situation for the proponents of statistical inference soon became complicated since—in particular after Word War II—the position of Neyman and Pearson also became known. There were a few texts that presented both theories,[42] not necessarily emphasizing their antagonism. The great mass, however, simply neglected the differences and the controversial issues and patched up the two antagonists into a single truth. This hybrid was presented as the desideratum of psychologists, as a mechanization of inference from data to hypotheses.

A typical example is J. P. Guilford, who in 1942 published the first edition of his *Fundamental Statistics in Psychology and Education*. This book probably became the most widely read and, in this sense, the most successful and influential statistics text. Here the theory of Neyman and Pearson was turned on its head, as when Guilford presented chapter headings like the "Direct Determination of the Probable Validity of a Null Hypothesis" or when he, through all successive editions, asserted, "We cannot prove the truth of the null-hypothesis; we can only demonstrate its improbability."[43] He shifted along with Fisher from the interpretation of the normal curve as a distribution of observable sample mean differences to true population mean differences. He even spoke a Bayesian language of "odds against the null hypothesis"[44] and declared the concept of power, still in the third edition in 1956, as "too complicated to discuss."[45]

The confusions in the statistical texts presented to the poor frustrated students were caused in part by the impossible attempt to make a single anonymous truth out of controversial issues, and in part by the attempt to sell this hybrid as the sine qua non of scientific inference.[46] In the years following World War II, there was more or less no mention of the controversial issues in the flood of texts on "statistics" that appeared. If the names of Fisher, Neyman, and Pearson were mentioned at all, as in the exceptional case of Hays's famous text, even then the controversial issues were not discussed, and the story suggested cumulative progress from Fisher to Neyman and Pearson: "The general theory of hypothesis testing first took form under the hand of Sir Ronald Fisher in the 1920s, but it was carried to a high state of development in the work of J. Neyman and E. S. Pearson, beginning about 1928."[47]

There is no mention of a deep controversy and extant problems in the few sentences that follow and describe the "high state of development." It should be added for the nonpsychologist that it was *not* customary to neglect controversial issues in the experimental psychology of the time. On the contrary, it was and still is general practice to discuss and even tolerate the existence of controversial theoretical issues.[48] Thus, the neglect of controversy and the presentation of a single hybrid theory of inferential statistics is all the more astonishing, and suggests how high the stakes were: namely, to safeguard the ideal of objectivity.

Neglect of Alternative Theories   It has been argued that only the Bayesian theories were truly theories of statistical *inference*. Alternative theories, such as Bayesian statistics, were completely neglected in psychological texts. Not until 1973 did the first textbook on Bayesian statistics for social scientists appear. In the same year Hays included a chapter on Bayesian methods in the second edition of his *Statistics for the Social Sciences*.[49] Even today, alternatives to the hybrid are only slowly penetrating the consciousness of experimenters. Traditionally, however, it was not even mentioned that an alternative view to the hybrid could exist.

Thus far the history of alternative theories. Since about 1960, Bayesian statistics have appeared in the psychological literature as models for the experimental subject's judgment under uncertainty,[50] that is, at the level of theory construction. However, this has not made Bayesian statistics any more attractive as a model for the experimenter's reasoning. It seems probable that once a form of mechanized inference has been institutionalized throughout the discipline, the stakes would be too high to risk any change in the rules that validated empirical results as knowledge.

Finally, this is a rather unique case of the neglect of alternative theories, which is *not* characteristic of the general attitude toward theoretical alternatives in psychology. This parallels the uniqueness of the neglect of extant controversies.

Anonymous Presentation of Ideas   Typically all the ideas (stemming either from Fisher or Neyman and Pearson) were presented in the texts *anonymously* as *the* corpus of inferential statistics. Neyman and Pearson are not mentioned at all in 21 out of the 25 textbooks I have looked through, and none of these mention the controversy. I personally was taught the hybrid without learning that some concepts came from Fisher, others from Neyman and Pearson; and I puzzled over why it was that some concepts, e.g., the Type II error (from Neyman and Pearson), seemed not to fit with others, e.g., the idea of null hypothesis testing (Fisher's idea). Again, the presentation of ideas without reference to their authors was and is unusual.

In my opinion, the anonymous presentation of a monolithic "inferential statistics" facilitated the neglect of controversial issues between the two positions as well as the neglect of alternatives. Thus, it facilitated the illusion of the "correct" mechanized inductive inference, apparently freeing the experimenter from his own subjectivity and responsibility.

Institutionalization of the Hybrid   Along with the textbooks, the anonymous hybrid entered the university curricula and the publishing policies of major journals. The following quotation is from the editorial of the *Journal of Experimental Psychology* in 1962, stating the criteria for the acceptance and rejection of papers submitted:

The next step in the assessment of an article involved a judgment with respect to the confidence to be placed in the findings—confidence that the results of the experiment would be repeatable under the conditions described. In editing the *Journal* there has been a strong reluctance to accept and publish results related to the principal concern of the research when those results were significant at the .05 level, whether by one- or two-tailed test! This has not implied a slavish worship of the .01 level or any other level, as some critics may have implied. Rather, it reflects a belief that it is the responsibility of the investigator in a science to reveal his effect in such a way that no reasonable man would be in a position to discredit the results by saying that they were the product of the way the ball bounced. At least, it was believed that such findings do not deserve a place in an archival journal. . . .

The same philosophy applied when negative results were submitted for publication, but here rejection frequently followed the decision that the investigator had not given the data an opportunity to disprove the null hypothesis, i.e., the sensitivity of the experiment was substandard for the type of investigation in question and was therefore not sufficient to persuade an expert in the area that the variable in question did not have an effect as great as other variables of known significant effect.[51]

This editorial shows the degree to which the hybrid had been institutionalized by 1962; the editorial is written as if there were no alternatives to significance testing of null hypotheses as the method of scientific inference, and as if there were no experimental researches without significance testing. What is explicitly said reflects some of the common confusions and errors spread by authors of the statistical manuals in psychology, the creators of the hybrid. They are not entirely to blame, since the real controversial issues as well as the equivocations and inconsistencies in the original writings made clear presentation of a single "inferential statistics" to the student a hopeless task. For instance, in the above quotation, it is stated (1) that the level of significance reflects the degree of confidence that the result is repeatable and (2) that the level of significance reflects the size of the effect of a variable. Both assertions, however, are incorrect.[52] Nevertheless, burdened with such confusions, the level of significance became an institutionalized measure for the quality of research. Note that the acceptance of the null hypothesis (a "negative result" in the language of the editorial) was considered as virtually a *sufficient* condition to reject an article, since it seemed to indicate poor experimental research.

The institutionalization of significance testing as *the* mechanization of inference was not at all restricted to the *Journal of Experimental Psychology*. Already in 1955 and 1956, the mean proportion of articles using significance tests was 81.5% in four leading journals from different areas within psychology. Significant results ($p < .05$) were reported in 97% of the studies.[53] In contrast, in unpublished work like congress reports and dissertations, the proportion of significant results was only between 60% and 70%—there was not so much selective pressure. In accordance with the above misunderstanding of "significance" as promising both confidence in replicability and size of effect, there was (1) not one single replication in a total of 362 experimental studies published in these four journals, and (2) measures for the size of the effect (so-called "practical significance") were rarely given.[54]

2.3 The Impact on Psychological Research

My general thesis is that certain classical ideals of natural science, adopted in psychology, influenced the use of probabilistic thinking in that discipline. Let us now consider a further step: How did the institutionalized dogma of inferential statistics itself influence the direction of experimental psychology? My answers are necessarily a kind of science fiction, since we can only speculate on how psychology might have evolved otherwise.

Shifting the Emphasis from Design and Hypotheses to Significance   Before the institutionalization there was much more sophistication in experimental design and hypotheses than in inference. Examples can be found in the work of Egon Brunswik and Jean Piaget. Contrary to common belief, designs with more than one independent variable, from two to five, were used long before analysis of variance presented this option to psychologists.[55] With the institutionalization of the hybrid, the focus of attention, however, shifted to what happened after the experiment rather than before. Researchers were rewarded for the exorcism of the null hypothesis, rather than for a careful experimental design or a phenomenological preanalysis.

Shifting the Emphasis from Theory-Guided Research to Inductive Research   A central feature of null hypothesis testing is that one can test many more main effects, interactions, or correlations than one has meaningful hypotheses for. Before the introduction of analysis of variance, researchers tested only as many effects as they had hypotheses. In the early applications of analysis of variance in the late '30s, researchers still did not use the possibility of testing every interaction, but tested only those few hypotheses they regarded as meaningful.[56] Soon, however, it was realized that analysis of variance, fortunately or unfortunately, generates and tests more hypotheses than the current conceptual position could handle, and psychologists availed themselves of the opportunity. This may have been a turning point, where the attention shifted from a priori hypotheses to a posteriori significant effects, and, thereby, from a more theory-guided research to a more inductive kind of research that waits for what turns out to be significant.

Theoretical development seems not to have gained from the enthusiasm about statistics. In addition, the fascination with sophisticated models of *how-to-transform numbers* was connected with an extremely simple view of the measurement problem, that is, the question of *how-to-get numbers*.[57] This marriage between sophisticated statistics and naive procedures for measurement has its historical parallel in the "correlation discipline" founded by Karl Pearson and Francis Galton. Here, the more or less complete lack of a theory of intelligence and of a related measurement theory went along with a fascination for sophisticated correlational statistics. The idea of research seemed to coincide with the idea of transforming numbers into other numbers, wherever they might come from in the first place.

It is possible that beliefs about objectivity disposed those who embraced the mechanization of inference in the form of statistics also to regard measurement as a straightforward, unproblematic matter. At least both views share a vision of knowledge independent of both the knower and his measurement instruments. In this sense, the cultivation of statistics may have actually inhibited the development of measurement theory.

Noncumulative Research?   Significance tests yielded yes-or-no answers instead of probabilities. This mirrors what was wanted in the original fields of application, agriculture and later education, from which the null hypothesis testing method made its way into the laboratory. Both the agriculturalist and the educational administrator had to make a yes-or-no decision of practical relevance, such as whether to introduce a new kind of manure or not, or a new curriculum. However, other than in these applied fields, there is generally *no* yes-or-no decision to be made in a scientific context. Here, acceptance or rejection of hypotheses is instead provisional, subject to reversal in light of further observations. The point is that the hybrid, based on yes-or-no decisions, suggests a different idea of what research is, than, for instance, Bayesian statistics. Since the hybrid excludes prior probabilities (which—as in Bayesian statistics—represent all the previous knowledge about the validity of a hypothesis), each new experiment started with zero knowledge and ended with a yes-or-no decision. The experiments were treated as isolated, without the benefit (or harm) of previous knowledge. Thus, the institutionalization of the hybrid suggested an idea of psychological research that consisted of a noncumulative, isolated presentation of yes-or-no decisions. Such an inflation of isolated results, each one conducted formally in a vacuum, characterizes the present state of knowledge in many areas of psychology. Bayesian statistics, on the contrary, suggests the idea of research as an ongoing, cumulative process of successive modification of the probabilities of hypotheses given new data. "Negative results" would not be omitted from the journals, as is the case with the publishing policies above quoted.

It should be added that prior probabilities seem to enter null hypothesis significance testing if they are very low. An example is ESP, which is given so low a prior probability that significant results are often taken as a proof of cheating.

Feelings of Guilt and Systematic Biases in the Literature   Since levels of significance became the institutionalized measure of quality of research, researchers often violated the logic of specifying the level *in advance*. They specified it *after* they had analyzed the data. This "cheating" may be understandable given the imperative "publish or perish" *and* the absence of reasonable criteria by which to select a particular a priori alpha risk for the decision. (As mentioned above, in contrast to applications like quality control, *no* real yes-or-no decision of practical relevance had to be made after the typical experiment, say, on the influence of motivational sets on the perception of incomplete pictures.) Other researchers, less sensitive to feelings of guilt, simply filled their data tables with a galaxy of stars, double stars, and triple stars, indicating different "levels of significance" (.05, .01, and .001), obviously determined a posteriori. A majority of authors were not only rejecting hypotheses; they were rejecting the logic of hypothesis testing too.

In general, it seems that the publishing policies above quoted generated systematic biases in what was reported in the journals. On the one hand, assume that the null hypothesis holds; then the journals are filled mainly with Type I errors. (By the way, this would explain the absence of published replications, since these should fail—"negative results.") On the other hand, assume that the null hypothesis does not hold; then the journals are filled with correct rejections. Not published in either case are correct and incorrect acceptances of the null hypotheses. If an experimenter would have allowed himself to think about this, he would have seen that this publishing policy makes sense only on the assumption that the null hypothesis is always incorrect. Under this assumption, however, significance testing of null hypotheses would be useless—except to give the illusion of an objective inference.

Finally, since no alternative is specified in null hypothesis testing (as opposed to Neyman and Pearson's theory), the Type II error (β) and the power of the test ($1 - \beta$) are unknown. Jacob Cohen, who analyzed a whole volume of the *Journal of Abnormal and Social Psychology* in 1960, specified such alternatives to determine the power of the experiments published there. This showed that the power of many studies was ridiculously low.[58] Since nevertheless nearly all studies reported significant results, either the effects must have been very great or these studies report Type I errors.

Inhibition of the Possibilities for Theory Construction Using Analysis of Variance
The atheoretical stance adopted by Fisher and the presentation of analysis of variance as *the* instrument of inference in psychology seems to have slowed down other interpretations of analysis of variance by psychologists. In particular, the modeling possibilities were recognized rather late. For instance, in 1967, Harold Kelley proposed analysis of variance (including the *F*-test) as a model for how people make causal attributions.[59] About the same time, Norman Anderson started to use analysis of variance as a model for the cognitive integration of multidimensional information in person perception and psychophysics.[60]

To summarize: My thesis is that experimental psychology pinned its hopes to two ideals of classical natural sciences—determinism and objectivity—in its struggle for scientific reputability. Probabilistic thinking seemed to offer a means toward objectivity in the sense of knowledge without knower—in particular, to offer a means for the mechanization of inductive inference. It is largely forgotten that Neyman and Pearson themselves argued that significance tests should carry only suggestive value, to be tempered by personal judgment.[61] The tempting illusion, which probabilistic thinking was made to serve, was the prospect of knowledge without personal coloration. Slowly, experimenters have now come to realize that this hope will be disappointed.

Finally, I want to emphasize that my thesis about the connection between inferential statistics and the ideal of objectivity deals with *one* function of inferential statistics, not with *the* function. Others have been mentioned. For instance, there is the early function of analysis of variance as the proof of the practical relevance of experimental psychology,[62] or even Abraham Kaplan's *law of the instrument*: "Give a small boy a hammer, and he will find that everthing he encounters needs pounding."[63]

3 Psychology without Individuals

In contrast to the natural sciences, psychology has to deal with two classes of human beings: the experimenters and the individuals investigated. Therefore there seems to exist—besides the experimenters—a second source of subjectivity that has to be eliminated. The idea of eliminating the experimenter may still be an irresistible temptation in many areas of research. The idea of eliminating the individual, however, was an innovation unique to psychology. Where did it come from and what does it mean?

3.1 The Ideal

From its inception, experimental psychology has been modeled in important respects upon classical physics. For instance, the question of what a psychological law and psychological experiment should mean was usually answered by pointing to examples from physics like Newton's law of falling bodies.[64] In the mid-'30s, psychological measurement was—under the influence of physical operationalism—reduced to physical measurement: "It *is* physical measurement. It always has been."[65] Classical physics as the image of science—this seemed to promise psychology the status of an "objective" science.

The point under consideration is the distinction between what I call *bipartite* and *tripartite* language.[66] The *bipartite* language is that out of which the laws of physics were made: the dualism of objects and variables. The *tripartite* language consists of *individuals*, who judge (react to) *objects* with respect to *variables*. It is the language of the typical "atomic sentence" obtained in psychological experiments; the experimenter observes that "individual X judges object Y as having the property Z." The ideal I am talking about is a psychological science based on a bipartite, physical language, thereby trying to eliminate the individuals, in addition to the experimenters. As illustration, I shall give two examples.

1. *Perception without individual perceivers*: Stanley Stevens tried to find psychophysical laws that consisted of variables (e.g., perceived loudness) and objects (e.g., pure tones). In order to measure perceived loudness, he asked individuals, for instance, to judge how much louder one tone is than a second, standard tone. The individual had to give numerical estimates like "4 times," "5 times," ("magnitude estimation"). These estimates were averaged and taken as the scale value of the tone (object) with respect to its subjective loudness (variable). The individuals who judged the tones were considered as *neutral measurement instruments* and played no role in the theory. The particular individual was of no interest, treated as a mere replication of the next individual; interindividual differences in perception or in the competence of fulfilling such complicated tasks as judging ratios of loudnesses were a priori eliminated by averaging. In short, the ideal was to consider perception without an individual perceiver. (Today, now that Stevens's dominating influence has weakened, we know that systematic individual differences exist even in elementary sensory perception, and that it makes little sense to average different perceptual strategies.[67])

2. *Judged personality without judges*: To illustrate the generality of the ideal, the second example is taken from a quite different area, from personality research. Raymond Cattell asked *individuals* to rate other *persons* on certain *traits*. The question he wanted to answer by means of those observations was, What are the *dimensions* of *man's* personality? As in Stevens's case, the "atomic sentences," i.e., the empirical observations, were tripartite, whereas the research question posed

was bipartite. (The corresponding tripartite question would have been, What are the *dimensions* with which certain *individuals* judge the personality of other *persons*?) In order to fit the observations to the bipartite question, the individuals were eliminated after averaging their judgments. These average judgments were transformed by correlation statistics and factor analysis, and the resulting factors were considered as the personality dimensions of the *ratees*. This bipartite enterprise has been called "objective" personality research.[68] In the 1960s it was criticized on various grounds. However, the bipartite, "physical" question still survives today. A major reason, in my opinion, is that it corresponds so well with the bipartite, "physical" structure of our commonsense language: "John is introverted" rather than "I construe John (for whatever reasons) as being introverted."

3.2 Probabilistic Thinking in the Service of the Ideal

Based on a physical image of science, the ideal dictated the elimination of the judging individuals, the raters, from psychological theories. In Stevens's case, the interest was in loudness (brightness, etc.) per se; in Cattell's case it was in personality per se. The subjectivity of the judges was unwanted. My argument is that probabilistic thinking played a considerable role as a means in this fight against the individual's subjectivity. My focus is the *random variable model*, which was given a special interpretation in experimental psychology: What is called "error" in the random variable model was equated with interindividual variability.

The Random Variable Model Laplace's philosophy of chance described each phenomenon in the physical and indeed the social world as governed by forces of two kinds: permanent and accidental. This dualism between "true" values and "errors" is the first assumption of what was later called the *random variable model*. The second is that the "errors" in repeated measurements are mutually independent and are random deviations that "oscillate" around the "true" value.

The Physical Image: Man Is Inert If man is to be modeled by the random variable model, something must be identified as "error." Like a physical object, man was considered as inert, an image that had been anticipated by Thomas Hobbes 300 years earlier: "That which is really within us, is, as I have said before, only motion, caused by the action of external objects."[69] With this "physical" image of inert, uniform objects in mind, two interpretations could be given to "error" variability: (1) repetitions over time, that is, repeated measurements on a single individual (intraindividual variability) and (2) repetitions over individuals (interindividual variability). From these two candidates for "error," however, the first one turned out to be dubious, because intraindividual repetitions often violated the second assumption of independence of "errors."[70] Thus, *both* the ideal and the independence assumption of the random variable model seemed to dictate a psychology without individuals. Let us consider a famous example.

Thurstone: perception without individuals: In 1927, Thurstone published his famous probabilistic measurement model known as the *law of comparative judgment*. My question is, what did Thurstone define as "error" in his normal distributional model? In his first paper in 1927, he suggested *intraindividual* repetitions be considered as "error."[71] In his second paper in the same year, he already offered *interindividual* repetitions as an alternative to intraindividual repetitions.[72] In the following applications and generalizations to attitude measurement, interindividual variability became the dominant interpretation for "error." Note that the two interpretations that Thurstone offered (1) were presented as if they were simply mere alternatives for data generation; (2) are both consistent with the above physical image of man; (3) but imply two different theories of perception.[73] Thurstone did not care about this latter ambiguity. Why then the change? I believe that the major reason was that intraindividual repetitions did not satisfy the independence assumption in the random variable model. Repeated measurements over time in a single individual often turned out to be dependent, thereby contradicting the assumption of independent "errors" around a single "true" value. There seemed to be a *process* initiated by repetitions over time. These time-dependent processes have been called exhaustion, learning, memory, loss of motivation, etc. Given those processes, intraindividual variability was ruled out as a candidate for "error."

In this sense, the random variable model dictated the interpretation of interindividual variability as "error." This was exactly what happened in experimental psychology in the period under consideration. Error terms were—in the applications of Thurstone's model, in analysis of variance, etc.—identified with interindividual variability. Studies and theories about the process of intraindividual change were rare (outside the specific area of learning and memory, which almost never dealt with interindividual differences). Even today, we are tempted to investigate 150 freshmen in Michigan for 20 minutes of their lifetimes, and think of them as interchangeable physical objects that do not change over time—which allows us to present our significant result as if it were about all mankind and all time. Accordingly, the *reactivity* of intraindividual measurement did not become a general research topic—a genuine "psychological topic" that would have contradicted the physical image of man. Rather it was considered as a source of trouble to be eliminated in careful experimental design.[74]

To summarize: in the thrall of a physical image of science and, as a consequence, a physical image of man, psychology was forced to eliminate the particular individual. The concept of "error" as defined in the random variable model could help here by identifying and conflating interindividual variability with "error." In addition, the independence assumption dictated the identification of interindividual variability rather than intraindividual variability with "error."

4 Conclusion

My thesis may be summarized as follows: Bound to determinism and objectivity as their ideals of science, experimental psychologists dealt with probabilistic thinking in a quite different way than their physicist contemporaries. Probabilistic thinking was suppressed on the level of psychological theory, maintaining the ideal of a deterministic theory and a Laplacian view of uncertainty as "ignorance." Probabilistic thinking was enthusiastically welcomed at the level of inference, generating the illusion of an objective, mechanical solution to the inference problem, the experimenter's subjectivity. Finally, on the level of measurement and data description, the random variable model was used to eliminate the individual as a further source of subjectivity, either by equating individual differences with error or by simple averaging over individuals. The irony is that the "harder" sciences, like physics and biology, have introduced uncertainty via probabilistic models into their subject matter, whereas the so-called "soft" ones, like psychology, were loathe to tolerate such uncertainty, and long clung to the ideals of classical physics.

Notes

1. Clark L. Hull, "The Problem of Intervening Variables in Molar Behavior Theory," *Psychological Review*, 50 (1943), 273–291, on p. 273.

2. See, for instance, the discussion of Egon Brunswik's probabilistic functionalism, published in the *Psychological Review*, 62 (1955), 193–242.

3. Werner Heisenberg, *Das Naturbild der heutigen Physik* (Hamburg: Rowohlt, 1955), p. 12.

4. German psychologists, at least, had no comparable enthusiasm for probability at either the inference or the theory construction level during this period. Hence, specifically national traditions may play an important role in the history of probabilistic thinking in psychology.

5. Pierre S. Laplace, *A Philosophical Essay on Probabilities* (1814, trans. by F. W. Truscott and F. L. Emory, New York: Dover, 1951).

6. Edwin G. Boring, "Eponym as Placebo." In E. G. Boring, ed., *History, Psychology, and Science: Selected Papers* (New York: Wiley, 1963), on p. 14.

7. David Krech, "Discussion: Theory and Reductionism," *Psychological Review*, 62 (1955), 229–231, on p. 230 (italics in the original).

8. An example of a theoretical position that was based on the principles of multiple cue mediation and cue substitutability instead of causal laws was Egon Brunswik's probabilistic functionalism (see note 2 and my chapter in this book, "Survival of the Fittest Probabilist: Brunswik, Thurstone, and the Two Disciplines of Psychology").

9. For an example of a "Darwinian" psychology that dispenses with prediction, see Howard E. Gruber, "The Fortunes of a Basic Darwinian Idea: Chance," in R. W. Rieber and Kurt Salzinger, eds., *The Roots of American Psychology: Historical Influences and Implications for the Future* (New York: The New York Academy of Sciences, 1977).

10. See Kurt Danziger, "Statistical Method and the Historical Development of Research Practice in American Psychology," this volume.

11. Wolfgang Marx, "Geni(t)ale Tiefenschau," *Psychologie heute*, 8 (1982), 30–33.

12. Kurt Lewin, *Dynamic Theory of Personality* (New York: McGraw-Hill, 1935).

13. Clark Hull, "Problem," p. 274 (note 1).

14. See note 2.

15. See Judith L. Scharff, "Skinner's Concept of the Operant: From Necessitarian to Probabilistic Causality," *Behaviorism* 10 (1982), 45–54.

16. B. F. Skinner, "The Concept of Reflex in the Description of Behavior," *The Journal of General Psychology*, 5 (1931), 427–458, on p. 446 (italics in the original).

17. B. F. Skinner, "Two Types of Conditioned Reflex and a Pseudo Type," *The Journal of General Psychology*, 12 (1935), 66–77.

18. B. F. Skinner, "Two Types of Conditioned Reflex: A Reply to Konorski and Miller," *The Journal of General Psychology*, 16 (1937), 272–279.

19. B. F. Skinner, *Science and Human Behavior* (New York: Macmillan, 1953).

20. Skinner, *Science*.

21. Leon M. Solomons, "A New Explanation of Weber's Law," *Psychological Review*, 7 (1900), 234–240.

22. L. L. Thurstone, "Psychophysical Analysis," *American Journal of Psychology*, 38 (1927), 368–389. For the issue of Thurstone's "independent discoveries" see my paper "Survival of the Fittest Probabilist," this volume, note 23.

23. Thurstone, "Psychophysical Analysis," p. 368.

24. Clark L. Hull, John M. Felsinger, Arthur I. Gladstone, and Harry G. Yamaguchi, "A Proposed Quantification of Habit Strength," *Psychological Review*, 54 (1947), 237–254. See also R. P. McDonald, "Hullian Theory and Thurstone's Judgment Model," *Australian Journal of Psychology*, 14 (1962), 32–38.

25. Hull, "Problem," p. 274 (note 1, italics in the original).

26. See my paper "Survival of the Fittest Probabilist," this volume.

27. Heisenberg, *Naturbild* (note 3).

28. Max Wertheimer, "Experimentelle Studien über das Sehen von Bewegungen," *Zeitschrift für Psychologie*, 61 (1912), 161–265.

29. E. Galanter, "Stanley Smith Stevens (1906–1973)," *Psychometrika*, 39 (1974), 1–2, on p. 1.

30. For a discussion of the discrepancy between magnitude estimation and category rating see Michael H. Birnbaum, "Controversies in Psychological Measurement," in Bernd Wegener, ed., *Social Attitudes and Psychological Measurement* (Hillsdale, NJ: Erlbaum, 1982). For a general discussion of the independence assumption in psychological measurement as well as for a conceptual perspective that considers measurement as modeling see my book *Messung und Modellbildung in der Psychologie* (Munich: Reinhardt, 1981).

31. See Anthony C. Rucci and Ryan D. Tweney, "Analysis of Variance and the 'Second Discipline' of Scientific Psychology: A Historical Account," *Psychological Bulletin*, 87 (1980), 166–184.

32. C. W. Telford, "The Refractory Phase of Voluntary and Associative Responses," *Journal of Experimental Psychology* 14 (1931), 1–36, on p. 18.

33. For a discussion of the differences between Fisher and Neyman and Pearson and the reception of the theories in psychology see Michael C. Acree, *Theories of Statistical Inference in Psychological Research: A Historico-Critical Study* (dissertation, Clark University, published by University Microfilms International, Ann Arbor, 1978).

34. R. A. Fisher, *The Design of Experiments* (Edinburgh: Oliver and Boyd, 1935), on pp. 18–20.

35. Fisher, *Design*.

36. Acree, *Theories*, on p. 398 (note 33).

37. R. A. Fisher and W. A. Mackenzie, "Studies in Crop Variation II: The Manurial Response of Different Potato Varieties," *Journal of Agricultural Science*, 13 (1923), 311–320. For the introduction of analysis of variance into psychology see A. D. Lovie, "The Analysis of

Variance in Experimental Psychology: 1934–1945," *British Journal of Mathematical and Statistical Psychology*, 32 (1979), 151–178, and Rucci and Tweney, "Analysis of Variance" (note 31).

38. For instance, in H. Sorensen's *Statistics for Students of Psychology and Education* (New York: McGraw-Hill, 1936) the chapter heading is "The Reliability of the Difference Between Means."

39. H. E. Garrett and J. Zubin, "The Analysis of Variance in Psychological Research," *Psychological Bulletin*, 40 (1943), 233–267, on p. 233.

40. G. W. Snedecor, *Statistical Methods* (1st ed. Ames, Iowa: Collegiate Press, 1937). Snedecor had also published a short book, *Calculation and Interpretation of Analysis of Variance and Covariance* (Ames, Iowa: Collegiate Press, 1934).

41. Lovie, "Analysis," on p. 169 (note 37).

42. An example is R. L. Anderson and T. A. Bancroft, *Statistical Theory in Research* (New York: McGraw-Hill, 1952).

43. J. P. Guilford, *Fundamental Statistics in Psychology and Education* (6th ed., with B. Fruchter, New York: McGraw-Hill, 1978), on p. 151, italics in the original.

44. Guilford, *Fundamental*, 1st ed., on p. 164.

45. Guilford, *Fundamental*, 3rd ed., 1956, on p. 217.

46. For further examples of confusions see Acree *Theories* (note 33) and Jürgen Bredenkamp, *Der Signifikanztest in der Psychologischen Forschung*. (Frankfurt: Akademische Verlagsanstalt, 1972).

47. William L. Hays, *Statistics* (New York: Holt, Rinehart and Winston, 1963), on p. 287.

48. Texts in psychology typically attempt to give an overview of the different theoretical positions existing in the field, including controversies. Examples for such an encyclopedic view of different positions are Melvin H. Marx, ed., *Theories in Contemporary Psychology* (New York: Macmillan, 1963), and *Psychological Theory* (New York: Macmillan, 1951). Another example is the early debate on intelligence and its measurement, published in the *Journal of Educational Psychology*, 12 (1921), 123–147, 195–216.

49. L. D. Phillips, *Bayesian Statistics for Social Scientists* (New York: Holt, Rinehart and Winston, 1973). The first article in a psychological journal on Bayesian statistics as an alternative to the hybrid seems to be Ward Edwards, Harold Lindman, and Leonard J. Savage, "Bayesian Statistical Inference for Psychological Research," *Psychological Review* 70 (1963), 193–242.

50. For a discussion see G. Gigerenzer and D. J. Murray, *Cognition as Intuitive Statistics* (Hillsdale, NJ: Erlbaum, 1987), and for an overview see D. Kahneman, P. Slovic, and A. Tversky, eds., *Judgment under Uncertainty: Heuristics and Biases* (Cambridge: Cambridge University Press, 1982).

51. A. W. Melton, "Editorial," *Journal of Experimental Psychology*, 64 (1962), 553–557, pp. 553–554.

52. For a discussion of these misunderstandings and the consequences of such "level of significance" policies see Bredenkamp, *Signifikanztest* (note 46).

53. T. D. Sterling, "Publication Decisions and their Possible Effects on Inferences Drawn from Tests of Significance—or Vice Versa," *Journal of the American Statistical Association*, 54 (1959), 30–34.

54. See Bredenkamp, *Signifikanztest* (note 46).

55. For instance, from the late 1920s on, Egon Brunswik and his research associates investigated the perception of stimuli (e.g., coins, stamps) varying in three and more dimensions (e.g., number, size, and value). See Egon Brunswik, *Wahrnehmung und Gegenstandswelt*

(Leipzig: Deuticke, 1934), pp. 131–166. Examples for multifactorial designs in the English literature are given by Lovie, "Analysis" (note 37).

56. See Lovie, "Analysis."

57. See my book *Messung* (note 30), as well as my article "Messung, Modellbildung und die kognitive Wende," in Manfred Amelang and H. J. Ahrens eds., *Brennpunkte der Persönlichkeitsforschung* Vol. I (Göttingen: Hogrefe, 1984).

58. J. Cohen, "The Statistical Power of Abnormal-Social Psychological Research: A Review," *Journal of Abnormal and Social Psychology*, 65 (1962), 145–153.

59. Harold H. Kelley, "Attribution Theory in Social Psychology," in D. Levine, ed., *Nebraska Symposium on Motivation*, Vol. 15 (Lincoln: University of Nebraska Press, 1967).

60. Normal H. Anderson, "Functional Measurement and Psychophysical Judgment," *Psychological Review*, 77 (1970), 153–170.

61. See Jerzy Neyman and Egon S. Pearson, *Joint Statistical Papers* (Cambridge: Cambridge University Press, 1967), e.g., on pp. 2, 56, and 204.

62. See Kurt Danziger's paper in this volume.

63. Abraham Kaplan, *The Conduct of Inquiry* (Scranton, PA: Chandler Publishing, 1964), on p. 28.

64. See my paper "Survival of the Fittest Probabilist" in this volume.

65. Douglas McGregor, "Scientific Measurement and Psychology," *Psychological Review*, 42 (1935), 246–266, on pp. 265–266 (italics in the original).

66. See Gigerenzer, "Messung" (note 57).

67. An example is the issue of binaural summation of loudness. It has been studied using averages over individuals as the unit and Stevens's scaling approach—e.g., G. S. Reynolds and S. S. Stevens, "Binaural Summation of Loudness," *Journal of the Acoustical Society of America*, 32 (1960), 1337–1344—and is now studied using the individual as the unit and axiomatic measurement theory—e.g., Gerd Gigerenzer and Gerhard Strube, "Are There Limits to Binaural Additivity of Loudness?" *Journal of Experimental Psychology: Human Perception and Performance*, 9 (1983), 126–136.

68. For a discussion of personality research with respect to the distinction between bipartite and tripartite questions see Gigerenzer, *Messung* (note 30), pp. 88–112, and Gigerenzer, "Messung," pp. 49–65 (note 57).

69. Thomas Hobbes, *Leviathan*, I, 6, 1651.

70. That intraindividual repetition often results in values that are better understood as a time-dependent process rather than as random deviations around a single true value has long been known to experimentalists. For instance, H. L. Hollingworth, "Variations in Efficiency during the Working Day," *Psychological Review*, 21 (1914), 473–491, analyzed repeated measures of an individual's performance during a working day. Some of the repeated measures showed a time-dependent steady increase (e.g., calculation, naming opposites); others (e.g., tapping) seemed to follow time-independent "random" variations around a single "true" value. In a preceding article, "Experimental Studies in Judgment. Judgment of the Comic," *Psychological Review*, 18 (1911), 132–156, Hollingworth analyzed both intraindividual and interindividual repeated presentations of jokes. The response to jokes that were told repeatedly (intraindividual repetition) showed a characteristic time-dependent process, depending on the stupidity of the joke, the particular individual, and other variables. In many areas, intraindividual repetitions were found to result in dependent values.

71. Thurstone, "Psychophysical Analysis" (note 22).

72. L. L. Thurstone, "A Law of Comparative Judgment," *Psychological Review*, 34 (1927), 273–286.

73. Thurstone's case is one of those where the *same* mathematical model has been applied to two *different* phenomena (repeated measures over time versus repeated measures over individuals), and the phenomena themselves came to be equated.

74. For instance, see Donald T. Campbell and Julian C. Stanley, "Experimental and Quasi-Experimental Designs for Research on Teaching," in N. L. Gage, ed., *Handbook of Research on Teaching* (Chicago: MacNally, 1963).

2 Statistical Method and the Historical Development of Research Practice in American Psychology

Kurt Danziger

Around 1950 the statistical techniques of analysis of variance came to occupy a dominant position in psychological research. This development can be understood in terms of certain changes in psychological research practice during the first half of the present century.

Traditionally, psychological research involved two divergent approaches to variability. In the "Wundtian" model the statistics of error were used merely to arrive at estimates of true values for certain magnitudes characteristic of individual functioning. In the "Galtonian" model, however, interindividual variability was the object of interest and the position of individual values was determined by reference to the characteristics of populations of individuals.

The Wundtian model of research practice was not effective in yielding information that was useful in practical contexts. But the Galtonian model was also limited because in its original form it made no provision for establishing causal effects by experimental manipulation.

A systematic survey of the psychological research literature shows the gradual emergence and increasing popularity of a new composite model of research practice, the "Neo-Galtonian." This substitutes experimentally constituted treatment groups for the natural groups of the original Galtonian model, but retains the latter's treatment of variability. It was the prior adoption of this model of psychological research practice that created technical needs that the analysis of variance was able to fill.

While this style of psychological research could produce knowledge that was potentially useful in certain restricted social contexts, it created problems for psychological theory, which was traditionally concerned with intraindividual processes. Behaviorism provided an early way of avoiding these problems, statistical inference a later strategy. The problems, however, have not gone away.

1 General Perspectives

The rapidity with which American psychologists adopted Fisherian analysis of variance techniques has been comprehensively demonstrated.[1] Although their initial understanding of these techniques was often defective,[2] psychologists embraced them with considerable enthusiasm. In the years immediately following the end of World War II the proportion of published psychological research studies that employed the analysis of variance techniques rose very steeply from an insignificant level to the point where the use of such techniques became a hallmark of psychological experimentation. It may be noted in parenthesis that Student's t-test seems to have entered experimental psychology at virtually the same time, enjoying a rapid rise in popularity that paralleled that of the Fisherian techniques.

How is this to be explained? If one examines the relevant literature of the time for clues, one will be told that the kinds of data experimental psychologists deal with lend themselves particularly well to analysis by means of the new techniques. This, however, only pushes the question one step further back. The active experimentalist may be forgiven the naive empiricism that allows him to experience his data as a piece of objective reality that he has merely exposed to view. The critical historian cannot permit himself this indulgence. He has to reckon with the fact that the data of the experimentalist are first of all a product of the experimenter's activity, both theoretical and practical. If the data present themselves in a form that makes them prime candidates for certain types of statistical analysis, this is not because Nature sends them to us already neatly packaged and labeled, but because the investigator has imposed certain practical and theoretical categories on them. If psychologists find that particular analytic techniques are appropriate for their work, this must mean that they have predefined the material they work with in such a way that they can recognize this appropriateness.

To illustrate this point we need only remind ourselves that not all psychologists saw anything of interest in the new statistical techniques. We shall look in vain for the psychoanalyst who suddenly becomes enamored by statistical tests of significance; we shall not find that the Gestalt psychologists embraced analysis of variance (ANOVA) as the answer to their prayers, nor shall we catch Jean Piaget employing the t-test. The group of psychologists who adopted Fisherian statistical techniques was by no means coextensive with the group of all psychologists. It was not even coextensive with the group of experimentalists. Those who found these statistical techniques congenial formed a subset of psychologists distinguished from other subsets by the way they defined their subject matter and the goals of their activity. Whether the statistical seed fell on stony or on fertile soil depended on the manner in which the soil had been prepared by the work of conceptualization that had already been performed on it.

This is not to say that the institutionalization of certain statistical methods within mainstream psychology remained without influence on the way in which research was done and on the kinds of data that were produced. Far from it! But historically this was a secondary effect. In order for methodological considerations to function as criteria for deciding on the production and publication of data, those considerations must have acquired an authority that only comes with institutionalization. It is the basis for this process of institutionalization that is at issue here, not the sociological mechanisms by which an already vigorous institutional form exerts its authority. The challenge is one of achieving a historical understanding of the origins of institutionalized forms. In the solution of this task the analysis of preexisting assumptions about the nature of the object of investigation plays an indispensable role.

We shall find, however, that these preexisting assumptions are closely linked to the purposes and aims of the investigator. If psychologists at different times and in different places have used a wide variety of definitions of their object of study, it is because their research interests have differed. The nature of the questions that are asked determines the form of the interrogation and hence the range of likely answers. Therefore, if we wish to understand the reasons for the adoption of particular methodologies, statistical or otherwise, we shall have to look at the research goals of the investigators.

By the mid-1930s there existed, at least in America, a vigorous school of psychological research that devoted itself to the assessment of the effects of specific treatment conditions on the performance of groups of experimental subjects. It was to this school of research that the early protagonists of ANOVA techniques were able to address themselves with the assurance that their message was likely to receive a sympathetic response. The fact that the use of the new technique would make it possible to increase the number of conditions that could be simultaneously investigated was certainly a great technical convenience, but it neither required nor led to a fundamental change of goals and assumptions on the part of the practical experimentalist.

Such a fundamental change had indeed taken place in the history of experimental psychology, though we must look for it in the years that preceded the introduction of the new statistical techniques. The data presented by Rucci and Tweney[3] show that, while ANOVA did not replace the use of correlational techniques in psychology, it did quite clearly substitute for the use of the Critical Ratio statistic. ANOVA appears to have been adopted as a replacement for that statistic because it could serve a similar function in the conduct of certain types of research. Its rapid success was connected with the fact that fundamentally new functions did not have to be invented for it; it simply did much more effectively what the existing technique was already attempting to do. Factorial designs were not unknown in experimental psychology, even in the absence of appropriate statistical analyses.[4]

We need to trace the development of the methodology of experimental psychology a little further back in order to detect clear signs of a more fundamental change. The practice of assessing the effects of experimental treatments by estimating the statistical significance of the difference between sample means had only just become well established among American experimental psychologists by the mid-1930s. Moreover, the adoption, by American experimental psychologists, of an early form of testing for the statistical significance of group differences was not in any way related to the influence of R. A. Fisher, who is virtually never cited in the American experimental literature of the time. Their adoption of this procedure was very much a function of certain technical needs created by the direction in which their own discipline was developing. I believe that the direction of this development was itself determined by the social demands to which the discipline was exposed and by the ideological commitments of the practitioners that corresponded to these demands. But before we can return to this question of historical causation, it is necessary to describe the nature of the change that had come upon the practice of psychological experimentation during the first three decades of the twentieth century.

2   The Wundtian Model and the Galtonian Model

Modern empirical psychological research originated in two entirely different and independent models, which may be referred to as the Galtonian model and the Wundtian model, respectively. The classical Wundtian experiment was designed to throw light on causal psychological processes operating in individual minds. It did this by systematically varying experimental conditions and observing the results. The relationships between experimental conditions and responses were not of primary interest in themselves, but were simply a means for the exploration of underlying psychological processes with causal properties. These causal processes existed in individual minds, and the experiment was designed to explore them in this individual context. The responses of experimental subjects represented the manifestations of an underlying "psychic causality." [5] Up to a point this was analogous to the way in which the responses of a physical body in an experiment represented the manifestations of an underlying physical causality.

From this point of view any increase in the number of experimental subjects above one constituted a replication of the experiment. It was sometimes dispensed with altogether, or at best limited to a handful of cases, the results of each one being reported individually. If the interindividual variability was large, this was considered prima facie evidence that the attempted experimental isolation of the determining causal factors had failed and that one was dealing with uncontrolled disturbing processes.[6]

The Galtonian model for psychological investigation was quite different. Here the aim of the investigation was the establishment of the distribution of qualities in a population and of the corelation of these qualities in the same or in different populations. The conception of causal explanation was Humean or positivistic.[7] (Strictly speaking, one should speak of a "Galton-Pearson," rather than a "Galtonian," model, because while Galton provided the inspiration, Pearson provided the philosophically and technically complete formulation. However, the term "Galtonian" is more convenient as an adjectival form as long as it is understood as a kind of historical shorthand.) What the collection and analysis of data was meant to accomplish was the demonstration of regular associations among observations, not the provision of evidence for the operation of underlying causal processes. Accordingly, experimental manipulation was limited to the construction of instruments for the collection of observations under standard conditions, mental tests and questionnaires, for the most part. These were administered to relatively large populations, so that the distribution and covariation of responses in these populations could be plotted. The properties of interest were population properties, and individual observations were relevant only in terms of their contribution or relationship to population parameters. Psychological investigations following this model therefore had to use large samples of subjects.

What I have referred to as the Wundtian and the Galtonian models of psychological research employed statistical methods in two very different ways. The first type of use is best exemplified by Fechnerian psychophysics, which was absorbed into the Wundtian model. Here, statistical methods are used as a way of combining observations in order to derive an assumed true value that characterizes a real natural process. This use remains very close to its origins in the methodology of physical observation where the Gaussian Law of Error is applied to errors of observation in a fairly straightforward manner. In the Galtonian model the practical meaning of "error" changes so that it is now taken to apply to the distribution of some characteristic in a population of individuals. The formal similarity of applying the Error Law to a population of observations and to a population of individuals must not be allowed to obscure the very real pragmatic difference that separates the two uses. An investigator using the Wundtian model would be using the statistics of error simply as a "Calculus of Error," while an investigator using the Galtonian model would be using them rather as a "Calculus of Exploration,"[8] where the aim is the description of the concomitance of attributes in natural populations.

This distinction is fundamental because it involves two kinds of research practice implying two basically different conceptions of causality. In the classical Wundtian experiment the establishment of a systematic relationship between two variables always involved an asymmetry that provided the basis for causal attributions. One of the variables was under the experimenter's control and could be "independently" varied. The question was whether this resulted in regular changes in the "dependent variable," i.e., whether a generative causal effect was involved. To take a classical Wundtian instance—did the direction of attention affect reaction times in a systematic way as predicted by the theory of apperception? The issue of errors of observation only arose in certain instances where a true value had to be assigned to a dependent variable that was subject to accidental fluctuations, e.g., a judgment of the presence or absence of a minute sensory difference. In this controlled laboratory situation the statistics of error play a relatively trivial role. They are in no sense constitutive of the underlying causal relationship that is being investigated.

The situation is quite different in the Galtonian model. Here the investigator attempted to impute relationships in situations over which he had no control. Such relationships had therefore to be symmetrical, because the investigator was unable to influence the situation so as to give the status of cause to one of the variables and the status of effect to the other. What he could hope for was a measure of concomitance or covariation. The descriptive statistics of the Galton-Pearson school provided him with such measures. Such statistics were therefore constitutive of the relationship described, whereas the calculus of error of the Wundtian model could not be said to constitute the causal relationships that were investigated. At the most fundamental level this involved a switch to a positivist conception of cause as mere concomitance without reference to generative processes. In this model the establishment of statistical correlations among attributes of natural populations became the goal of research, while in the Wundtian model statistics had been at best a means for ensuring the reliability of certain observations.

The reasons for the switch from the Wundtian to the Galtonian model of psychological investigation emerge quite clearly from contemporary discussions bearing on the question of the direction that was to be taken by the new discipline. The Wundtian model had helped to confer a certain aura of scientific respectability on the new field, but it suffered from the grave defect that its range of application was extremely limited. Wundt himself saw only a restricted scope for experimental psychology in his sense, and he actually opposed both the independence of psychology from philosophy and the application of scientific psychology to real-life problems.[9] This attitude was emphatically rejected by some of the younger German psychologists, and it was never accepted by any but a tiny handful of American psychologists. The desirability of making psychology relevant to real life was already proclaimed by William James,[10] and the question of how to do this remained a major preoccupation among the leaders of early American psychology.[11] The future of the discipline in their society plainly depended on this.

The early years of the twentieth century were therefore a period during which many psychologists were attempting to break the narrow bounds that the Wundtian model had prescribed for them. For some, the way toward the experimental study of more complex psychological processes involved the systematic elaboration of the technique of introspection in a manner never sanctioned by Wundt.[12] This proved to be a blind alley. Others, especially in America, adopted a style of research that deviated fundamentally from the Wundtian model. The alternative Galtonian model provided them with a heuristic that was singularly appropriate for their purposes. Indeed, there were very real parallels between the situation faced by Galton and that which confronted early twentieth-century American psychologists. Galton's abiding interests lay in the creation of a science of heredity, but his problem was one of establishing such a science in the absence of controlled experimentation. The psychologists' problem was one of promoting a socially relevant science involving areas of human functioning that were not accessible to precise experimental study under controlled conditions. Descriptive statistical measures of group attributes had seemed to provide a solution to Galton's problem. His example proved attractive to the many psychologists who followed in his footsteps.

From the point of view of the establishment of psychology as a profession, the adoption of the Galtonian model proved to be a huge success. It provided psychologists with a readily marketable social skill, namely, intelligence testing, which demonstrated the social utility of psychological knowledge and firmly established the psychologist as a recognized technical expert. The mass testing of recruits for the American army in World War I greatly accelerated this process of professionalization, which had just begun to get under way.[13] Even prior to the war, American psychologists were attempting to interest businessmen and various social agencies in their services, but the scene of their most impressive successes was the educational system. Their involvement with the training of teachers and educational administrators was already intense in the early years of the century, and in the following decades the relationship of psychology and education might, with some justice, be described as symbiotic. The education industry not only provided psychologists with jobs and a recognized social function; it also came to exert a major influence on psychological research, in terms of both subject matter and methodology. Certainly it greatly favored the Galtonian style of research over the Wundtian, for the goals of educationally relevant research were set by educational administrators whose interest was in large-scale characteristics of populations and programs, not in specific causal processes in individual minds.

3   Neo-Galtonianism and the Treatment Group

The original Galtonian groups were natural groups; that is to say, they were drawn from actual socially defined populations, men and women, sophomores and freshmen, nine-year-olds and ten-year-olds, etc., and the only problems that could arise about their constitution were relatively technical problems of the representativeness of the sample. But in due course psychologists began to constitute research groups on the basis of psychological measurements, the prototypical example being groups differing in intelligence. These groups were then compared in terms of their performance on some other psychological measure. It was not uncommon to give a completely unwarranted causal interpretation to the group differences obtained in such studies—for example, to regard a difference in intelligence as the cause of differences in other psychological characteristics. This is indicative of the dilemma created by the Galtonian research paradigm. While it permitted psychologists to gain a foothold in the practical world of administrative decisions, it confined that foothold to the relatively restricted area of selection procedures. Yet there were other areas of practical decision making for which psychological knowledge could only become relevant if it took the form of a knowledge of causal relationships. Therefore, there was always the temptation to give an illegitimate causal interpretation to the descriptive data produced by the application of the Galtonian model.

A further development in research methodology provided a way out of this apparent dead end. One of the early results of American psychologists' interest in finding practical applications for their work could be seen in studies of the effects of training and fatigue on functions like motor skills and rote learning. Although these topics were of little significance in terms of the then prevalent theoretical goals of psychology, their relevance to practical situations in education and industry was obvious. In due course, younger psychologists like Thorndike gave these topics a new theoretical significance, but what requires attention in the present context are the consequences that this type of content had for research methodology. The most apparent immediate effect is the tendency to report the results of such studies in the form of group data. While individual results are often still given in the published reports, there is a slowly growing tendency to group the results from a number of subjects exposed to similar experimental conditions in aggregate values. It is these latter that come to be treated as the real results of the experiment, the individual departures from the aggregate values being treated as "error." Eventually, the actual performances of the individuals in the experiment are not reported at all, and only group values are given and judged worthy of scientific discussion.

At first, these aggregate values are simply used to summarize the performance of a single group of subjects taken through a particular program of practice or exercise. But in due course the limits of the kind of information obtainable on this basis become apparent. As in the case of the original Galtonian studies on natural groups, the data from these artifically created laboratory groups are essentially descriptive. To be able to make causal inferences it is necessary to introduce a comparative perspective and to study the difference in the performance of two or more groups exposed to different conditions. Thus is born a fundamentally new entity in psychological research, namely, the *treatment group*. Starting from insignificant beginnings early in the century, this unit of study was beginning to occupy a noticeable place in the literature of the 1920s and was eventually to provide the dominant form for psychological research.

The invention of the treatment group provided a way of dealing with problems in drawing causal conclusions from correlational data in natural human populations. Psychologists, who had been taught to appreciate the value of laboratory methods by Wundt and his pupils, were able to see that the source of the trouble lay in the absence of controlled conditions that could be systematically varied in an experiment. They therefore modified the original Galtonian model by applying it, not to natural groups, but to deliberately constituted experimental groups. Because these groups had been created by the manipulations of the experimenter, it was assumed that any relevant differences in their attributes could be seen as the effects of these manipulations. There emerged a style of research that is a hybrid product of the Wundtian and the Galtonian models. One may call it the *neo-Galtonian* model because the Galtonian component is the dominant one. For while the shift from natural to artificial groups had left the essentials of the Galtonian model intact, the shift from individual to group data had destroyed the basis of the Wundtian model and left only the external trappings of controlled experimentation. What was lost was the core concept of psychic causality, which had provided the link between the experimental manipulations and the theoretical interpretations of their effects.

This was the research context that gradually generated an interest in statistical significance among experimental psychologists. At first, experimenters were content to negotiate agreement about the significance of differences between treatment groups by informal demonstration—the scores of one group were pointed to as being "obviously" superior to those of a comparison group. But where differences were not all that obvious, this left room for disagreement about their genuineness. This led to informal methods of improving the demonstration— pointing to the limited degree of overlap between the two distributions, for example. While the use of neo-Galtonian research designs was still relatively novel, such methods might have been acceptable; but by about 1930, treatment group designs were becoming more and more widely used, and the old demonstrations could no longer be relied on to produce agreement among investigators. More formal ways of expressing the significance of group differences began to be explored.[14] In the immediately following years the use of the Critical Ratio became standardized and widely adopted by experimental psychologists. While the earlier statistical texts written for psychologists give only brief mention to the analysis of group differences,[15] the texts on psychological statistics of the mid-1930s generally contain a considerably expanded treatment of this topic.[16] This development takes us to the threshold of the period marked by the rapid adoption of the *t*-test and analysis of variance.

4   Practical Context and Theoretical Problems

It seems clear that historical shifts in psychologists' use of statistics must be seen in terms of changes in research practice that preceded these shifts. However, to appreciate the full significance of these historical developments one has to be aware of the fact that they were intimately tied up with fundamental changes in the goals of psychological research and in the scope of psychological theory.

As has already been noted, the change from a Wundtian to a Galtonian model for psychological research was related to a professional interest in producing data of practical value. This is confirmed by an analysis of research practice in empirical studies published in various psychological journals during the period 1914–1936. What one finds here is that the Galtonian research paradigm predominates in organs like the *Journal of Educational Psychology* and the *Journal of Applied Psychology* from the very beginning, while it gains only gradual acceptance in journals devoted to "pure" research, like *the, Journal of Experimental Psychology* and the *American Journal of Psychology*. The impetus for the adoption of this paradigm seems to come from the area of practically useful psychological research. The same picture emerges in the German psychological literature when one compares publications in the *Zeitschrift fuer angewandte Psychologie* with those in other psychological journals. The difference is that German academic psychology seems to have resisted the research style of applied psychology much more strongly than American academic psychology.

When one speaks of applied psychology in this context one must be clear about the fact that a very specific kind of application of psychological knowledge is involved. It is an application that takes place in an administrative context, and the individuals to whom such knowledge is useful are administrators. The kind of research that is involved here is not designed to enhance the self-understanding of individuals, but to provide administrators with a data base from which to make more effective decisions about the programs and individuals under their control. The psychological studies initiated in this context were generally characterized by the following features: (1) the goals of the research were determined by issues of specific and immediate social concern rather than by issues of general psychological theory; (2) the questions asked were questions of output, performance, and efficiency, rather than questions involving internal psychological processes; (3) the subjects on whom this research was carried out, and to whom its results applied, were assumed to be *minors*, either in legal fact or in the more general sense that they were persons without valid insight who were not free to determine their own fate but were objects of social control by those in authority.

It appears that the neo-Galtonian model of psychological research represented the routinized and quasi-formal expression of these features. The model's substitution of the treatment group for the individual as the primary object of study, as well as its concentration on the analysis of quantified performance measures, was a crystallization of the research requirements that arose in certain practical institutional contexts. Historically, the requirements of educational institutions, or rather, of their administrators, appear to have played the dominant role. But insofar as they were faced with analogous decisions, administrators of other social agencies and of business corporations could at times find the results of this style of psychological research of interest. One should not underestimate the potential merits of this style of research from a certain point of view. It was and remains capable of yielding results of some utility in situations where its limitations are acceptable. These are primarily situations in which human individuals appear in the role of minors with curtailed insight and autonomy, where decisions hinge on quantitative measures of group performance, and where limited institutional goals disallow critical reflection on the range and definition of relevant treatment conditions.

However, the shift of the data base of psychology, from the reports of individuals to statistical aggregates referring to groups of individuals, quickly produced tensions at the level of psychological theory. Traditionally, psychology had been devoted to the elaboration of theoretical statements about processes taking place in the minds of individuals and had used methods determined by this theoretical goal. But with the shift to aggregate data, a yawning chasm appeared between the specific results of psychological investigation and the traditional theoretical goal of understanding processes and structures in individual minds. This led to major shifts in psychological theory, chiefly among British and American psychologists, whose work was more quickly and profoundly affected by Galtonian research models.

One strategy was to use a Galtonian statistical base to develop models of mental ability by means of the technique of factor analysis. This meant going back to a model of the mind—faculty psychology—that had been discredited for a century and had represented the very antithesis of a scientific approach to the pioneers of experimental psychology. As time went on it became clear that the use of statistical analysis would not produce the desired consensus about the nature and arrangement of human faculties. A statistical pattern of scores on a psychological test and the processes characterizing an individual mind at work belong to two entirely different universes of discourse. The mathematical factors used to describe a certain pattern of cell entries in a matrix permit no convincing conclusions about the individual psychological patterns that generated the observed statistical pattern. Not only is the latter crucially dependent on the nature of the population tested and on the devices of the tester, but the outcome of the analysis in turn depends on subjective decisions about which criteria are to be employed in any particular case. Thus, the use of factor analysis remained limited to a relatively small community of believers and produced no generally accepted solution to the basic problem of Galtonian psychology.

A second, rather more widely accepted, theoretical stance, which often accompanied the adoption of the Galtonian research paradigm, took the form of methodological behaviorism. This must not be confused with J. B. Watson's metaphysical behaviorism, which was much less prevalent among American psychologists than is commonly believed.[17] A kind of methodological behaviorism, expressed in the phrase "Intelligence is what intelligence tests measure," did, however, become the dominant research philosophy in American psychology. Basically, this involved a renunciation of psychology's traditional theoretical task and its redefinition in terms of the kind of information produced by the Galtonian research paradigm. It meant giving up any attempt at exploring the psychological basis of individual action and restricting oneself to gathering the kind of statistical cata that might be useful in practical administrative contexts. Theoretical statements were to be limited to summarizing such data, though in practice this often involved the reification of statistical artifacts.

Finally, there was a third way of handling the discrepancy between the statistical data base and the traditional theoretical goals of the discipline, which developed much more slowly than the other two. In fact, during the period of inital adoption of the Galtonian research paradigm there were only isolated examples of this third approach. What is involved here is a kind of modeling in which the statistical structure of data based on the responses of many individuals is assumed to conform to the structure of the relevant psychological processes operating on the individual level. This approach achieved importance through the work of Thurstone,[18] but there had been earlier attempts in this direction. For example, as early as 1909 Thorndike published a psychophysical study showing a high degree of variability in a group of 37 subjects.[19] At the time, it was most unconventional for questions of psychophysics to be investigated in this Galtonian style, but Thorndike was a pioneer of this approach, and the conclusions he drew from his data were quite radical. There is no such thing as a general psychophysical law, he claimed, but only a multitude of "specialized tendencies to response." The variability that was observed was not to be regarded as "error" around some underlying parameter—it was a direct reflection of the fact that individuals are aggregates of specific response tendencies, a certain number of which are sampled by the experimental situation. It does not require a major theoretical leap to add that such a sampling may also take place due to fluctuations in the internal as well as the external evironments of these specific response tendencies. The point is that the pattern of group data produced by certain experimental operations is taken to represent a pattern of psychological processes in the individual. It is clear that the scope of such an approach will depend on the sophistication of the statistical model employed. A major reason for the slowness with which this approach was accepted undoubtedly lay in the crudeness of the earlier models.

It was suggested above that analysis of variance and related techniques became popular in psychology because psychological research practice had switched to models that yielded aggregate data of potential practical usefulness. (In this connection it may be noted that agricultural metaphors did not enter psychological research with Fisherian techniques, but had already been used by educational psychologists like Thorndike who pioneered the new approach to psychological research.) On the level of research practice the new statistical techniques served to alleviate problems of reaching a consensus about the significance of aggregate data, especially in cases where more than one factor was being investigated.

It is now possible to see that Fisherian techniques were also adopted because it was hoped that they would help to solve the theoretical problem that was becoming more acute with the irresistible spread of the Galtonian research paradigm from areas of applied research to areas of pure research. As long as research was limited to specific practical predictions, some form of methodological behaviorism was quite workable, but where research was expected to throw light on psychological processes in individuals the gap between the goal and the statistical data base became a problem. The earlier treatments of statistical significance in texts of psychological statistics had designated it as a reliability problem and thus spoke to the concern about reaching consensus about data on the level of research practice. But when the new perspective suggested that the problem be approached as one of statistical inference, it seemed to speak to the theoretical problems of psychologists as well as to their practical problems. It seemed that the hypothetical distributions of the statistical analysis could be identified with characteristics of real psychological systems, thus permitting a bridging of the gap between data that referred to groups and theoretical constructs that referred to individuals.

This idea proved to be very attractive to psychologists because it finally enabled them to develop a unified research practice. Until the large-scale adoption of the concept of statistical inference, psychological research had shown an extremely divergent pattern. On the one hand, there was essentially statistical research originating in problems of administratively applied psychology, as we have seen. But, on the other hand, there was research on individual subjects designed to investigate issues posed by major theoretical systems like those of the Gestalt school, Piaget, or various systems of clinical psychology. The notion of statistical inference and its attendant technology made it possible to insist that group data provided the only adequate empirical basis, not only for making practical decisions and evaluating theories developed in a practical context, but also for testing *any* psychological theory that claimed scientific status. Thus, the adoption of the methodology of statistical inference marked the emergence of a common research paradigm for the discipline as a whole.

This had extremely far-reaching consequences for psychological theorizing, which I have discussed elsewhere.[20] Suffice it to say that the requirements of the methodology came to impose very severe limitations on the structure of acceptable theories. For instance, it came to be accepted that a scientific psychological theory had to be such that it would be capable of generating predictions on the level of group data. This tended to rule out theories that made a fundamental distinction between surface or phenotypical patterns and underlying or genotypical patterns. At any rate, there are good grounds for asserting that the adoption of the statistical inference model had theoretical consequences for psychology that far transcend its purely technical significance.

Notes

1. A. J. Rucci ard R. D. Tweney, "Analysis of Variance and the 'Second Discipline' of Scientific Psychology: A Historical Account," *Psychological Bulletin*, 87 (1980), 166–184

2. A. D. Lovie, "The Analysis of Variance in Experimental Psychology: 1934–1945," *British Journal of Mathematical and Statistical Psychology*, 32 (1979), 151–178

3. Rucci and Tweney, "Analysis."

4. Lovie, "Analysis."

5. W. Wundt, *Logik: Eine Untersuchung der Principien der Erkenntnis*, 2 Vols. (Stuttgart: Enke, 1880–1883); K. Danziger, "Wundt's Psychological Experiment in the Light of His Philosophy of Science," *Psychological Research*, 42 (1980), 109–122.

6. E. B. Titchener, "The Type-Theory of the Simple Reaction," *Mind*, 4 (1895), 506–514.

7. K. Pearson, *The Grammar of Science* (London: Black, 1892).

8. L. Hogben, *Statistical Theory* (London: Allen and Unwin, 1957).

9. W. Wundt, *Die Psychologie im Kampf ums Dasein* (Leipzig: Kröner, 1913).

10. W. James, "A Plea for Psychology as a Natural Science," *Philiosophical Review*, 1 (1982), 146–153.

11. K. Danziger, "The Social Origins of Modern Psychology," in A. R. Buss, ed., *Psychology in a Social Context* (New York: Irvington, 1979); J. M. O'Donnell, *The Origins of Behaviorism: American Psychology, 1870–1920* (New York: New York University Press, 1985).

12. K. Danziger, "The History of Introspection Reconsidered," *Journal of the History of the Behavioral Sciences*, 16 (1980). 241–262

13. F. Samelson, "World War I Intelligence Testing and the Development of Psychology," *Journal of the History of the Behavioral Sciences*, 13 (1977), 274–282.

14. H. M. Walker, "Concerning the Standard Error of a Difference," *Journal of Educational Psychology*, 20 (1929), 53–60.

15. E. L. Thorndike, *An Introduction to the Theory of Mental and Social Measurement*, 2nd ed. (New York: Columbia University, 1913); H. D. Rugg, *Statistical Methods Applied to Education* (Boston: Houghton Mifflin, 1917); T. L. Kelley, *Statistical Method* (New York: MacMillan, 1923); A. S. Otis, *Statistical Method in Education Measurement* (New York and Chicago: World Book Co., 1925); H. E. Garrett, *Statistics in Psychology and Education*, 1st and 2 nd eds. (New York: Longmans, Green, 1926 and 1937); K. J. Holzinger, *Statistical Methods for Students in Education* (Boston: Ginn, 1928).

16. Garrett, *Statistics;* H. Sorenson, *Statistics for Students of Psychology and Education* (New York: McGraw-Hill, 1936).

17. F. Samelson, "Struggle for Scientific Authority: The Reception of Watson's Behaviorism, 1913–1920," *Journal of the History of the Behavioral Sciences*, 17 (1981), 399–425.

18. L. Thurstone, "A Law of Comparative Judgment," *Psychological Review*, 34 (1927), 273–286.

19. E. L. Thorndike, "A Note on the Accuracy of Discrimination of Weights and Lengths," *Psychological Review*, 16 (1909), 340–346.

20. K. Danziger, "The Methodological Imperative in Psychology," *Philosophy of the Social Sciences*, 15 (1985), 1–13.

3 Survival of the Fittest Probabilist: Brunswik, Thurstone, and the Two Disciplines of Psychology

Gerd Gigerenzer

Egon Brunswik (1903–1955) made the uncertainty of perceptual cues the basis of his Darwinian view of perception, which he called "probabilistic functionalism." The reasons for the almost unanimous rejection of Brunswik's views during his lifetime are investigated with respect to the division of psychology into two more or less unrelated disciplines, the "experimental" and the "correlational." Perception was the topic of the experimental discipline, and its method was experiment. Intelligence was the original topic of the correlational discipline, to be studied with Karl Pearson's correlational statistics. These bonds between topic and method were of historical rather than substantive origin. Brunswik proposed to study perception in the natural environment by correlational statistics, rather than in the laboratory by the principles of experimental isolation and control. In addition, he considered the perceptual system as an "intuitive statistician" of the Pearson school. Thus he integrated what many of his contemporaries thought could not be, which resulted in incomprehension and rejection. Ironically, the "intuitive statistician" became a major metaphor in the cognitive revolution that occurred after Brunswik's death— although equipped with the new inferential statistics institutionalized in experimental psychology and not with correlational statistics. Brunswik's failure is compared with the first success of a probabilistic conception, that of L. L. Thurstone, who was the fitter of the two in adapting to the two rigid disciplines.

In 1963, Edwin G. Boring, dean of the history of psychology, could summarize the relationship between psychology and probability in only two words: "Determinism reigns." [1] And it reigned over two empires: Man was governed by deterministic laws, and psychology, the science of man, was itself also governed by deterministic laws, by the "Zeitgeist," with the help of local "Ortsgeister" (rather than by the incalculable "Great Man"). Since the history of psychology was for a long time written by a stenographer of the "Zeitgeist," it is not all that surprising that the history of probabilistic ideas in psychology has until now been barely touched.

My focus of interest is the Viennese psychologist Egon Brunswik (1903–1955), who postulated uncertainty (of perceptual cues) as the fundamental property of human and animal perception, and drew rich implications from his "probabilistic functionalism" for methodology and experimental design. His answers to the fundamental questions "What to look for?" and "How to proceed?" were not those of his contemporaries. If Brunswik's revolutionary ideas had triumphed over Boring's *Zeitgeist*, experimental psychology would have gone a different route, but his probabilistic revolution failed at the time. "Brunswik was a brilliant man who wasted his life." [2] This was the judgment of Boring, the determinist, on Brunswik, the probabilist, an authoritative judgment shared by the majority at the time—and a judgment Brunswik might have anticipated when he finally committed suicide in 1955.

My question is the following: Why did probabilistic functionalism meet with almost universal rejection? Is the answer a general dislike of and unfamiliarity with probability by Brunswik's contemporaries? Or even that his writings were difficult to read, as is repeatedly claimed? There is some, but only a little, truth in the last point, if one refers to the inconsistencies in his writings, as I shall show. The first explanation is, as I shall argue, false; probabilistic thinking was not generally rejected.

I shall argue that the division of psychology into two more or less unrelated and even opposing disciplines lies at the heart of the rejection. Brunswik integrated the study of perception, *the* topic of experimental psychology, with the methods of the second discipline, correlational psychology. He thereby connected what was understood by his contemporaries to be unconnected, and they reacted with incomprehension and rejection. The case of Louis Leon Thurstone, who had the first success with a probabilistic conception in psychology, is used as a control for the argument. First, however, I shall present some background information and describe the major ideas of the unsuccessful and the successful probabilists.

1   The Background

Egon Brunswik was born in Budapest, 1903; his father was Hungarian, an engineer, his mother Austrian. Louis Leon Thunström—the Swedish name of his parents was changed later to Thurstone—was born in Chicago in 1887. There are some parallels in their lives. Both received a degree in engineering before they finally turned to psychology. Since engineering in Chicago meant something different from engineering in Vienna, this first parallel might be considered as a first line of divergence, leading to Thurstone's pragmatic view, his interest in application rather than in theoretical structure, as opposed to Brunswik's systematic view, concerned with consistency in theory and methodology.[3] Brunswik received his doctorate in 1927 in Vienna, Thurstone in Chicago in 1917. Both developed their probabilistic ideas in the same field, in perception and psychophysics, and generalized them to other areas of psychology. Brunswik studied with Karl Bühler and Moritz Schlick, and started his career as an assistant to the former in Vienna. The key to his probabilism is his experiments on perceptual constancy, first published in the late 1920s. Thurstone's concern with probability did not develop until he returned to Chicago in the mid-1920s, after seven years in the newly established Division of Applied Psychology at the Carnegie Institute of Technology in Pittsburgh. He himself wondered why he did not undertake any fundamental theoretical research until his breakthrough in 1927. Thus, in spite of Thurstone's being elder, the framework for their probabilistic ideas was laid out at about the same time.

Edward C. Tolman seems to have been responsible for Brunswik having spent a year as a Rockefeller fellow in the Department of Psychology at the University of California at Berkeley in 1935–1936. In 1937 Brunswik permanently joined the Berkeley faculty, and remained a member until his death. The occasion for their contact and later friendship was that Tolman, at the time one of psychology's major figures, spent a sabbatical year in Vienna, where he recognized the similarities between his purposive behaviorism and Brunswik's intentionalistic view of perception.[4]

Thus both Brunswik and Thurstone lived in the same country from 1937 on, made their major contributions in the same time interval, between 1925 and 1955 (the year in which they both died), and shared interests in probability, psychophysics, and the development of an "objective psychology." Nevertheless, Thurstone seems never to have mentioned Brunswik in his writings; Brunswik at least makes a few short remarks concerning Thurstone.

2 The Ideas

2.1 The Vienna Brunswik

The key to Brunswik's probabilistic functionalism can be found in his early experiments on perceptual constancy.[5] Perceptual constancy was then being much studied in Europe, but Brunswik approached this topic from a somewhat different point of view. His approach was committed to the methodological physicalism of the Vienna Circle and to Karl Bühler's call for an "objective" psychology.[6] For Brunswik, this meant reducing introspection to the very last unavoidable minimum, thereby maximizing inter- and intrasubjective agreement of observation and communication. Brunswik distinguished between the *Gegenstand*, a property of an object measured in physical units, and the *Gegebenheit*, the same property as perceived by an individual, also to be measured in physical units. For example, in the study of size constancy, the *Gegenstand* was the actual size of a distant object, and the *Gegebenheit* was the perceived size. To measure the perceived size of a target object, a standard series of objects differing in size was presented to the individual. The individual was then asked to select that object in this series that was perceived to be of the same size as the target object. The unavoidable minimum of introspection that was left was these subjective judgments of equality.

There were three basic findings concerning the perception of size, brightness, area, volume, shape, and other physical properties of objects. I shall use size constancy as an example. First, if the distance of the object to the individual is varied, then the perceived size (*Gegebenheit*) remains relatively constant despite the retinal projection varying according to the laws of optics. For instance, the size of a man (*Gegenstand*, or distal stimulus) is perceived as about the same at a distance of three meters and at six meters, although the retinal projection (proximal stimulus) is twice as great at a distance of three meters. This is the basic phenomenon to which the expression "perceptual constancy" or "object constancy" refers. The second general result was that perceptual constancy is *not* perfect. If constancy were perfect, Brunswik would say that perception "achieves the distal stimulus." In principle, perception may "achieve" either the distal stimulus, the proximal stimulus, or some compromise (*Zwischengegenstand*). In fact, perception does not "achieve" the distal stimulus, nor—and to an even lesser extent —does it "achieve" the proximal stimulus, when instructed to judge these stimuli. That is, judgments of both physical size and retinal projection are incorrect. Perception tends to achieve merely some *Zwischengegenstand*, a compromise between these two "poles of intention." A third general result was that perceptual achievement was a function of the individual's learning history, age, sex, and "attitude," i.e., whether he or she had been instructed to judge the distal or the proximal stimulus. Brunswik's conclusions from these results were the following:

1. *Perceptual cues are ambiguous and can be replaced by other cues.* The observation was that size constancy is imperfect and that other achievements, like brightness, color, and shape constancies, are even more imperfect. Brunswik's question was, Why is constancy imperfect? Bühler had offered his *duplicity principle* as a general model of the mechanism of perceptual constancy. The principle states that invariance (constancy) is mediated through the variance of two kinds of information, which are processed together. In size constancy, these are the retinal projection and at least one distance cue. Now, why is perceptual constancy imperfect? Brunswik's answer was, Because the information necessary—like the distance to the organism—has to be *estimated* by the organism. There exist many cues providing information for distance, like binocular parallax, convergence of eye axes, accommodation, vertical position of the object, and number of in-between objects. *None* of these cues, however, he emphasized, is a perfect cue for distance. They are all *uncertain* cues, some more and some less ambiguous.

This ambiguity of perceptual cues is the very point where probability is located in Brunswik's system: There exists an irreduceable uncertainty in the relationship between the cue available and the distal object to be perceived.

To reduce the uncertainty, the perceptual system uses multiple cues, depending on what is available in the experiment. The experiments showed that in principle every cue could be *substituted* for by other cues. This principle of *cue substitutability* implied that in a concrete situation there may be an immense number of possibilities for how many and what cues are used in the mediation process.[7]

2. *Perception is intentionalistic.* The observation was that perception tends to "achieve" the distal stimulus rather than some proximal stimulus immediately given on the organism's receptor surface. Brunswik's interpretation was a Darwinian or functional view of perception as an "intentionalistic achievement": In order to survive, the organism must "know" the distal stimuli and cannot be content with the immediately given proximal stimuli. Perception was seen as an adaptation. Note that he distinguished this intentionalistic achievement of perception, measured objectively as described above, from the *conscious* intention of the intellect, as "measurable" by introspection: What is intended consciously may not be achieved perceptually; and what is achieved perceptually may be different from what the intellect "thinks" is achieved (e.g., the "real world" rather than *Zwischengegenstände*).

3. *Perceptual cues are learned.* The fact that the degree of constancy depends on age, sex, training, and knowledge led him early to conclude that perceptual constancy must be learned.[8] Learning of perceptual cues as a means of achieving perceptual constancy may itself be partially substituted by learning of verbal cues, which then allows for "conceptual constancy" by means of words.[9]

4. *The program: inventory of perceptual achievements.* Brunswik called this program *Psychologie vom Gegenstand her* ("psychology in terms of objects"). This

expression emphasizes objectivity, connected with a pure consideration of perception as an *achievement*. "Achievement" here means how well perception captures the physical world, measured quantitatively. The very last aim is an empirical one: an inventory of the environment as it is "achieved" by perception and its comparison with the actual physical world.

5. *The generalized program*.[10] Perceptual constancy is concerned with the question whether a property like perceived size remains constant under variation of *external* conditions like the distance to the observer. Brunswik's generalized problem of constancy was concerned with the question of whether a property like perceived size remains constant under variation of other properties of the *same* object like its shape or color. This generalization allowed for a broad research program, which he himself called "multidimensional psychophysics." Examples are experiments on coins varying in value, area, and number; none of these properties ("poles of intention") could be judged independently of the other two. For instance, the area of coins was overestimated if the value was higher; what was perceptually "achieved" turned out to be a compromise (*Zwischengegenstand*) between area and value.

To summarize in Brunswik's own words: The study of perception asks questions about the "internal difference between two methods of observation,"[11] between *Gegebenheit* and *Gegenstand*, perception and physical measurement. The philosophical conclusion drawn from these experimental results was straightforward: The immediately given world of perception and the physical world do not coincide. The portrayal of the physical world by intuitive perception "is in a certain way imperfect, not completely isomorphic; an investigation of this incompleteness is the very task of psychology."[12]

2.2   The Berkeley Brunswik

The basic ideas were those of the Vienna Brunswik: The principles of cue ambiguity, cue substitutability, and goal-orientation of perception were now summarized as the principle of *vicarious functioning*.[13] Simple cause-effect relationships (a Newtonian view of the world) were considered as principles of the inorganic world, whereas vicarious functioning (a Darwinian view of the world) was considered as a significant principle of the organic world. Therefore, psychology is in the neighborhood of biology, not, as it was and still is hoped, of physics. Such a biological orientation was essentially the point of view of Karl Bühler, Brunswik's teacher, and common within psychology in German-speaking countries between World War I and World War II.[14]

Whereas all this can be traced back to the writings of the Vienna Brunswik, it was only the Berkeley Brunswik who drew the consequences from this program for a suitable methodology. These formed his idea of a "representative design," which was a challenge to the existing belief about what proper experimentation in psychology should be. Proper experimentation, as it was assumed at his time, should follow the ideas of *isolation, control*, and *one-to-one causality*: Isolate one variable, vary it systematically, hold all other variables constant; then concomitant variation in the dependent variable will indicate the causal relationship in question. This belief has been handed down to psychology from such famous origins in physics as Galileo's reputed study of the fundamental laws of falling bodies. Brunswik called this idea of an experiment the "systematic design," as opposed to his "representative design."

The methodological implications Brunswik derived from his Darwinian view of perception can be restated as follows: If perception (or behavior, respectively) works via vicarious functioning, then the systematic design is an inappropriate way to study perception, *since it does not allow for vicarious functioning.* For instance, according to the principles of isolation and control the organism can only use the cue varied in the actual experiment and no other cues. The discovery, then, that he actually utilizes it, i.e., that the variation of the cue has an effect on the individual's response, thus does not sanction the conclusion that he will use this cue when vicarious functioning is not prevented by the systematic design of the experiment. Generalizability of experimental results to the natural environment of the individual will accordingly be the exception rather than the rule. As a consequence of isolation and control, Brunswik argued, variables become "artificially" tied and untied. Therefore he compared the systematic design with a Hollywood-style black-white dramatization of the environment, similar to cliché literature: "Old-fashioned opera plots or soap melodramas that 'drip with generosity' share with cowboy movies what David Hume would have called 'inseparable' associations of noble character, overpowering strength, fairness, courage, youth, final success, and so forth, or of their opposites." [15]

Let me illustrate the point of artificial tying of variables by an exaggerated example. Imagine the question is, What cues are used for judging the intelligence of others? Three variables are selected: sex, beard, and spectacles. These are systematically varied, everything else being held constant. Such a systematic tying and untying would not be representative for our environments; bearded women that do or do not wear spectacles may lead to significant effects of variables or interactions that vanish in a different systematic design, and so on. Actually, the variation of results as a function of the specific design was and still is, unfortunately, the undeniable experience in our work as experimental psychologists. In 1975, Lee J. Cronbach's judgment was; "Generalizations decay." [16] As Brunswik put it, "The main function both of art and of systematic experimentation, then, is to shake and mold us by exaggeration and extreme correlation or absence of correlation." [17]

Natural Environment and Correlation instead of Isolation and Control   What, then, is a proper methodology, if not isolation and control? Brunswik's answer was to study perception in the *natural environment* rather than under the restricted conditions of the laboratory. This allowed for vicarious functioning, for the introduction of all relevant cues, and for the "natural" correlations between them rather than the artificial tying and untying of cues according to the principles of isolation and control. Since one could not study the whole natural environment, a *representative sample of objects* had to be taken from the environment.

The actual development of such a "representative design," as he called his methodology later, was gradual. In 1939 he published the first study on maze learning, using rats, where reinforcement was probabilistic rather than "always" or "never."[18] The artificial miniature environment he used for the rats was a simulation of an environment where the "laws" to be learned are probabilistic rather than deterministic. In the next year he published a short report on the first study of perceptual constancy in a natural environment.[19] He accompanied his subject in her usual environment and took a random sample of objects on which she focused during her daily activities. Asking her for judgments of size and distance, and comparing them with the actual size and possible distance cues, Brunswik tried out an experimental idea that psychology had never seen before.

The use of correlation analysis[20] in such a representative design allowed for the analysis of (1) the *achievement*, that is, the degree of perceptual constancy, (2) the *ecological validities*, and (3) the *cue-utilization* of an individual. *Achievement* was defined as the correlation between the perceived stimulus (judged size, brightness, etc.) and the distal stimulus (physical size, albedo, etc.). This correlation was called the *functional validity* of an individual's perception with respect to a distal variable. The correlation between a cue (e.g., retinal projection) and a distal stimulus (e.g., physical size) was called the *ecological validity* of a cue in a given environment. It indicates the extent to which a cue is capable of standing as a probability cue for the distal stimulus. Finally, the ecological validities of cues could be compared with the actual *cue-utilization* of an individual, as defined by the correlation between judgment and cue. Cues that show high ecological validity need not be used ("trusted") by an individual, and vice versa.

Probabilistic Functionalism   The term "probabilistic functionalism" defines the program of the Berkeley Brunswik: To study how an organism comes to terms with an environment that presents itself via ambiguous cues as uncertain. This program includes the study of functional validities, ecological validities, and cue-utilization. One can see how far Brunswik went beyond the traditional view of experimental psychology: analysis of achievement and correlation hypotheses (cue-utilization) rather than one-to-one cause-effect laws; natural environment instead of isolation and control; use of correlation statistics rather than tests for mean differences; sampling of both individuals *and* objects rather than sampling of individuals alone; and finally the study of the "texture" of the environment in terms of ecological validities. This last idea was completely new, since it did not arise in the systematic design, where the "texture," i.e., the tying and untying of variables, was generated by the experimenter. From the point of view of systematic design, the "texture" of the environment was the subject matter of textbooks on experimental design, something to be constructed artificially, not to be studied experimentally.

Just at the time of Brunswik's challenge, however, the systematic design became linked with new statistical developments. In particular, R. A. Fisher's analysis of variance allowed for generalizing the ideas of isolation, control, and one-to-one cause-effect relationships to two or a few more independent variables. The new inferential statistics was soon institutionalized in experimental psychology, leaving the Darwinian view of perception no ecological niche, as it were.[21] Although Brunswik's challenge to the purpose and to the methodology of experimental psychology was widely recognized, his program was completely rejected during his lifetime.[22] Let us now see how to make a probabilistic idea successful.

2.3 The Success Story: Thurstone's "Quantitative Rational Science"

Brunswik's contribution to psychology can be reconstructed as the consistent and uncompromising extension of his basic experimental findings in perceptual constancy. In contrast, Thurstone's voluminuous work focuses on distinct areas of research. There seems to be little in common in terms of a guiding psychological principle like Brunswik's idea of *cue ambiguity*. The common denominator seems to be Thurstone's concern with psychology as a "quantitative rational science," what he called "mathematical psychology"; for him this meant to make psychology both "objective" and "applicable" by developing measurement models for quantifying psychic phenomena—in this very broad sense of quantification, which connects his "law of comparative judgment" to his work on multiple factor analysis and intelligence measurement. In the following I shall consider only that work of Thurstone in which probabilistic ideas were of essential importance, that is, his work on psychophysics published in 1927, which he himself regarded as the best work he ever produced.

In 1927, Thurstone published three papers suggesting a new solution to an old problem.[23] The old problem was, How to measure a psychological magnitude, for instance, perceived brightness? The old solution was found by Fechner—as he tells us, when he lay in bed on the morning of the 22nd of October 1850 puzzling over how to fight the materialism of his time and to support his monism and spiritualism. In order to demonstrate the essential unity of mind and body, indeed of all existence, Fechner had to find the relationship between physical and psychic events. For this, in turn, he needed a measure of sensation. His solution was to determine the just noticeable differences (jnds), to assume that they are all psychologically equal, and then simply to count the number of jnds from the absolute threshold to the sensation being measured. This number would indicate the quantitative measure. The relationship he finally found was his famous logarithmic law, in which he saw a principle of cosmic importance.

Thurstone faced the same problem, measurement. His solution, however, was based on the normal distribution. The use and abuse of the normal distribution was already quite common, as exemplified by Francis Galton's enthusiasm: "I know of scarcely anything so apt to impress the imagination as the wonderful form of cosmic order expressed by the 'Law of Frequency of Error'. The law would have been personified by the Greeks and deified, if they had known of it." [24] Whereas Galton found in the normal curve a piece of cosmic order such as Fechner found in his logarithmic law, Thurstone's approach to the probability distribution was entirely *pragmatic*.

In his first paper of 1927, "Psychophysical Analysis," Thurstone frankly asserts that "this analysis has nothing really to do with any psychological system"(!), that he is "not now interested in the nature of the process" (he never became so), and that he does not want "to disturb the main argument with systematic irrelevances ... regarding the psychic or physiological nature of the psychophysical judgment."[25]

The solution he proposed was the following. First, he assumed that the repeated presentation of an object does not always result in the same psychological value (e.g., of brightness, loudness, etc.), but in a normal distribution of such psychological values. Second, imagine that two objects a and b are presented and the individual is asked which one is louder, brighter, etc. Then the probabilities p_{ab} that a is judged louder, brighter, etc., than b will depend on the overlap of the two normal distributions. With some additional assumptions on the variances and covariances of the two distributions, an easy way to state this function is

$$p_{ab} = N(x_a - x_b) \qquad \text{for all } a, b \in A, \tag{1}$$

where N is the cumulative normal probability function, A is a set of n stimuli, and x_a and x_b are the means of the respective distributions, which are considered as the psychological values sought. For instance, $p_{ab} = .5$ if $x_a = x_b$. Thus far the model. To test the model, repeated presentations of all stimulus pairs from a set of n stimuli provide relative frequencies as estimates for the p_{ab}s. Since equation (1) implies certain restrictions on the values of the probabilities (relative frequencies), the model can be tested. If it is accepted, then the equations can be solved for the unknown scale values x_a, x_b, etc..

The fascinating aspect about this use of probability is that Thurstone made measurement an empirical question. The desired psychological scale values are formulated within a mathematical model, which is tested against empirical relative frequencies, and if the model holds, scale values can be derived. This integrative view of mathematical modeling and measurement was real progress as compared to the earlier approaches as well as to the later psychophysical scaling methods of Stanley S. Stevens, where measurement was regarded as separate from modeling, as a preliminary step to model building and testing only.

Note that Thurstone starts, contrary to Brunswik, without a psychological interpretation of his probabilistic ideas. He asserts that his choice of the normal distribution (as opposed to a different probability distribution) was not motivated by some theory about how perception works, but by the simple pragmatic ground "that its presence can be experimentally tested."[26] This lack of psychological interpretation enabled Thurstone and his followers to apply the function broadly. Let us consider this reluctance to ground his probabilistic concepts in a precise interpretation.

With respect to his first publication in 1927, the reader is left with an impression, not stated clearly, that the probability curve is used as a model for *intraindividual* variation of responses to the repeated presentation of the same stimuli. This would imply certain theoretical restrictions on how perception works under repeated stimulus presentation (e.g., deviations from an assumed "true" psychological value must remain independent over replications) as well as some practical restrictions. An important practical restriction would be that the scale could be constructed only for a narrow range, since probabilities different from $p_{ab} = 1$ or $p_{ab} = 0$ only occur when the difference between the stimuli compared is small. Probabilities that differ from one and zero are, however, a necessary condition for differentiation and therefore for scaling the stimuli. In particular, in his attempt to measure attitudes, Thurstone soon must have realized that probabilities of one or zero might be the rule rather than the exception if one starts with an *intraindividual* interpretation of the probability distribution. The way he solved this practical problem was to consider the probability distribution as *interindividual* variation, since lack of agreement between individuals generated data where both $a > b$ and $b < a$ appeared. Accordingly, in his second publication of 1927, in which he promoted his probabilistic idea stated in equation (1) to a "law of comparative judgment," he then distinguished two cases of application: the normal curve as a model for intraindividual variation and as a model for interindividual variation. The choice seemed to be a matter of convenience for the experimenter, rather than one of psychological theory. What the choice meant in terms of a theory of perception, a theory of attitudes, or of something else never was made explicit by Thurstone. He thought of his probabilistic model simply as a testable formal model. This first probabilistic measurement model in psychology was influential not only as the basis for rewriting psychophysics around the normal probability curve and for measuring attitudes; it also laid the foundation of today's probabilistic measurement theories and anticipated signal detection theory.

A final observation illustrates Thurstone's ability to give up ideas quickly, if they seemed to be unsuccessful, and to adapt to different ideas if they proved successful. A few years before his breakthrough in 1927, Thurstone enthusiastically rejected any kind of stimulus-response psychology on the grounds "that we have come to forget the individual person altogether." He favored Freudian ideas, drives, motives, and the primacy of the organism over the stimulus: "I suggest that we dethrone the stimulus."[27] A few years later, he made his breakthrough with a purely probabilistic stimulus-response model, not hesitating to treat individual differences as "error." Now, his turn came to forget the individual person (and, as far as I can see, he never quoted these earlier articles again).

3 Why Failure, Why Success?

The question is why Brunswik's and Thurstone's probabilistic ideas received such different responses from the scientific community—remember: different responses at about the same time, in the same country, and with respect to the same topic, namely, perception and psychophysics. Looking at Brunswik alone, one might hypothesize a general dislike of and unfamiliarity with probability and uncertainty in a scientific community sharing a deterministic, "nomothetic" view. The case of Thurstone, however, demonstrates that it cannot be that simple.

The key to my answer is the general misunderstanding of Brunswik's ideas rather than a general negative evaluation of them. My answer is as follows. In Brunswik's time psychology was divided into two major research programs, which were called the "two disciplines of scientific psychology" by Cronbach in 1957.[28] Each was a different arrangement of ideas about purpose, subject matter, and methodology; each was based on a different metaphor of man. The connection of ideas within each discipline was fairly *arbitrary*, historically determined, rather than connected by inherent logical or psychological necessity (as arbitrary indeed as the division into two disciplines itself was). I see Brunswik's program as an *attempt at a consistent integration* of ideas from the two disciplines about purpose, subject matter, and methodology. This means that he tried to untie the arbitrarily connected ideas within each discipline and thereby unify psychology. Moreover, he went beyond the scope of the two disciplines with respect to his metaphor of *man as an intuitive statistician.*

For psychologists, the two disciplines functioned as cognitive schemata, acquired by professional conditioning. These two schemata consisted of the two respective clusters of ideas. Their function was to free one from asking the problematic questions "What to look for?" and "How to proceed?" and to provide time for empirical work. An understanding of a new program, however, depended on whether it could be *assimilated* into one of the existing schemata. Understanding by means of assimilation, in turn, depended on the existence of similar clusters of ideas about purpose, subject matter, and methodology. Otherwise, forced "assimilation" would work like a Procrustean bed. The only alternative then would have been, to use Piaget's terms, to *accommodate* the existing schemata. Accommodation, however, would have demanded the questioning of the respective clusterings of ideas (as well as of the underlying conception of man), which seemed to be a horror rather than a challenge for Brunswik's contemporaries. What remained was the attempt to assimilate what could not be assimilated, that is, to understand Brunswik's program in terms of one of the two existing cognitive schemata. This Procrustean situation resulted in fundamental misunderstandings and a general rejection of what Brunswik wanted. In contrast, Thurstone left the autonomy of the two disciplines undisturbed.

4   The Two Disciplines

The first discipline, the "Tight Little Island"[29] of "experimental psychology," had its origins in the ideas of Fechner and Wundt. The traditional topics were perception, memory, and learning. In Brunswik's time it became largely restricted to a behavioristic and mechanistic conception of man, thereby eliminating mental concepts from the subject matter. The second discipline, the "Holy Roman Empire" of "correlational psychology" or "differential psychology," had its origin in Galton's and Pearson's concern with individual differences in "natural ability," closely linked to their eugenics program. This duality originated for historical reasons rather than as a straightforward consequence of the subject matter.

4.1 The Experimental Discipline

In Brunswik's words, the experimental discipline was nomothetic, systematic, and reductionist. Hull, Hilgard, Krech, Lewin, Postman, and other leading figures all subscribed to a nomothetic psychology. In Postman's words, "The essence of the nomothetic position is that it looks for uniform laws of behavior, e.g., for stimulus-response correlations which are unity, or, at least, theoretically conceived to be unity." [30] Nomothetic psychology included both stimulus-response psychology without any interest in mediating processes, like Skinner's operant conditioning view, and theories focusing on mediating processes, whether molar (e.g., Postman) or molecular, reducing psychology to physiology (e.g., Krech). As Hull stated it, the ideal of a nomothetic psychology was "a full-blown natural science." [31] The task of psychology was seen as finding deterministic laws—in Hull's case, laws between inner processes stated in a quasi-physiological language—whose model was the law of falling bodies. The methodology for finding these laws was also thought to be Galilean, or "systematic," to use Brunswik's language; the basic elements were isolation and control. From the late 1930s on, this methodology could be generalized to more than one independent variable by using Fisher's statistical work, developed in agriculture. As I mentioned earlier, Fisher's statistical models were fully integrated into the deterministic experimental discipline: There were true values and error only, the later being attributed to the experimenter, e.g., as measurement error or lack of control over relevant conditions, and not to the subject matter itself.

What was the underlying conception of man in a reductionist experimental psychology? Since both the conception of law and the methodology were taken over from physics, some essential aspects that distinguish psychology from physics—from a nonreductionist view—were cut off in a Procrustean bed: Laws should consist of *variables* and *objects* only. The third component that distinguishes psychology from physics, the *individual* who tries to come to terms with *objects* and *variables* by means of "hypotheses," did not appear in these laws. Individual differences were of no interest and were treated as error in the analysis of variance. The implicit, underlying conception of man was that of a physical object that could learn, that is, whose quantitative values in a few basic variables—like Hull's "habit strength"—could be modified by external manipulation. Henceforward I shall call this the "physical metaphor of man."

4.2 The Correlational Discipline

In contrast, the second discipline was from the beginning, with Galton's interest in natural ability and eugenics, genetically oriented and thereby not concerned with learning by external manipulation. This original interest explains the obvious lack of theory in this second discipline (which was meant to be a genetic rather than a psychological theory of natural ability, or intelligence). Instead of finding genetic laws, not available at the time, the purpose was to find a measurement instrument for natural ability to allow for an "objective" registration of individual differences.

This was a necessary step toward a successful eugenics program. The methodology was based on tests and correlation analysis, a means to an "inductive" inventory of man's inner life—in terms of "factors" of intelligence or personality. Besides those attempts to prove the basic assumption, i.e., that intelligence or other traits are inherited, the purpose was to quantify individual differences with respect to these "factors." The underlying conception of man was Darwinian rather than a physical one: Although man was again, like a physical object, construed as a bundle of variables (traits), men differ from one another, and these differences were made the basis for social and political arguments. However, Brunswik's principal opponents were the advocates of a physical rather than a Darwinian metaphor of man, the latter not taking part in the debate at all.

4.3 The Two Disciplines as Arbitrary Connections of Elements

The first point I want to make is that the elements within each discipline (elements concerning *purpose*, *subject matter*, and *methodology*) had *no* necessary logical or psychological affinity to one another. Their connection was historical rather than substantive. I shall present some examples of these arbitrary connections.

The study of perception, memory, and learning was historically associated with the goal of law finding, as the study of intelligence and personality was historically associated with the goal of quantifying individual differences. Aside from these historical reasons, there was no substantive reason to consider the determination of individual differences as the goal when studying intelligence, and to consider individual differences as an annoyance when studying perception. Only in the last two decades, however, since the dominant view of Stevens's approach to psy-chophysics (based on averaging across individuals and a "one independent variable–one dependent variable" design) has been overcome, have we come to realize that individual differences in perception cannot be simply eliminated by averaging, since different individuals behave as if they use different perceptual strategies.[32]

With respect to the methodology, both stimuli (e.g., test items) and individuals were sampled in the correlational discipline, whereas only individuals were sampled in the experimental discipline. Nevertheless, even in the latter case generalization was intended over individuals *and* stimuli. Note the asymmetry in the logic of induction! An example, as recent as 1977, is a review of research on the effect of the sex of the experimenter on the performance of the subjects. In a total of 63 studies reviewed, *thousands* of individuals were used, whereas in about 40% of the studies only a single (!) experimenter of each sex was used as stimulus, and in the other studies not many more than two of each sex.[33] It is not too surprising that the generalizations drawn in these studies ("sex of experimenter has/has no influence") are con-tradictory and do not allow a definite conclusion.

Although the clustering of subject matter, purpose, and methodology was arbi-trary, the two disciplines became distinct from one another. The journals of one discipline had little influence on the journals of the other;[34] their programs were presented in different textbooks, or in more or less unrelated chapters of the same textbook. The psychologist's values were affected by the schism; in 1954, Thorn-dike[35] reported that a psychologist's esteem for the experimenters was correlated −.80 with his esteem for the colleagues of the correlational discipline! The issue of the separation had been repeatedly taken up in the presidential addresses before the American Psychological Association. In the 1938 address, John F. Dashiell complained about the independence and direct antagonism of the two subgroups, and optimistically discerned some signs of a developing rapproche-ment.[36] In the 1957 address, however, Lee J. Cronbach's judgment was that "psy-chology continues to this day to be limited by the dedication of its investigators to one or the other method of inquiry" and even talked about a "standstill."[37] (As late as 1975, he judged the "theoretical progress to have been disappointing"[38] and gave a survey of his own program for a combined discipline, a science of "aptitude × treatment interactions.") Also in 1957, Boring argued that the schism was partly healed in the 1940s; first because the "experimentalists could not longer look down their noses at applied psychology," since mental testing had worked well during the war, and second, since all psychologists "needed to use the newer statistical methods as well as the Fisher techniques for assessing the significance of results."[39]

Finally, given these historically evolved, arbitrary bases for our research programs, is it not strange that we psychologists tend to neglect our history, believing in simple cumulative advance? Or must these two go hand in hand: the neglect of history and its tyranny over our ways of thinking?

5   Man as an Intuitive Statistician

Brunswik's program was based on a metaphor of man that was quite different from those in the two disciplines. His conception of probability and his metaphor of man were two sides of the same coin. However, Brunswik did not spell out the underlying metaphor until shortly before his death. Thereby he himself may have been partially responsible for the misunderstandings concerning the role probability plays in his psychology. I shall present a brief overview of the development of Brunswik's conception of man in his writings.

In his book *Wahrnehmung und Gegenstandswelt*, which summarizes the program of the Vienna Brunswik, man is a *perceptual being* (*Wahrnehmungswesen*). This perceptual being is only part of the whole personality; it extends over two steps of Bühler's three-step model, instinct and conditioning, not including intellect. Within the whole personality, the perceptual being is autonomous; it functions inductively by conditioning. Conditioning means that relationships between proximal cues and distal stimuli are learned from their joint occurrence. However, adaptation by experience is slow compared with the quick adaptation by means of the intellect. To summarize in Brunswik's words, the perceptual being is like a "stupid animal" that acts under the principle "better uncertain than nothing" without being conscious of this uncertainty. The inductive uncertainty in the perceptual inference, e.g., the inference about the size of an object, is *not* noticed. We do not see the size of a tower oscillating according to the uncertainty of the cues available. We have the illusion of certainty, i.e., of a unique perceptual impression.

Under the influence of Tolman, but while still in Vienna, Brunswik described organisms (both rats and human beings) as venturing *hypotheses* about the relationships between cues and objects. In his early writings in English, during the transition time between Vienna and Berkeley, it is often not clear that his physicalism was only methodological and not thematic, thereby sometimes suggesting a physical metaphor of man. To quote him in 1937: "In problems of reception (i.e., perception and thinking) as well as of action, psychology in terms of objects [as he called his program at the time] would turn out to be a physical and biological natural science."[40]

Similarily, in the same article he refers to his basic idea of multiple and variable causal mediation, symbolized in his "lens model," as a process that "can be completely understood in terms of a certain type of *physical* process." And he adds; "The only question remaining open for a physical explanation is as to how natural or artificial tools (or organismic "institutions") like collecting lenses . . . might have developed at all. This general genetic question . . . belongs to the field of theoretical biology."[41]

These few quotations show that part of his writing could be understood not only as a methodological physicalism but also as a variety of thematic physicalism. The reduction of mediating processes to mere physical processes is not far from the reductionist psychology and the ideal of a thematic unity of sciences to which he was otherwise deeply opposed.

In the same and in subsequent publications Brunswik introduced for his perceptual being the analogies of the instinctlike Id of psychoanalysis, the creative thinking in Gestalt psychology, the intuition of essences (*Wesensschau*), Helmholtz's unconscious inference, and the finite mortal individual who can do nothing but make a posit, or wager (Reichenbach's terms).

Only very late, in the Berkeley discussion and in two papers published posthumously, did Brunswik find the concise formula for his underlying conception of man: the perceptual being as an "intuitive statistician." This formula spells out the essential link between probability and Brunswik's conception of man. Based on past experience, the perceptual being must guess his world by means of ambiguous cues, similar to the statistician who inductively must estimate unknown parameters. In constrast to the reasoning of the professional statistician, the process is *intuitive* rather than conscious: Neither the data (cues), the weighting of the cues (correlational hypotheses), nor the statistical character of the outcome (perception) is "known." The intuitive statistician does not perceive the "confidence intervals" of his estimates. Above all, the intuitive statistician infers his world by calculating correlations; he is a statistician of Karl Pearson's school, and not of those statistical schools—like Fisher's—that became dominant at this time in experimental psychology.

6   Misunderstood Probabilism

The foremost prominent experimental psychologists at the time participated in the two debates on Brunswik's probabilism. The first debate was arranged between

Clark Hull, Kurt Lewin, and Egon Brunswik at the Sixth International Congress for the Unity of Science in Chicago, September 1941. The second debate was held at the Berkeley Conference for the Unity of Science at the University of California, July 1953. The second debate was entitled "Symposium on the Probability Approach in Psychology"; the participants were Herbert Feigl, Ernest R. Hilgard, David Krech, Leo Postman, and Egon Brunswik. The debates were published in the *Psychological Review* in 1943 and 1955. I shall use these debates to make my point that the two disciplines of psychology acted as cognitive schemata for Brunswik's contemporaries. His probabilism was distorted in the experimentalists' schema.

6.1 The Attempt to Integrate

I consider Brunswik's probabilism as an attempt at a consistent integration of elements from the two disciplines. Let us consider this in detail.

His major area of research, perception, was the traditional topic of the experimental discipline. His purpose, to study the achievement of an individual in his natural environments, was, however, quite different from that of his experimental colleagues. With this focus on what he called "what-problems," the achievement, rather than "how-problems," the study of mediation, he shared the inventory program of the Galton-Pearson discipline. Moreover, with his focus on adaptation to a natural environment rather than to an artificial set of intelligence test items he followed in Darwin's footsteps even farther than Galton and his followers had. They gave up the natural environment in favor of measurement in controlled settings, as exemplified by intelligence testing. Here Brunswik joins Whitman, Heinroth, Lorenz, and other ethologists who followed Darwin's naturalistic, observational, and descriptive approach. Quite distinct from the ethologists' program, however, his program was the unique attempt to integrate the study of adaptation to the natural environment with objective measurement in the sense of methodological physicalism—to unite Darwin with the Vienna Circle.

Brunswik's methodology, representative design, allowed all variables to vary simultaneously. Consequently, the role of variables could only be ascertained after the fact, by ex post facto correlation analysis. Working backward via partial correlation analysis, however, was the methodological principle of the correlational discipline, which could hardly be assimilated into the experimental psychologists' schema at the time. This difficulty in cognitively unknotting the ideas of isolation and control from perception, on the one hand, and correlational analysis from intelligence, on the other, is exemplified in Hilgard's reaction to Brunswik. When Hilgard rejected Brunswik's methodology, he talked about correlational analysis as tied to the nature-nurture controversy (which concerns intelligence); he ignored Brunswik's novel integration of perception and correlation. And let him speak the final word: "Correlation is an instrument of the devil."[42]

Let us compare this with Thurstone, who is of particular interest since he worked both on perception and intelligence. When studying perception, Thurstone's probabilistic law followed the principles of isolation and control with its "one independent variable–one dependent variable" approach. Besides measurement, his purpose was law finding; he regarded his "law of comparative judgment" as the third psychophysical law in addition to Fechner's and Weber's laws. Thus, when he studied the traditional subject matter of the experimental discipline, he did this with the traditional purpose and methodology. However, when Thurstone's interests returned from perception to intelligence, a topic on which he had published before his breakthrough, at the same time he shifted to correlation analysis and factor analysis as the methodology. From this moment on, isolation, control, and law finding were passé, thereby leaving the historically established ties between topic, methodology, and purpose intact.

6.2 The Failure to Understand

In addition to his attempt at integration, Brunswik's metaphor of man provided a second obstacle to understanding his ideas. Probability and the conception of man were, as I have argued above, two sides of the same coin. Consequently, it turned out to be difficult to understand where probability entered into his system for all those whose theoretical position was based on a different (implicit) metaphor of man.

Brunswik's experimentalist colleagues tried to understand the role that probability plays in his system in analogy to the role that probability plays in physics. Note that this physicalist analogy was wholly in keeping with the experimentalists' emulation of physics as the ideal science. It would not have come to mind as easily from a Darwinian point of view, where uncertainty is represented by such metaphors as the "irregular tree," or from the point of view of telecommunication, which would have suggested analogies like "noisy communication" and "redundancy" in order to understand Brunswik's probabilism.

After Brunswik had advanced his program in the first discussion in Chicago, both Hull and Lewin criticized it using examples from physics and an understanding of law finding in psychology as a direct offshoot of law finding in classical physics. Hull argued that he and Lewin believed in uniform laws of behavior, which correspond to correlations of 1.00. However, since progress in isolating such a law was very laborious and time-consuming, "All of us may as well give it up, as Brunswik seems already to have done." [43] Hull went on to construe Brunswik's "probability laws" in terms of his metaphor of science, his beloved law of falling bodies. "Probability laws" therefore turned out to be somewhat imprecise versions of laws of the Galileo type. At least at the time of this first debate, Brunswik seems not to be innocent of such misunderstandings. In his paper, on which Hull commented, he opposed the "nomothetic laws," favored by everyone in the debate besides him, to his "probability laws" (a term he borrowed from Reichenbach), and described the difference as follows: "Imperfect correlations fill in the gap between law and isolated fact. Laws allow prediction with certainty, statistics (correlations) predictions with probability." [44] This led to the impression that his "probability laws" (e.g., the ecological and functional validities) lie somewhere in between the deterministic law on the one hand, and the isolated, singular fact, on the other. Moreover, it suggested quantitative differences in precision only, whereas Brunswik in fact used the term "probability law" for something qualitatively different from Hull's understanding of what a "law" is.

Hull went on to the laws of thermodynamics in gases as a further attempt to understand and to evaluate Brunswik's probabilism. By means of this analogy, he could make the point that even if one accepted chance variability in the microscopic or molecular conditions, uniform quantitative laws for the central tendencies would nevertheless emerge. Therefore, in a psychology concerned with the molar rather than the molecular conditions of behavior, there would be no need and no place for probability and uncertainty. Again, this analogy of randomness at the microscopic level, which is then tamed at the molar level by the "ideal gas law," is completely inapplicable to Brunswik's probabilistic system. Brunswik even thought that psychological "laws" become more and more probabilistic and subject to various interacting factors, the more *molar* the behavior in question is.

The analogy of thermodynamics was repeated in the Berkeley discussion, twelve year later, by Hilgard, together with other analogies, such as Mendelian genetics, to argue for what Hilgard called the status quo: "I should wish to be as nomothetic as I can be, while as probabilistic as I have to be."[45] Hilgard started his discussion with a friendly example that was meant to demonstrate the usefulness of Brunswik's suggestions. This example is revealing since it demonstrates the complete assimilation of Brunswik's ideas into the straitjacket of the experimentalist's schema. In short, the example is about *one* independent variable (age of weaning) and *one* dependent variable (neurotic behavior in later childhood), and two studies that found contradictory relationships. Hilgard's solution was that the two studies had investigated two different ranges in the weaning age, one up to one year, and the other between one and four years. Always to investigate the whole range of the independent variable—this "wise suggestion" was then the vestige of Brunswik's ideas that survived the forced assimilation into Hilgard's schema: "I think we would have learned the major lessons to be taught us by Brunswik."[46]

As another analogy, the indeterminism of quantum mechanics was used to evaluate Brunswik's program. Krech put the issue into this simple formula: "Brunswik, to use a cliché of Einstein, thinks that 'God *does* gamble'. I think He does not. ... I *prefer* nomothetic laws. I have faith (with Einstein) that unambiguous regularities will be discovered."[47] In his defense, published shortly before his death, Brunswik rejected this analogy for the first time rigorously: "I *do* believe that 'God does *not* gamble," and "I fully realize that the impasse of quantum physics is irrelevant to our case." The crucial point was, "While God may not gamble, animals and humans do, and ... they cannot help but to gamble in an ecology that is of essence only partly accessible to their foresight."[48]

Uncertainty is located neither in the environment, as Krech and Hilgard reconstrued Brunswik's thinking, nor in intraorganismic processes, as Hull's analogy semmed to suggest. As Brunswik stated in 1955, "Uncertainty is a feature of the relationships between the organism and the distal environment," and "Uncertainty is not seen as a necessary feature of intraorganismic processes."[49] Again, there seems to be an evolution in Brunswik's thought. The early Berkeley Brunswik could not spell out the essential place where probability enters his program. He even put forward similar "misconstructions" as Hull, Hilgard, and Krech had. For instance, in 1944 he wrote that probability is "due to the ecological ambiguity of cues if not to more intrinsically organismic imperfections as well."[50] And shortly before, in the Chicago discussion, he had proclaimed his revolution on the grounds of "the probability character of the causal (partial cause-and-effect) relationships in the environment."[51] No wonder his colleagues were confused.

To summarize, the attempts to understand and to evaluate Brunswik's probabilism by means of physics were entirely misleading. Brunswik's probabilism was neither a somewhat unreliable law of falling bodies nor the "microscopic chaos" of statistical mechanics nor the indeterminism of quantum physics. A tentative partial explanation for Brunswik's own rather late clarification of the issue might be the close link between Brunswik's ideas and those of the logical positivists of the Vienna Circle. Since leading members of the Vienna Circle, e.g., Otto Neurath, shared the probabilistic world view of quantum physics, such a link could have rendered more difficult Brunswik's own cognitive separation of the respective ideas.

Let us again compare this with Thurstone's probabilistic "law." All the probabilistic concepts in the "law" were presented as mere technical assumptions rather than as having psychological content. Therefore, the "law" could be easily understood as a mere "tool" for quantifying psychological properties. In fact, it was used as a means of establishing nomothetic laws. For instance, Hull used reaction latencies as a measure of his "habit strength." Since reaction latencies showed variability, he could use Thurstone's law for the quantification of habit strength.[52] Based on this variability of reaction latencies, Hull developed his concept of oscillation of habit strength, which could be "tentatively attributed to the random spontaneous firing of the individual nerve cells."[53] From such remarks it could be inferred that if Hull, the determinist, could have been converted into a probabilist, then Hull's theory of behavior would have become an analogy to statistical mechanics. The analogy would lie in the probabilistic conception of *man as a bundle of spontaneous firing nerve cells*, where spontaneity is tamed by molar behavioral laws corresponding to the "ideal gas laws." Such a probabilistic interpretation would have left the physical metaphor of man untouched. Brunswik, however, found himself between and beyond the two disciplines, alone in his attempt to integrate what was considered to be unrelated.

7 What Happened to the Ideas?

Brunswik's ideas are scattered throughout present-day psychology, but the program as a whole has not survived. It is as if the body of ideas had been cut into parts, which then reappeared in other bodies.

Since the cognitive revolution around 1960, many theories in experimental psychology view man as an intuitive statistician, but without the correlational statistics of the other discipline and the Darwinian emphasis on natural environments.[54] Like Brunswik, R. L. Gregory understands perception as inductive inference; however, Gregory's "betting machine" works on the basis of the statistical testing methods used in experimental psychology, such as Fisher's null hypothesis testing method.[55] Harold H. Kelley's view that the mind draws causal inferences by intuitively calculating an analysis of variance became one of the most influential research programs in experimental social psychology.[56] In their study of judgment under uncertainty, D. Kahneman and A. Tversky measure the reasoning of the intuitive statistician against the laws of probability theory.[57] The new intuitive statistician, however, is investigated using the principles of isolation and control. As a consequence, the process of information search and selection is usually not dealt with, and the study of reasoning under uncertainty is reduced to the question of how the mind combines the information given. In contrast, a Brunswikian approach would demand not specifying the pieces of information (the independent variables in the experiment) in advance. Rather, it would begin with taking a representative sample of the free floating information in a particular natural environment. The interest would be in the selection and utilization of cues from an environment, rather than in the combination of given isolated cues in the laboratory.

The study of perception in natural environments is defended by J. J. Gibson, who was influenced by Brunswik, but rejects Brunswik's idea of fundamentally uncertain cues and his methodology.[58] Brunswik's concern with the generalizability beyond the laboratory has become a central issue—but within the framework of systematic design.[59] As far as I can see, there is today not a single research group carrying out the program of probabilistic functionalism in its original form, including those psychologists who are strongly committed to Brunswikian ideas.[60]

In contrast, both of Thurstone's major programs, the law of comparative judgment in perception and psychophysics and the correlational and factor analysis of intelligence, were quite successful and stimulated much further research. Thurstone left undisturbed the division of psychology into two disciplines. When he studied perception, he adapted to the purpose of law finding and to the methodological principles of isolation and control; when he studied intelligence, he turned to individual differences and correlational analysis. Brunswik's "intuitive statistician," however, died in the desert between the two rigid disciplines of psychology.

Notes

1. Address of the Honorary President of the 17th International Congress of Psychology at Washington, D.C., in 1963. Reprinted as "Eponym as Placebo," in Edward G. Boring ed., *History, Psychology and Science: Selected Papers* (New York: Wiley, 1963), on p. 14.

2. Edward G. Boring, in 1962 (quoted by Kenneth R. Hammond, "Introduction to Brunswikian Theory and Methods," *New Directions for Methodology of Social and Behavioral Science*, 3 (1980), 1–11, on p. 9). A hallmark of Boring's deterministic history of a deterministic psychology is his neglect of probabilistic ideas, both Brunswik's and Thurstone's. Thurstone's famous probabilistic "law" is mentioned neither in *Sensation and Perception in the History of Experimental Psychology* (New York: Appleton-Century-Crofts, 1942) nor in *A History of Experimental Psychology* (New York: Appleton-Century-Crofts, 2nd ed., 1957). Brunswik's work in Vienna is given credit in the first book, which was written before Brunswik developed his "probabilistic functionalism" in Berkeley. In the second book, which was published afterward, both the Vienna Brunswik and the Berkeley Brunswik are eliminated from Boring's history.

3. For Thurstone's published life see Louis L. Thurstone, "Autobiography," in H. S. Langfeld et al., eds., *A History of Psychology in Autobiography.* Vol. 4 (Worcester, MA: Clark Univer-

sity Press, 1952, pp. 295–321); Dorothy C. Adkins, "Louis Leon Thurstone: Creative Thinker, Dedicated Teacher, Eminent Psychologist," in Norman Frederiksen and Harold Gulliksen, eds., *Contributions to Mathematical Psychology* (New York: Holt, Rinehart and Winston, 1964); Harold Gulliksen, "Louis Leon Thurstone, Experimental and Mathematical Psychologist," *American Psychologist*, 23 (1968), 786–802. In Brunswik's case, there exists neither an autobiography nor a biography. Some information about his life is contained in the obituaries—see Albert Wellek, "Nachruf," *Psychologische Rundschau*, 7 (1956), 155–156; Edward C. Tolman, "Egon Brunswik, Psychologist and Philosopher of Science," *Science*, 122 (1955), 910; Edward C. Tolman, "Egon Brunswik: 1903–1955," *American Journal of Psychology*, 69 (1956), 315–324.

4. See Edward C. Tolman and Egon Brunswik, "The Organism and the Causal Texture of the Environment," *Psychological Review*, 42 (1935), 43–77.

5. See Egon Brunswik, "Zur Entwicklung der Albedowahrnehmung," *Zeitschrift für Psychologie*, 109 (1929), 40–115; Egon Brunswik, "Über Farben-, Grössen- und Gestaltkonstanz in der Jugend," in Hans Volkelt, ed., *Bericht über den XI. Kongress für experimentelle Psychologie in Wien, 1929* (Jena: Gustav Fischer, 1930), pp. 52–56; Egon Brunswik, "Die Zugänglichkeit von Gegenständen für die Wahrnehmung und deren quantitative Bestimmung," *Archiv für Psychologie*, 88 (1933), 377–418.

6. Karl Bühler, *Die Krise der Psychologie*, 2nd ed. (Jena: 1929).

7. Brunswik, "Zugänglichkeit," on pp. 411–418 (note 5).

8. Brunswik, "Albedowahrnehmung," on p. 113 (note 5).

9. Brunswik, "Farben-, Grössen- und Gestaltkonstanz," on p. 54 (note 5).

10. The program is proposed in Brunswik's first book, *Wahrnehmung und Gegenstandswelt: Grundlegung einer Psychologie vom Gegenstand her* (Leipzig: Deuticke, 1934). Altogether, Brunswik published between 1929 and 1938 two books and about a dozen, mostly experimental, articles in German language.

11. Brunswik, "Wahrnehmung," on p. 4.

12. Egon Brunswik, "Psychology as a Science of Objective Relations," *Philosophy of Science*, 4 (1937), 227–260.

13. In search of a definition of the subject matter of psychology, the term "vicarious functioning" was introduced by W. S. Hunter, *Human Behavior* (Chicago: University of Chicago Press, 1928).

14. See Eckart Scheerer, "Organische Weltanschauung und Ganzheitspsychologie," in C. F. Graumann, ed., *Psychologie im Nationalsozialismus* (Berlin: Springer, 1984).

15. Egon Brunswik, "Representative Design and Probabilistic Theory in a Functional Psychology," *Psychological Review*, 62 (1955), 193–217, on p. 215.

16. Lee J. Cronbach, "Beyond the Two Disciplines of Scientific Psychology," *American Psychologist*, 30 (1975), 116–127, on p. 122.

17. Brunswik, "Representative Design," p. 215 (note 15).

18. Egon Brunswik, "Probability as a Determiner of Rat Behavior," *Journal of Experimental Psychology*, 25 (1939), 175–197.

19. There are two early reports of this first realization of representative design, Egon Brunswik, "A Random Sample of Estimated Sizes and Their Relation to Corresponding Size Measurements," *Psychological Bulletin*, 37 (1940), 585–586; Egon Brunswik, "Perceptual Size-Constancy in Life-Situations," *Psychological Bulletin*, 38 (1941), 611–612. The experiment is fully described in Egon Brunswik, "Distal Focussing of Perception: Size-Constancy in a Representative Sample of Situations," *Psychological Monographs*, 56 (1944), 1–49.

20. In 1940, the "constancy ratio," developed by the Vienna Brunswik as a quantiative measure of constancy, was replaced by correlation statistics; see Egon Brunswik, "Thing Constancy as Measured by Correlation Coefficients," *Psychological Review*, 47 (1940), 69–78.

21. See my chapter "Probabilistic Thinking and the Fight against Subjectivity" in this volume.

22. Kenneth R. Hammond visited Brunswik during the spring of 1954: "My emotions are hard to describe. Before my eyes one of the most significant figures in psychology was pointing with pride to one assistant professor and one graduate student who were doing the sort of research he advocated. Apparently there was no one else" (Hammond, "Introduction," on p. 9 note 2).

23. Louis L. Thurstone, "Psychophysical Analysis," *American Journal of Psychology*, 38 (1927), 368–389; Louis L. Thurstone, "A Law of Comparative Judgment," *Psychological Review*, 34 (1927), 273–286; Louis L. Thurstone, "Three Psychophysical Laws," *Psychological Review*, 34 (1927), 424–432.

These articles contain no reference to the fact that the basic idea had been stated earlier by L. M. Solomons, "A New Explanation of Weber's Law," *Psychological Review*, 7 (1900), 234–240. This case of an "independent" discovery parallels that of "multiple factor analysis," which has been attributed by many authors (and by Thurstone himself) to Thurstone. In fact, however, it is due to Garnett's work in 1919. Again, Thurstone mentions no predecessors—see James H. Steiger and Peter Schönemann, "A History of Factor Indeterminacy," in Samuel Shye, ed., *Theory Construction and Data Analysis in the Behavioral Sciences* (San Francisco: Jossey-Bass, 1978).

In Thurstone's obituary, J. P. Guilford noted that Thurstone's "independence of thinking showed itself in the fact that he did not read widely in the psychological literature, as he was quite willing to admit" (*Psychometrika*, 20 (1955), 264). Deal Wolfe noted in his obituary of Thurstone, "He once told me that he was forced to work on psychological problems that others had not tackled because he had such a poor memory he could not remember what had been done before" (*American Journal of Psychology*, 69 (1956), 132).

24. Francis Galton, *Natural Inheritance* (New York: Macmillan, 1889), on p. 66.

25. Thurstone, "Analysis," pp. 368–369 (note 23).

26. Thurstone, "Analysis," p. 373. However, Thurstone was wrong. A direct test of the normal assumption turned out to be impossible, whereas support can be given today by theories about how the central nervous system estimates information from the stream of nervous impluses; see Gerd Gigerenzer, *Messung und Modellbildung in der Psychologie* (Munich: Reinhardt, 1981), on pp. 286–299.

27. Louis L. Thurstone, "The Stimulus-Response Fallacy in Psychology," *Psychological Review*, 30 (1923), 354–369, on p. 364; see also Louis L. Thurstone, "Influence of Freudianism on Theoretical Psychology", *Psychological Review*, 31 (1924), 175–183.

28. Lee J. Cronbach, "The Two Disciplines of Scientific Psychology," *American Psychologist*, 12 (1957), 671–684.

29. Cronbach, "Two Disciplines," on p. 671.

30. Leo Postman, "The Probability Approach and Nomothetic Theory," *Psychological Review*, 62 (1955), 218–225, on pp. 220–221.

31. Clark L. Hull, "The Problem of Intervening Variables in Molar Behavior Theory," *Psychological Review*, 50 (1943), 273–291, on p. 273.

32. In contrast to Stevens's scaling approach, perception is studied today using the individual as a unit rather than through averages. This holds both for the "functional measurement" approach—see Norman H. Anderson, *Foundations of Information Integration Theory* (New York: Academic Press, 1981)—and the "axiomatic measurement" approach—see Gerd Gig-

ernzer and Gerhard Strube, "Are There Limits to Binaural Additivity of Loudness," *Journal of Experimental Psychology: Human Perception and Performance*, 9 (1983), 126–136.

33. Donna K. Rumenik, Deborah R. Capasso, and Clyde Hendrik, "Experimenter Sex Effects in Behavioral Research," *Psychological Bulletin*, 84 (1977), 852–877.

34. R. S. Daniel and C. M. Louttit, *Professional Problems in Psychology* (New York: Prentice Hall, 1953).

35. R. L. Thorndike, "The Psychological Value System of Psychologists," *American Psychologist*, 9 (1945), 787–790.

36. John F. Dashiell, "Some Rapprochements in Contemporary Psychology," *Psychological Bulletin*, 36 (1939), 1–24.

37. Cronbach, "Two Disciplines," pp. 671–678 (note 28).

38. Cronbach, "Beyond" (note 16).

39. Boring, "A History," pp. 577–578 (note 2). Boring nevertheless dares to suggest personality differences that maintain the schism: "The tester . . . usually liked other people."

40. Egon Brunswik, "Psychology as a Science of Objective Relations," *Philosophy of Science*, 4 (1937), 227–260, on p. 236.

41. Brunswik, "Psychology," on p. 234.

42. Ernest R. Hilgard, "Discussion of Probabilistic Functionalism," *Psychological Review*, 62 (1955), 226–228, on p. 228.

43. Hull, "Problem," on p. 274 (note 31).

44. Egon Brunswik, "Organismic Achievement and Environmental Probability," *Psychological Review*, 50 (1943), 255–272, on p. 269.

45. Hilgard, "Discussion," on p. 228 (note 42).

46. Hilgard, "Discussion," on p. 228.

47. David Krech, "Discussion: Theory and Reductionism," *Psychological Review*, 62 (1955), 299-231, on p. 229.

48. Egon Brunswik, "In Defense of Probabilistic Functionalism: A Reply," *Psychological Review*, 62 (1955), 236–242, on p. 236.

49. Brunswik, "Representative Design," on p. 210 (note 15).

50. Brunswik, "Distal Focussing," on p. 47 (note 19).

51. Brunswik, "Organismic Achievement," on p. 261 (note 44).

52. C. L. Hull, J. M. Felsinger, A. I. Gladstone, and H. G. Yamaguchi, "A Proposed Quantification of Habit Strength," *Psychological Review*, 54 (1947), 237–254.

53. Hull, "Problem," on p. 276 (note 31).

54. For an account of recent cognitive theories based on the statistical metaphor see Gerd Gigerenzer and David J. Murray, *Cognition as Intuitive Statistics* (Hillsdale, NJ: Erlbaum, 1987).

55. R. L. Gregory, *Concepts and Mechanisms of Perception* (New York: Charles Scribner's Sons, 1974).

56. Harold H. Kelley, "The Process of Causal Attribution", *American Psychologist*, 28 (1973), 107–128.

57. See D. Kahneman, P. Slovic, and A. Tversky, eds., *Judgment under Uncertainty* (Cambridge: Cambridge University Press, 1982), and R. Nisbett and L. Ross, *Human Inference: Strategies and Shortcomings in Social Judgment* (Englewood Cliffs, NJ: Prentice-Hall, 1980).

58. J. J. Gibson, *The Ecological Approach to Visual Perception* (Boston; Houghton-Mifflin, 1979).

59. See D. T. Cambell and J. C. Stanley, "Experimental and Quasi-Experimental Designs for Research on Teaching," in N. L. Gage, ed., *Handbook of Research on Teaching* (Chicago: MacNally, 1963).

60. See Kenneth R. Hammond and Nancy E. Wascoe, eds., "Realizations of Brunswik's Representative Design," *New Directions for Methodology of Social and Behavioral Science* 3 (1980), whole number.

4 A Perspective for Viewing the Integration of Probability Theory into Psychology

David J. Murray

Since the 1950s, probabilistic models of learning and decision making (including sensory detection tasks) have tended to predominate over deterministic models. This chapter traces some of the sources of dissatisfaction with the latter by outlining their history from the seventeenth century to the present—the main models considered are those of Herbart, Fechner, Hull, and Luce. It is shown that deterministic models have problems associated with measurability, nonfalsifiability, and overcomplexity. Nevertheless they remain tempting so long as one makes assertions about "internal scalable magnitudes." The contrast between the two kinds of models is thrown into relief by examining the problem of what is meant by "memory strength."

The main purpose of this chapter is to offer a framework from which to view the incorporation of probabilistic concepts into psychology. In two previous articles,[1,2] I reviewed, in strict chronological order, uses made of probabilistic theory between about 1800 and 1954. An article bringing this chronological account up to date still needs to be written. But in the present essay the purpose will be to evaluate the circumstances under which psychologists are obliged to use the language of probability. One way in which this evaluation may be achieved is to survey the widespread use of deterministic (nonprobabilistic) models in psychology and to show that they were associated with certain difficulties that probabilistic models escaped to some extent. The main body of this essay will therefore consist of an account of deterministic models; this will lead us to a general position where we are skeptical of the value, for modeling purposes, of the notion that organisms possess psychological qualities that can be described as "internal scalable magnitudes." Instead, psychological *events* can be viewed as occurring with given probabilities. The essay will finish with an examination of a particular contemporary problem that seems to epitomize the general problem of the quantification of psychology as it has occurred over the last century, namely, the problem of how to represent the processes of human memory.

1 Deterministic Models versus Probabilistic Models

The contrast between exact prediction and probabilistic prediction has often been encapsulated in the distinction between 'deterministic' science and 'probabilistic'

While working on this project, the author was a guest at the Center for Interdisciplinary Research, University of Bielefeld, where he was supported by a grant from the Volkswagen Foundation. Support in Canada came from Natural Sciences and Engineering Research Council of Canada Operating Grant A0126 and a grant from the Queen's University Advisory Research Council. I particularly wish to thank Lorenz Krüger and Michael Heidelberger of Bielefeld for their encouragement of the research, and Peter Dodwell of Queen's for commenting on the first draft of this chapter.

science. The word 'determinism' has a complicated history, as Hacking[3] has pointed out, and may not have appeared with its present connotations until the late nineteenth century, but the gist of deterministic doctrine has long been held to be inherent in the writings of Laplace, who believed that perfect knowledge of the universe at one instant would allow us to predict its state at the next instant. This is the sort of foreknowledge usually only achieved given a knowledge of chemical compounding and mechanics of a kind predicted by equations concerning the behavior of physical objects.

In contrast, a probabilistic view can take one or both of two approaches. In one, the terms of the equations are themselves probabilities; the probabilities may be unknowns or derived from observations of relative frequencies in the external world. These equations allow us to predict the relative frequencies of future events, for example, the suicide rate at some time in the future, but do not allow us to state with certainty what will happen to an individual in the population to which the prediction applies. Such probabilistic "laws," or statements of relative frequencies, which were discussed at great length by Windelband[4] and others in the late nineteenth century, have the disadvantage that one can only move with uncertainty from the general to the particular, but they nevertheless also assume that the universe is determinate. This is the kind of use of probability most often found in psychology, where it is often assumed that a given event can be the outcome of many smaller unobservable events that determine that some of the elements in the equations are random variables. The second approach to probabilism assumes a chance or random element in nature that is irreducible to a description of our ignorance. This is a rare form of probabilism in psychology.

1.1 Thomasius

We note immediately that those who first sought to incorporate numbers into psychology did so with expressions we would now term 'deterministic'. The pioneer who first attracts our attention was Christian Thomasius, a professor of law at Leipzig who started the fashion for lecturing in the vernacular as opposed to Latin. In 1690, because of theological and academic controversies, he moved to Halle, where a new and innovative university was founded a short time later. In 1691 and 1692 he published two monographs[5] in which he claimed to be able to give a profile of an individual's personality by rating the person on a scale between 5 and 60 (probably arbitrarily chosen limits) on four dimensions. These dimensions were rational love (*vernünftige Liebe*), sensuousness (*Wollust*), ambitiousness (*Ehrgeiz*), and acquisitiveness (*Geldgeiz*). He gave comparisons between his ratings of a person on these scales and those of two other raters and showed that although the actual numerical ratings fluctuated between raters, the relative weights assigned to the four dimensions were the same for each rater—all estimated the person in question to be dominated by sensuousness and, to a lesser extent, acquisitiveness, followed by social ambitiousness, with the lowest rating being on rational love. Variability as a feature of psychological data and as a source of error was a problem from the very beginning, but Thomasius's claim was that the variability was not so severe as to vitiate the validity of his general claim that psychological characteristics could be measured by ratings. As psychology progressed, ratings came to play an increasing role in attempts at quantification, as we shall outline.

1.2 Herbart

The next major figure, J. F. Herbart,[6] introduced determinism into psychological modelbuilding. He wished to predict the ebb and flow of presentations (*Vorstellungen*) into consciousness. Nowadays some modelers of short-term memory processes take the position that conscious contents are "items" that enter and leave short-term memory with certain "probabilities." Herbart adopted a quite different terminology, which is deterministic by virtue of the fact that *Vorstellungen* are conceived of as entities with varying strengths. His mechanics and dynamics of consciousness depend on a few simple assumptions concerning the strength of a weak item after it has been inhibited by a stronger item; one assumption is that the reduction in strength in question is predictable from a knowledge of the initial strengths of the two *Vorstellungen* in question.

The question is immediately raised as to whether the system is useless because these 'strengths' are difficult, perhaps impossible, to measure. We shall see that this problem becomes a serious one for psychophysics as well. Restricting our remarks to Herbart's case for the moment, we note that there are behaviorally derived indices that may correlate with what Herbart called the strength of presentations. The more recently an item has been presented, the faster it is recognized when it is presented again.[7] So one possible of strength in consciousness is recognition time. Another possible index is a rating of the strength of an item in consciousness, but I have never seen this used.[8] If recognition time is taken as an index, we find that Herbart's system is not so remote from present-day preoccupations as might be thought. For example, if we ask subjects to recognize items from a list of very recent items, mean recognition time per item increases linearly with list length.[9] But this finding can be directly predicted from Herbart's model, because he says that the degree to which each of a set of n presentations coexisting in consciousness is inhibited is a direct function of n and all we need assume is that recognition time varies inversely as the strength of the item being recognized.[10]

But the objection can be raised that recognition time is too 'indirect' a measure of the strength of a presentation. The problem with indirect measures represents an epitome of a more general problem concerned with propositions about causes and effects in psychology, which in turn affects our attitude to the value of 'indirect' measures. Imagine we have a response that can be thought of as an effect E. We hypothesise that E was caused by some prior event C1, which in turn was caused by some prior event C2, and so on. Let E, for instance, be a keypress. C1 might then be postulated to be a nervous message to the finger that presses the key. C2 might be the motor command initiated by the subject. C3 might be a judgment made about a stimulus. We turn to the question of the falsification of a claim that "C1 causes E" as opposed to the falsification of the claim that "C3 causes E." In the first case, if we vary C1, the kind of message sent to the finger, we claim that E will vary (to be precise, will not occur unless just the right kind of message is sent). If E does occur every time C1 occurs, I can make a claim that the proposition "C1 causes E" is true. If E does not occur if C1 does occur, I can make a claim that the proposition "C1 causes E" is false. Now take the case of C3. If E occurs every time C3 occurs, I can make the claim "C3 causes E" is true. But if E does not occur every time C3 occurs, I am *not* entitled to infer that "C3 causes E" is false. It might be that after C3 occurred, C2 or C1 did not occur and hence that E did not occur. If we accept that C1 is a 'direct' cause and that C3 is an 'indirect' cause, we see that the problem resides in the fact that a proposition about direct causes can be falsified readily, whereas a proposition about indirect causes cannot be falsified readily.

We can now apply this line of reasoning to deterministic models like Herbart's that rely on the assumption that something in consciousness has a certain 'strength'. If we could measure the strength directly, we could readily falsify propositions about the thing in consciousness. But we cannot; we have only indirect measures such as reaction times or ratings. To take the case of reaction times: I might predict that an item strong in consciousness will be associated with a rapid reaction time. I assert that "item with strength C will yield reaction time T." But because this is indirect I have to admit that other events can intervene between C and T. For example, the subject may wish to double check his belief that the item actually occurred. The double checking will slow down his reaction time, so we find that the item with strength C did not yield reaction time T, but a slower one. But I cannot on the basis of this evidence be sure that I have falsified my original assertion. I am simply in a sort of experimental impasse that is familiar to anybody who has worked with reaction times and that has led some people to be skeptical of reaction times as a 'measure' of any cognitive activity. This impasse, which can be related to the problem of falsifiability, has meant that deterministic models that postulate that a psychological experience can have a 'strength' or magnitude have been difficult to test because the experimenter is forced to use indirect measures.

1.3   Indirect Measurement as a Problem in Fechner's Model

The problem of indirect measures has sorely afflicted the next deterministic model that arose after Herbart's. In 1846 Weber summarized the results of a number of experiments in the following words:

It appears from my experiments that the smallest difference between two weights which we can distinguish by way of feeling changes in muscle-tension is that difference shown by two weights roughly bearing the relation of 39 to 40, i.e. when one is about 1/40 heavier than the other ... If we go by the feeling of pressure exerted by the two weights on the skin, we can actually distinguish a weight-difference of 1/30, i.e. when the weights are in the relation 29 to 30 ....
I have shown that results for weight-judgements are the same whether ounces or half-ounces be used: for the results do not depend on the number of grains making up the extra weight, but upon whether the extra weight is 1/30 or 1/50 of the weight to be compared with the second weight. It is the same with the comparisons of the lengths of two lines or the pitches of two tones.[11]

This was essentially the main statement by Weber of what has come to be known as Weber's law. If S is the physical magnitude of a stimulus, then the fraction ΔS that needs to be added to S to make it just discriminable or just noticeably different from S is a fraction of S that is always the same no matter what the absolute value of S. Expressed as an equation, $\Delta S = kS$, where k is a constant. We immediately note that this is a relation between physical magnitudes that is required if a psychological event, the experiencing of a just noticeable difference, is to occur. The actual detection or estimation of when that event has occurred is prone to experimental error, and the first use of probability theory in psychology was Fechner's application of the equation for the normal distribution in order to estimate the variability seen in ΔS from measurements distributed about that value. In fact, Fechner was one of the first to appreciate the value of probabilism as a means of representing Nature (see the paper by Heidelberger in volume 1). This problem aside, it may also be noted that at low values of S, the fraction $\Delta S/S$ has values different from those for the higher values of S, and that Gregory[12] has suggested that at these low values the discrimination of ΔS from S is being influenced by neural noise. Otherwise the reliability of Weber's law is such that it has been taken as a given property of nature both in deterministic models such as Fechner's and in modern models of how the probability of a given choice varies with the set of elements from which one has to choose.[13] In the latter instance, it is assumed that the probability that a will be chosen over b remains the same if to a is added an increment Δa and to b is added the same proportional increment Δb.

For Fechner, Weber's law represented the link between mind and matter for which his metaphysics had led him to look. Like Herbart, Fechner assumed without too much discussion that psychological experiences could be assigned a measure of strength or magnitude. But whereas Herbart talked of the strength of presentations in a context of consciousness, Fechner restricted his efforts to the study of the presumed strength or magnitude of sensations. He was driven in the direction of his law of sensation strength by a number of influences. One was his admiration of a suggested law of the strength of subjective value of money; the law was suggested by Daniel Bernoulli in the previous century.[14] Another influence arose from astronomy, where there had been discussion of the relationship between the apparent brightness of a star as it appeared to the eye and the amount of light actually being received from the star. Pliskoff[15] has shown that by 1856, four years before the publication of Fechner's book, astronomers were aware that to each arithmetic increment in apparent brightness there corresponded a multiplicative increment in light flux. The main influence on Fechner, however, was his desire to work from a known fact relating mind and matter, Weber's law, to a broader law relating sensation magnitude to stimulus magnitude. This he was able to do by the assumption of what he called his "fundamental formula": He assumed that if we started with a sensation strength R, a just noticeable difference in sensation strength ΔR was a function of the Weber ratio, i.e., $\Delta R = C\Delta S/S$. From this it can be inferred that R increases as the logarithm of S. Later, Plateau's work suggested a possible change in the fundamental formula: Instead of assuming that the absolute value of the increment ΔR was a function of the Weber ratio, it might be surmised that the relative value of the increment, $\Delta R/R$, was a function of the Weber ratio.[16] This leads to the assertion that R increases as a power of S. Stevens and others[17] have put much effort into showing that, if we have a measure of R, a plot of that measure against S is better fitted by a power function than by a logarithmic function.

But there is the problem again, exactly as we had it for Herbart: How do we measure R? There seem to be no studies where reaction times have been used to explore R, even though reaction time decreases, the more intense the stimulus.[18] Fechner tried to think of R as a scale in which the units were just noticeable differences, which we have seen are difficult to measure. Stevens made the important innovation of using ratings of sensory magnitude and was able to add substantially to our insight into psychophysics when he demonstrated that ratings along one dimension (say, visual intensity) could be monotonically related to ratings along another dimension (say, sound intensity). Stevens's claims for the power relationship between R and S were largely based on ratings of magnitudes. But even with ratings, such as assigning a numerical value to the sensation associated with a given stimulus, the problem of indirectness has not been solved. We have a response such as a rating; this is presumed to reflect in some way an internal state we call R; and in turn R reflects in some way a stimulus whose strength is S. We can talk about this chain in terms of direct and indirect causes; a more satisfactory way perhaps is to follow Shepard[19] and assert that if $R = f(S)$ and if the measure (e.g., a rating) $= g(R)$, then the latter measure $= g(f(S))$. There seems no way of separately estimating the effects of the two functions, which in our terminology is tantamount to saying that one cannot falsify a proposition of the form "S causes R" or "R causes the measure." It must be noted that we assume here that R has a single absolute value; if R is measured relative to another value of R, certain conceptual problems are simplified, but we prefer to defer discussion of this until we have outlined further problems with the hypothesis that R has a magnitude that can be defined in terms of a single number or value on a scale.

A more profound objection to the idea that a sensation can have a centain strength was put forward by von Kries in 1882.[20] Von Kries first asked what 'measurement' meant. It meant, he said, that you could take two nonidentical objects (*Stücke*) and evaluate their equality. This can be done for objects in space, where we can see that two objects are of equal length, or for processes in time, which take up equal durations. In detail, von Kries spelled out how from units of length and time combined with units of mass (a more subtle measure that requires a postulate) we arrive at measures of speed, force, work, temperature, and measures concerned with electricity. Matters become difficult, however, if we try to measure stimulus intensity, because we need further postulates and a greater knowledge than we have of nervous processes. The question becomes intractable if we try to measure sensation intensity, because we cannot derive such a measure from length, time, and mass; and we cannot properly define 'equality'. To define equality means that we can say that the change from $R1$ to $R2$ is the same as the change from Rk to Rl or that Rm is so-and-so as many times as great as Rn. But for von Kries the difference between a pressure of 2 and 3 pounds on the skin simply is not "equal" to the difference between 10 and 15 pounds on the skin, even though by Weber's law the difference ostensibly reflects the same ratio. Moreover, he argued, one cannot reasonably say that one pain is 10 times as intense as another. There is, he admitted, something special about ratios between stimulus intensities that do influence our judgements about them: If we have an object of objective brightness X on a background of objective brightness Y, these two brightnesses may continue to yield the same subjective contrast to each other even when the overall illumination changes or the pupil varies in size. Thus objective brightnesses A and B can be experienced also as nA and nB or mA and mB. This demonstration of Weber's law leads us naturally to judge that sensation intensities change geometrically. But we still cannot assign a meaningful measure to sensation intensity itself. At the end of his paper, moreover, von Kries noted that when we say, in everyday conversation, that two sensations are "equal," we are likely to be referring to extremes of sensation—we might say two different sensations are equally strong if they stand at the pain threshold, or that two weights feel equally heavy if they require the same muscular exertion. We also use relative terms, as when we judge a Raphael painting to be 'more' beautiful than a genre painting. But being able to use language in this casual way still does not imply that a sensation is the sort of thing that can be measured in the proper sense of the word as defined earlier. Even Fechner agreed that if this argument was correct, the foundations of psychophysics would be undermined. A deterministic psychophysics whose basic unit was sensation strength would, by this argument, be rendered implausible; and there is reason to believe that von Kries himself was led to consider the problems of probability in the light of this apparent failure of determinism.[21]

The indefinability problem as applied to the concept of sensation strength was not the only route that led to the incorporation of probabilistic concepts in psychophysics. In 1892 Fullerton and Cattell,[22] apparently unhappy with the fact that Weber's law relates a physical ratio to the unstable and difficult-to-measure 'just noticeable difference', suggested instead that we measure only the error of observation in a set of difficult discrimination tasks. This is a physical measurement—a standard deviation or a probable error—we may denote $\Delta'S$. Using the fact established by probability theory that the algebraic sum of a number of errors is the average error multiplied by the square root of the number, they argued that one could consider S as a sum of partial units and that therefore $\Delta'S$ would be a function of the square root of S. Weber's law said that ΔS was a linear function of S. A resolution of the conflict was suggested when Woodworth, Guilford, and others in the early twentieth century[23] offered a deterministic account of the relationship between $\Delta'S$ and ΔS; the point of interest here, though, is that Fullerton and Cattell offered a relationship between physical measures only and used variability as an index reflecting, not the strength of a sensation, but a degree of uncertainty. It should be noted that Stevens, at a later period, would maintain that direct measures of sensation strength were possible, even though they may be subject to variability.

It was not only Fullerton and Cattell who introduced variability as an underlying factor in accounting for the relationship between ΔS and S at threshold. In 1900 Solomons[24] wrote a short paper in which he argued that Weber's law *had* to take the form it did because, at threshold, a comparison stimulus $\Delta S + S$ is judged to be just noticeably different from the standard S only if the nervous excitation aroused by $\Delta S + S$ equaled or just exceeded the most extreme deviation in excitation due to random changes in nervous activity (Solomons called this 'irritability'). We must imagine that a stimulus of strength S arouses excitation that is super-imposed upon a fluctuating background level of nervous irritability I whose greatest deviation is ΔI. Only if the excitation from ΔS equals (threshold) or exceeds ΔI will $\Delta S + S$ be judged as different from S. Solomons assumed that I was independent of S except in certain cases. When I is independent of S, the sensation magnitude R is assumed to be a function of IS. At threshold, it is assumed that the variations in R due to an increment in S (with I held constant) are equal to the variations in R due to the increment in I (with S held constant). That is, $R = I \Delta S = S \Delta I$, whence we infer that threshold $\Delta S = S \Delta I / I$. This is Weber's law in which the constant k is now given an explicit value $\Delta I / I$. In this form, R is not mentioned. If, however, we assume that $R = cIS$, where c is a constant, then the effect upon R of a variation in the value of I must be proportional to S, a way of stating Weber's law in which we do mention R. Solomons also argued that when very few cells were active, as might be the case for a very low value of S, S might directly influence I and cause the variability I to be greater than was the case where many more cells were in action. This, he held, accounted for the change in the value of the constant in Weber's law for low levels of S. Solomons did not speculate, as Fullerton and Cattell did for the relationship between $\Delta'S$ and S, on the relationship between ΔI and I. However, he did phrase the general outcome of his theorizing in terms of probabilities: "The probability of a given difference being perceived is the probability of the difference between the two values of I being less than the difference between the stimuli. One hundred per cent. of correct judgements will only be reached when the difference between the stimuli is more than twice the greatest variation of I" (p. 236).

1.4   Overcomplexity as a Problem in Hull's Model

Solomons's ideas about the variability of the background against which a stimulus was changed were predecessors for similar probabilistic notions put forward in a psychophysical context by Thurstone in 1927. But, wishing to continue the discussion of deterministic models, we move forward to the next major model of this type, that of Hull. Hull's system was reported in three texts, spanning the period 1943–1952,[25] and is widely acknowledged to be the most ambitious theory of learning available. But it has suffered from the criticism that it sees 'learning' as a process whereby response tendencies are cumulatively strengthened with experience; the probabilistic approach, in contrast, sees learning as being better described in terms of increasing probabilities of response without the assumption that 'something' is being strengthened. From the present point of view, however, Hull's is the best example of a deterministic model whose apparent drawbacks persuaded researchers of the viability of the alternative probabilistic viewpoint. It is therefore worth closer investigation.

Hull accepted from the outset the notion that responses could only be given if the antecedent neural activity upon which the response depended exceeded a resting background level by some threshold value. He also accepted that any chosen response was not always the same in every respect but showed variability in strength, latency, or probability. He referred to this dimension as 'behavioral oscillation'; in the *Principles of Behavior* he assumed the variability to be normally distributed about a mean value, though an analysis of latency data led him to argue, in the *Essentials of Behavior*, that the distribution was actually leptokurtic (sharply peaked). Before a given response can be evoked, he argued, there could be a small number of trials in which the antecedents to the evocation of the response failed to exceed the threshold either because they were not intense enough or because the natural variability of the response yielded sometimes superthreshold and sometimes subthreshold values. But it is difficult to describe this phenomenon without being more precise about these 'antecedents'; essentially Hull considered that a response had a certain tendency or potential to be evoked and that this 'effective reaction potential', to give it its proper name, was a theoretical construct that had a magnitude, measurable on some definite scale.

The variables that determined the magnitude of the effective reaction potential were both external and internal; external variables would include the number of trials on which the response had been reinforced; internal variables included other theoretical (or 'symbolic') constructs, each of which was scalable in magnitude. These included the animal's drive strength, the degree of habit strength as determined by the training sequence, and the degree of inhibition set up by the effort involved in making the response. In the version described in the *Principles of Behavior*, there are thirteen symbolic constructs, that is, unobservables held to play a part in the chain of events relating stimulus to response. Three more were added in the *Essentials of Behavior*. The transition from a deterministic model of sensation, with two unknown functions, to a model explaining how a stimulus can yield a particular response out of a large repertoire has increased the number of unknown functions by an order of magnitude. The question is whether the model is tractable enough to be useful.

Hull was quite confident that it was. He believed that the magnitude of each symbolic construct was determined by measurable stimulus events and that the magnitude of the final construct in the chain, effective reaction potential, was reflected directly in response measures, such as response probability, latency or strength, or the resistance of the response to extinction. That is, the constructs involved in all the stages before the effective reaction potential is set up are functions of measurable events, and in turn response measures are functions of the effective reaction potential. The bulk of Hull's three books is devoted to enlightened conjectures as to what these functions are; habit strength, for instance, is held to be related to the number of reinforcements N by the expression

$$\text{habit strength} = 1 - 10^{-.0305N},$$

with other equations of this sort being derived from experimental data. Effective reaction potential is related to the median reaction latency M by the expression

$$\text{effective reaction potential} = 2.845(M)^{-.483}.$$

In the final version of the model in *A Behavior System* there are seventeen postulates relating either symbolic constructs to external measures or symbolic constructs to each other. An example of the latter, from the *Principles*, is the assertion that habit strength and drive combine multiplicatively to produce the symbolic construct labeled reaction potentiality, which in turn will be transformed after inhibition has taken its toll into effective reaction potential. In turn effective reaction potential will be modified by behavioral oscillation to produce the momentary effective reaction potential, the final antecedent for the response.

Because each symbolic construct could be related to measurable external events, Hull believed that his system was scientifically as creditable as a physics in which from the behavior of atoms we make postulates about subatomic events or a chemistry where from the behavior of compounds we make postulates about molecular structures. As Hull put it, "All well-developed sciences freely employ theoretical constructs wherever they prove useful, sometimes even sequences or chains of them. The scientific utility of logical constructs consists in the mediation of valid deductions; this in turn is absolutely dependent upon every construct, or construct chain, being securely anchored both on the antecedent side and on the consequent side to conditions or events which are directly observable. If possible, they should also be measurable" (*Principles of Behavior*, p. 382). Hull's confidence arose from the fact that the mathematical forms relating external measures to constructs and constructs to constructs were almost all based on experimental findings rather than guesswork. In fact his caution in this respect is shown by the conservative nature of the first two books, most of whose texts is concerned with making inferences from data to the model. Only in the last book does Hull make inferences from the model to data, i.e., predictions. But a great deal of research effort in the 1940s and 1950s went into trying to confirm the postulates and expanding from them.[26]

Hull's effort also instigated a discussion among philosophers of science about the value of 'intervening variables' in science; in 1948 MacCorquodale and Meehl[27] subsumed what Hull called theoretical constructs under the general heading of "intervening variables" because they seemed so securely anchored with reference to measurable external items and contrasted them with "hypothetical constructs," concepts that needed extra explanation and could not be inferred from laws about external measures. Hull, for example, hypothesized that as a rat approached a goal box containing food, it would make anticipatory eating responses; these 'fractional anticipatory goal responses' had to be hypothesized—they were not inherent in equations relating running speed to effective reaction potential, which was in turn related to drive and habit strength.

The impression given by the ambience of learning theory in Hull's time is one of optimism that the system is tractable. The charge that one cannot falsify a claim relating a response measure to a function early in the sequence because other functions may have influenced the outcome was countered by the argument that one *can* investigate one function at a time by holding everything constant except the variable in question. Hull never wanted to claim that from a single set of response measures he could infer what was happening to habit strength and drive strength and inhibition all at the same time; but he did claim that knowing two of these, he could infer the other, because he had the relevant information in his empirical equations. Although he may have had sixteen unobservable theoretical constructs in his system, a *known* function related each construct to an observable measure, whereas in the case of sensory magnitude discussed earlier, the function relating sensory magnitude to observable measures was unknown. Many of Hull's graphs, for instance, plot a theoretical construct magnitude as a function of observable measures. The problem with Hull's system was not so much its untestability as its complexity. A system that restricted itself to making deterministic statements about the relationship between stimuli and responses with no reference to unobservable intervening processes would be simpler, and one of the reasons that research inspired by Skinner is more prevalent in the 1980s than is research inspired by Hull lies in the former's greater simplicity of both conception and execution. And we have it on Estes's testimony that one of the reasons Estes developed his probabilistic stimulus sampling theory was his desire for a mathematical model of relationships between reinforcement contingencies and response probabilities in Skinner's *simple* operant conditioning paradigm.[28]

1.5 Luce's Measures of Preference Level

Hull's theoretical constructs, such as habit strength and effective reaction potential, were scalable magnitudes from which such response measures as response amplitude, response latency, and resistance to extinction could be derived reasonably readily. It turned out to be more complicated to predict response probability; the factor of behavioral oscillation had to be taken into account, and Hull gave a detailed numerical example illustrating how the "probability of reaction evocation is a normal probability (ogival) function of the superthreshold magnitude of effective reaction potential" (*Principles of Behavior*, p. 330). For researchers studying situations in which choices have to be made, the usual behavioral evidence consists only of probabilities of choice or latencies of choice. Restricting our discussion for the moment to the former, we note that it is possible to predict a choice probability either by having a knowledge of previous choices and setting up an equation relating these to a future choice or by setting up both a model of internal processes presumed to determine choices and also equations relating these processes to future choices. The latter method is essentially that of Hull, who made the probability of choice contingent on habit strength and other theoretical constructs.

Although after 1950 there was a surge in models that predicted future choice probabilities from past choice probabilities with little reference to any internal scalable magnitudes (e.g., the stimulus sampling model of Estes, the linear operator model of Bush and Mosteller), deterministic models were also given a new lease on life when it was discovered that one could make predictions from a thesis about choice probabilities to the existence of an internal scalable magnitude presumed to underlie those choices. To be precise: in 1959, Luce[29] showed that from a particular axiom about how individuals make choices between alternative objects, one could infer the existence of a ratio scale describing the degrees of preference allotted to the several alternatives. The axiom can be stated in several ways, of which a drastically simplified one is the following. Imagine we are in a restaurant and must choose a dish from a set of main courses. The main courses consist of meat dishes, poultry dishes, and seafood dishes. The choice axiom says that the probability of choosing a particular dish (say, steak) from the main course is the probability of choosing a meat dish from the main course times the probability of choosing steak from the meat dishes. Or, to state it another way, the probability of choosing steak from a main course menu is conditional upon the probability of choosing a meat dish.

The question is whether, from a knowledge of choice behavior, we can infer the existence of an internal scalable magnitude that might be labeled "food preference." Luce claimed we could if the axiom were correct. His formal argument will be found in *Individual Choice Behavior*, p. 23; in the words of the foregoing example, we can write (where P is probability)

$$P(\text{choose steak from meat dishes}) = \frac{P(\text{choose steak from main course})}{P(\text{choose meat dish from main course})}.$$

The ratio on the right-hand side is unchanged if one multiplies both numerator and denominator by the same positive number k. The numerator can be replaced by a function v (steak), which is equivalent to the probability shown multiplied by the number; the denominator can be replaced by a function that is the sum of the probabilities of choosing each meat dish times the number. Each of these latter terms can be represented by v(meat dish 1), v(meat dish 2), and so on. The probability on the left-hand side can therefore be represented by the ratio v(steak)/the sum of v(meat dish 1), v(meat dish 2), and so on. This is a single number, which could represent a value on a scale of preference for various types of meat dish. There was a precedent for this kind of reasoning in work by Bradley and Terry,[30] who argued that an observed probability of choosing a over b in a paired-comparison choice task was a function of $v(a)/v(a) + v(b)$. Luce went on to show that such a scale was unique except for its unit and introduced the whole matter by writing (p.3),

The method of attack is to introduce a single axiom relating the various probabilities of choices from different sets of alternatives. It is a simple and, I feel, intuitively compelling axiom that appears to illuminate many of the more traditional problems, in particular the question of whether or not a comparatively unique scale exists which reflects choice behavior. Such a scale, unique except for its unit, is shown to exist very generally. It appears to be the formal counterpart of the intuitive idea of utility (or value) in economics, of incentive value in motivation, of subjective sensation in psychophysics, and of response strength in learning theory.

Luce's monograph had implications for both deterministic theory and probabilistic theory. With respect to the former, he offered an alternative to Hull's response strength theory in which he suggested that a learning model could be one in which the aim was to predict the probability of choosing a particular alternative from a set of alternatives offered on a given trial. It was assumed that this probability could be predicted if one knew the strengths of the preferences for each alternative; according to the model, the probability of choosing alternative i on trial n is given by the ratio of $v(i)$ to the sum of the v values corresponding to each of the other response alternatives. The task of the model then becomes to predict the probability of response on trial $n + 1$. It is assumed that an event such as a reinforcement serves to weight each v value by an amount β; β will be greater than one if the response is rewarded, and between zero and one if the response is nonrewarded or punished. The probability of giving response i on trial $n + 1$ will then be given by the ratio of $v(i)$ to the sum of the other v values. Luce gave reasons for preferring this 'Beta model' to other models in which probabilities of response were predicted on the basis of probabilistic assumptions; Sternberg[31] in 1963 pointed out that the Beta model made predictions that were not easily discriminated from those made by Hull and reported some results suggesting that for avoidance learning, predictions from a linear operator model were superior to those from the Beta model, whereas the latter was good at predicting reversal learning following overlearning. In general, stimulus sampling models have been more widely developed and tested than has the Beta model, but Herrnstein has developed a variation of the Beta model that has been tested in operant conditioning situations.[32]

The implication of Luce's monograph for further developments of probabilistic models was also important. In the area of decision making Luce showed that v-scale values allotted to choice alternatives could be weighted by numbers reflecting estimates of the probability that each alternative would actually occur, yielding estimates of utilities. In psychophysics v-scale values could be integrated with the assumptions of signal detection theory to yield exact predictions of performance in stimulus detection tasks. Perhaps the most fruitful outcome emerged when Luce compared his work with that of Thurstone. Both men assumed that a psychophysical discrimination took place when the internal response to one stimulus B was somehow 'greater' than that to another stimulus A. For Thurstone, the choice depended on the assumption that to each stimulus there was associated an array of normally distributed internal states; the means of differences between means of any two distributions of this kind are also distributed normally; if the difference between the means associated with B and A were sufficiently large, the two were perceived as discriminable. For Luce, a function needed to be found according to which discrimination depended on the difference between $v(A)$ and $v(B)$. This function turned out to be one in which the differences between v values were logistically distributed rather than normally distributed. Since the logistic and normal distributions have very similar shapes when plotted graphically, both being bell-like, it was difficult to decide whether experimentally obtained data fitted the Thurstone model better than it fitted the Choice Axiom model; a clear discussion of this problem is presented by Laming.[33] But one can go even further and ask whether it is possible to find a probability distribution of internal states such that the distribution of differences between means associated with such distributions is logistic. Following work by Holman and Marley, Yellott was able to show in 1977 that if Thurstone's array of internal states had the double exponential distribution, then the resulting differences between means were logistically distributed, and moreover that the double exponential distribution is the only distribution with this property.[34] In addition Yellott pointed to the fact that the double exponential distribution is one of three possible distributions for the limit, as n tends to infinity, of the maximum Z_n^* of n independent and identically distributed random variables $Z_1, \ldots, Z_n$.

For the case of sensory discrimination, therefore, we have a model that combines probabilistic with deterministic reasoning as follows. Sets of random variables with independent and identical distributions combine in such a way that their maxima are reflected in an array of mental states. There are many such arrays, each being distributed according to a double exponential function, and each has a mean M. Discrimination between mental states occurs when two means $M1$ and $M2$ differ by an amount suggesting that they are not samples from the same distribution (array). The distribution of such differences, in turn, is logistic, and this is consistent with the choice axiom assumption that the means $M1$, $M2$ can be allotted numbers that reflect internal scalable magnitudes. This scale can be derived from choice data such as paired comparison data. This mixture of the deterministic and probabilistic is also found in the Beta model, and it is worthwhile to note that Sternberg, in his discussion of learning models,[35] wrote, "It is probably a mistake to think of deterministic and stochastic treatments of a stochastic model as dichotomous. Deterministic approximations can be made at various stages in the analysis of a model by assuming that the probability distribution of some quantity is concentrated at its mean" (p. 40).

At a more general level, we note that v-scale values are obtained when stimuli are arranged by subject in a particular way. In tests of discrimination, two stimuli might be presented and the subject must choose which seems to be the more intense of the two. Or stimuli might be presented as a set of several items that the subject must rank in order of intensity. Luce discussed the problem of relating paired comparison data to ranked data; we bypass this issue to remark that in both cases we have an instance where the derivation of an internal scalable magnitude arises from a study of the relations *between* responses. The picture is that a stimulus variable, e.g., stimulus magnitude, affects the internal magnitude, which is then reflected in a pattern of choice responses, from which in turn we work back to the internal magnitude. This means that assertions about the final step in the stimulus-response chain, the mapping of the internal magnitude onto responses, may be erroneous if the choice axiom does not hold; and assertions about the first step, from stimulus magnitude to internal magnitude, are conjectural but nevertheless falsifiable by examination of the derived internal magnitude values in question, provided the choice axiom does hold.

Considerable effort has therefore gone into answering the question of whether the choice axiom is valid; Luce's opinion in 1977[36] was that the axiom had received support from several experimental studies of choice behavior, but that it received less support when there were high degrees of similarity between the choices; moreover, the choice axiom might be a special case, valid for paired comparisons where the alternatives had little in common, of a more general principle, offered by Tversky in 1972,[37] according to which choices are made by eliminating all alternatives but one. As for the question of how stimulus magnitude relates to internal magnitude, Luce made a number of important conjectures in the monograph to try to account for the fact that the relationship seems to change from context to context, being sometimes a logarithmic function and at other times a power function; but by 1977 he seemed less convinced of the success of these efforts.

1.6 Shepard

Before summarizing these remarks on deterministic models, we should note that psychophysicists are still very concerned with the question of whether sensation can be represented by an internal scalable magnitude. A recent idea, put forward in 1981 by Shepard,[38] but formalized in axiomatic fashion in 1972 by Krantz on the basis of a draft paper by Shepard, is that the internal magnitude is based not so much on a one-to-one mapping of stimulus onto sensation as on a mapping of relations *between* stimuli. Studies of cross-modal matching, where a subject must make, say, two brightnesses equivalent to two sounds differing by a certain amount, and of magnitude estimation led Shepard to conclude that psychological equivalence occurs when we have equal ratios of stimulus magnitude. Clearly his argument was strongly influenced by the pioneering work of Stevens, who had asserted that equal physical ratios gives rise to equal sense ratios. Shepard argued that it was possible to construct a series of stimuli such that the psychological magnitude of the difference between successive pairs remained constant; this would occur if the stimuli formed a series S1, cS1, c^2S1, c^3S1, ... Shepard could not deny that an isolated stimulus set up an internal state to which a number might be attached, but preferred to believe that the system operated relationally, by judging ratios between stimuli. Shepard outlines the importance of this concept for our understanding of internal scalable magnitudes in the following words (p. 50):

Naturally, then, when someone speaks of measuring a sensation, it suggests that the sensation has likewise been fixed on a numerical scale, and that one can thereby determine whether one sensation is twice another, is equal to the sum of two others, and so on. According to the analysis presented above, however, the operations of magnitude estimation and cross-modality matching, upon which Stevens proposed to base psychophysical measurement, do not determine any more than an ordinal structure on the psychological magnitudes .... So, although the subject himself can tell us that one such inner magnitude is greater than another, the psychophysical operations that we have considered are powerless to tell us anything further about *how much* greater the one is than the other ... what the psychophysicist measures is an important constant governing how a subject transduces any stimulus from a particular sensory continuum ....

Shepard points to the greater adaptiveness, for evolutionary purposes, of a system that responds to ratios of intensities and extensities rather than to single values of these. It is to be noted that this argument, if correct, makes the probability of response dependent on a relation between two stimuli, just as Weber's law does. It does not make the probability of response a function of an internal scalable magnitude, as Fechner's law does. It thus escapes the criticisms of von Kries.

1.7 The Rise of Probabilistic Models

It is now time to summarize our findings on deterministic models and to ascertain why at the present time probabilistic models have tended to supersede them.

First, the notion of an internal scalable magnitude of a given strength has given rise to a number of problems that can be organised in terms of an S-R (Stimulus-response) framework. Herbart's concept of the strength of a *Vorstellung* led us to ask how this was reflected in a response measure, with a resulting discussion of some of the inherent difficulties. Fechner's concept of the strength of a sensation led to difficulties in separately estimating the effects of a stimulus on the sensation, and the effects of the sensation on the response; Von Kries also argued that 'sensation strength' had no scientific meaning. Hull's concept of several theoretical constructs avoided some of these difficulties because he was so explicit about how the constructs were anchored either to the stimulus or the response, but their sheer number led to a complexity of the model that drove other learning theorists to simpler expressions relating probabilities of current responses to probabilities of previous responses. Luce's concept of a v scale was based on relations between responses, but left the effects of the stimulus on v-scale values still conjectural. Weber and Shepard bypassed the notion that there is an internal scalable magnitude of sensation that corresponds one-to-one with stimulus intensity by arguing that the probability of a response of a particular kind (e.g., 'greater than') depended on the relation between stimuli.

Since this raises the issue of relations between stimuli, and relations between responses, it may be asked whether we could argue that we could obtain indirect evidence supporting models incorporating internal magnitudes by studying relations between stimuli as they are reflected in relations between responses. Aside from Shepard's claim that we can, we might note that other models have, it is claimed, received support from evidence in which a relation between stimuli is reflected in a (predicted) relation between responses. For example, Herbart predicted that in a list of items ABCD . . . , A was closely associated with B, less closely with C, even less closely with D, and so on. In 1885 Ebbinghaus[39] did an experiment to test this prediction, and the experiment was successful. Does this not offer indirect support for Herbart's model? Hull made many predictions about the learning of serial lists based on a model involving internal levels of excitation and inhibition; for example, he claimed that reminiscence[40] would be greater for the middle items of a serial list than for the end items. These, and other predictions, will be found in the *Mathematico-deductive Theory of Rote Learning*,[41] and, later, in *A Behavior System*, other predictions are made about discrimination learning and problem solving by animals. When the predictions are successful, do they not bias us in favor of the model? The answer must be that once again we find ourselves in a sort of impasse, where we must argue that successful predictions of relations between responses from relations between stimuli constitute indirect evidence in favor of the model, but because the evidence is indirect, it is always possible that another model could do as well. Nevertheless, a failure to predict successfully a pattern of responses to a pattern of stimuli would seem to falsify the model. But there is always the experimental problem that our failure to confirm the model arose from a failure to control all the variables that needed to be controlled. Again, we are confronted with the difficulty of falsifying a complex model.

The answer to this impasse in the 1950s was Estes's stimulus-sampling model for learning, in which it was assumed that on trial n, a given subset of stimuli from an array of stimuli was associated with (conditioned to) a given response from a repertoire, this association taking place with a fixed probability.[42] Performance on trial $n + 1$ could then be predicted and the probability of association could be estimated from the data. This bald statement gives, however, no indication of the versatility of the stimulus-sampling class of models. The versatility can be described as having three aspects.

First, the model itself possessed versatility by virtue of the variety of conceptions of the sampling procedure. A stimulus array could be seen as a "population" from which a sample of stimulus elements could be drawn on a single trial: this was the picture first explored by Estes in 1950. But by 1965, when Atkinson and Estes[43] wrote their detailed account of stimulus-sampling theory, this model took a variety of forms, including a simple form, in which only one element of the stimulus display was sampled on each trial, and a complicated form, in which different patterns of stimulus elements were sampled on each trial. A classification of these various alternatives will be found in the 1970 text of Coombs, Dawes, and Tversky[44] (p.281). The one-element model was particularly successful when applied to human paired-associates learning, and the best-known equation for the learning curve to emerge from this literature was that of Bower.[45] In 1961 he adduced that, in this context,

$$P_n = 1 - (1 - g)(1 - c)^{n-1},$$

where P_n is the probability of a correct response on trial n, c is the probability (estimable from the data) that a correct response will be mentally hooked up with the corresponding stimulus on a given trial, and g is the probability that a response will be correctly guessed. Equations for pattern models were also derived by Atkinson and Estes and were often quite elaborate.

A second way in which these models were versatile was their ability to predict not only the probability of a correct response on trial n, but also such statistics as the number of errors to be expected before an item was learned, the distribution of errors between the first (probably chance) correct response and the next (probably learned) correct response, and standard deviations of the number of expected errors. Third, versatility was also exemplified by the number of learning situations to which the models could be applied—these included paired-associates learning, as mentioned, discrimination learning by humans and animals, various schedules of reinforcement, responding to compounds of stimuli, and investigations of stimulus generalization. One of the most interesting outcomes of the approach was a rational explanation of 'probability matching', first explored by Brunswik in 1939.[46] If a response is rewarded on $P\%$ of the trials, the animal will be giving that response on about $P\%$ of the trials when learning has reached its asymptote. This result was predicted by stimulus-sampling theory.

Another type of probabilistic model to appear in the 1950s was Bush and Mosteller's linear operator model.[47] Here, the notion is that the probability of a response on trial n is changed to a new probability applying on trial $n + 1$ by virtue of a factor called an 'operator'. The operator can increase the probability, as might be expected if the response on trial n were rewarded, or decrease it if the response on trial n were punished or were fatiguing. Luce's Beta model is in fact a type of (nonlinear) operator model. Bush and Mosteller's 1955 book clearly indicated that operator models shared versatility with stimulus-sampling models: different conceptions of the learning process could be represented by different ways of combining operators; statistics concerning distributions of response probabilities could be derived; and the model could be applied to a range of learning situations, including free-recall verbal learning, avoidance learning, imitation learning, discrimination experiments, and runway experiments.

By 1970, texts on mathematical psychology, in their chapters on learning theory, dealt almost exclusively with stimulus-sampling theory and operator models. Hull's work was mentioned, but it was generally to point out that Hull's concept of an internal magnitude of 'response strength' that grew with reinforcement was less compatible with experimental data than was a model in which a correct response was the result of an all-or-none hookup with a stimulus, this hookup taking place with a fixed probability (on this, see particularly Restle and Greeno, 1970, chapter 1).[48]

Second, it was realized by Solomons that variability in the system may be consistent with certain outcomes that seemed deterministic, such as Weber's law. Thurstone in 1927 realized that an internal scalable magnitude might have values that fluctuated from moment to moment and that only by considering certain limits, such as means, could he arrive at an account of sensory discrimination. His approach was revived in the 1950s, when the signal detection theorists suggested that the human observer based a detection response on a 'likelihood ratio'; that is, an estimation is made as to whether a given internal state might have arisen from a distribution of states corresponding to 'signals' as opposed to noise.[49] One's emission of the response in part depended on the criterion set by oneself for accepting that a signal was indeed present; there were therefore two aspects to a judgment, a sensory aspect depending on the degree of sensitivity of the system and a decision aspect depending on the payoffs associated with success or the penalties associated with failure. In both the systems of Thurstone and the signal detection theorists, the distributions of internal states corresponding to stimulus situations were supposed to be normal, but we have seen that different distributions might be involved if Luce's choice axiom is valid. Note that in both the signal detection literature and the choice axiom literature, underlying distributions must be inferred from discrete responses, inferences that have required considerable mathematical elaboration of both models. Note also that other examples exist in the psychological literature where underlying distributions are inferred from discrete response data.

In 1965, McGill[50] gave a detailed account of reaction time data in which he showed how various patterns of systematic changes in reaction time as a function of trials or stimulus conditions were consistent with underlying probability distributions presumably representing processes in the nervous system. Variability in nature and variability in response patterns therefore demand probabilistic representation in psychology.

Third, as noted earlier, probabilistic and deterministic models are not necessarily dichotomous. Differences between means of distributions can be labeled as scale values in a manner most explicitly outlined by Luce. From premises about internal scalable magnitudes and random variability in the system, one can make predictions about response probabilities, as shown by Hull. One can now add that from probabilistic arguments, one can make inferences about internal events. This is perhaps best shown in the following model, due to Atkinson and Shiffrin in 1968.[51]

Imagine that one's task is to predict the probability of recall of the number that had been paired with a certain letter, after a few other letter-number pairs have been presented for learning. The model assumes that the information about the target pair enters a sensory register where some information may be lost; it then goes with probability p_1 into a short-term store with a fixed number of slots r, where new information may knock out old; it then goes with probability p_2 into long-term store, where some information may decay at a rate t. The probability of a correct response will depend on the four parameters p_1, r, p_2, and t as well as on the number of pairs perceived between the presentation of the target pair and the cue to retrieve it. We can plot a retention curve (percentage correct as a function of the number of pairs intervening between presentation and probe) and from this somehow deduce the four unknown values. The falsifiability problem would prima facie seem as great in this case as if we were dealing with a system with four unknown internal magnitudes. But a computer can do a search and find values of the four unknowns that combined into a single equation can be shown to fit the observed retention function acceptably well. If the computer had failed to find values of the four unknowns that, when combined, acceptably fitted the observed data, the model might have been falsified; as it is, the model is indirectly confirmed, though it has not been shown to determine the data uniquely.

The success of this and other similar models in memory and learning has led to reduced interest in models where learning is conceived of as a growth in strength, or forgetting as a diminution in strength, of some internal magnitude. Instead, learning is represented as a sequence of finite states with information being transferred from state to state with a certain fixed probability. Bower's model is a simple two-state model, with c representing the probability of an item going from the 'unlearned' to the 'learned' state; other more complicated models incorporating forgetting factors are available.[52] In these finite state models we have equations in which the elements are probabilities, and they therefore deserve the name 'probabilistic' models. But they represent a determinate view of nature. They also represent a simpler system than one in which four internal magnitudes *interact*, where estimation of single values fitting an observed curve would be very difficult.

This consideration brings us to a crossroads in our evaluation of deterministic and probabilistic models. Many of the problems that have arisen with deterministic models, such as the measurement of internal scalable magnitudes and the falsifiability of assertions about them, arise because internal scalable magnitudes represent continuous functions that can take on many values. The alternative offered by probabilistic models is that one can take the same *behavior* as might be seen as the outcome of the interplay of internal scalable magnitudes and see it instead as a discrete event determined by other discrete events. If the model is simple enough, it can be tested experimentally and falsified; it is worth noting that in 1965 Atkinson and Estes described 'multi-element' stimulus-sampling models that were so complex as to be difficult to falsify,[53] so the appeal of simplicity in certain probabilistic models should not be underestimated. In such models, a response is given not because a hypothetical response potential has reached a particular strength, but because another internal event, such as a learned hookup, has occurred. Here the probabilistic modeler must *assume* that a description of internal sets in terms of fixed entities going from state to state, or in terms of a limited number of prior internal responses, is adequate for his purposes. The problem for the probabilistic modeler then becomes that of integrating continuous variables into the model.

Attempts have been made to do this. For example, Atkinson and Estes in 1965, in the course of describing stimulus-sampling theory, argued that the stimulation effective on any one learning trial could be seen as a sample from a total population of stimulus elements. They then suggested that "just as the ratio of sample size to population size is a natural way of representing stimulus variability, sample size per se may be taken as a correspondent of stimulus intensity" (p. 124). The probability that a stimulus element could lead to an internal event, such as a learned hookup with a response, *could* then be influenced by the number of elements in the sample and a continuous variable (stimulus intensity) thereby integrated into a probabilistic model.

Another example, already hinted at in the discussion on Luce's work, is the derivation of continuous measures from data analyzed according to probabilistic principles. A notable instance of such a model is that of Wickelgren and Norman,[54] who in 1966 and subsequently have derived a measure they call 'memory strength' from the analysis of recognition data using the methods of signal detection. They showed, for example, that memory strength, as measured by the signal detection parameter d', diminishes with the number of items intervening between the original presentation of the stimulus and the recognition test, and may depend on the amount of attention paid to the stimulus when it was first presented. As with genetics and other instances in science, a probabilistic account at one level of analysis (here, distributions of mental states) has led to a deterministic account at a later level of analysis (here, memory strength). We must conclude, then, that models are possible in which intensity of a stimulus or perhaps an internal magnitude are incorporated into models describing transitions between finite states; and models deriving internal scalable magnitudes from probabilistic measures are also possible. But the matter of choice between continuous magnitude representations and finite state representations of psychological processes still leaves us at a crossroads, to which we shall return in the next section when we come to discuss further the concept of 'memory strength'.

2 The Dilemma Epitomized: The Problem of Memory

So far, we have taken a historical approach to the dilemma of whether to use deterministic as opposed to probabilistic models in psychology. We now focus on a particular problem, because it epitomizes the general quandary faced by psychologists who wish to mathematize the subject. Absolutely central to any general psychology is the problem of memory; it is because of the power of memory that we can learn language, think in that language, answer examination questions, and carry on conversations. Without memory we are reduced to the plight of the patient with Korsakoff's psychosis, usually a result of alcoholism, who cannot answer more than the simplest questions about his life and who cannot integrate new information into his general corpus of knowledge.

Historically speaking, the study of memory has undergone four phases. First, there was the period of philosophical speculation, starting with Aristotle's *De Memoria*[55] and continuing through the Middle Ages to the associationistic writings of the nineteenth century. Then there was the early experimental period (about 1879–1960) pioneered by Ebbinghaus.[56] The associationistic tradition was continued in the sense that it was believed that memory retrieval was the result of finding the correct route to a particularly memory 'trace'. It was in this period, particularly its later stages, that it was shown that the learning of lists of words greatly depended on the meaningfulness and frequency in the language of such words. The predominant model was Hullian: Memories were the results of 'traces' that were 'strengthened' by practice and experience. Even in this period, however, there was one probabilistic model, produced in 1930 by Thurstone,[57] that claimed that one could evolve a theory of learning by seeing the successful performance of an act to be learned as having a particular probability p, with the probability of an error then being $(1 - p)$. This model, however, had little influence on experimenters concerned with 'verbal learning'. Hull's model was more influential.

The third phase took place in the context of the 'cognitive revolution' (about 1960–1980). Memory was now seen as an aspect of the flow of information through the organism, and there were so many models of roughly the same kind,

sensory memory ⟶ short-term memory ⟶ long-term memory,
(STM) (LTM)

that many analogies were made between human memory and other kinds of storage systems.[58] Some of these were listed by Roediger in 1980.[59] The important point was that this kind of model could be described in terms of events—for example, the event of entering sensory memory, the event of entering STM, the entering of LTM, and such other events as forgetting from a store or retrieval from a store. Hence stochastic probabilistic models came to the fore, with Atkinson and Shiffrin's model, outlined above, being the best example. (Note, however, that in the context of STM, the deterministic concept of 'memory strength' persisted in Wickelgren and Norman's model.) In the context of LTM, the list of variables presumably enhancing the 'memorability' of information was extended to include imagery value and the ease with which the information could be 'organized' into a coherent package.

The fourth phase is focused on Tulving's new book, *Elements of Episodic Memory* (1983), although Semon in the 1900s anticipated some of Tulving's ideas.[60] Tulving argues that there is good experimental evidence supporting the division of human memory into three kinds: procedural memory (e.g., memory for skills such as skiing, typing, tying shoelaces); episodic memory (essentially memory that is tagged to particular moments in time, as are almost all memories of personal experience; it also includes experiments of the type where a list of familiar words is presented and the subject must recall which words, out of his total vocabulary, were those presented in the experiment); and semantic memory, which is similar to what the philosophers of the first phase called 'knowledge' but is not related to particular points in time, and includes such knowledge as that of one's own language, that of one's academic discipline, and general facts about the universe. Amnesias can selectively influence these types of memory. Within each type, encoding into memory is a particular event; retrieving from memory is a particular event; and there is little discussion of what is meant by 'memory strength', because this is held to be a misnomer—all memory events are determined from the conjunction of an encoding event *and* a retrieving event. Moreover, in this model we avoid the notion of 'trace strength' because it is held that if A is left as a trace, and a stimulus event like A occurs, then the effect of the stimulus is not to strengthen the trace of A (as Hull and the associationists had assumed) but is to lay down a new trace, which, by a sort of sympathetic resonance, may 'remind' one of the old trace of A. Multiple representations of similar events may therefore be laid down in memory; what we conventionally call a 'strong' memory may be one with many encodings, and therefore many retrieval access routes, rather than a single strong 'trace' as in the models of the first three phases.

This short survey leads us directly to the conclusion that the deterministic concept of 'memory strength' may have been a misleading concept resulting from a particular image or metaphor of memory laid down in Greek times. But before rejecting it out of hand, we must ask whether in the Tulving model, the term 'memory strength' can be adapted to cohere with the model, and whether probabilistic models have any advantage in the Tulving model. First, as mentioned above, the casual use of the term 'a strong memory' *may* be meaningful; we refer to a particular readiness to retrieve such memories, and there is nothing in the model that denies that some memories are more likely to be accessed in the course of a lifetime than are others. But with Tulving the metaphor has changed; it is not only denied that there is an entity (such as a trace) that has an 'internal scalable magnitude', but also it is asserted that it is a property of the world that some events are more likely to reoccur than are others. Moreover, the reasons a memory is 'strong' in episodic memory will probably have to do with the emotions, whereas the reason a memory is 'strong' in semantic memory may have to do with the frequency with which it is experienced.[61] Any attempt to mathematize Tulving's model must therefore be prepared to explicate the concept of memory 'strength' in such a way that it is related to events *outside* the observer as well as to events inside him—and then there is the very real question of whether such an exercise is worth it. There is, however, one aspect of the total storage-retrieval situation that might be quantifiable; we have suggested that the 'strength' of a memory might be a circuitous way of referring to the number of access routes to that memory. Any experience that leads to many associations will, almost by definition, have numerous access routes. We should consider seriously therefore the possibility that a measure of accessibility could be formulated in terms of the number of associations to which an event (the laying down of a trace) gives rise. The best hope for a determined determinist is therefore to abandon the notion that a trace can be talked about in isolation. "Connectivity" would be the sort of measure in a future model, based on Tulving's, in which magnitudes would play a part.

But would not a probabilistic approach be better, particularly in view of the practical difficulties of establishing a measure of the number of associations/access routes a trace has? Tulving has described how the encoding procedure at the time of storage should be seen as an event, and so should any later retrieval; correct retrieval is thus a joint function of encoding and retrieval conditions, but it is impossible to state in advance that an encoded item will necessarily be retrieved, nor is it possible to retrodict, from a retrieval probability less than one, whether an item was for certainty encoded originally. The uncertainty here is very similar to the uncertainty expounded by Shepard (p. 29) in the context of a deterministic account of sensation strengths; in both cases we can only talk of relationships between stimuli and relationships between responses (sensory judgments in the case of Shepard, retrieval acts in the case of Tulving). There are two models of memory currently available that incorporate probabilistic reasoning. In 1967 Bower defined memories as having different numbers of components, with each component having a finite probability of being forgotten over the course of time.[62] In this model recall and recognition patterns can be related to variations in the number of stimulus components, and clearly the 'probability of forgetting' in Bower's terminology can be transformed to a 'probability of retrieval' in Tulving's terminology.

In the second model of long-term memory, the concept we earlier called 'connectivity' is evoked in the notion of a 'retrieval structure'; the model was first put forward in 1981 by Raaijmakers and Shiffrin and has since been extended (1984) by Gullind and Shiffrin.[63] A 'retrieval structure' is a matrix in which the elements are the 'strengths' of association between an 'image' in long-term memory and a cue that could be used in a test situation, such as the item itself (a recognition test) or another item that could serve as a retrieval cue. Retrieval cues can include cues from other items in the list, preexperimental cues, and context cues. From this table of strengths a few equations (incorporating Luce's choice axiom) specify the retrieval probabilities. The strengths themselves are determined by four main parameters, some of which reflect the amount of attention and processing given at the time of encoding. In order to fit the model to the data, the method of using a computer search to find values of the parameters best fitting experimentally obtained data may be employed, or enlightened guesses may be made of the value of the parameters and data thereby generated, which can be matched to experimentally obtained data. The model, called "Search of Associative Memory" (SAM), has been successfully used to predict the effects of list length on free recall, the effects of partial cuing of recall, the differential effects of word frequency on recall and recognition, and a variety of other results. 'Strength of association' is a deterministic concept, but the model assumes a probabilistic sampling mechanism in the course of search, and predicts therefore the probability that an item will be recovered given a cue. As in Tulving's model, therefore, there is no assumption that a trace has a 'strength' in isolation, but it is here assumed that the connectivity between items can be represented by finite strengths. There *are* differences, concerning the nature of recognition, between Tulving's views and those of the SAM model, but we suspect a reconciliation will not be long in coming. We therefore find that, in contemporary theory, currently the most ambitious model of memory utilizes a mixture of deterministic and probabilistic concepts.

We may close by noting that models of the nervous system are essentially deterministic,insofar as a nerve sends messages to the brain that reflect the intensity of the stimulus in question. For a particular sensory receptor, a very weak stimulus may elicit no response, but once the stimulus reaches a 'threshold' intensity, a nervous impulse is transmitted down the fiber from the receptor. The intensity of suprathreshold stimuli is encoded by the frequency of the impulses traversing the fiber.[64] Thus the brain receives an analog message about intensities in the external world, and it becomes a meaningful question to ask whether mental entities such as 'sensations', 'habits', and 'memories' have strengths that are also analog reflections of the external world. Our survey has led us to the conclusion that a quantification of psychology is more likely to be successful if we think of chains of sensory or mnemonic events rather than of 'internal scalable magnitudes', though we have noted that measures of 'associative strength' may be feasible. We leave it as a question for the future as to how the intensity-reflecting mapping of the nervous system can be translated into the event-reflecting terminology of modern cognitive psychology.

Notes

1. D. J. Murray, "The Use of Probability Theory in Psychology prior to 1930," in M. Heidelberger and L. Krüger, eds., *Probability and Conceptual Change in Scientific Thought* (Bielefeld: B. Kleine, 1982), pp. 141–163.

2. D. J. Murray, "Probability in the History of Psychology, 1930–1954," in M. Heidelberger, L. Krüger, and R. Rheinwald, eds., *Probability since 1800: Interdisciplinary Studies of Scientific Development* (Bielefeld: B. Kleine, 1983), pp. 214–235.

3. I. M. Hacking, "Nineteenth Century Cracks in the Concept of Determinism," in M. Heidelberger and L. Krüger, eds., *Conceptual Change*, pp. 5–34 (note 1).

4. W. Windelband, *Die Lehren vom Zufall* (Berlin: A. W. Schade, 1870).

5. C. Thomasius wrote two monographs on his system: *Das Verborgene des Herzens anderer Menschen auch wider ihren Willen aus der täglichen Conversation zuerkennen* (Halle: Christoph Salfeld, 1691) and *Weitere Erleuterung durch unterschiedene Exempel des ohnelangst gethane Vorschlags wegen der neuen Wissenschaft Anderer Menschen Gemuther erkennenzulernen* (Halle: Christoph Salfeld, 1692). These are summarized by P. McReynolds and K. Ludwig, "Psychometrics in the Seventeenth Century: the Personology of Christian Thomasius," paper presented at the Ninth Annual Meeting of CHEIRON, the International Society for the History of the Behavioral and Social Sciences, University of Colorado, Boulder, June 9–11, 1977. The idea of applying mathematics to psychology was also proposed by Christian Wolff

(1679–1754). His use of the term *psychometria* is discussed in A. Metraux, "An Essay on the Early Beginnings of Psychometrics," in G. Eckardt and L. Sprung, *Advances in Historiography of Psychology* (Berlin: VEB Deutscher Verlag der Wissenschaften, 1983). See also G. P. Brooks and S. K. Aalto, "The Rise and Fall of Moral algebra: Frances Hutcheson and the Mathematization of Psychology," *Journal of the History of the Behavioural Sciences*, 17 (1981), 343–356.

6. J. F. Herbart put forward a sketch of his theory in his *Lehrbuch zur Psychologie* (Königsberg: Unzer, 1816), translated by M. K. Smith as *A Textbook in Psychology* (New York: Appleton, 1891). The theory was extended in his *Psychologie als Wissenschaft neu gegründet auf Erfahrung, Metaphysik und Mathematik* (Königsberg: Unzer, 1824–5; reprinted Amsterdam: E. J. Bonset, 1968). His use of mathematics appears to be original, though his system owes much to Leibniz.

7. See, for example, B. Forrin and K. Cunningham, "Recognition Time and Serial Position of Probed Item in Short-Term Memory" *Journal of Experimental Psychology*, 99 (1973), 272–279, or R. Ratcliff, "A Theory of Memory Retrieval," *Psychological Review*, 85 (1978), 59–108.

8. Ratings of subjective confidence in recognition responses have, however, been used to derive a measure of 'memory strength' in short-term recognition experiments. See note 54.

9. The first paper to report this was by S. Sternberg, "High Speed Scanning in Human Memory," *Science*, 153 (1966), 652–654. The linear 'set-size' effect has been confirmed in many studies since, and is discussed in the papers described in note 7.

10. This statement is based on the results of an exercise in which I inserted imaginary values into Herbart's equations describing the reduction in strength of each item in a set of *n* items competing in conciousness. The equations are given in the opening chapter of *Psychologie als Wissenschaft* (note 6).

11. The quotation is from the translation of Weber's *Der Tastsinn und das Gemeingefühl* (1846) by H. E. Ross and D. J. Murray in E. H. Weber, *The Sense of Touch* (New York: Academic Press, 1978), pp. 220–221.

12. R. L. Gregory, *Eye and Brain: the Psychology of Seeing* (New York: McGraw-Hill, 1966), pp. 85–89.

13. J. C. Falmagne and G. Iverson, "Conjoint Weber Laws and Additivity," *Journal of Mathematical Psychology*, 20 (1979), 164–183; L. Narens, "A Note on Weber's Law for Conjoint Structures," *Journal of Mathematical Psychology*, 21 (1980), 88–91.

14. D. Bernoulli, "Specimen theoriae novae de mensura sortis," *Commentarii academiae scientarum imperialis petropolitanae*, 5 (1730; pub. 1738), 175–192.

15. S. S. Pliskoff, "Antecedents to Fechner's Law: The Astronomers J. Herschel, W. R. Davies and N. R. Pogson," *Journal of the Experimental Analysis of Behavior*, 28 (1977), 185–187.

16. J. A. F. Plateau, "Sur la mesure des sensations physiques, et sur la loi qui lie l'intensité de ses sensations à l'intensité de la cause excitante," *Bulletins de l'Academie Royale de Belgique*, 2nd series, 33 (1872), No. 5.

17 See, e.g., S. S. Stevens, "The Psychophysics of Sensory Function", in W. A. Rosenblith, ed., *Sensory Communication* (Cambridge, MA: MIT Press, 1961), or S. S. Stevens, "To Honor Fechner and Repeal His Law," *Science*, 133 (1961), 80–86.

18. For example, a curve of reaction time plotted against sound intensity will be found in R. Chocholle, "Variation des temps de réaction auditifs en fonction de l'intensité à diverses frequences," *Année Psychologique*, 41–42 (1945), 65–124.

19. R. N. Shepard, "Psychological Relations and Psychophysical Scales: On the Status of "Direct" Psychophysical Measurement," *Journal of Mathematical Psychology*, 24 (1981), 21–57.

20. J. von Kries, "Ueber die Messung intensiver Grossen und über das sogenannte psycho-physische Gesetz," *Vierteljahrsschrift für Philosophie und Soziologie*, 6 (1882), 257–294.

21. M. Heidelberger (personal communication). Von Kries went on to write his treatise on probability, which appeared four years later: *Die Principien der Wahrscheinlichkeitsrechnung* (Tübingen: Mohr, 1886).

22. G. S. Fullerton and J. M. Cattell, "On the Perception of Small Differences," *Publications of the University of Pennsylvania, Philosophical Series*, 2 (1892), Reprinted in A. T. Poffenberger, ed., *James McKeen Cattell: Man of Science*, Vol. I, pp. 142–251 (Lancaster: The Science Press, 1947).

23. R. S. Woodworth, "Professor Cattell's Psychophysical Contributions," *Archives of Psychology*, 40 (1914), 60–74; J. P. Guilford, "A Generalised Psychophysical Law," *Psychological Review*, 39 (1932), 73–85. A short account of this literature will be found in D. Lewis, *Quantitative Methods in Psychology* (New York: McGraw-Hill, 1960), pp. 433–438.

24. L. M. Solomons, "A New Explanation of Weber's Law," *Psychological Review*, 7 (1900), 234–240.

25. C. L. Hull, *Principles of Behavior* (New York: Appleton-Century-Crofts, 1943); *Essentials of Behavior* (New Haven: Yale University Press, 1951); *A Behavior System* (New Haven: Yale University Press, 1952).

26. See notably K. W. Spence, *Behavior Theory and Conditioning* (New Haven: Yale University Press, 1956), and W. H. Thorpe, *Learning and Instinct in Animals* (London: Methuen, 1st ed. 1956, 2nd ed. 1963).

27. K. MacCorquodale and P. E. Meehl, "Hypothetical Constructs and Intervening Variables," *Psychological Review*, 55 (1948), 95–107.

28. In the following reference, Estes describes original work on operant conditioning that led to the stimulus sampling model. In referring to Hull's method, he remarked, "[Hull's method] would have … the drawback of leaving us with a number of arbitrary functions intervening between our learning theory and the testable predictions derivable from the theory … our unwillingness to face an unending sequence of *ad hoc* assumptions about response measures in each new situation, together with the fact that we have found strong reasons to regard behavior as essentially probabilistic suggests a possible way out. Suppose that we take as our basic theoretical dependent variable the probability of occurrence associated with any given operationally defined response class. The desired unification can then be achieved by formulating laws of learning in terms of response probability as a dependent variable": W. K. Estes, "The Statistical Approach to Learning theory," in S. Koch, *Psychology: A study of a Science*, Vol. 2, p. 393 (New York: McGraw-Hill, 1959).

29. R. D. Luce, *Individual Choice Behavior* (New York: Wiley, 1959).

30. R. A. Bradley and M. E. Terry, "Rank Analysis of Incomplete Block Designs. I. The Method of Paired Comparisons," *Biometrika*, 39 (1952), 324–345.

31 S. Sternberg, "Stochastic Learning Theory," in R. D. Luce, R. R. Bush, and E. Galanter, eds. *Handbook of Mathematical Psychology* (3 Vols.) (New York: Wiley, 1963–5).

32. R. J. Herrnstein, "On the Law of Effect," *Journal of the Experimental Analysis of Behavior*, 13 (1970), 243–266. The connection between operant conditioning and Luce's v scale is made most explicit in R. J. Herrnstein and D. H. Loveland, "Matching in a Network," *Journal of the Experimental Analysis of Behavior*, 26 (1976), 143–153.

33. D. Laming, *Mathematical Psychology* (New York: Academic Press, 1973), Chapter 2.

34. The work of E. Holman and A. A. J. Marley is described by R. D. Luce and P. Suppes, "Preference, Utility, and Subjective Probability," in R. D. Luce, R. R. Bush, and E. Galanter, eds., *Handbook*, Vol. III, p. 338. The main paper is J. I. Yellott, Jr., "The Relationship

between Luce's Choice Axiom, Thurstone's Theory of Comparative Judgment, and the Double Exponential Distribution," *Journal of Mathematical Psychology*, 15 (1977), 109–144. By 'double exponential' is here meant a distribution of the form

$$P(X < t) = \exp[-\exp(-at + b)].$$

Luce, "Choice Axiom," p. 217 (note 36), notes some terminological confusions about the meaning of 'double exponential'.

35. S. Sternberg, "Stochastic," in R. D. Luce, R. R. Bush, and E. Galanter, eds., *Handbook* (note 31).

36. R. D. Luce, "The Choice Axiom after Twenty Years," *Journal of Mathematical Psychology*, 15 (1977), 215–233.

37. A. Tversky, "Choice by Elimination," *Journal of Mathematical Psychology*, 9 (1972), 341–367.

38. R. N. Shepard, "Relations" (note 19); D. H. Krantz, "A Theory of Magnitude Estimation and Cross-Modality Matching," *Journal of Mathematical Psychology*, 9 (1972), 168–199.

39. H. Ebbinghaus, *Über das Gedächtnis: Untersuchungen zur experimentellen Psychologie* (Leipzig: Duncker and Humblot, 1885), esp. Chapter 9. Translated by H. A. Ruger and C. E. Bussenius as *Memory: a Contribution to Experimental Psychology* (New York: Dover Press, 1964).

40. If a measure of retention is plotted against the amount of time elapsed since the item was learned, the curve usually decreases. But sometimes retention after an interval exceeds that associated with a shorter interval; this is known as "reminiscence."

41. C. L. Hull, C. I. Hovland, R. T. Ross, M. Hall, D. T. Perkins, and F. B. Fitch, *Mathematico-Deductive Theory of Rote Learning* (New Haven: Yale University Press, 1940). This work introduced Hullian scientific methodology and paved the way for the works listed in note 25.

42. W. K. Estes, "Toward a Statistical Theory of Learning," *Psychological Review*, 57 (1950), 94–107.

43. R. C. Atkinson and W. K. Estes, "Stimulus Sampling Theory," in R. D. Luce, R. R. Bush and E. Galanter, eds., *Handbook*, Vol. II, pp. 121–168 (note 31).

44. C. H. Coombs, R. M. Dawes, and A. Tversky, *Mathematical Psychology: An Elementary Introduction* (Englewood Cliffs, NJ: Prentice-Hall, 1970).

45. G. H. Bower, "Application of a Model to Paired-Associate Learning," *Psychometrika*, 26 (1961), 255–280.

46. E. Brunswik, "Probability as a Determiner of Rat Behavior," *Journal of Experimental Psychology*, 25 (1939), 175–197.

47. R. R. Bush and F. Mosteller, "A Mathematical Model for Simple Learning." *Psychological Review*, 58 (1951), 313–323. Also R. R. Bush and F. Mosteller, *Stochastic Models for Learning* (New York: Wiley, 1955).

48. F. Restle and J. G. Greeno, *Introduction to Mathematical Psychology* (Reading, MA: Addison-Wesley, 1970).

49. L. L. Thurstone, "A Law of Comparative Judgment," *Psychological Review*, 34 (1927), 273–286. Signal detection calculations have to take into account instances where subjects report detecting a signal when in fact none was present ("false alarms"). Among the first to report the analysis of false alarms were M. Smith and E. A. Wilson, "A Model of the Auditory Threshold and Its Application to the Problem of the Multiple Observer," *Psychological Monographs*, No. 359, 1953. The report by W. W. Peterson and T. G. Birdsall, "The Theory of Signal Detectability," Electronic Defense Group, University of Michigan, Technical Report

100 David J. Murray

No. 13, Sept. 1953, was followed by the widely read paper of W. P. Tanner, Jr., and J. A. Swets, "A Decision-Making Theory of Visual Detection," *Psychological Review*, 61 (1954), 401–409.

50. W. J. McGill, "Stochastic Latency Mechanisms," in R. D. Luce, R. R. Bush, and E. Galanter, eds., *Handbook of Mathematical Psychology*, (3 Vols.) (New York: Wiley, 1963–5).

51. R. C. Atkinson and R. M. Shiffrin, "Human Memory: A Proposed System and Its Control Processes," in K. W. Spence and J. T. Spence, eds., *The Psychology of Learning and Motivation*, Vol. 2. (New York: Academic Press, 1968).

52. On this, see D. Laming, *Mathematical Psychology*, pp. 260–270 (note 33).

53. R. C. Atkinson and W. K. Estes, "Stimulus Sampling Theory," in R. D. Luce, R. R. Bush, and E. Galanter, eds., *Handbook*, Vol. II, p. 264 (note 31).

54. W. A. Wickelgren and D. A. Norman, "Strength Models and Serial Position in Short-Term Recognition Memory," *Journal of Mathematical Psychology*, 3 (1966), 316–347; W. A. Wickelgren and K. M. Berian, "Dual Trace Theory and the Consolidation of Long-Term Memory," *Journal of Mathematical Psychology*, 8 (1971), 404–417.

55 Aristotle, *De Memoria*, in R. McKeon, ed., *The Basic Works of Aristotle* (New York: Random House, 1941).

56. See particularly H. Ebbinghaus, *Über das Gedächtnis* (note 39).

57. L. L. Thurstone, "The Learning Function," *Journal of General Psychology*, 3 (1930), 469–491.

58. A book that summarized early contributions to cognitive psychology, including Broadbent's 1958 stress on the difference between short- and long-term memories, was U. Neisser, *Cognitive Psychology* (New York: Appleton-Century-Crofts, 1967).

59. Roediger, H. L., III, "Memory Metaphors in Cognitive Psychology," *Memory and Cognition*, 8 (1980), 231–246.

60. For references to Semon's works, see D. L. Schacter, J. E. Eich, and E. Tulving, "Richard Semon's Theory of Memory," *Journal of Verbal Learning and Verbal Behavior*, 17 (1978), 478–485, and E. Tulving, *Elements of Episodic Memory* (New York: Oxford University Press, 1983).

61. This suggestion is based on research in progress by the author.

62. G. Bower, "A Multicomponent Theory of the Memory Trace," in K. W. Spence and J. T. Spence eds., *The Psychology of Learning and Motivation: Advances in Research and Theory*, Vol. 1 (New York: Academic Press, 1967).

63. J. G. W. Raaijmakers and R. M. Shiffrin, "Search of Associative Memory," *Psychological Review*, 88 (1981), 93–134; G. Gillund and R. M. Shiffrin, "A Retrieval Model for Both Recognition and Recall," *Psychological Review*, 91 (1984), 1–67.

64. An early report that conveys the excitement of this discovery is E. D. Adrian, *The Basis of Sensation* (London: Christophers, 1928).

II SOCIOLOGY

5 The Two Empirical Roots of Social Theory and the Probability Revolution

Anthony Oberschall

Aside from statistical inference, probability theory has had little impact on social theory, which is concerned with the conditions for social order and the origins and change of institutions. Nineteenth-century theorists borrowed organic and mechanical analogies and adopted the deterministic outlook of the natural sciences. Two bodies of social data were used. The first consisted of ethnographic accounts of non-European peoples. Theorists imposed an evolutionary scheme on them to explain institutional origins and change, e.g., of the family. The second consisted of contemporary moral statistics, e.g., birth, deaths, crime, suicide, which became abundant in the era of statistical enthusiasm. Regularities in moral statistical rates caused wonderment, especially in the case of voluntary actions such as suicide and marriage. Theories and controversies were set off about the micro foundations of the macro social order. A synthesis of the two empirical roots within a single theory addressing the base of social order was attempted by Durkheim. Suicide exemplified the force of a moral order in the group upon its members, a force transcending particular individual choices. Durkheim and his circle became absorbed with the study of collective representations, especially religion, on which the moral order was anchored. Because of their evolutionary approach, they came to neglect moral statistics, quantitative techniques, and the study of contemporary Europe. Halbwachs, the principal exception, found it impossible to reconcile Durkheimian theory with individualist and probabilistic thinking gaining currency in other sciences. The micro-macro gap in sociological theory persists, but methodological individualism shows some promise of bridging it.

1 The Eighteenth-Century Legacy

Social theory starts with questions about social order. Norms and belief systems provide stable expectations necessary for sustaining lasting social interactions. According to Evans Pritchard, "It is evident that there must be uniformities and regularities in social life, that society must have some sort of order, or its members couldn't live together. It is only because people know the kind of behavior expected of them, and what kind of behavior to expect from others . . . that each and all are able to go about their affairs. They can make predictions, anticipate events, and lead their lives in harmony with their fellows. . . ." [1] The habits, needs, and preferences, and the particular rules and institutions prevailing in a group or society tend to be accepted as self-evident, like one's native language, or else are believed to be divinely ordained and not to be questioned.

As Europeans penetrated into the Americas, Asia, and Africa, the discovery of peoples and societies with different rules and institutions required explanation. Why monogamy here and polygamy there? Why is the same behavior a crime in one group and permitted in another? Is there some lawfulness in the temporal and spatial patterns of institutional variations? It is questions such as these that eighteenth-century social theorists raised.

By the mideighteenth century an impressive volume of descriptions from travelers, explorers, missionaries, adventurers, and merchants had accumulated about peoples everywhere, from great empires such as China to small societies of hunters in North America. This information considerably enriched the stock of knowledge about the diversity of customs, religions, political regimes, and laws that men lived and had lived under. It was noted that some of the institutions of contemporary preliterate peoples resembled those recorded by ancient historians and geographers such as Tacitus, Herodotus, and Pliny, and indeed those in the Old Testament. It appeared plausible that Europeans had once lived much like contemporary "primitives" or "savages," but had managed to develop and progress from these origins. Savages, on the other hand, were a case of arrested development. Referring to American Indians, Adam Ferguson wrote that "It is in [their] present condition that we are to behold, as in a mirror, the features of our own progenitors." [2] A study of their institutions would reveal the laws of change of social institutions and of mankind itself. This idea was at the root of eighteenth-century, pre-Darwinian social evolutionary thinking and led to comparative analyses of customs, institutions, and societies, the essence of a wholistic, macro-level approach to social theory.

More information about one's own society also leads to puzzles. Events and actions such as births, deaths, marriage, and crime manifest regular patterns in the aggregate. Yet these patterns result from the uncoordinated activities and choices of a multitude of people, each pursuing a private end, and not from an imposed design. How is such a spontaneously generated orderliness at all possible?

Adam Smith, Malthus, and other political economists sought to answer this question with assumptions about individuals' wants, preferences, motivations, and goals, from which aggregated deductions were made for explaining the lawfulness and self-equilibrating properties exhibited by the market and the population.

The political economy approach stimulated interest in quantitative information about the entire population. Thus Malthus outlined what amounted to a comprehensive research program on the working classes in the *Essay on Population*:

... the histories of mankind which we possess are in general the histories only of the higher classes. We have not many accounts that can be depended upon of the manner and customs of that part of mankind were ... [the population] movements chiefly take place .... This branch of statistical knowledge has of late years been attended in some countries, and we may promise ourselves a clearer insight into the internal structure of human society from the progress of these inquiries. But the science may be said to be in its infancy, and many of the objects on which it would be desirable to have information have been either omitted or not stated with sufficient accuracy. [3]

These two kinds of information about people and institutions are the two empirical roots of social theory: one root is the ethnographic and historical record of customs, beliefs and institutions; the other is the record of events and actions in contemporary populations. It is my belief that a preoccupation with either kind of information channels the theorist toward a different kind of social theory and methodology. One approach is wholistic, starts with institutions and groups, and views individual actions as the product of social influences and contraints that transcend and overwhelm people. It employs a macro-theoretical method. The other approach is individualistic, starts with the motivations, incentives, beliefs, and choices of individuals, and views norms and institutions as the product of the intersection of a multitude of individual actions. It uses a micro-theoretical method. Whichever approach one espouses, one runs into the difficulties of moving from the micro to the macro level, or vice versa. It is evident that institutions constrain individual actions, but it is also evident that if many people change their actions in concert, the institution itself will change.

The individualistic, micro-theoretical method coupled with quantitative data on actions extends easily to the study of the properties of distributions, of averages and dispersions, and of the mathematical principles underlying the mechanisms for generating distributions, i.e., probability theory. Wholistic methods, on the other hand, have a built-in resistance to distributional and probabilist thinking. Instead, organismic analogies putting the accent on structure and complementary parts and development are compatible with those of social analysis. Thus the choice of method in social theory predisposed theorists' reception of probabilistic ideas.

My essay seeks to explain how it was that social theory in the nineteenth century took off in a wholistic, macro-theoretical direction, and came to rely largely on historical and ethnographic data rather than on statistical data on contemporary society.[4] In particular, I shall concentrate on the French tradition of social theory, with Durkheim at the center, because contemporary social theory bears his characteristic imprint more than that of any other social theorist.

2   Statistical Enthusiasm: New Data on Contemporary Society

In comparison to the second half of the eighteenth century, a great deal more information about contemporary European society became available in the first half of the nineteenth, especially numerical data on social problems. Why did this happen, and what impact did this data explosion have on social theory?

The rapid diffusion of data and of the statistical method was due to change in both demand and supply. Important on the demand side was the rapid growth of cities, of the working class, and of social problems. Unskilled workers, women, children, and migrants from impoverished rural districts crowded into hastily constructed, congested dwellings. They formed the first urban wage-earning, unskilled lower class, which exhibited many social pathologies from the vantage point of the privileged, and constituted a threat to political stability, safety, and health. Whether or not the lower class was more unhealthy, ignorant, criminal, poor, drunken, irreligious, and socially disorganized in cities than in the villages and small towns that they had come from is not the issue. The point is that they were far more visible, concentrated, and physically proximate to the affluent, and could not be ignored. Social problems needed attention, but before steps could be taken, their dimensions and causes had to be established. Middle class and aristocratic reformers, philanthropists, professionals, civil servants, and humanitarians, gathered in statistical societies and voluntary associations, set out to do so by means of social surveys and investigations and pressured governments to do likewise. Public agencies obliged by collecting and releasing summaries of administrative records, conducting official inquiries and parliamentary hearings, and setting up statistical agencies to monitor social trends continuously.

In Paris, the 1820s and 1830s witnessed the creation of highly competitive, daily-circulation newspapers that fed the middle class reading public a daily dose of crime stories and criminal and public health statistics. These newly invented social barometers were carefully scrutinized for indications of incipient social unrest.[5] The demand for social statistics was thus not only sustained by a small circle of elite reformers, but by the attentions of a wider public using it for a variety of purposes ranging from entertainment to politics.

On the intellectual plane, the spirit of the era has been aptly characterized as "the will to quantify everything, to measure everything, to know everything."[6] It was a continuation of the intellectual program of the eighteenth-century encyclopedists, for whom knowledge based on observation was a precondition of social engineering and reform carried out by an enlightened intellectual and scientific elite. The faith in the numerical or statistical method, the sense of participating in the creation of a new science, and the spirit of reform are perhaps best expressed by the physician Parent-Duchatelet in the opening pages of his monumental work on prostitution: "During the collection and analysis of all my data, I made the greatest effort to obtain numerical results on all matters that I chose to dwell upon. It is only with the help of [the numerical] method that one advances science, and that one provides for the government the means of progressing with confidence to a higher state of perfection. This method, which I will call statistical ... will in a short time become widely adopted."[7]

Immersed in numerical facts, the statistical enthusiasts sought to create a social science as rigorous and objective as the natural sciences. It would base social reforms on "the laws which govern men's habits and the principles of human nature, upon which the structure of society and its movement depends," as Lord Brougham expressed it to the members of the Social Science Association in 1857.[8]

On the supply side, the 1820s and '30s experienced a revolutionary change in the means of production of social statistics. Whereas earlier the typical social investigation was conducted by a lone scholar or reformer, with limited means and time, statistical societies pooled the resources of their members, hired interviewers, tabulators, and assistants, and were able to complete large-scale research projects in a short time. Scientific societies and academies gave out grants and awarded prizes for statistical research.

Governments had greater means yet for responding to reformers' and the public's thirst for information. In France, government agencies, departments, and commissions that used to keep confidential records and to conduct inquiries and surveys whose results were restricted to government circles now released that information to a wider public. Starting with 1827, a volume of national criminal justice statistics entitled *Compte General de L'Administration de la Justice Crimi-*

nelle was issued annually, and starting in 1831, every three years nationwide statistics on primary schooling were published on top of an annual report on the "state of primary education in the entire kingdom." In Britain, Parliamentary Commissions conducted numerous inquiries and published reports. Further, social and administrative reforms created officials and inspectors (factory, poor law, public health, sanitary, etc.) who were charged with monitoring new reform legislation. With each reform and expansion of the governmental apparatus, there would be released the annual reports of these agencies as well as special reports of their investigative activities that were meant to gain public support.

Cullen has provided a detailed picture of the link between reform, bureaucratic expansion, and production of information in Britain.[9] National criminal statistics were first published by the Home Office in 1810 as a consequence of the controversy over capital punishment, with reform MPs seeking a reduction in capital offenses. Both sides demanded this information for their arguments. The Voluntary Association for Education of Working People pressured the government to collect and release national figures on the schooling of the nation's children and on school facilities, figures that religious bodies and schools were not able to provide accurately Ecclesiastical registers of births and deaths did not include Catholics and dissenters. Consequently a national system of birth and death registration was instituted at the General Registrar's Office. This office, under William Farr, then produced statistical studies on the links between mortality and social environmental factors. The poor law medical officers engaged in investigations of rubbish collection, water purity, and sewage disposal that enabled Edwin Chadwick to release reports and studies of sanitary conditions filled with statistical data.

The civil servants who headed and staffed these new agencies and directed these studies were often closely associated with nongovernmental associations such as the Statistical Society of London (later Royal SS), the Central Society of Education, and the Health of Towns Associations. It is this social circle of prominent civil servants, physicians, politicians, professionals, scholars, and other reformers joined together through overlapping memberships and informal networks and located mostly in London that provided a powerful mechanism for triggering rapid diffusion of the statistical method from center to periphery.

There is an element of faddishness in any such process. During the bandwagon phase, exaggerated claims were voiced and unrealistically high expectations were entertained about the benefits of the numerical method to science and reform. When the limitations of statistics were eventually recognized, disappointment and rejection tended to be exaggerated as well.

Whether an innovation remains a permanent feature of the intellectual landscape will depend on the extent to which the scientific and intellectual world has become institutionalized, i.e., the extent to which it has a resource base and controls its own development from within, independent of shifting currents of public opinion and political support. In the particular instance of social and moral statistics, there did not yet exist in the nineteenth-century an autonomous social science rooted in distinct institutions and professions, such as universities and scientific societies are today. Thus there were discontinuities in the adoption of the statistical method in the emerging social sciences once the era of statistical enthusiasm faded in the second half of the nineteenth-century.[10] Nevertheless, the census agencies created everywhere during the era of statistical enthusiasm and social reform developed a bureaucratic momentum of their own and survived, even prospered, because governments, business, and policy makers everywhere found useful the information they routinely provided. Nowhere was this more true than in the German states that pioneered social insurance for which information on mortality, morbidity, industrial accidents, savings, and the standard of living of the industrial working class was necessary. Thus in the later nineteenth-century government and administrative statistics, including social and moral statistics, kept growing. But the figures no longer elicit wonderment and comment by a large public, nor are they scrutinized for quantitative laws and regularities as frequently as they had been in the 1820s and '30s by reformers and scientists.

3   In Search of Social Theory: Quetelet and Moral Statistics

Though Quetelet is largely omitted from histories of sociology, his "social physics" was the most influential nineteenth-century effort to provide, as Malthus had hoped, "a clearer insight into the internal structure of human society" by means of "statistical knowledge." As for moral statistics, it ceased to exist as a distinct scientific activity by the end of the century. Yet from the 1820s to the 1840s, moral statistics was the designation for a vigorous field of intellectual activities that many hoped would become the foundation of a quantitatively and empirically grounded social science. Starting with Graunt's *Observations on the Bills of Mortality*, statistical regularities had been discovered and analyzed for mortality, for the sex ratio at birth, and for some other demographic variables. With the nineteenth-century social and moral statisticians, the search for empirical regularities, like some sort of an intellectual California gold rush, was extended to crime, suicide, illegitimacy, church attendance, pauperism, schooling, philanthropic acts such as donations to the poor, and other moral actions. The intellectual challenge of the age was the analysis of these figures for the purpose of detecting regularities and relationships in them, and of accounting for numerical lawfulness in some systematic fashion.

A case in point was the brilliant *Essai sur la Statistique Morale de la France* submitted to the Academie des Sciences in 1833 by the Paris lawyer André Guerry. It is a sophisticated work of criminology, based on careful analyses of the recently published French crime statistics. It is replete with tables, graphs, and maps. In it, Guerry establishes differences in crime rates between the principal social and life-cycle categories (gender, age, ...), regional development (degree of urbanization, degree of literacy, ...), and seasonal variations, separately for crimes against persons and crimes against property, which to this day have remained basic categories for expressing crime rates. In the conclusion, Guerry stated the premise of moral statistics: "... the majority of moral facts, considered in the aggregate rather than in each individual, are determined by regular causes whose variations are contained within narrow limits, and which can be subjected to direct and numerical observations just as those of the material world."[11]

Like many moral statisticians, Guerry proceeded cautiously before erecting an intellectual edifice on his figures and empirical generalizations. As he himself stated in the preface, "No system has directed our work; we have avoided leaning on any theory."[12] Such was not the case for the Belgian astronomer Quetelet, who had a far wider vision and scientific ambitions.

More than any other man in the 1820–1850 period, Quetelet sought to lay the foundations for a science of man and society, spanning the entire spectrum of human phenomena and actions, from the physical to the moral, social, and intellectual. Though he failed in this endeavor, Quetelet expressed the fundamental question about the social order most vividly. In *Système Social*, he wrote in a section on marriage,

In the realm of human action there is none in which free will intervenes as directly as it does in marriage. It is one of the most important life events, which people as a rule undertake with the utmost caution.... If one observes marriages specifically, one fact will especially astonish us: we shall discover that not only does the total sum of marriages remain constant from year to year, in cities as well in the countryside, but that the ratios of marriages between single men and single women, and between widows and widowers, also remain constant.... It appears as though, from one end of the kingdom to the other, the population has agreed to enter into the same number of marriages, and agreed to do so for each province, for the cities and countryside, and for single men, women, widows, and widowers ....[13]

Since the entire population does not get together yearly in order to decide how many of them are going to be married and who will marry whom, according to some agreed-upon quota or plan, how to explain the regularities observed in these and other moral statistics? Quetelet's "physique sociale" and his theory of "l'homme moyen" were meant to provide an explanation for the numerical patterns he discovered in physical, demographic, social, and moral data. On numerous occasions he commented on how the frequency distribution of military draftees' heights was a good fit to the Gaussian or "normal" error distribution, known in astronomy. He then made a bold assumption, a leap of faith as it were, by assuming that the generating mechanism for errors in astronomical measurements was also applicable to physical, social, and moral processes that produce the observed stability of rates and ratios, and the lesser fluctuations from year to year. He assumed a constant, intense force—which he termed "penchant"—similar to gravity for celestial bodies that fixed them in their true orbit, and a large number of independent, small, perturbing forces, acting in all directions, which produced the random scatter of actual astronomical observations about the true planetary position, and by analogy the deviations about the mean for frequency distributions of physical, social, and moral phenomena.

For a physical trait, such as height, there did exist an observed frequency distribution, whose mean value could be calculated, and deviations from the mean measured. For moral attributes, such as crime, suicide, and marriage, this was not the case, since most people never get arrested during their lifetimes and do not commit suicide. In what sense could Quetelet then write about a normally dis-

tributed "penchant au crime," "penchant au suicide," "penchant au marriage,"
which were analoguous to the constant force of gravity acting on celestial bodies so
as to keep them to their orbital paths?

Quetelet made another bold assumption. Noting that some age-specific crime
rates peaked for men in their midtwenties and were progressively lower as one
observed the younger and the older age cohorts on either side of the maximum, and
that a similar pattern was true for the age-specific marriage rate, he applied the
error law and model to moral phenomena. In the *Système Social*, he made an
extended analogy between the laws of classical mechanics and what he called moral
laws.[14] Influences on moral actions were like forces. Forces have varying magnitudes and directions, and so do moral influences. Instincts are powerful, active
forces, found in mankind in inverse proportion to reason. They are continuously
acting forces (as gravity is); examples are egoism, family pride, self-preservation,
"consciousness of kind," and so on. Then there are, as in mechanics, impulse forces,
for instance, the human will, which usually has zero intensity and is seldom used.
The balance of all these forces acting on an individual produces a point of equilibrium, called the moral center of gravity. As one deviates from the center of gravity
under the influence of an impulse force, powerful counterforces tend to push one
back to the center of gravity. For instance, when an ambitious soldier advances in
the ranks aiming to become a general, the obstacles will grow as the square of the
velocity of his advancement because the envy of his rivals will grow and impede his
efforts. And so on.

The analogies between mechanics and social processes run amok in Quetelet.
Applied to the crime, suicide, and marriage rates, these ideas boil down to the
following: The "penchant au crime," etc., is the sum of these continuously acting,
powerful drives or instincts, such as self-preservation, egoism, and sense of duty,
and of the weaker, short-term, accidental, independently acting forces, such as the
human will and the influence of peers. Drives or instincts create the central tendency
of the distribution; the human will and social influences account for the dispersion
about the central tendency. The set of means of all these normally distributed
physical and moral traits and dispositions Quetelet sums up in the concept of the
"Homme Moyen," or average man. The statistical regularities in moral and social
statistics are then easily explained. Since drives and instincts rooted in human
nature are by definition unchanging or only very slowly changing, and possess, also
by definition, far greater intensity than the accidental forces, the rates and ratios
must be constant, or nearly so. The stability of social order is thus ultimately rooted
in the invariant traits and dispositions of the "Homme Moyen."

But was this an explanation, or merely an analogy, and not a helpful one at that?
And not the relegation of human will in moral actions to a secondary, accidental
force a direct attack on free will? Quetelet's ideas were bound to raise a great deal of
controversy, and did.

Already in 1833, Guerry objected to the inference of a criminal disposition or
propensity from crime rates. Noting the higher crime rates of men compared to
women, he wrote that "It would be an error to think that these numbers represent,
for each sex, the degree of energy of their criminal disposition, to think, for

example, that in crimes against persons this disposition is really five times stronger in man than in woman."[15] And he mentioned the obvious: For many crimes (corruption, financial fraud, personal violence) women have fewer means and opportunities to commit them than men.

Drobisch, the most comprehensive critic of Quetelet's theoretic ideas, argued against the concept of the "Homme Moyen."[16] Taking marriage as a good example of a freely made choice, he examined the very low probability of marriage in various age classes—for instance, Quetelet had calculated the "disposition to marriage" of Belgian men, age 25–30, as a mere .0884—and comments that these probabilities in no way express the true inclination to marry, but reflect inhibitory factors such as parents' consent, partners' readiness to marry, lack of resources to form a household, and so on. Variations of marriage rates between groups and over time can be attributed to changes in the social and economic conditions that favor or block marriage. Drobisch then subjects the notion of a disposition to crime and to suicide to similar criticisms. And he concludes that when such rates exhibit certain regularities, it is because in most societies these social and economic conditions tend to remain fairly stable in the short run. The upshot of all this is that Quetelet's concept of the "average man," the sum of all these average dispositions, is nothing but "an abstract, mathematical fiction," and that the aggregate rates (e.g., suicide rates) exercise no causal force on any individual suicide choice.

An equally penetrating examination of Quetelet's ideas was provided by the philosopher and mathematician Wilhelm Lexis, who devised various mathematical tests for establishing whether fluctuations and differences in observed rates might be generated through a random mechanism similar to the drawing of black and white balls from an urn, as would be true if human action were caused by a great many, independent causes each subject to random variations, as Quetelet claimed.[17] Lexis, however, demonstrated that of all the usual moral actions (suicide, crime, marriage, etc.) none could pass his test for a random generating mechanism. Indeed, only the sex ratio at birth passed Lexis's test, not a variable where human will or intention could have any effect on the outcome. These results were a blow to Quetelet's often voiced hope for a social physics grounded in the mathematical theory of probability, as Laplace had done for celestial mechanics.

Although Durhkeim reiterated Drobisch's critique of Quetelet, he was fascinated by the orderly aggregate patterns resulting from the uncoordinated actions of a multitude of people and wrote that, though flawed, Quetelet's theory "has remained the only systematic explanation of this [question]."[18] In *Suicide*, Durkheim strove to provide a better explanation.

Fundamental objections as well might be raised about Quetelet's application of astronomy and mechanics to the human and social sciences. In a thinly veiled reference to Quetelet, Max Weber commented on this issue in his most important methodological essay as follows: "Again and again the idea recurs that the ideal that all knowledge, including the cultural sciences, should strive for, even if only in a distant future, must be a system of propositions from which reality can be 'deduced'. A prominent natural scientist has thought, as is well known, that the ideal for cultural analysis is to provide an astronomical analysis of life events.[19] And he goes on to warn against the naive immitation of the method and language of the natural sciences in the social and cultural sciences—what Hayek was later to characterize as "scienticism."

In 1874, the ardent German moral statistician and theologian Alexander von Oettingen drew up a balance sheet on Quetelet: French statisticians such as Guerry, Dufau, and Moreau de Jonnes made little use of Quetelet's ideas, and the leading British statisticians, such as Nieson, Porter, and Farr, ignored him altogether. In Germany, he had more of an impact and generated some philosophic controversies over free will and determinism, but his greatest success was among "physicians, jurists, theologians, and philosophers," and not the leading academic social scientists, like Knapp and Schmoller, who were highly critical.[20]

By all accounts then, Quetelet's "social physics" must be judged a failure. What, then, explains Quetelet's high reputation in the eyes of his contemporaries, and for the large amount of controversies his ideas elicited in nineteenth-century scientific and intellectual circles? Three explanations come to mind. First, despite Quetelet's ambivalent disclaimers that his theory did not undermine free will in moral choices, this was not how others read him, and thus the predictable storm about free will and determinism, which has got to be one of the most confusing controversies in the already crowded annals of confusing controversies. Second, Quetelet certainly posed an important, intriguing question about the statistical properties of moral actions. As long as no one came up with a better answer, one had to wrestle with Quetelet's analogies, terminology, and ideas, however much one disagreed. Third, I suspect that many social scientists are at their most confusing when they utter profound truths about causality and methodological principles. Fortunately, they often do not apply these ideas to the solution of concrete, substantive problems. Quetelet was a case in point. If one skips the pages on celestial mechanics, the average man, social physics, and the center of gravity, Quetelet's writings are filled with a wealth of observations, comments, suggestions, and facts about crime, marriage, suicide, and dozens of other topics, which are quite instructive. Many must have enjoyed reading him for this reason and not because of his philosophic and methodological pronouncements.

4   Opposition to Social Statistics

After 1850, the statistical method becomes the target of criticism. For the founders of the statistical societies, collecting numerical facts and applying the statistical method would settle once and for all questions about cause and effect in human behavior and institutions through the discovery of social laws. These laws in turn would unambiguously guide an effective social reform. The prospectus for the creation of a statistical society of London announced in 1834 that "The Statistical Society will consider it to be the first and most essential role of its conduct . . . to confine its attention rigorously to facts—and, as far as it may be found possible, to facts which can be stated numerically and arranged in tables."[21] Eventually it became more and more evident that facts do not speak for themselves, that covariation is not causation, and that most research is more likely to require painstaking, follow-up work and to raise new questions than to settle age-old questions about the cause of poverty and criminality in a definitive fashion.

The more sophisticated statisticians knew this all along. When Guerry discovered a positive relationship between the standard of living and property crime rates in French departments, which contradicted the conventional wisdom, he cautioned that "the question of the influence of prosperity or of misery on morality is more difficult to resolve than one would have suspected at first."[22] But the general membership of statistical societies and reform publics had had no such reservations when they embraced statistics. For France, the historian Chevalier concludes that the Second Empire, after 1851, "does not benefit from the same extensive documentation that accumulated in the Paris of the July Monarchy .... The results of these statistical researches do not receive in the Second Empire the same dissemination that they received in the first half of the nineteenth-century, and do not awaken a comparable, strong interest."[23]

Abrams's account of the British reform and statistical movement traces the growing disillusionment with statistics.[24] With the exception of Manchester and London, the statistical societies went out of existence. Instead a moralizing perspective took hold. There was a great growth of voluntary associations concerned with moral improvement and, above all, of temperance associations.[25] Whereas society and social conditions are seen earlier as the causes of social pathologies, whose chief victims were working class people, in the Victorian era the consumption of alcoholic beverages becomes the principal cause of crime, pauperism, disease, and insanity. Social reform now consists of moral uplifting and of converting the victims to better moral habits.

In the two surviving British statistical societies there is a decline in survey work on social problems, and an increase in papers published on economic topics using published statistics. The quality of social research performed and the technical competence of statistical analysts also decline markedly.[26] Social reform and ameliorism proved too shaky a foundation on which to build a quantitative social science. Fifty years after the start of the era of statistical enthusiasm in the 1820s, the same impressionistic and arbitrary eyeballing techniques were used to argue for a positive or negative relationship between two variables. No permanent cumulation in techniques of data collection and analysis had been made. Until the twentieth century, this is typical of empirical social research everywhere.

Not only did rank-and-file interest and support for moral statistics and for an applied social science decline, but opposition is voiced by scholars and social scientists as well. In midcentury France, the most prominent social scientist and critic of statistics was Frederic LePlay, one of the originators of social science fieldwork, an advocate of the method of direct observation, the inventor of the method of family budgets, and the creator of an empirically grounded sociology of the working class and peasant family, on which he based a broader theory of social change and a conservative social policy.

Trained as a mining engineer at the Ecole Polytechnic and the Ecole des Mines in the 1820s, LePlay distinguished himself early and became editor of the *Annales des*

Mines. As editor and mining consultant in the 1830s and 1840s, LePlay crisscrossed most of Europe on assignments for governments and wealthy mine properietors during which he perfected the method of workers' family monographs based on participant observation that he had first conceived as a student in 1829 on his first field trip in the forests and mines of the Harz region of Germany. During his lifetime, LePlay did detailed studies of about a hundred families (in later life he made use of assistants), most of which were published in *Les Ouvriers Européens* (1855) and the four volumes of *Ouvriers des Deux Mondes*(1857–1862).

LePlay had a brilliant career. He served on numerous government commissions, headed a government statistical agency responsible for mineral and industrial statistics, occupied the chair of metallurgy at the Ecole des Mines, became general manager of the 1855 and the 1867 Universal Expositions in Paris, was appointed "conseiller d'Etat" and advised Napoleon III, and ended his public career in the French Senate. In addition to being a successful engineer, administrator, scholar, and politician, LePlay was also at the center of a conservative social reform movement of his own creation, which sought to formulate social policy and reforms grounded on scientific principles.

An indication of the new antistatistical climate of opinion is the fact that an empirically and quantitatively trained and oriented scientist of LePlay's stature should repudiate the moral statistical tradition represented by Quetelet. Although in his 1840 statistical treatise *Vues Générales sur la Statistique*, LePlay advocated government collection of social and moral data, the failure of the National Assembly's nationwide social survey of the condition of the working class during the 1848 revolution converted him from supporter to critic of government-collected statistical information.[27] By the time of his first major book in 1855, *Les Ouvriers Européens*, which ironically won the Montyon prize for statistics conferred by the Academie des Sciences, LePlay had become a severe critic of the statistical tradition. In its stead he advocated his observational method.

In a typical passage LePlay writes that even though "ingenious savants" (probably an oblique reference to Quetelet) have occasionally surmounted the difficulties inherent in the statisical method, "... The statisticians' method is not the direct observation of facts; it is a compilation and interpretation ... of facts collected from a standpoint quite different and foreign to that of scientific concerns. Despite their seeming generality and seductive uniformity, statistical documents have contributed but feebly to the progress of social science."[28]

In contrast, LePlay advocated an alternative method in which observation is "not entrusted to a multitude of agents" untrained in the topic, but to a "few specialized men who are knowledgeable about the subject matter." He differentiated his methods from that of the statisticians' in the sharpest terms: "... For discovering the nuances described in the [workers' family] monographs that this book is made up of, the author had to maintain an intimate contact with the populations he studied. By means of an extended stay in the dwellings of the families that he chose for detailed description, he became slowly initiated to their language, their habits, their needs, their sentiments, their passions, and their prejudices."[29]

With the decline of the quantitatively minded social reform movement and the opposition of prominent men like LePlay, the earlier linkages between social theory, applied social science, and moral statistics were loosened. By the end of the century, moral statistics meant voluminous compilations of statistical data without an attempt to make sense of its contents in other than a superficial descriptive fashion. Such was the situation when Durkheim sought once more to wed moral statistics to sociological theory, and to apply the new social science to a growing social problem.

5 Durkheim's Sociology and Moral Statistics

By the end of the nineteenth century Durkheim had become the foremost champion of sociology in France, and was well on his way to founding the first university-based school of sociology. Durkheim has left his characteristic stamp on socio-logy to this day. As a social theorist, he was a functionalist and evolutionist. But he was also conversant with Quetelet and the moral statistical tradition. He himself undertook several quantitative studies utilizing moral statistics. If there was going to be a fruitful, lasting integration of social theory and the empirical, quantitative study of contemporary society, Durkheim surely was in a pivotal position to accomplish it. To what extent he did is the topic of this section.

Durkheim once defined sociology succinctly in a letter to his collaborator Bouglé as follows: "The object of sociology as a whole is to determine the conditions for the conservation of societies." [30] How does moral statistics fit into such a conception?

For Quetelet the stability of the social order rested on the stable statistical properties of various actions, such as marriage and crime rates. Durkheim and the sociologists, building on the older social evolutionary tradition of the Scottish moralists, wanted to explain in addition why certain actions are regarded as criminal in some societies, but not in others; what accounted for differences in the intensity and manner of punishment; why some societies had no specialized agen-cies of social control, whereas others did have criminal justice systems distinct from other institutions. Moreover, they searched for laws that would account for the change from one type of social control institution to the other. It would be surprising if a satisfactory theory of crime rates, to stay with this one example, would at the same time answer these other questions that sociologists considered central for the new sociology. Thus Durkheim wrote in the preface to *Suicide*, his major statistical work,"... we think we have established a certain number of propositions concerning marriage, widowhood, family life, religious society, etc., which, if we are not mistaken, are more instructive than the common theories of moralists [about] these conditions or institutions." [31]

Sociology is not just all about marriage rates, divorce rates, the differential mortality of males and females. It deals more comprehensively with marriage, widowhood, and family life.

Durkheim objected to the then current explanation of social bonds, of solidarity, and of the origin and change in social institutions, such as the family, legal systems, and religion. His point of view comes across most clearly in the sections of *Les Règles de la Mèthode Sociologique*, where he is critical of Spencer. Spencer, like Quetelet before him, postulated a drive, instinct, or sentiment rooted in human nature corresponding to various social institutions: for instance, fear of death was at the root of religion, and fear of other men at the root of government. If this were true, the study of social relationships, solidarity, and social institutions would be but a "corollary of psychology."[32]

Durkheim maintained instead that beliefs, sentiments, and drives, such as egoism and altruism, are socially rooted, originating in and shaped by social influences exterior to the individual, which are "endowed with a dominating and coercive power that gets imposed on [the individual] whether or not he wants it."[33] Because these sentiments and drives are socially determined, one has to study the social bonds and relations of mutual dependence among men for grasping how social solidarity, moral authority, and shared beliefs cement social relationships into stable institutions. Weaken the social bonds and mutual dependence among men, and you will weaken the moral authority and constraints that regulate behavior, to the point that some will even commit suicide—destroy themselves. And this despite the instinct for self-preservation recognized as the most powerful of all instincts. If suicide rates were correlated inversely with the intensity of social bonds and moral authority within groups, it would constitute a convincing demonstration of the importance of social influences, norms, and supraindividual variables for explaining human behavior.

That is precisely what Durkheim set out to do in *Suicide*, in a quantitative fashion, using the accumulated information on the topic, which was huge, and had been classified, sifted, and analyzed by many investigators from all points of view. And he did demonstrate how the suicide rate varied systematically for different religious groups, social categories (age, sex), occupation and professions, minority and majority status, marital and domestic groups (single, married, widowed, with and without children), and so on. From these statistical analyses, he inferred the following, typical generalization: "Suicide varies inversely with the degree of integration of domestic society,"where integration was not an individual trait or disposition, but the attribute of a group or category or social unit, i.e., density of social bonds and attachments among its members. Thus he remained true to his own methodological rule of explaining social facts, the suicide rate, with reference to other social facts, moral integration, without introducing physiological, psychological, biological, or any other variables and principles of explanation.

Durkheim's achievement was not methodological. He used the same technique of comparing rate and ratio differences that was then in common use. Nor did he discover any startlingly new statistical relationships between suicide and other variables. What he did do was to supply a consistent explanation for all the known statistical relationships, and this from a new, sociological perspective.

In the closing pages of *Suicide*, Durkheim claimed that his sociological theory of suicide based on the moral integration (or the moral constitution) of the group cleared up Quetelet's puzzle concerning the statistical regularities of moral actions, i.e., both the pattern of stable group differences and the pattern of stability over time: "From this point of view there is no longer anything mysterious about the stability of the suicide rate. . . . Since this inclination to suicide has its source in the moral constitution of groups, it must differ from group to group, and in each of them remain for long periods practically the same."[34]

Beyond the issue of suicide and crime rates, it is instructive to examine just what the points of similarity and difference were between Quetelet's social physics and Durkheim's sociology.

First, Quetelet and Durkheim were both strongly committed to determinism, and to the discovery of laws of society. To be sure, Quetelet entertained at times the idea of a chance mechanism in human events, such as the accidental causes that produce dispersion about mean values. Yet his biographer, Lottin, wrote appropriately that "ever since his first works . . . Quetelet was pursued by the idea that the entire universe is subject to laws, that nothing is left to chance."[35]

As for Durkheim, in a short essay on sociology in France in the nineteenth century in which he traces its antecedents, he notes that the decisive step was the idea that social phenomena are subject to laws, because "the sentiment that there are laws is the crucial factor of scientific thinking."[36] In his writings Durkheim had no hesitation in referring to his empirical generalizations on suicide and on the evolution of punishment as "laws." He further wrote that he had chosen to study the subject matter of suicide because "real laws are discoverable which demonstrate the possibility of sociology better than any dialectical argument."[37] There is no hint in Durkheim's writings that he ever thought a chance mechanism at the individual level might generate laws at the macro level.

Second, they were both concerned with social laws at the aggregate or societal unit of analysis, and not with explaining individual behavior. It is these social laws, and not individual behavior, that account for the social order, or as Durkheim expressed it, account for the "conservation" of society. Nor did either of them think that it is possible to arrive at social laws by starting with assumptions about individuals and deriving aggregate consequences from their actions, as was done by political economists. On the contrary, only by studying the properties of aggregates can theorists arrive at valid social laws, because the numerous idiosyncratic causes that impact differently on individuals, called "accidental" causes by Quetelet, cancel each others' effects in the aggregate.

On this score, Quetelet was explicit:

We must . . . lose sight of man as an isolated individual, and consider him only as a part of the species. In stripping him of his individuality, we shall eliminate all that is merely accidental . . . one must . . . study aggregates in order to eliminate from observations everything that is merely fortuitous and individual. The probability calculus indicates that . . . one approaches closer to the truth or to the laws one wants to grasp to the extent that observations include a greater number of individuals. These laws . . . can be applied to particular individuals only within certain limits . . . it is the social body that we are aiming to study, and not the pecularities that distinguish the individuals that make it up.[38]

In his various writings, Durkheim uses concepts and expressions that are remini-

scent of Quetelet. Currents of opinion—we would today call these "social influences"—which influence people toward marriage, suicide, fertility, and so on,
Durkheim writes, are certainly social facts:

At first they appear indistinguishable from the forms they assume in each
particular case. But statistics provides us the means of isolating them. They are
... expressed ... by the rates of fertility, of marriage, of suicide .... Since each of
these numbers includes all the particular cases without distinction, the individual
circumstances ... become neutralized and consequently do not contribute to the
rates. What it expresses is a certain collective state .... This is what social
phenomena are when they are stripped of all foreign elements.[39]

More so than Quetelet, Durkheim underlined that the properties of society,
groups and institutions are different in kind, not just magnitude, from the properties of the individuals that compose them. In his Bordeaux inaugural address, he
stated, "... when collected into a social unit by means of lasting bonds, men form a
new being possessing its own nature and laws. It is a social being. The phenomena
that occur in it certainly have their ultimate roots in the consciousness of the
individual. But collective life is not simply a blown-up image of individual life. It
presents a character *sui generis* that inductions based on psychology alone would
not allow one to predict."[40]

There are differences as well. For Quetelet, the precise shape of the distribution
about the average mattered. The normal distribution of soldiers' heights, Quetelet
thought, indicated that the mean was the "true value," and hence rooted in human
nature, or in some fundamental character of a nation or people, which he referred to
as a "penchant" when applied to moral and intellectual qualities. Durkheim
rejected this interpretation.[41] Yet Durkheim used the notion of deviation from the
average as an objective criterion for distinguishing the normal from the pathological and to give scientific validity to value judgments.[42] For Durkheim, the most
frequent or common, with some exceptions, is the normal, and deviations from the
most frequent are pathological.[43] Thus crime as such is a normal phenomenon since
it is found in every society, at every stage of development, although excessive crime
rates can be pathological.

These similarities between Quetelet and Durkheim are mainly methodological;
the differences between them are fundamental and theoretical. Durkheim's sociology was tied to an evolutionary and functionalist way of thinking alien to
Quetelet's social physics.[44] One made use of mechanical and astronomical analogies, and the other of biological analogies, but beyond that, they possessed a very
different notion of the key intellectual problems of social theory, whether it be
called social physics, sociology, or something else.

Durkheim provided an overview of his conception of sociology and of its
antecedents to nonspecialized audiences on two occasions, yet on neither occasion
does he refer to Quetelet or to moral statistics as antecedents.[45] Instead he reviewed the main ideas of Montesqieu, Saint Simon, Comte, Spencer, Espinas, and
Schaeffle, the German historical school of economists, and comparative juris-
prudence. According to Durkheim, the key conceptual breakthroughs were two.

First, society and institutions evolve spontaneously, and were not created by Human design; they are thus subject to laws just as all other natural phenomena, which humans cannot arbitrarily alter.[46] Second, society is analogous to an organism, made up of interdependent parts with specialized functions that contribute to its survival. Like other organisms, society (and institutions) evolves according to laws.[47]

The first priority of sociology is the study of collective representations, "common ideas and sentiments that successive generations transmit to each other and that assure the coherence as well as continuity of collective life,"[48] e.g., language, religious traditions, political beliefs, and judgments that have a practical aim and a binding force, like norms of conduct and laws.

And further, "[Societies] are nothing if they are not systems of representation ... the essential aim of sociology is to research how collective representation are formed and are combined."[49]

The central Durkheimian idea that society is an organismlike entity whose orderliness is founded on these collective representations is entirely lacking in Quetelet's social physics. For Quetelet, the social system is stable because the complex bundle of drives, instincts, dispositions, and qualities, physical, moral, and intellectual, of the "average man" are in equilibrium; i.e., human nature itself assures social order.

To be sure, there is a limited place in Durkheim's sociology for actions and events in the moral statistical sense. The clearest statement of it is expressed in a 1909 essay where Durkheim wrote,

The principal problems of sociology are the study of how a political, legal, moral, economic, religious institution has been formed ... , the causes that occasioned it, to what useful ends it corresponds. Comparative history now is the only instrument that sociologists possess for resolving such questions. . . . There are nonetheless occasions when the raw materials of sociological comparisons must be sought in a discipline other than history. That happens when one studies not how a law or moral norm has come about ... but why it is that it is more or less adhered to by the groups that practice it. For instance, instead of asking where the norm prohibiting homicide comes from, one researches instead the diverse causes that make people, groups of all kinds, more or less inclined to break it. . . . To resolve such questions, one must essentially turn to statistics.[50]

This position was quite a retreat from his earlier, far more positive assessment of moral statistics' contribution to sociology in the preface to *Suicide* in 1895.

6 Durkheim, Primitive Religion, and Evolution

Although not in principle opposed to quantitative research and the statistical method, after *Suicide* Durkheim became interested in questions that could not be answered using those techniques. Even though the programmatic statement in the first issue of the *Année Sociologique* stated the purpose of the new journal as "the construction of sociology starting from the research findings of specialized sciences,

120   Anthony Oberschall

the history of law, of custom, of religions, moral statistics, economic science ... , "
the last two were neglected in the Durkheim circle, as can be seen from the main
intellectual pursuits of its members. Mauss studied sacrifice, magic, reciprocity in
gift giving; Bougle, the Indian caste system and the history of egalitarian ideas;
Hubert, celtic archeology and the conception of time in religion and magic; Davy,
the history of contract and of Ancient Egypt; Granet, Chinese civilization; Beuchat,
American Indians and Eskimos; Fauconnet, social responsibility; Levi-Bruhl was a
philosopher, Hertz an anthropologist studying Alpine cults, and Doutte an arabist.
That leaves of course Halbwachs and Simiand, major intellectual figures to be sure,
but a small minority within the circle.

There are other indications of this shift in interest as well. According to Mauss,
Durkheim had lectured on ethics in Bordeaux from 1898 to 1900, and had in that
connection undertaken statistical analyses of crime rates that were comparable to
his work on suicide. Yet Durkheim left these studies for his students to finish.[51]
Whether they did is doubtful since they were not published, and do not survive. On
the other hand, Durkheim did publish in 1901 in the *Année Sociologique* a study of
the genesis and evolution of punishment, based on historical sources. That was his
priority.

A similar fate befell Durkheim's studies of the family and kinship, which he had
intended to study in their statistical, demographic, and contemporary aspects, but
ended up neglecting at the expense of a study of primitive kinship based on the
writings of ethnographers.[52] What accounts for these changes?

We know from various sources that Durkheim had experienced in 1895 a major
intellectual brainstorm about the central place of religion for his theory of collective
representations. In a 1907 letter, he expressed it as follows: "... it is only in 1895 that
I had the clear sentiment of the capital role played by religion in social life ... it was
for me a revelation ... it was entirely due to the studies I had undertaken in the
history of religions and in particular my reading of the works of Robertson Smith
and his school."[53]

Following this insight Durkheim started on an intellectual odyssey that took him
to the study of kinship and religion among Australian aborigines. At the end, he
had laid the foundations for a sociology of knowledge and culture that turned out to
be of more lasting significance for social science than his study of suicide.

About the role of religion as the basic building block of culture and of collective
representations, Durkheim concluded, "The most essential notions of the human
mind, notions of time, of space, of genus and species, of force and causality, of
personality, ... those which philosophers have labeled categories, and which dominate the whole of logical thought, have been elaborated in the very words of
religion. It is from religion that science has taken them."[54]

Granted all that, there is no reason, in principle, why the consensual religious and
ideological roots of collective representatives and of moral authority could not be
studied in contemporary France. Or, for that matter, why not study early Christian-
ity, Islam, Buddhism, the Protestant Reformation, as Max Weber was to do? Why
this obsession with primitive peoples and ethnographic sources? The consequences
for Durkheim's sociology were evident. The shift to the comparative study of

religion and to ethnography, especially of the most "primitive" peoples, led to a loss of interest in moral and statistical data on contemporary France and Europe as the means of confronting sociological theory with bodies of evidence.

Durkheim and most other sociologists and anthropologists of his time were evolutionists. In this view, social units could be classified in some ascending order of development, from the primitive horde to modern society, according to their degree of complexity and differentiation resulting from combinations of elementary units. The social units and institutions of a society are interdependent in complex ways, like the organs of a biological organism. Savages, primitive peoples, what we refer to today as preliterate tribal societies, are cases of arrested development, like children or embryos that did not make it to adulthood. Contemporary complex societies, however, have made it through all stages. The mechanism of social evolution is adaptation of the group under the pressure of human wants and competition resulting from wars, migrations, population increase, and the like. A successful adaptation is one whose effects or consequences are beneficial for the maintenance of the whole system. Survivals sometimes remain so long as they do not jeopardize the successful adaptation of the whole social organism. Family structure and religion, social classes and castes, the state and the economy are not the conscious, planned creations of individuals, as is the case for tools, artifacts, and laws enacted by a legislature. They evolve from simple to complex. The goal of science is to discover the laws of social evolution.

For Durkheim, too, social organization evolved from simple horde to simple polysegmented societies, to polysegmented societies simply compounded, and so on, up to contemporary European societies, with due allowances made for variations in structure at each stage due to the "incomplete coalescence of the united segments," and further allowances made for "pathologic" forms, such as the "anomic" division of labor.[55] To be sure, Durkheim's functionalism and social evolutionary views were not of the crude scientist variety. But one consequence that did follow from them was that the proper scientific methodology consisted of studying first the most elementary manifestation of a social institution, whether it is the family, religion, categories of thought, moral authority, or religion, and that in order to understand their contemporary manifestations, the entire evolutionary chain would have to be studied and understood. In *Les Règles de la Méthode Sociologique*, he had formulated the following rule: "... one cannot explain a social fact of some complexity without following its integral development through all social types."[56] Applying the rule to the religious roots of collective representations and moral authority, he wrote, "Primitive civilizations constitute ... privileged cases, because they are simple cases.... [They offer us] a means of discerning the ever present causes upon which the most essential forms of religious thought and practice depend,"[57] and "A science in its infancy must pose problems in their simplest form, and only later make them gradually more complicated. When we have understood very elementary religions, we will be able to move on to others."[58]

Social evolutionism thus accounts for Durkheim's turn to comparative ethnographic and historical studies, especially of the most simple tribal societies. At the time, this meant armchair reading, not field work by trained professionals based on direct observation, which was just then starting in anthropology. Ironically, it is not true that family and kinship are "simpler" in primitive societies than in ours. And as far as their religions and myths are concerned, they rival in richness, complexity, and levels of meaning anything written by Genet and Pirandello. Durkheim and his followers ended up complicating for themselves an already very difficult intellectual task.

At least one member of the Durkheim circle had serious reservations about such an approach to sociological theory. Lapie was one of the original collaborators of the *Année Sociologique*. He had taught in North Africa and had studied Arabs firsthand. In a letter to Bouglé, who was more approachable than Durkheim, Lapie wrote,

You know my opinion about books that deal with savages. When will the *Année Sociologique*, instead of repeating the same errors as Spencer did for the [Foulani] ... and the Bushmen, undertake investigations about social facts that surround us and concern French society, whose laws it is perhaps simpler to establish, though they are less well known that those of primitive societies? Why, for example, should one not undertake an investigation of the forms of religious sentiment in France? ... An intelligent statistical investigation ... might advance social science appreciably.[59]

Lapie's advice went unheeded. He eventually dropped out of the Durkheim circle.

Social evolutionary approaches to sociological theory and the use of ethnographic data on non-European peoples predominated elsewhere as well, to the point that it is arbitrary to draw a distinction between anthropology and sociology at the turn of the century. In the United States, colleges and universities started offering courses in sociology in the 1880–1920 period. Many departments were at first designated "sociology *and* anthropology," including the most important one at Chicago. The major theoretical works of the early American sociologists were a mixture of social evolutionism, ethnography, comparative religion, philosophy of history, *Volkerpsychologie*, and moralizing: such was Sumner's *Folkways* and *The Science of Society*, and Thomas's *Source Book of Social Origins* (1908), whose subtitle was "Ethnological Materials, Psychological Standpoint, Classified and Annotated Bibliographies for the Interpretation of Savage Society."

In Britain, this was even more true, from Spencer's *Principles of Sociology* to Westermarck's *The History of Human Marriage* and *The Origin and Development of Moral Ideas*, and to Hobhouse's *Mind in Evolution* (an evolutionary study of animal psychology) and *The Material Culture and Social Institution of the Simpler Peoples*.[60] These was no difference to speak of, in definition of subject matter, method, data, and theory between the first British academic sociologists and anthropologists. The sociology syllabus at the London School of Economics, the first academic course of sociological study in Britain, was basically a combination of evolutionary philosophy and comparative anthropology.

These comparisons with the Durkheim circle indicate that at the same time as an academic sociology was becoming institutionalized in Britain, France, and the

United States, social evolutionism and a concentration on ethnographic data on non-European societies prevailed among academic sociologists. Compared to the 1830s and 1840s, the quantitative study of group processes and social action had declined. Thus not moral statistics, but comparative ethnography and social evolutionary theory became the takeoff point for twentieth-century sociology.

7 Halbwachs's Ambivalent Verdict

The most important French sociologist between the two world wars was Maurice Halbwachs. He would have to be the key figure in any renewed synthesis of social theory and research on contemporary society. Though he has been described as "an authority on methodological questions and the Durkheimians' most sophisticated statistician,"[61] he was cautious and ambivalent regarding the marriage of sociology with quantitative techniques. Indeed, the two proved to be difficult to join in the Durkheimian framework.

Halbwachs was born in 1877 and died in a German concentration camp in 1944. He studied at the Lycée Henri IV and came there under the influence of Bergson.[62] He went on to study philosophy at the Ecole Normale, where he became a socialist and a Dreyfusard. Halbwachs had an extraordinarily diverse and rich professional and intellectual career. Under the influence of his friend and mentor Simiand he gradually moved from philosophy to sociology, economics, and demography, and joined the *Année Sociologique* team in 1905. His law thesis, a study of real estate prices in Paris, and his Ph.D. dissertation, which was based on family budgets and dealt with the standard of life of the working class, were topics virtually ignored by others in the Durkheim circle except for Simiand. During the war he spent two years at the Ministry of Armaments headed by the socialist Albert Thomas, where his main responsibility was workers' salaries and the cost of living. Earlier he spent two years in Germany and Austria on a lectureship and fellowship in 1904 and 1909, where he studied among other things Marxism and political economy. After he wrote on the repression of a strike by the Berlin police for the socialist paper *L'Humanité* in 1909, he was expelled in what became a minor international incident.

His subsequent academic career was spent principally at the University of Strasbourg, where he became France's first professor of sociology in 1922, and after 1935 at the Sorbonne. Among other things, he introduced to a French audience the thought of Max Weber, Pareto, Veblen, Keynes, and Schumpeter in lengthy reviews. In 1930, he was a visiting professor at the Sociology Department of the University of Chicago, then the leading American center of sociology, whose activities he described in detail in a 1932 *Annales* report. His major works and books were creative departures from Durkheim rather than mere extensions. A case in point was his 1930 book *Les Causes du Suicide*. In the preface of that book, Marcel Mauss wrote that Halbwachs "felt compelled to undertake new research, pose new problems, and present facts from another perspective" than Durkheim had.[63] Similarly, Halbwachs's major book on collective representations, *La Mémoire Collective*, elicits the following comment from Karady, the recent reeditor of

Halbwachs's essays: "... the Durkheimian sociology of knowledge was basically genetic, and drew upon "primitive" facts within an evolutionary framework. As for Halbwachs, he utilizes historical and contemporary data and observations to study mental processes here and now."[64]

Halbwachs was also one of the first social scientists to monitor the British mathematical statisticians and to devote a great deal of thought to the application of probability and mathematical statistics to sociology. Already in 1912 he had published a book on Quetelet and moral statistics titled *La Théorie de l'Homme Moyen*. He taught statistics in Strasbourg jointly with the mathematician Frechet and coauthored with him *Le Calcul des Probabilités à la Portée de Tous* in 1924. In a series of articles in the 1920s and '30s, he discussed in turn whether statistical analysis is an appropriate substitute for experiments (which leads him to reanalyze data from Mendel's genetic experiments and Galton's laws on human inheritance), the concept of "law" in sociology, problems of multivariate analysis and the use of standardized rates, and the applicability of the probability calculus to social processes, among other topics.

In all these writings, Halbwachs keeps repeating the same basic objection. Probability theory can be applied only when each event or combination of events is independent of the others, as would be the case for successive rolls of dice or drawings of colored balls from an urn. But there is no analog to chance mechanisms in social and moral processes; far from acting independently, human beings influence each other; they are influenced by shared norms, beliefs, and a common past, and the institutions of society are interdependent. In the conclusion of his critique of Quetelet's theories, Halbwachs writes, "The law of large numbers ... can be applied only if the forces or causes that are combined in various ways ... are rigorously 'independent'.... It is especially in the social world that such a hypothesis is inadmissible,"[65] and again "Society and the moral actions of its members are perhaps of all phenomena the domain in which it is least possible to consider an individual and his actions in isolation from all others': that would be omitting the most essential matter. It means that this domain is where the calculus of probability is least applicable."[66]

In point of fact, this judgment turned out to be mistaken. It is possible to describe social influence processes by means of stochastic models if one is willing to alter the assumptions of the basic coin-tossing or urn experiments slightly, as Lazarsfeld has shown.[67]

In the Durkheimian conception of society and social processes that Halbwachs, shared, there was no room for chance in human affairs and for "independence" among human beings. Durkheimian sociology does not start with social action and the individual decision maker, and build from that elementary entirty the concept of social units, such as group and family, and of institutions, such as marriage and property. Durkheimian sociology uses groups and institutions and belief systems as the basic units of analysis. If the social actor and human choice are studied. as suicide was, it was to demonstrate the existence and concrete effects of these groups and institutions, which cannot be directly studied, for they are a system of relationships between human beings, existing independently of any particular person, and not a collection of physical attributes or entities such as a house that is made of bricks. Since groups and institutions exercise a pervasive influence on all their members, including the choices they make, there is no room for actors making these choices independently of each other.

The Durkheimian bias against randomness and chance events in social processes is manifest in one of Halbwachs's most interesting statistical essays on the consequences of the large number of French male deaths in World War I upon postwar marriage. Commenting on the narrowing age gap between the spouses, he writes, "One might surmise, it is true, that these changes can be explained by a mechanical and blind game of competition between age cohorts, analoguous to sexual selection in the animals world as Darwin depicted it." [68]

Indeed, the data can be interpreted in that manner. But Halbwachs immediately draws back from such a heretical thought into the safety of a Dukheimian mold: "If marriage were nothing but the union of sexes, that would be true. But marriage is much else besides."

Sentiments of maturity and of marital responsibilities are collective representations shaped and promoted by all, not just by those who are single at the marriageable ages. Thus "... beyond the matrimonial steps taken by individuals, there exists as it were a collective marriage market whose direction and rhythm are regulated by the evolution of society." [69]

Halbwachs rejects blind chance in human affairs, and with it probability theory, but not statistical analysis for getting at empirical generalizations, what he calls "statistical laws." His ambivalence can best be summarized in the conclusion of a long 1923 essay: "... statistical laws do not follow from the laws of chance ... but the statistician is obliged at every moment to make use of the calculus of probability." [70]

Elsewhere, in a 1935 essay, Halbwachs is critical of the use of multivariate techniques introduced by Yule and other British mathematical statisticians. Commenting on the standardization of the age pyramid in making cross-national comparisons of mortality rates, he writes,

How long would the French live if, remaining French, they lived under the same physical and social conditions as the Swedes? How long would the Germans live if they lived under the same conditions as the French? This boils down ... to the question of how long the camel would live, if it were transported to the polar regions while still remaining a camel, and how long a reindeer would live if it were transported into the Sahara, and still remained a reindeer.... Like the *homo economicus, such as homo demographicus* is an abstraction too carefully detached from reality for us to learn anything useful from it. [71]

Several important consequences followed from Halbwachs's ambivalence. [72] Sociology became but precariously institutionalized in France between the two world wars. It needed allies, both within and without academia to expand numerically and to invigorate it intellectually, just as in the early nineteenth century moral statistics and empirical research on social conditions rode a crest of popularity and scientific interest through its association with liberal social reform. Durkheim's theoretical system had two potentially fruitful, built-in, empirical tendencies. One led to anthropology, ethnography, and comparative history, by way of Mauss and Bougle. The other might lead to the quantitative, empirical study of contemporary society, if only an advocate could be found. Halbwachs was the obvious candidate. Numerically the Mauss group was by far the most important, but Halbwachs and Simiand, his kindred spirit, had a more central position at the University of Paris and the Collège de France. In the early part of this century, the natural ally of the quantitatively oriented, positivist legacy of Durkheim would have been the rapidly growing professional statisticians whose main responsibility was the quinquennial census conducted by the Statistique Générale de la France (SGF). They were much concerned with narrow methodological questions (e.g., of data collection, classification, and processing) of no interest to most of the Durkheim group, since most did not use the SGF data. The statisticians, on the other hand, had little use for the "literary" interests of the Durkheimians. With his mathematical knowledge, his wartime experience in a statistical agency, his interest and work in demography and urban ecology, his contemporary empirical orientation, and his standing among the Durkheims, Halbwachs might have bridged the growing gap between the statisticians and the sociologists. Yet we have described his ambivalence about such an intellectual marriage.

In Halbwachs's case, there were additional reasons, perhaps of a personal kind. He was somewhat of a lone scholar, not an academic entrepreneur. Within the Durkheim group, he and Simiand were intellectual marginals. And as far as the statisticians were concerned, they preferred dealing with the two at arm's length. Despite the fact that he had coauthored with Halbwachs an article in the *Encyclopédie Française* of 1936 on population over 100 pages long,[73] Alfred Sauvy, the grand old man of French demography, recalled about the 1930s, ". . . we were on guard against the sociologists, because they committed almost always statistical errors. Even Simiand was shunned, especially on account of the obscurity of his language. We were on very good terms with Halbwachs but contested his statistical knowledge."[74]

The context makes clear that it is not Halbwachs's mathematical skills that Sauvy contested, but his knowledge about data collection and use of official data. The upshot of all of this was that an integration of social theory with empirical social research on contemporary society failed to be made; the fate of the Durkheim circle in this regard paralleled that of moral statistics and Quetelet.

8 Conclusion

The Durkheimian legacy to sociology and to social theory was the study of institutions, norms, laws, religious beliefs, and collective representations more generally as social facts that had a reality of their own, obeying evolutionary laws of their own, independently of the motivations, dispositions, and purposes of the human beings that participated in them or shared them. If these institutions and collective representations were not created by human design, according to a carefully laid out master plan, how could they have originated and how do they change? The answer was provided by a functionalist explanation: Institutions that contribute to adaptation in the environment and to maintaining the social system are favored over those that do not.[75] Most social theorists at the turn of the century were functionalists, and most looked to ethnography, history, and religion as the raw materials for an inductive study of institutions and the laws of their evolution.

However, the political economists had shown that the laws of supply and demand in the market could be deduced from the actions of utility- and profit-maximizing consumers and firms. Might it not be the case that other institutions, such as marriage and property rights, could be accounted for by an individualist method rather than by wholism and functionalism?

The issue is not just "academic." The usual objection to Durkheimian and functionalist sociology has been their intellectual blinders in coming to terms with power, conflict, and the processes of social change, as well as their "oversocialized" view of man. A functionalist explanation for slavery, race segregation, the persecution of witches, or authoritarian government can all too easily become a legitimation of the institution, since its very existence is proof of its capacity to adapt to a social environment and to contribute stability to a social system. Durkheim provided himself with an escape from this dilemma. He would argue that some institutions are "pathological" or abnormal, yet this too would lead to difficulties, as George Friedmann pointed out: "Durkheim would have been obliged to consider 'abnormal' most of the forms taken by labor in modern society. . . ."[76] With a wholistic theory like functionalism, one is unable to move beyond "functional adaptation" to the concept of "social welfare." With an individualist method, it follows in a straightforward manner that what benefits some, including social system stability, may be at the expense of others, possibly even the majority. An individualist method can clarify the paradoxes of collective action, e.g., "the prisoner's dilemma" and "the tragedy of the commons." It can explain why public goods or institutions fail to be brought about by voluntary choice and cooperation even though they are functional and preferred by all.[77]

The alternative to wholism and functionalism is methodological individualism. Its sociological version is the action theory advocated by Max Weber.[78] Though not successful in reorienting social theory as practiced by his contemporaries, Weber put his views forward forcefully in several of his theoretical writings, especially in a 1913 methodological essay (which served as a draft of the opening pages of the posthumous *Wirtschaft und Gesellschaft*). Weber wrote that "'Verstehen' . . . is in the end also the reason . . . why sociology must treat the individual human being and his actions as the basic unit, 'its atom'. . . . Concepts such as the 'state,' 'association,' feudalism, and similar entities represent for sociology categories of certain kinds of human interaction . . . it is the task [of sociology] to reduce them . . . to the actions of its constituent human beings."[79]

In this view, social order is grounded on the mutual expectations of social actors about each others' behavior and the resulting choices based on calculation of one's chances of success, given what one expects others to do. Elsewhere he writes in the same vein that "the stability of custom rests essentially on the fact that he who does not take it into account in his actions is maladapted, i.e., must pay the cost of many inconveniences ... so long as the actions of the majority in his environment take that custom into account when they act."[80] As the contemporary reader will recognize, Weber conceptualizes custom as the stable outcome of strategic interaction in the game theory sense. Whereas for Durkheim, expectations about order are based on conformity to norms and on consensus about collective representations, for Weber such expectation can also result from a multitude of purposive actors who pursue their private ends:

Many ... especially notable uniformities ... of social action are not determined by orientation to any sort of norm that is held to be valid, nor do they rest on custom, but entirely on the fact that ... social action is best adapted to the normal interests of the actors as they are conscious of them. This is above all true of economic action, for example, the uniformities of price determination in a 'free market', but is by no means confined to such cases. ... This phenomenon ... can bring about results that are very similar to those that an authoritarian agency, very often in vain, has attempted to obtain by coercion. ... Observation of this has, in fact, been one of the important sources of economics as a science. But it is true in all spheres of action as well.[81]

Had Halbwachs been mindful of this possibility, he would not have dismissed out of hand an explanation for the post–World War I French marriage market based on the "mechanical and blind play of competition."

There is actually no contradiction between the concept of a competitive marriage market and the Durkheimian view of marital choice based on norms and collective representations. Both sets of variables must enter a satisfactory sociological explanation of marital choice. The impact of norms and collective representations can be entered into a model of individual choice as costs and constraints. Methodological individualism does not imply rejection of the social dimensions in human behavior, which Durkheim had concluded from his reading of nineteenth-century utilitarians and economists. And it too, in addition to wholism and functionalism, can be employed to account for institutions and institutional change, e.g., the origins of the state, systems of property ownership, and various forms of family.[82]

Notes

1. E. E. Evans-Pritchard, *Social Anthropology* (London: 1951), p. 49.

2. Adam Ferguson, *An Essay on the History of Civil Society* (1767), quoted in J. W. Burrow, *Evolution and Society* (Cambridge: Cambridge University Press, 1966), p. 12.

3. Thomas Malthus, *Essay on the Principle of Population* (1798), Book 7, Chapter 2.

4. The intellectual consequences of this trend are dealt with in the conclusion of this essay.

5. Louis Chevalier, *Classes Laborieuses et Classes Dangereuses* (Paris: Plon, 1955).

6. Chevalier, *Classes Dangereuses*, p. 23 (note 5).

7. A. J. B. Parent-Duchatelet, *De la Prostitution Dans la Ville de Paris* (Paris: 1836), p. 21.

8. Quoted in Stephen Cole, "Continuity and Institutionalization in Science," in: Anthony Oberschall, ed., *The Establishment of Empirical Sociology* (New York: Harper and Row, 1972), p. 75.

9. M. J. Cullen, *The Statistical Movement in Early Victorian Britain* (New York: Barnes and Noble, 1975).

10. On these and other discontinuities, cf. Oberschall, *Establishment* (note 8).

11. Andre Guerry, *Essai sur la Statistique Morale de la France* (Paris: Crochard, 1833), p. 69.

12. Guerry, *Essai*, p. III (note 11).

13. Adolphe Quetelet, *Du Système Social et des Lois Qui le Régissent* (Paris: 1848), p. 138.

14. Quetelet, *Système Social*, pp. 74–81 (note 13).

15. Guerry, *Essai*, pp. 20–21 (note 11).

16. Moritz Drobisch, *Die moralische Statistik und die menschliche Willensfreiheit* (Leipzig: Voss, 1867), pp. 20–57.

17. Wilhelm Lexis, *Zur Theorie der Massenerscheinungen in der menschlichen Gesellschaft* (Leipzig: 1877).

18. Emile Durkheim, *Suicide* (Glencoe: Free Press, 1951 [1897]), pp. 300, 300–305.

19. Max Weber, "Die Objektivität sozialwissenschaftlicher und sozialpolitischer Erkenntnisse," *Archiv für Sozialwissenschaft und Sozialpolitik*, I (1904), 47ff.

20. Alexander v. Oettingen, *Die Moralstatistik* (1874), pp. 26–34.

21. Quoted in Cole, "Continuity," p. 75 (note 8).

22. Guerry, *Essai*, p. 43 (note 11).

23. Chevalier, *Classes Dangereuses*, p. 21 (note 5).

24. Philip Abrams, *The Origins of British Sociology, 1834–1914* (Chicago: Chicago University Press, 1968).

25. Abrams, *Origins*, pp. 38–39 (note 24).

26. For Manchester, David Elesh, "The Manchester Statistical Society: A Case Study," in Oberschall, *Establishment* (note 8); for London, Cole, "Continuity" (note 8).

27. Catherine Bodard Silver, ed., *Frederic LePlay* (Chicago: Chicago University Press, 1982), pp. 45–46.

28. Frédéric LePlay, *Les Ouvriers Européens* (Paris: 1855), pp. 11–13.

29. LePlay, *Ouvriers*, pp. 11–13 (note 28).

30. Emile Durkheim, *Les Règles de la Méthode Sociologique* (Paris: P. U. F., 1981 [1895]), p. ix.

31. Durkheim, *Suicide*, p. 37 (note 18).

32. Durkheim, *Régles*, p. 101 (note 30).

33. Durkheim, *Suicide*, p. 300 (note 18).

34. Durkheim, *Suicide*, p. 305 (note 18).

35. Joseph Lottin, *Quetelet, Statisticien et Sociologue* (Louvain: 1912), p. 276.

36. Emile Durkheim, "La Sociologie en France au XIX Siecle" (1900), in *La Science Sociale et L'Action*, (Paris: P. U. F., 1970), p. 113.

37. Quoted in Steven Lukes, *Emile Durkheim* (London: Lane, 1973), p. 192.

38. Adolphe Quetelet, *Sur l'Homme et le Développement de ses Facultés* (Paris: 1835), pp. 4ff.

130   Anthony Oberschall

39. Durkheim, *Règles*, pp. 9–10 (note 30).

40. Emile Durkheim, "Cours de Science Sociale" (1888), in *Science Sociale*, p. 86 (note 36).

41. Durkheim, *Suicide*, p. 304 (note 18).

42. Lukes, *Durkheim*, p. 28 (note 37).

43. Durkheim, *Règles*, pp. 55–56 (note 30).

44. Durkheim, *Règles*, pp. 57–63 (note 30)

45. Durkheim, "Cours" (note 40) and "La Sociologie" (note 36).

46. Durkheim, "Cours," pp. 75–86 (note 40).

47. Durkheim, "Cours," pp. 85–100 (note 40).

48. Durkheim, "Cours," p. 101 (note 40).

49. Durkheim, "La Sociologie," pp. 124–125 (note 36)

50. Emile Durkheim, "Sociologie et Sciences Sociales" (1909), in *Science Sociale*, pp. 153ff. (note 36).

51. Lukes, *Durkheim*, pp. 256–257 (note 37).

52. Lukes, *Durkheim*, pp. 179–180 (note 37).

53. Letter quoted by Karady in "Les Durkheimiens," *Revue Française de Sociologie*, 20 (1979), 72.

54. Quoted in Lukes, *Durkheim*, p. 445 (note 37).

55. Durkheim, *Règles*, Chapters 3 and 4 (note 30).

56. Durkheim, *Règles*, p. 137 (note 30).

57. Quoted in Lukes, *Durkheim*, p. 181 (note 37).

58. Quoted in Lukes, *Durkheim*, p. 458 (note 37).

59. Letter printed in "Les Durkheimiens," *Revue Française de Sociologie*, 20 (1979), 36.

60. Abrams, *Origins* (note 24).

61. Philippe Besnard, ed., *The Sociological Domain* (Cambridge: Cambridge University Press, 1983), p. 264.

62. My account of Halbwachs's career relies heavily on the introduction in Victor Karady, ed., *Maurice Halbwachs, Classes Sociales et Morphologie* (Paris: 1972).

63. Quoted in Anthony Giddens's preface to Maurice Halbwachs, *The Causes of Suicide* (London: Routledge and Kegan Paul, 1978), p. XXIX.

64. Karady, *Halbwachs*, p. 18 (note 62).

65. Maurice Halbwachs, *La Théorie de l'Homme Moyen* (Paris: Alcan, 1912), pp. 145–146.

66. Halbwachs, *L'Homme Moyen*, p. 174 (note 65).

67. Paul Lazarsfeld, "Notes on the History of Quantification in Sociology," *Isis*, 52 (1961), 308–309. The technical point is that it is quite appropriate to change the probabilities of the outcomes at each drawing depending on prior outcomes and then applying the basic urn model to derive a distribution of outcomes. One then obtains "contagious" stochastic distributions.

68. Karady, *Halbwachs*, p. 267 (note 62).

69. Karady, *Halbwachs*, p. 268 (note 62).

70. Karady, *Halbwachs*, p. 307 (note 62).

71. Karady, *Halbwachs*, p. 337 (note 62).

72. The next paragraph relies heavily on Alain Desrosiers, "Un Essai de Mise en Relation des Histoires Récentes de la Statistique," unpublished paper (Paris: INSEE, October 15, 1982).

73. Adolphe Laundry, *La Révolution Démographique* (Paris: INED, 1982 [1934]); cf. the Introduction by Alain Girard, p. 21.

74. Quoted in Desrosiers, "Essai," p. 4 (note 72).

75. Peel defines the functionalist method as follows: "The essence of this method is to explain institutions not in terms of motives or purposes, either of the actors whose actions compose the institutions, or of the creators of the institutions, but in terms of the functions they fulfill, that is of their effects or consequences for the whole system of which they form a part within the environment to which they are adapted"—in J. D. Y. Peel, *Herbert Spencer* (London: Heinemann, 1971), p. 183.

76. Quoted in Lukes, *Durkheim*, p. 124 (note 37).

77. Cf. Edna Ullman-Margalit, *The Emergence of Norms* (Oxford: Clarendon Press, 1977).

78. On Weber's action theory, see Paul Lazarsfeld and Anthony Oberschall, "Max Weber and Empirical Social Research," *American Sociological Review*, 30 (1965). According to Ullman-Margalit, *Norms*, p. 14 (note 77), "Methodological individualism is the view according to which statements about social collectivities can be reduced to statements referring solely to individual human beings, their actions, and relations among them."

79. Max Weber, "Über einige Kategorien der verstehenden Soziologie" (1913), in *Soziologie, Analysen, Politik* (Stuttgart: Kröner, 1956), p. 110.

80. Max Weber, *Wirtschaft und Gesellschaft*, Vol. I (Tübingen: Mohr, 1956), p. 22.

81. Weber, *Wirtschaft*, pp. 21–22 (note 81).

82. There is now a considerable literature on these topics and the new institutional economics. The March 1984 issue of the *Zeitschrift für die gesammte Staatswissenschaft*, Vol. 140, No 1 (the English title of this journal is *Journal of Institutional and Theoretical Economics*) contains the papers from a 1983 symposium on these topics in which the principal figures participated.

III ECONOMICS

The Probabilistic Revolution in Economics—an Overview

Mary S. Morgan

Counting the people and their incomes has a tradition in political economy that goes back to the late seventeenth century, though rapacious rulers have always excelled at this task, as evidenced by the Domesday Book of eleventh-century England. The tradition flowered afresh in the social sciences of the nineteenth century when the statistical science of demography developed separately from economics.[1] In one respect economists' measurements of income were similar to those of early census takers' measurements of population—both aimed at a complete measure of society's resources. But because of the circular flow of economic life, the "total" income of society does not equal the sum of the individual incomes.

In other areas of measurement in economics, however, a sampling process was inevitable. It is clear that prices in the market place for any one commodity generally show considerable variation. The problem of obtaining one representative measure of prices and their changes cannot be resolved simply by taking a census of all price observations because accurate measurement requires that each price be weighted by its relevant quantity. For practical reasons, samples must be taken and index numbers constructed from these. This process is far from simple, and probability tests were used to determine the most accurate measurement method. Almost all economic variables are measured either in complete form or are representative samples in index number form. Horváth's chapter discusses the theoretical development of both types of measurement, which occurred in the late nineteenth and early twentieth centuries and formed an important precondition for any statistical consideration of economic laws.

Despite the observed variation in prices, economic theory since the eighteenth century has been formulated on the basis of one unique market price for each good.[2] But in the 1870s, the labor theory of value, in which there is one price for labor, was replaced by the marginal theory of value, in which wages vary. It was soon discovered that the empirical distribution of incomes (and certain other variables, such as the size of firms) was similar in different countries, and consequently, desultory attempts were made to provide theoretical probabilistic models that might have generated the observed distributions. Economic theories featuring variables with a range of values explicitly represented by a probability distribution have become more commonplace only in the most recent years.

While statistical methods began to be applied to measuring and discovering economic relationships in the late nineteenth century, the real development of such methods took place in the 1920s and 1930s. In this work, economists rejected the use of probability mathematics because it was thought to be inapplicable to economic data. Morgan's chapter suggests that this only changed in the 1940s with T. Haavelmo's reassessment of the practical domain of probability theory in testing economic theories, which followed the work of R. A. Fisher in biology and parallel developments in the theory of statistical inference by J. Neyman and E. S. Pearson.

In the field of theory building, we have a strange picture—although several eminent economists of the nineteenth and early twentieth centuries wrote on probability theory (for example, Cournot, Jevons, and Keynes), theoretical economics remained persistently deterministic until very recently. Ménard's chapter maintains that this is because economists were, and remain, strongly influenced by classical mechanics. Theories were formulated within the idealized model of perfectly competitive market forces and self-equilibrating mechanisms. Equilibrium solutions depend crucially on the presence of a mystical "invisible hand" (under the late eighteenth-century classical model of Adam Smith) or economic agents having perfect information or foresight (under the twentieth-century neoclassical version of the model). Uncertainty must exist in the absence of such perfect knowledge, and probability has been introduced into the theory to try and deal with this.

The effective starting point for this introduction of probability into the theory of *individual* economic behavior might well be dated by Frank Knight's book *Risk, Uncertainty and Profit* (1921). Knight distinguished very clearly between risk (uncertain knowledge about the individual outcome but certain knowledge about the probability distribution of outcomes) and uncertainty (uncertain knowledge about the probability distribution of outcomes). Risk can be insured against—and insuring against wrong decisions takes away the costs of uncertainty and removes content from the decision. Genuine uncertainty according to Knight cannot be insured against.

Particularly since the 1930s, this problem of economic behavior and decision taking in an uncertain world (for example, when a firm must take decisions in ignorance of a rival firm's actions), has become a preoccupation of economic theorists. The basic model involves the application of probabilities to the possible outcomes and the gain or loss associated with each outcome. The probabilities used to measure the expected utility of various outcomes can be either objective/frequentist or personal. (The use of probability in the measurement of utility goes back to Bernoulli in the eighteenth century, but had been lost to economics in the interim.) Approaches to the decision problem vary: in game theory (following von Neumann and Morgenstern), both risk and uncertainty are covered within an *optimizing* framework, while H. Simon's approach involves a search for *satisfactory* decisions in the absence of knowledge of the distribution of outcomes. Knight's distinction between risk and uncertainty has become blurred in this more recent theoretical treatment, and it seems that uncertainty has only partly been tamed by the attempts to make it conform to a probabilistic model.

The same thing appears to have happened when we look at the way probability enters economic theories at the level of *society*. According to a suggestion by the Russian economist E. E. Slutsky in 1927, probability, in the guise of random shocks, may provide the impetus responsible for the Brownian-motion movements observed over time in economic variables. Further, some theoretical explanations of the business cycle (and the 1930s depression) involved fallible expectations—that is, individual agents' plans, based on their expectations, prove to be incorrect when aggregated with those of others. More recent economic models of expectations have been based on the theory that the subjective probability distribution of economic agents regarding some economic outcome will equal the objective probability distribution in the system as a whole, even though individuals' expectations prove incorrect. According to this theory of "rational expectations," there will be no systematic error in the interactions between expectations and resulting outcomes.

But this still leaves the economy open to real shocks of a random nature (such as unanticipated policy changes or acts of God), and once again it appears that probability ideas have been used successfully to plug one gap in the system of economic theory only to find that uncertainty remains as considerable as ever.

Notes

1. R. A. Horváth, "The Rise of Demography as an Autonomous Science," *Universität Bielefeld Institut fur Bevolkerungsforschung* (IBS Materialien, No. 12, 1983), pp. 1–51.

2. E. Streissler, "On Probabilistic Models in Economic Theory," in *Probability and Conceptual Change in Scientific Thought*, ed. M. Heidelberger and Lorenz Krüger (22nd Report, Forschungsschwerpunkt Wissenschaftsforschung, Universität Bielefeld, 1982), pp.165–184.

6 Why Was There No Probabilistic Revolution in Economic Thought?

Claude Ménard

Mathematical Economics developed, from the beginning of the nineteenth century, under the influence of the rational mechanics of Lagrange. For almost one and a half centuries economists have ignored probability. This is a very paradoxical fact indeed. How is it that an instrument, originally conceived to a great extent for the use of the emerging "social sciences," was neglected by the very individuals who should have used and developed it? This chapter suggests that theoretical economics has been, and is still, thought of in the deterministic terms of the physics of the end of the eighteenth century. The probabilistic revolution in economic thought is still no more than a subject for a "scientific research program."

Before expounding my thesis, there is some need, I think, to explain the title of this chapter and give you some idea of my understanding of the question. The problems it raises, which I would like to submit for open consideration, will be looked at from the point of view of an economist mainly interested in the theoretical foundations of his discipline.

Rather than develop a specific point in the history of economic thought, I would like to give an overview of the evolution of economics, and try to find some explanation for a very paradoxical fact: Why is it that there was such a long time (almost one and a half centuries) before probability was taken seriously as a tool of investigation in the economics profession? And, even more precisely, how is it that a new mathematical instrument, elaborated at the end of the eighteenth century for the special use of the developing social sciences, particularly "political economy," was openly subordinated or ignored by the greatest theoretical economists for almost six generations?

Why indeed is it that there was no probabilistic revolution in economic thought? Why did this new mathematical instrument, apparently so well adapted to approximate social facts, not transform the concepts and the methods of modern economists? Is it because economists, even the more inventive ones, were insensitive to the novelty of probabilistic analysis, or is it because probability was not as convenient as had been hoped for the examination of economic facts?

But these are enough questions. It would be pretentious of me to suggest that I shall be giving you the answers. So you have to be prepared to be somewhat disappointed.

Let me add one other limitation. All my examples will be taken from theoretical economics. I do not intend to explore the contribution of probability to the modern analysis of economic data. This, of course, is a severe limitation. I assume that taking into consideration those aspects would not alter my main thesis: that is, that there has been no deep transformation of economic thought by the probabilistic approach.

I shall not try to outline the history of probability theory. Others can do that better than I. But I would like to go into the history of my own discipline, and particularly the history of a subfield that originated in the nineteenth century: mathematical economics. I would like to ask why the most prominent contributors to economics—from Cournot to Keynes—did not use probability, even though they (at least the two thinkers mentioned) well knew its possibilities.

Let us first look at Cournot, who had the honor to be the first model builder in economics, and to be the first mathematician who took modern economics seriously, seriously enough to publish a book on economics, now a classic: *Researches into the Mathematical Principles of the Theory of Wealth* (1838).

Cournot was in a good position to generate a radical transformation of economic thinking. He was trained, and well trained, as a mathematician in the tradition of Laplace. He was much impressed by the great masters of the late eighteenth century: Condorcet, Lagrange, Laplace, and Monge. And he was very close to Poisson, the leading probabilist of that time. Poisson was his protector; he found Cournot a job, and Poisson also played a major role in Cournot's theoretical orientation. It was Poisson who invited Cournot to work on probability as applied to insurance problems; and Poisson gave him the idea of writing an essay on probability, the well-known *Exposition de la théorie des chances et des probabilités* (1843), which was a very influential book in the nineteenth century. So we can say that Cournot, as a practitioner, was in a very good position to evaluate the possibilities of probability as a mode of investigation.

There is also another point that I would like to emphasize. It is the ideological context within which Cournot wrote his books on economics and probability. After the chaos of the Napoleonic wars and the ending of the French Revolution, the Restoration of the 1820s did not succeed in restoring an equilibrium between the social forces of French society. On the contrary, there were many social tensions, which led to outbursts in 1830, 1834, and so on. This period of unrest, however, gave rise to numerous projects for social reform. We must remember that Saint-Simon, Auguste Comte, etc., became very influential during the period, and that the 1830s saw the birth of many socialist utopias. Cournot was not a socialist, nor was he a utopian, but in his economic research he pretended to contribute to a better ordering of economic and social processes. In those years, this sounded like the work of a reformer.

We can therefore summarize the situation as follows: In 1838, Cournot understood the need for a new formulation of economic theory, and he was convinced that it must be done in mathematical terms. By the standards of his time, he was very well trained as a probabilist; and he was an admirer and follower of Condorcet, Laplace, and Poisson.[1]

What is the result of this highly favorable conjunction for a probabilistic revolution in economic theory? Cournot's leading book, *Researches into the Mathematical Principles of the Theory of Wealth*, did not use any probability at all. On the contrary, the first rigorous model in economic theory was a rejection of the idea that probability could be a useful tool at the core of economic analysis.

In adopting this attitude, Cournot was inspired by nineteenth-century physics, prior to Boltzmann and Gibbs. Cournot considered physics to be the exemplary science, and from it he drew analogies based on Adam Smith's exposition of economics. Probability was invoked, but only to the extent that its principles could restore (despite the apparent discontinuity of social phenomena) the properties of the universal law of continuity that Cournot believed to be the cause of the success of mathematical analysis in mechanics.[2]

In the mathematization of economics, which began with Cournot, and was to dominate economic theory from the end of the nineteenth century until now, probability did not play a central role. This was not because the probabilistic method was inadequate; after all, probability mathematics was not all that rudimentary. It is rather because the ambitions of the first model builder in economics, in formulating his idea of the new economic science to be created, led him to seek solutions in rational economics, understood to be a counterpart of "rational mechanics." The law of profit was compared to the work performed by machines; in both cases, productivity had to be maximized.[3] To accomplish this reformulation, there existed a very sophisticated instrument: mathematical analysis as perfected by Laplace, Lagrange, and Cauchy. Thus was elaborated the first economic model according to the image of classical physics.

We now come to the second stage of my little story, the so-called marginal revolution of the 1870s, which happened simultaneously in three different countries: Great Britain, France, and Austria.[4] Here again there are some paradoxical facts that I must summarize, because they shed light on the resistance to probability theory in economic thought.

If we do not consider the Austrian tradition—for the simple reason that unfortunately I am not competent to do so—we can identify the marginal revolution with the names of Jevons, Walras, Edgeworth, and, a little bit later, Marshall.

All of them, though to a lesser degree Marshall, were well trained in mathematics. But, and this is even more important for the story, all of them were informed of the potential of probability. Walras had been trained as an engineer at the *Ecole des Mines* in Paris, where probability was taught. Moreover, he was an admirer of Cournot; and his correspondence with Cheysson, another French mathematical economist, indicates a certain understanding of this form of reasoning.[5] This familiarity with probability was also true of Jevons and Edgeworth. Jevons was a disciple of De Morgan, the author of an *Essay on Probabilities and Their Applications* (1838). Certain correspondence from Edgeworth to the Royal Society—for example, the one published in *Nature* in 1889—shows that he had a subtle understanding of what was possible in statistics and probability.

Notwithstanding these favorable conditions, none of them used probability as an instrument in developing theoretical economics. We might even say that they explicitly showed resistance to the idea of using probability in economic theory, as instanced in a paper published by Edgeworth for the *Economic Journal* in 1910, which is entitled "Applications of Probability to Economics."[6] There is no use of probability at all in this paper!

To understand this curious fact, it would be useful to consider very briefly the position of Leon Walras. He was, and he is still today, the most influential theoretician of the general equilibrium approach, which is the main current in modern mathematical economics.

Imbued with his readings of Descartes, Newton, Lagrange, and Cournot, Walras set himself a goal: the creation of a pure economics worthy of rational mechanics. His rapport with these works provided the theoretical standard: Economy was to be studied according to the methods and principles of the science of falling bodies; general equilibrium of markets was compared to universal gravitation; "virtualy interchangeable quantities" of goods were compared to "virtual velocities"; and so on. In the same way, the theory of production and exchange was to be based on the mathematics of the classical physics.[7]

This can be clearly seen in Walras's main criticism of Cournot, as well as in the criticism of Jevons by Edgeworth. Cournot and Jevons were said to be mistaken in employing empirical considerations in their analysis of the demand function so that verification would have required statistics, i.e., the confrontation of estimated frequencies with observed frequencies. On the contrary, Walras, like Edgeworth, attempted to formulate equations of general equilibrium, that is to say, of the interaction of all prices under pure competition based on rationally determined demand curves, deduced from the relationship between the utility of a given commodity and its scarcity. More rigorously stated, this system had to be based on the principle of marginal utility, or the satisfaction of marginal need under the threat of scarcity. This calculation required infinitesimal calculus, not probability. We have had to wait until Savage, in the 1950s, to see how the probabilistic approach could be useful in microeconomics.

Here again the development of mathematical economics was almost immune to the possibilities of probability, at the same time—and we should emphasize this point—that statistics and probability were deeply modifying biology and physics. As an illustration, there is no article at all on probability in Palgrave's universally known *Dictionary of Political Economy*, published at the end of the nineteenth century (but there are four pages on descriptive statistics).[8] The image of the *homoeconomicus*, already well defined at that time, strengthened the idea of an economics based on classical physics, that is to say, based on *a purely deterministic approach*.

I must, however, mention one exception. There was one area, at least, where the economic models inspired by classical physics were clearly not adequate: this was the case of oligopoly, where producers or sellers are few and must take into account the behavior of their rivals. This situation of interdependence, when a strategy by the economic agents is involved, cannot be fully described using a mechanistic model. Cournot, once again, had formulated "the question"; and both Edgeworth and Marshall were quite concerned by this problem.[9] Pigou, one of the most influential economists at the beginning of this century, at least before Keynes arrived, even published a paper in 1908 saying that probability *may be* of some help in understanding this situation.[10] But he did not go further, and we have had to wait for von Neumann and Morgenstern before this approach was elaborated.[11] Probably one reason why the possibility was ignored is epistemological in nature. As Edgeworth summarized it in 1897, all these theoreticians were convinced that an oligopolized market would be the source of economic chaos: instability of prices and uncertainty in determining the quantities to be produced.[12]

Oligopoly, where the strategic behavior of rivals had to be probabilized, was, for the economists of the late nineteenth and the beginning of the twentieth centuries, the greatest economic evil, to be eradicated by antitrust laws rather than be analyzed by new theoretical models of a probabilistic type.

Let me introduce very briefly the last stage of this development, which concerns the period between the two world wars. I shall only allude to that period, which was one of transition.

In one sense we can say that nothing happened in those twenty years or so. After the pioneering work of Cournot and the developments in the late nineteenth century—with the major contribution of the so-called "marginal revolution"—the interwar period was at first a time of perplexity and then a time of essential transformation in economic analysis with the "Keynesian revolution." But this was not a revolution at all if one considers the instruments that were used, nor if one looks at the kind of model building that was done. Economic theory continued to be defined using a deterministic vocabulary—and, more precisely, a vocabulary coming from the old classical physics.

Let us consider Keynes. Long before he published the *General Theory* (1936), he was the author of a very good *Treatise on Probability*.[13] But, as was true for Cournot, there is no place in Keynes's economic model for a probabilistic approach. We much go so far as to say, and this is clear in Keynes's correspondence, that he did not think mathematical probability could be a useful tool for economists. He was even suspicious of the much less ambitious program of the newly born Econometric Society. It was only after World War II that some Keynesians tried to introduce probabilistic aspects into their models. And yet, even today, most of our macromodels are of a purely deterministic type.[14]

There were, however, some dissenting voices in the interwar period. Surprising though it might be, one was Wesley Mitchell, who, in his Presidential Address published in the *American Economic Review* in 1925, stated that economic theory was too closely related to the old mechanical conceptions of Lagrange, which implied the notion of sameness, certainty, and invariant laws.[15] Mitchell argued that economic thought should be reconstructed in the light of modern physics, which involved the notions of variety, probability, and approximations. But he was an institutionalist, that is, on the margin of the theoretical economics profession. He was much respected as an empirical analyst; but when he urged "radical changes in economic theory," there was no significant response to his proposal. And we must recognize that Mitchell himself never used probability in a significant way.

The same must be said of the beginning of the Econometric Society and of its journal, *Econometrica*. If we look at the issues published before World War II, we must admit that the upsurge of interest in econometrics did not amount to a probabilistic revolution in economic thought.[16]

What are the conclusions of this historical survey of mathematical economics from the beginning of the nineteenth century to the 1940s? Clearly that there was no probabilistic revolution in economic thought, and that the most influential theoretical economist during this period did not even seriously consider that probability could be helpful in economics.

Such a conclusion is paradoxical. An instrument conceived to a great extent originally for the use of the emerging "social sciences" was neglected by the very individuals who logically should have used and developed it. Instead probability found acceptance in another network of analogies and in other disciplines.

It must be remembered that, at the beginning of the nineteenth century, statistics and probability constituted two relatively unconnected areas of mathematics aimed at revealing the laws of social behavior. Cournot, like his teacher Poisson, contributed greatly to the linking of these kinds of knowledge in his *Exposition de la théorie des chances*, which was translated into English, German, and Spanish.[17] His innovative aims, however, were coupled with considerable pessimism about the social sciences. As we have seen, Cournot felt that probability could not establish precise forecasts and could not be, as Condorcet had once hoped, a revolutionary technique for the social sciences.[18]

The second half of the nineteenth century confirmed Cournot's view: The links between statistics and probabilities were multiplied—but these theoretical developments took place outside economics. Probability gained its second wind as an instrument of knowledge on grounds other than those defined in the books of the great founders of modern mathematical economics.[19]

Already in 1850, as Keynes has shown, Maxwell had affirmed that "the true logic for this world is the calculus of probabilities."[20] This insight, confirmed in his work of 1859, found its demonstration and generalization in the work of Boltzmann. However, it was Gibbs who played the most significant role in extending this thesis beyond the field of kinetic theory and in propagating the idea.[21] And it was through the influence of Maxwell and Gibbs that theoretical economists—namely, Edgeworth and especially Samuelson in his *Foundations of Economic Analysis*, published in 1947—rediscovered that probability could be of some help in economics.[22] But even these economists only indicated the possibility: the preliminary steps in a research program, not the program itself.

There is also another point that I would like to emphasize. I have underlined the fact that the analogy with classical mechanics was mainly responsible for the ignorance of the possibilities offered by probability in economic theory. Let me just mention that economists interested in mathematics were greatly impressed by the advances of biology, which was revitalized by the use of statistics and probability at the end of the nineteenth century. Edgeworth drew attention to this, and Bowley, who should be recognized as a founder of modern econometrics, often quoted the works of Galton and Pearson.[23] Moreover, we should not overlook that the term *econometrics* was adopted in the 1930s from *biometrik*.

This would suggest that mathematical economics was elaborated under the influence of the rational mechanics of Lagrange. For almost one and a half centuries economists have ignored probability. The biological analogies no doubt played an essential role in the reworking of the idea of economics. This practice in the 1930s and there after explains the renewed interest in probability and statistics.[24]

Nonetheless, theoretical economics is still, it seems to me, thought of in the deterministic terms of late eighteenth-century physics. The probabilistic revolution in economic thought is still no more than subject for a "scientific research program." But the research has not yet begun. And I am not sure that it soon will.

Notes

1. Antoine Augustin Cournot, *Souvenirs* (Paris: Hachette, 1913).

2. Cournot, *Researches into the Mathematical Theory of Wealth* (Paris: Hachette, 1938), Chapter 4. See also Claude Ménard, *La formation d'una rationalité économique* (Paris: Flammarion, 1978), Chapter 2.

3. See Cournot's translation of Kater and Lardner, *Elements de mécanique* (Paris: Mathias, 1834); Addendum, and Leon Walras, "Economique et mécanique," *Bulletin de la société vaudoise de sciences naturelles* 45 (1909), 313–327.

4. Mark Blaug, *Economic Theory in Retrospect* (Cambridge: Cambridge University Press, 1968, 2nd ed.), Chapter 8.

5. Leon Walras, *Correspondence and Related Papers*, ed. William Jaffe (Amsterdam: Koninklijke Nederlandse Akademie van Wetenschappen, 1965), Vol. 2 (letters of January, Febuary, April, and May 1886).

6. Francis Y. Edgeworth, "Applications of Probability to Economics," *Economic Journal*, 20 (1920), 286–295 and 441–465.

7. Walras, "Economique et Mécanique"; see also Walras, *Elements of Pure Economics* (1874), translated by William Jaffe London: (George Allen & Unwin, 1954) Preface to the 4th ed.

8. Palgrave, *Dictionary of Political Economy* (London: Macmillan, 1926).

9. F. Y. Edgeworth, *Mathematical Psychics* (London: Kegan Paul, 1881), and also his article "La Teorie Pura del Monopolio," *Giornale degli economisti*, 15 (1897), 13–31, 307–320, and 405–414. See also Alfred Marshall, *Principles of Economics*, 1920 (8th ed. London: Macmillan, 1969), Appendixes C–E.

10. A. C. Pigou, "Equilibrium under Bilateral Monopoly," *Economic Journal*, 18 (1908), 205–220.

11. Oscar Morgenstern and John von Neumann, *Theory of Games and Economic Behavior* (Princeton: Princeton University Press, 1944), Chapter 1, No. 2.3; Chapter 2, No. 11; Chapter 3, No. 1.

12. F. Y. Edgeworth, "La teoria pura," pp. 441ff. (note 9).

13. John Meynard Keynes, *A Treatise on Probability* (London: Macmillan, 1921).

14. Spiro Latsis, ed., *Methods and Appraisal in Economics* (Cambridge: Cambridge University Press, 1976); see particularly Herbert Simon's contribution.

15. Wesley C. Mitchell, "Quantitative Analysis in Economic Theory," *American Economic Review*, 15 (1925), 1–12.

16. This is particularly clear in the "scientific program" developed in the first issue, *Econometrica*, 1 (1933).

17. A. A. Cournot, *Exposition de la théorie des chances et des probabilités* (Paris: Hachette, 1843), p. 261.

18. A. A. Cournot, Addendum to Herschel, *Traité d'astronomie* (Paris: Paulin, 1836), pp. 504ff.

19. Westergaard, *Contributions to the History of Statistics*, (London: P. King & Son, 1932).

20. Keynes, *Treatise*, p. 172n2 (note 13).

21. J. W. Gibbs, *Elementary Principles of Statistical Mechanics* (New York; 1982).

22. Paul, Samuelson, *Foundations of Economic Analysis*, 1947 (New York: Atheneum, 1967), Chapter XII and 1964's Foreword.

23. A. L., Bowley, *Elements of Statistics* (London: 1921).

24. See, for example, K. Wagemann, *Konjunkturlehre* (Berlin: 1928).

7 The Rise of Macroeconomic Calculations in Economic Statistics

Robert A. Horváth

Probabilistic thinking in economics broke through relatively recently, no earlier than the 1930s in the form of econometrics, even though the origins of statistical verification of theoretical economic results go back to the turn of the century. The basic idea of the present chapter is that this development had an important precondition: the rise of macroeconomic calculations in economic statistics, which started one generation earlier, in the late 1860s. Without this important step it would have been difficult to link the individualistic marginalist economic theory with probabilistic thinking and the law of large numbers.

This chapter deals first with the pioneering contributions to national income calculations by De Bruyn Kops and Dudley Baxter in 1869 and those of Fellner from 1901 to 1903. It then concentrates on the contributions of Marshall and Edgeworth to the statistical verification of theoretical economic concepts on a deterministic probabilistic basis and on their work in developing index-number theory as a tool for such analysis. This work was completed in the post-World War I period in the field of national income calculations by several scholars, especially Stamp and Fellner, and in "index-number theory" and applications on the deterministic sampling basis by I. Fisher. The paper concludes with the 1930s, that is, before the era of modern probabilistic sampling.

1

The importance of macroeconomic calculations in economic statistics was already emphasized in a previous paper of mine, given at the ZiF colloquium in September 1982, and a rough sketch of their development was included in the context of a discussion of verbal versus mathematical economic and real econometric descriptions of the economic system.[1] Now I would like to enter into the details of the problem from the point of view of economic statistics proper and discuss how the probability calculus influenced the statistical study of macroeconomic calculations.

As a starting point, I would like to refer to "A Brief Survey of the Development of Economics in the Last Century" given in a paper by Ragnar Frisch.[2] In his assessment—written in 1965—Frisch stressed the fact that in this development "... neither the classicists nor the neo-classicists did much to verify their theoretical results by statistical observations. The reason was partly that the statistics were poor, and partly that neither the classical nor the neo-classical theory were built up with a systematic statistical verification in mind." After mentioning the tentative attempts by the German historical and the American institutionalist schools, Frisch continued with the statement that "in the first part of the 20th century the picture changed ... the theoreticians themselves took up a systematic work of building up the theory in such a way that the theory could be brought into immediate contact with the observational material."

My point in this context is that this statement is totally acceptable as concerns developments in theoretical economics, including early forerunners of mathematical economics, but that, in my view, this process began in economic statistics at least one generation earlier. It was especially at the Queteletian Congress in The Hague in 1869 that the idea of national income calculations, to be carried out by the modern national official statistical services, was launched and the quantification of the theoretical economic concepts was undertaken as a necessary part of this new development. The combination of economic theory, mathematics, and statistics in Frisch's view was the second breakthrough in this context since the work of John Stuart Mill, "... the first being the bringing back into economic theory of the subjective element by the marginalist school in the form of the study of human wants."

The launching of the idea of national income calculations in economic statistics was an essential step toward the quantification on a global level of economic analysis, if indeed the nation was to be considered as a global unit of the economic system, as Frisch formulated it.

To prove my thesis by historical analysis, that this "second breakthrough" happened one generation earlier than in Frisch's timing, is the purpose of the present chapter.

2

The origins of the initiatives in the field of economic statistics at the 1869 The Hague Congress go back to the earlier 1867 Vienna Congress, where resolutions were adopted for a more thorough study of the problems of the state budget in order to establish an International Financial Statistics.

The 1867 proposition was made by the Belgian statistician Heuschling and adopted unanimously with the proviso of Baron De Hock that these statistics should include data on state incomes, outlays, state funds, and property.[3] Even complementary proposals that the budget data should also include data on local government and all institutions of public interest, such as corporations, etc., were adopted. From this enlargement of the concept it was only one step to an international comparison of all charges ensuing from the financing of public activity, and the rapporteur of this problem at the 1869 The Hague Congress, the Dutch statistician De Bruyn Kops, made this final decisive step.

He pointed out that the 1867 enlargement of the basis of comparison "could not provide a final global evaluation of economic work done by a nation—this could only be done by utilizing the concept of national income." Only by elaborating such a concept—showing the approximate dimensions of the national economy rather than its accurate numerical values—might it be possible to proceed to global statistical comparisons. First, one could compare the growth or diminution in the material well-being of a whole nation from one year to another, i.e., over a period of time, and second, one could compare the actual state of the incomes of two nations for the same date, i.e., make point estimates at one time over geographical space. De Bruyn insisted on the methodological aspect of this kind of statistical approximation; this approach provided a "methodological" framework that revealed, in particular, several practical economic problems.

But from the scientific historical point of view, it is clear that the quantification of national income within this framework was not possible at the end of the 1860s without the corresponding theoretical economic concepts. These were either already available, or in the absence of such notions, had to be formulated within the methodology of the political economy of the epoch. De Bruyn, with an exemplary intellectual determination, proceeded in this direction, using the relatively complete income-tax statistics and the data on property inheritance available in England, France, and Belgium. First, he raised two fundamental questions of economic methodological interest: first whether the national income concept should comprise all individual incomes, and second how to distinguish between gross income and net income.

For these problems, the definition on an individual basis as elaborated by classical political economy certainly had no validity; the aggregation of all individual incomes implies several double counts, and so the sum total surpasses the real national income on the personal basis, even if we are not counting those small individual incomes that are free of taxation (i.e., under the minimal taxation level). De Bruyn proposed to leave out of consideration all private revenues that represented a transfer among individuals ensuing from a redistribution of the national income, as well as those production costs that were not closely connected with the replacement of the working capital for the year of count, but that arose from the utilization of capital accumulated in earlier years. So the point was to deduct the costs of social production of the current year from the new value created and thence to determine the remaining net value of the national income available for consumption or for saving—a kind of net "value added" according to modern terminology. This concept of national income developed by De Bruyn comprises not only material production but also services, creating any kind of individual income, and all revenues, even those not taxable because of the minimum-level limitation.

The quantification of national income for economic statistical purposes thus shows a marked difference from the statistical computation for taxation, i.e., for fiscal purposes—the latter surpassing by far the magnitude of the macroeconomic national income concept. This was the result of De Bruyn's investigations on the basis of individual incomes.

The perspicacity of this author is best demonstrated by the fact that, by thinking over the whole computation of national income, he discovered a possible alternative methodology of approximation by concentrating on the problem of the aggregation of material production. This second approximation—to find out the new value produced, measured by "the exchangeable material products"—made it necessary that a further amount of individual incomes should be eliminated from the aggregation, either totally or partially. Only those "primary incomes" should be counted whose product is material, leaving out all "secondary incomes," which represent a transfer from the value of primary incomes. From this point of view individuals may be taken in the physical sense as members of a population, but also in the legal sense as associations of persons or institutions with either a private or a public legal character. This hypothesis permitted him to consider not only economic, but state or self-governing bodies in the income calculation on the productive, or material, basis. The incomes of such bodies—providing they do not arise, from productive activity or from the interest of capital or from the use of state and other public property—are clearly to be eliminated from this concept of national income. The only exception are taxes, which may be considered as that part of the general costs of production that are paid for the cooperation of the state in providing a legal and social framework by its very existence—a very modern idea indeed, corresponding to the up-to-date economic theory of the state as the fifth factor of production.

But logically, if these taxes are counted as state incomes, their amount should be deducted, of course, from personal incomes as the price paid for the state's cooperation in productive activities, as also should the rents and payments paid to state employees as secondary incomes derived from taxation. The same logic applies to any incomes originating from loan capital in the form of interest, to avoid any double counting of the interest paid by the debtor as production cost and by the creditor as income on capital. From the rigorous exposition of the problems of this alternative methodology, De Bruyn arrived at the very modern conclusion that an unequivocal national income concept based on production requires the elimination of all incomes on services, i.e., deriving from the so-called "immaterial production," and retaining only the aggregation of the incomes of all primary producers—individual or collective. In this case, he insisted, the computation of national income should be theoretically equal to the material net product measured in terms of economic exchangeable goods.

Expressed differently, there are two different methods of approximation possible, each of them justified from the methodological point of view: first, aggregating individual net incomes from the point of view of the total sum at disposal for consumption or saving purposes; and, second, adding up the net values of the goods produced in the different sectors of material production in the economic system as a global unit, measuring the net result of labor in producing each year the new amount of wealth of the nation. These results of De Bruyn's analysis of macroeconomic statistical computation methods deserve to be highly appreciated by historians of science. His work anticipated the whole subsequent development of this problem in economic theory and economic statistics, with the dualism of the methodology as one of its most characteristic features.

The merits of De Bruyn's activity are, however, not restricted to this basic methodological accomplishment. He also anticipated (though not as clearly) a second major problem of statistical national income calculations: the role of money. His rather optimistic basic assessment of this problem was that by limiting the national income calculations strictly to one year, the role of money could be eliminated as a disturbing factor and restricted to an accounting unit only, and thus none of its genuine economic functions would have to be taken into consideration. The consequence of this standpoint was the elimination from national income of the savings of earlier years made through the credit system, as well as those investments that could be made from the savings of future years with the help of credit. But if we want to calculate the national income for a longer period (which De Bruyn explicitly referred to without exploiting it), this implies that the changes in the purchasing power of money and all the long-run effects of credit operations in the changing economic environment have to be taken into consideration. In this case the assumption of a static price level should clearly be abandoned, and the need for a double calculation, once on current and once on stable prices, emerges.

The final proposition of De Bruyn—to adopt the national income concept based on income tax for Western Europe, i.e., the first methodological approach—was effectively supported in another pioneering work by Dudley Baxter presented at the same time at The Hague Congress, and included in the annexes to its papers.[4] This happened despite the counterarguments of von Mayr that for several European countries, e.g., Bavaria, this basis was totally inadequate since income taxation was only subsidiary—it was only applied in those cases where no net product item was available. The analysis of Dudley Baxter's paper shows that he was also aware of the dualism in the methodology of the computation of national income, but he argued in favor of the facilities concerning the first method; he emphasized that for the material product basis of computation, the United Kingdom had no reliable data, not even approximate figures.

The method of calculation adopted in the first approach is delineated rather sketchily by Baxter as "too long to describe here" and performed with "many precautions and rectifications," giving no further details. In fact, as sources he used the income tax returns of all annual incomes above £100 Sterling and the statistical material contained in the official "Blue Books" and "in the papers of many private enquirers." Among the latter he only named the data of Professor Leone Levi. In detail, he used the schedules of the five income tax classes from A to E, covering the tax return of (1) owners and (2) occupiers of lands and houses, (3) those of English and foreign fundholders, (4) the returns of trades, professions, and foreign property, and finally (5) those of public officers. The use of these schedules in many cases required a conflation of the income for one person, if it was drawn from multiple sources or from company participation. But finally, the total number of all persons was ascertained from the schedules and from complementary census data. The data for persons below the minimum taxable income level were taken from the census by number and occupation and, with the help of the average earnings of the latter, an estimate of the aggregate national income was inferred. On this basis—as a by-product—Baxter also made a comprehensive calculation of the tax burden in the United Kingdom by referring to the categorization of Smith and McCulloch to determine the "final tax payer." His findings showed the burden of gross taxation to be somewhat lower than that of the net taxation, the relative quotas being 10.5% and 13.5%, respectively. This additional calculation, made in an absolutely clear and simple manner, fulfilled the proposals of earlier congresses concerning the international comparison of state budgets and the burden of taxation in the different countries.

These successful attempts of De Bruyn and Baxter were, as it turned out, judged completely negatively by von Mayr—despite the fact that De Bruyn had clearly shown that calculations based on the second, material product, approach should theoretically produce another aggregate sum of the national income, which could serve the same statistical analytical purpose. But, due to details concerning the structure and composition of national income, the results of the two approaches would not allow any international comparison—to this extent, the counterarguments expressed at the congress by von Mayr were, of course, valid.[5]

3

At the 1869 The Hague Congress, nobody noticed that the two Hungarian delegates, Keleti and Körösy, who represented the newly founded official statistical service of their country, submitted a list of recent Hungarian publications of international interest, among them a paper by Elek Fényes. This paper, published in 1867, was entitled "A Comparison of the Economic Strength of the Austrian and Hungarian Empires" and represented the text of a lecture held at the Hungarian Academy of Sciences.[6]

In it, the author, one of the best Hungarian statisticians of the previous three decades, compared the national wealths and the national incomes of the two constituent parts of the newly founded Austro-Hungarian Dual Monarchy, enabling him to determine the appropriate share of the "common" budget of the associated states. In the coming decades, this problem aroused a major controversy between the statisticians and economists of the two countries, and, finally, the so-called Hungarian "quota" was determined along the lines proposed by Fényes, i.e., on the basis of the taxation system with crude approximations, but using a scientifically sound methodology using the net material product basis of national income.

This precedent enabled another Hungarian scholar, Frigyes Fellner, to concentrate at the turn of the century on developing scientifically coherent methodologies of national wealth and national income calculations for those countries whose mode of taxation was based on the net material product. In his first paper on national wealth in 1901,[7] a careful analysis of earlier work made Fellner realize that most economic statisticians had concentrated their efforts on establishing a numerical estimate of national wealth, but that they had made no progress in using this as a basis for a complementary estimate of national income. This was the case for a considerable number of scholars between the years 1863 and 1899—some twenty scientists. He mentioned another twenty names prior to the 1860s and even some earlier pioneers, such as Petty, Vauban, King, and Krug. It is noteworthy that Fellner also quoted Baxter, who seemed to him to be the only exception.

Fellner's paper was first presented at the Budapest Session of the International Statistical Institute in 1901 in French, under the title "The Estimation of National Wealth." In it Fellner denominated the individual income basis of estimation as the "subjective method" of the statistical approach to national income and wealth, and the other one as the "objective method." Fellner was eager to demonstrate the advantages of the second approach, especially its capacity to lay bare the structure and contributions of the branches of material production to the creation of the new yearly global national product and the accumulation of national wealth. Fellner analyzed also the "subjective method," stressing its underlying disadvantage: its reliance on individual income declarations.

From the point of view of the "objective method," the connection between the calculation of national wealth and that of national income was a crucial step, and Fellner quoted the representatives of this line of thought, especially Rümelin and Schall. Their studies, of 1863 and 1884, respectively, had already recognized this important link.[8] He also listed and criticized a third method proposed and elaborated by the French statistician de Foville in this context. The latter proposed to measure the annual change in wealth by inheritance, and its multiplication by the generational mean duration of life. Theoretically, the result of this multiplication was equal to the sum of the national wealth. Fellner also mentioned a fourth kind of methodological approximation for national wealth calculations, developed in the United States, to establish national wealth on the basis of official census property data.[9]

These theoretical considerations led to Fellner's attempt to establish the approximate value of the Hungarian national wealth. He soon discovered that, despite the firm guiding principles indicating the use of the direct taxation of the net product basis, he could rarely find the appropriate data. To complement these, he had to use the data of the subjective method, i.e., mixed data, including also the successoral methodology (based on inheritance) of de Foville. On the basis of many hypotheses and assumptions for all three approximations, he proposed to accept the arithmetic mean of the three methods as the result. Fellner was aware that the kind of economic statistics developed by him would have to solve two major methodological problems in the future, the first being to provide for the fluctuation and changes in the relation of national income to national wealth. Fellner was led to the conviction that national income was better suited for international comparisons than national wealth since the former was wholly calculable in monetary terms. Thus the final result of Fellner's thoughts on national wealth lead him directly to research exclusively in national income.

That this period of economic statistical efforts influenced theoretical economic thinking in neoclassical economics before the second "breakthrough" of Frisch may best be demonstrated by examining the system of Marshall.[10] In it, the first "breakthough," the renewal of the mid-nineteenth century political economy by the "restitution of the role of the human element and that of subjective value consider-ations," was characterized either by individually different appraisals of economic values and satisfactions or by individually different considerations of prices and the purchasing power of money. The starting point of this marginalist approach was definitely the economic individual and the formation of his income; but it was heading increasingly toward the macroeconomic level, and national income was aggregated simply as the sum of individual incomes. Here the application of probability began to play an important role, as the law of large numbers permitted the elimination of individual and group differences. The per capita income is the average of the national level and served in the Marshallian system as a general indicator of national wealth, as well as of economic prosperity in general. However, in Marshall's thinking, the problem of wealth was a secondary notion; it was dependent theoretically on the primary notion of income, and therefore the expo-sition of wealth was introduced only after the notion of income. "A fortiori," he emphasized that the measure of wealth could seldom be direct—it is mainly based on the estimate of income with reference to labor inputs, or to the using up of capital with recourse to the general rate of interest in capitalizing wealth.

And so, in this context Marshall referred to statistical estimates of national wealth, and enumerated the best estimates before, and in later editions after, the turn of the nineteenth century, practically in the same way as Fellner. So his standpoint, as that of a theoretical economist, was much more comprehensive than that of von Mayr, and paved the way for the recognition and reception of macroeconomic statistical calculations, especially that of national income itself.

No wonder Fellner felt that there was an important gap to be filled, not only in economic statistics but also in theoretical economics, by pursuing his investigations in this direction. The result was his paper "The Evaluation of National Income," presented at the 1903 Berlin Session of the International Statistical Institute, in which he also made a practical attempt to estimate the national income of Hungary.[11] But as he also included his ideas on the theoretical foundations of the problem and a critical appraisal of the work done before him, this paper represented the direction for future development.

The starting point of Fellner's second paper was his insistence on the unsolved character of national income estimation in economic statistics, in which economic theory failed to come to the rescue. However, there were two possible theoretical approaches to the computation of this important concept, which are identical with those of national wealth estimation, as Fellner quite casually recognized: namely, the so-called "objective" and "subjective" methods, now renamed by him in German the "real" and "personal" methods. To uncover the theoretical roots of the national income concept, he enlisted all major economists from Smith and Ricardo. The tenets of these classics, concerning the concept of national income, rested on the material product by deducing production costs. In addition, he also reviewed those economists who wanted to determine national income by the aggregation of personal incomes, from J. B. Say and Roscher to Marshall. It was clear to him that, from the very beginning, many authors had been obliged to combine the two concepts to make them at least theoretically operative. He also insisted that the aggregate personal income concept was essentially a private or microeconomic notion and that its aggregation necessarily gave a greater sum than the "real" national income through the inclusion of a part of the distribution process. The deduction of the latter, including that of the immaterial services, gives a macroeconomically correct approach to the real national income.

To illustrate this point and the statistical work done before him, he summarized the calculations of Schall for Württemberg in 1884 and those of von Czoernig for Austria in 1861 by the so-called real or objective method, and then he also included the results of Dudley Baxter's estimate for the United Kingdom and those of Soetber for Prussia in 1879.[12] As regards Baxter's work, Fellner criticized it, finding it strange that the income tax returns of the United Kingdom only gave the incomes of 1.3 million persons, since some 12.5 million people remained under the £100 Sterling income limit. The data utilized for the people under the tax limit were rather crude estimates. The same was true for Soetber's calculation with respect to the data of some 3.5 million people. Thus, in the calculation for Hungary, on the basis of the objective method, in conformity with the Hungarian tax system, Fellner insisted on using the combination of the two methods: the complementary use of the personal method was necessary to obtain a "conjectural statistics" on the objective basis.

One wonders, on the basis of this rudimentary analysis of the work done by Fellner, why his pioneering attempt was not really appreciated—by either the representatives of economic statistics or those of theoretical economics of his epoch. Maybe Colin Clark was not far from the truth when, in the preface of the second (1951) edition of his world-famous work, *The Conditions of Economic Progress*, he expressed the opinion that, before 1940, neither of the two above-mentioned scientific communities had grasped the tremendous importance of national income calculations.[13] Fellner's efforts, during the following decade, to determine the national income of the Austro-Hungarian Empire[14] also went unheeded; and the peace-negotiators after World War I failed to accept that only this realistic economic basis would make it possible to dismember the Hungarian economy fairly so as to create the new so-called "successor states" and to determine realistic reparation payments. The revival of these kinds of macroeconomic calculations was slow and intermittent, even between the two world wars.

4

However, after World War I, the strong criticism by Keynes of the economic consequences of the peace treatise[15] and the grave problem of the German reparations motivated one of the finest British scholars, Sir Josiah Stamp, in 1920–1921, to take up the problem of national wealth and national capital,[16] while fully recognizing the close connection to the calculation of national income.

Despite the fact that, in principle, he was against a so-called point estimation in time and in favor of an analysis for a longer period, the latter was made impossible by postwar inflation. So Stamp proceeded with one-year estimates of national wealth and national income, recognizing the utmost necessity even for such a restricted calculation. He compiled a comprehensive survey of the earlier British efforts in this field—e.g., of Crammond Bowley (1914), Mallet (1916), Strutt (1915), and himself (1916)—relating to national wealth, and those of *Bowley* (1912) and Chiozza-Money and Whitaker (1912 and 1920) relating to the national income. He was aware from the beginning, as shown in his postwar essay, that the duality of methodologies comprising the alternative personal income and production basis methods had, through a confusion of these two basic ideas, "played havoc" in the estimation of national wealth; and he found this confusion was difficult to clarify completely when dealing with the methodology of national income calculations.

Instead of two basic methods, he spoke of three by separating (1) the income tax basis and (2) the complementary occupational census approximation, and confronting these two with the third method, (3) the net output or production census basis. As additional methods, two other complementary approaches were mentioned by him: (4) the capitalizing of interest yields on capital property and (5) the so-called Australian income census method. This chosen terminology was not much help in making the whole methodological issue clearer. But later, in his subsequent publications, Stamp freed the problem from all ambiguity, and spoke of the production-based "Hungarian" method as one possible basic approach and of the aggregated personal income, or "British," method as the other fundamental one. As a third possibility he mentioned "mixed" methods: using these two combined, or any other complementary method as an alternative for some components of national income.

The elements of another type of possible confusion also emerged in this first postwar paper of Stamp. In particular, when he took up the problem of taxation capacity and its economic limitations, he was really closer to the problem of the distribution of national income. In this respect, Stamp was confronted first with the effect of changes in the value of money, that is with the effect of changing price levels on incomes, profits, or wages, and second, with the laws of their diminishing utility from the personal point of view, i.e., the subjective valuation of the purchasing power of money, already present in the system of Marshall. This problem became more crucial in the "welfare economy" system of Pigou[17] and complicated the work of economic statisticians in the period between the two wars.

For Pigou, the concept of welfare included the notion of "economic value" of every purchasable service even if not connected with productive economic activity, since he supposed that such kind of service is apt to augment "human welfare," or, more precisely, "human happiness." So the latter was meant not only in the strict "economic" but also in the wider "ethical" sense. There was no obstacle to including, in the aggregate value of the national income, "services" such as defense and public health, which were "produced" to augment the "national dividend" and, through it, the "national welfare," as Clark rightly pointed out. But a more orthodox marginalist, the Hungarian Farkas Heller, was skeptical about whether such a micro-economic marginalist concept of "personal income" could be extended for macroeconomic use.[18] According to him, starting out from the nation as a macroeconomic productive unit, the inclusion of services was possible, if at all, only on the basis of productive services, but certainly not on the redistribution basis (i.e., including redistributed incomes). From this point of view, intricate difficulties arose regarding the basis of the aggregated income.

The material product method fared better, with its theoretical problems at the strictly methodological level. This is clearly shown in Fellner's later efforts, i.e., in his calculations for postwar Hungary on the basis of the averages of the years 1927–1929. On the one hand he counted the value of productive services *net* and not gross, and on the other hand he insisted on retaining the net value of the balance of foreign trade, which he augmented by those items of the international balance of payments that represented an increase in value issuing from the foreign trade turnover of material goods. In the early 1930s a more rigorous and systematic approach to this problem was used only in Soviet economic statistics, where, even at that time, the concept of the economic balance of national income calculations was being developed, and the conceptually heterogeneous items within the production-based national income concept could be more easily identified and sorted out when confronted with an equivalent balancing item of material consumption.[19] The first initiative in this development of Western economic statistical thinking was provided by the rudimentary abstract national income schemes in Keynes's basic treatise of 1936, *The General Theory*,[20] which, under pressure from British economic efforts during World War II, led to the emergence of the so-called "national budget" computation. Within this concept, national production and consumption figures were compared, in gross and net form; i.e., items of the national income that did not derive from the state budget were also included on the basis of statistical data—if available—in the calculations and forecasting. The first attempt at such a calculation appeared in the 1941 White Paper, but only the 1945 White Paper, *National Income and Expenditure of the United Kingdom*, which basically represented the conceptual ideas of Keynes and, from the economic statistical point of view, those of Stone, could be considered a methodologically fully developed version.[21] This framework was important and pioneering insofar as it provided a clear-cut distinction between such concepts as the national income on the material product basis and the corresponding final material consumption concept, or alternatively an aggregated income concept, with which to measure all final consumption. The latter was considered to include both material and im-material products, as in the traditional British income tax basis, and in accordance with the national income concept of the welfare economists, which took account not only of objective but also of subjective valuation. These distinctions, however, sometimes confused rather than clarified the subject, as, in 1956, Joan Robinson rightly remarked, but no doubt they were necessary stages for future development.[22]

In this respect, also, pioneering work was done in the United States, by Simon Kuznets. In 1947, he elaborated a tripartite balance of national income by adding a third intermediate balance, between the national income product balance side and that of the final material consumption, which exhibited the intermediary distribution of national income.[23] This was an antecedent of the final distribution or consumption balance, the sum total of which should theoretically be equal to the other two previous balances. One cannot fully appreciate the importance of this theoretical performance, since the requirement had already been stated in the early analysis of De Bruyn and Fellner.

Even if we exclude the problem of changes in the purchasing power of money (i.e., if one calculates the national income on a one-year basis of current prices only), another underlying complication emerges if the pattern of income distribution—primary or final—is not stable, but subject to considerable variations. Bowley tackled this problem when he found a startling constancy of many proportions and rates of movements in his investigations. These seemed to point to a fixed system of causation, and the distribution of income had an appearance of inevitableness—according to a quotation by Stamp. The latter commented on this problem in a very deterministic way, concluding that "when we look at the distribution in all the civilized countries that we know and we find this peculiar characteristic distribution exhibited by the graph (i.e., the Pareto-curve) with slight differences, one wonders whether it is in the nature of the universe that it should be so, and our minds go back to the Middle Ages, when the community seems to have consisted of very rich barons and a lot of wealthless serfs, and we wonder whether the present kind of distribution could have obtained at that age." The similarity of this interpretation, and even its wording, to that of Quetelet is striking; the latter stated that the limits of variation of social elements is decreasing with the spread of civilization.[24] In this context, Quetelet was speaking of the diminution of price fluctuations between their maximal and minimal deviations in the midnineteenth century. We want to stress this point in order to show that in this part of economic statistics, i.e., the national income calculations, the stochastic aspect of the problem was not yet present in these short-term year-to-year calculations. This aspect came into play only with the long-term considerations, when the changes in the purchasing power of money, or, roughly speaking, the changes in prices, no longer had to be eliminated from the valuation of national income, even when the subjective valuation, either on the Marshallian level or that advocated by Pigou, was not taken into consideration.

5

Thus we come to the final basic problem of our analysis of the development of macroeconomic calculations: the index number problem, i.e., a basic methodology to account for the fluctuations of the purchasing power of money. This problem was not new; but from the 1860s on, it was seen from a new angle, that of statistical methodology. This did not detract from the work of early pioneers up to the 1860s, such as Carli, Dutot, Smith, Young, Shuckburgh-Evelyn, Lowe, Poulett-Scrope, and Tooke, but from then on, the methodological development was accelerated and brought increasingly into harmony not only with macroeconomic calculations but also according to Schumpeter, with probability considerations.[25] The new approach was first used mostly by German scientists, especially Laspeyres, Paasche, and Drobisch, within the context of both the theory of statistical averages and the theory of economic value fluctuations—measured by price changes of subsets of chosen commodities as compared to an initial level. On this basis, Laspeyres solved the problem of the basic value indices, introducing a formula with the help of the arithmetic mean and on the basis of augmenting prices.[26] Paasche solved the same problem on the basis of augmenting physical values with a similar formula named after him.[27] Drobisch, in a sense, complemented their work,[28] by using the so-called weight-crossing. Some two decades later, in 1887, Marshall realized[29] that from the seemingly static index numbers, representing mostly short-term changes, one could easily derive a whole statistical series of index numbers by choosing for every short-term period (most conveniently for five or ten years) an ever-changing base: that of the previous year. The "chain-index" thus produced would eliminate the problem of long-term changes in the purchasing power of money.

At the same time, in the light of the work of the above-mentioned scientists, Jevons proposed a technically and methodologically similar solution,[30] but based on the geometric mean. In his view changes in economic variables are never linear;

they move and differ in time. During this period most of the underlying technical and methodological problems of index numbers were solved from the economic statistical point of view, i.e., the measurement problem of the ratio of prices—mostly in percentages—of a given set of commodities at a particular date. On this basis they expressed the change in prices of the same items over a period chosen as the standard—usually one year. The main construction problems of such composite index numbers were (1) the choice of commodities, (2) the quotation of prices, and (3) the choice of the time basis. The discovery of the interdependence of the three basic index formulas—the value index, the price index, and the volume index—belongs to this period of perfection of technique that began in the late 1860s, although it occurred some two decades later with the work of the Danish statistician Westergaard.[31]

This period of the technical improvement of index numbers overlapped with the development of the probabilistic aspect of the index number calculations, introduced into economic statistics by English neoclassical economic thinking rather than by the work of the above-mentioned German statisticians. Edgeworth and Marshall were among the first, in the late 1880s and the early 1890s, to initiate this new development. In Marshallian thought, the classical Queteletian probabilistic approach was continued. With the help of "a priori" deterministic probabilities on the basis of the law of large numbers, Marshall obtained the same result as Smith[32] and Quetelet[33] in eliminating subjective valuation differences between economic individuals. Although he started out from the widely differing appreciation of the purchasing power of money between the rich and the poor or other groups, his conclusion was that "... the same price [for the individual] measures different satisfactions because money may be more plentiful for him or because his sensibility may vary, but differences may be neglected by considering an average of large numbers of people." It is also characteristic of this passage that Marshall refers to Edgeworth's *Mathematical Psychics* from the year 1881 for a proof.[34]

According to Marshall's line of thought, "an external standard of measurement" must exist in economics with which the strength of individual motives can be measured by a money-price ratio; otherwise we could only speak of "subjective impressions." With the help of the external standard, "facts could be grouped and statements could be made about them approximately accurately and numerically." These quotations did not mean that Marshall completely abandoned the basis of subjective valuation or utility theory. On the contrary—when he made a finer distinction between the production of national income and its final, mostly private, consumption, he came closer to the right theoretical solution than the later, neoclassical welfare economists, such as Pigou. Marshall, in particular, suggested the measurement of demand, one of the later preoccupations of the econometric era of the 1930s, when he stated "... though the theory of demand is yet in its infancy, we can already see that it may be possible to collect and arrange statistics of consumption in such a way as to throw light on different questions of great importance of public well-being." But in his view these individual wants would compensate each other, and so the main result of his investigations was that the law of general demand may be induced by aggregation on the basis of price statistics, even when the "disturbing causes" reemerge from the Queteletian system.

Marshall was of the opinion that prices are "highly conjectural and liable to large error." So despite his positive efforts to create a statistically operative economic theory, he partly encouraged those voices that expressed a deep distrust of statistical measurement as applied in economics, and which were raised especially against the use of index numbers. The strongest objection was perhaps raised by the Dutch statistician and economist Pierson,[35] with Walsh at the opposite end overemphasizing their importance.[36] Compared with these views, Marshall and Edgeworth[37] represented a balanced judgment, insisting on the need for a numerical increase in statistical observations and on improvement in their quality, together with their methodological development. From our analysis, it seems to be clear that, in this field, Marshall was still close to the Queteletian line but at the same time he surpassed it. This is best shown in his treatment of the problem of economic development, where he was well aware of the difference between simply applying the law of large numbers with "a priori" reasoning and the problem of representativeness in the sense of the more modern sampling method. For example, when he was speaking of the central statistical values of grouped frequency distributions of prices, or of the representativeness of costs, the so-called "normal costs," or of the "average firm," he was aware that "the latter term is of a particular sort, not taken at random, but selected after a broad survey." This idea clearly indicates the direction of modern stratified sampling.[38]

Regarding this question, however, Edgeworth was even closer to modern probabilistic thinking and to modern statistical methodology, e.g., when he dealt with the problem of the measurement of probable error in one of his early papers in 1897. Working on the problem of the advantages or disadvantages of bimetallism in economic theory,[39] he first tried to solve this problem with the help of physical analogies and the application of "a priori" probability and the law of normal error according to De Morgan. But he soon discovered that the test of the fitness of a more accurate measure of the dispersion about the mean is the probable error.

In his subsequent work, Edgeworth made an even more conscious effort to differentiate "a priori" or "unverified" probability applications in economics from those of an "a posteriori" or "statistical" probability character. For this purpose he applied the probable error test more systematically and also developed the sample basis of statistical material. This particular work began with his "Defence of Index-Numbers" of 1896,[40] in which he insisted that index numbers rest ultimately on speculative assumptions, the probability characteristics of which are connected with the Laplacian law of error. This reasoning is a consequence of the fact that "a common trend comes out in the average, but the particular movements are independent," and that this is "generally fulfilled both in physics and social phenomena." Later, in his 1910 paper on "Applications of Probability in Economics,"[41] this tenet was modified insofar as "... of course, this kind of 'a priori' presumption is liable to be superseded by specific evidence as to the shape of the curve" (as suggested by Marshall, Colson, and even Dupuit in 1844), because "there is required, I think, in a case of this sort, in order to override the 'a priori' probability either very definite specific evidence, or the consensus of high authorities."[42] Edgeworth also carried out pioneering work in his 1910 article on index numbers in

Palgrave's Dictionary of Political Economy. Here he provided a systematic foundation of all statistical methodological problems involved in constructing index numbers and thus paved the way for the contributions of many scholars working in this field, especially those of Irving Fisher, familiar to all economists and economic statisticians.

6

When giving a careful analysis of the recent development of the construction of index numbers as a tool of economic analysis, one has to agree with Schumpeter,[43] who observed a marked development at the turn of the century, due to the contributions of the above-mentioned representatives of economics and economic statistics but including also the important contributions of W. C. Mitchell.[44] Schumpeter stressed the fact that the new era of the probability approach was no doubt best connected with the name of Irving Fisher and his famous work *The Making of Index-Numbers,* published in 1922.[45] It is true, that Fisher's studies went as far back as 1892, with his paper on *Mathematical Investigations in the Theory of Value and Prices,* and also included as a connecting link in his thought his *The Purchasing Power of Money* of 1911;[46] but as Schumpeter rightly emphasized, "every later work was based on this [1922 work]" and his analysis of "tests."

In his major analysis of index numbers, Fisher paid ample tribute to all his predecessors, whether their contributions were independent results or were consciously applied and deliberately criticized by him. Thus, the postwar work represented an impressive synthesis of the subject from all scientifically relevant angles and justly merited its outstanding position in the development of this field of economic statistics. Fisher aimed at developing a set of instruments for the measurement of real wages, trade and money rates, and especially the purchasing power of money, by elucidating several varieties of index numbers, their tests, and their reliability. According to the prefatory note of the editor, W. T. Foster, by developing such instruments, the traditional economics, based on verbal, philosophical methods and opinions, became a thing of the past, and the present already belonged to statistical measurement. The editor also stressed the fact that, for the formulas examined by Fisher, the so-called instrumental error of index numbers was below the 1/800th level, which once and for all rendered invalid any arguments of the kind raised by Pierson. Fisher himself in the preface to the first edition stressed the inductive character of his study, but pointed out that this had not excluded the theoretical element from his investigations. As he put it, these investigations had at the same time the character of both "art" and "science." He modestly stated that he had examined more than 100 formulas, but by consulting his text one gets a more precise picture—he examined some 170 in grand total.

In his methodology, Fisher went into great detail, starting with simple indices or price relations for individual goods or articles and showing the average percentage change of prices between various points in time. He defined the "real index numbers" as showing the prices of a number of commodities as an average of their "price relatives" (a relation of the form P_1/P_0, where the subscripts refer to time periods). Their main characteristics consist primarily in the indication of price movements over time and only secondarily in comparisons between two places. His method of exposing the simple and aggregated formulas was threefold, giving numerical, graphic, and algebraic presentations to their characteristics. These characteristics derived from the tests introduced by him in chapter IV, including the so-called "reversal tests," being either time (T_1) or factor reversal tests (T_2). The third alternative, the commodity reversal test, later named by Walsh the "circular test," is only a preliminary condition according to Fisher that should be taken for granted, and perhaps this was the reason why it had never been formulated before him. These tests state the possibility of interchanges in the formulas of the two sets of magnitudes—the two time periods for T_1 and the two factors (P and Q, price and quantity) for T_2—without there being any ensuing alteration in fairness and reliability of the resulting index numbers. Fisher himself was of the opinion, according to his preface, that among these tests the time reversal tests were by far the best, the factor reversal test also yielded good results, and the third, or "circular," test might be rejected on the grounds that "its conception is erroneous and theoretically it may not be fully fulfilled." In principle these testing procedures were designed to produce very similar results for all good formulas and for all bad ones very dissimilar results. Fisher's main preoccupation was to find out the best formulas according to these criteria (which included simplicity and rapid computability) in chapter XI. Among the formulas, he did not find more than eight that more or less satisfied all the requirements; by far the best of them was the famous formula 353, the "ideal formula." This was constructed upon a crossed formula basis invented by Fisher himself, who found it strange that people had used crossed weights within the formulas since Drobisch, but "curiously enough the possibility of formula crossing was overlooked." The ideal formula is the geometric mean of the crossed weight aggregates of the two basic price and volume indices, as developed by Laspeyres and Paasche:

$$IF_{353} = \sqrt{\frac{Sp_1q_0}{Sp_0q_0} \times \frac{Sp_1q_1}{Sp_0q_1}}$$

(where S is the sum of the prices-times-quantity terms and their subscripts refer to time periods).

This type of index had already been mentioned by Bowley in 1899, and by Walsh in 1901, as Fisher realized in the early 1920s, but these authors did not advocate the formula with Fisher's systematic derivation and argumentation: ensuring the "middle course" of the original two formulas whose weights were crossed and improving accuracy through forming their average, which leads to the smallest possible bias in measure, not to speak of the simplicity and rapidity of its computation. According to Fisher's calculations the accuracy of the ideal formula is so high that its bias or error of measurement is *"within less than an eight-of one percent!* . . . or a hand's breadth on the top of the Washington Monument, or less than three ounces in a man's weight, or a cent added to a \$8 expense." Or, put differently, as physicists or astronomers would say, "The '*instrumental* error' is negligible." Fisher held that "a good formula for one purpose is a good formula for all known purposes," and in terms of practical application "... all methods (if free of freakishness and bias) *agree!*"

Another great merit of Fisher should be mentioned, that with great simplicity he solved the problem of the probability basis of the making of index numbers.[47] His starting point was that "when a large number of [price] relatives are used there is usually little asymmetry in any case," i.e., the distribution is nearly normal, and, second, that "any asymmetry displayed on the ratio chart at least in the distribution of the price relatives taken forward is reversed when we have to consider the relatives taken backward." In other words, by using cross weighting and formula crossing or ratio reversibility, according to the Fisherian tests, the two-way characteristics of the weighted and crossed index result in symmetry. From this quotation, it follows that, in the long run, there is no asymmetry in any direction. For this reason, all formulas may be arranged in order of their remoteness from, and their closeness to, the ideal formula, and any probable error of the latter may be taken as a standard for their measure of deviation ranged in 5 classes as done by Fisher $(0, \pm 1, \text{ or } \pm 2)$.

At the same time, he arranged them in qualitative classes, such as "worthless, poor, fair, good, very good, excellent, and superlative index numbers." The top 29 formulas, according to his analysis, are within an 0.5–1.0% accuracy range from the ideal index number, and a statistical measuring instrument is as good and precise as a physical apparatus. In consequence, there is no need to differentiate between index numbers according to subject matter, as Mitchell proposed; any possible further improvement of index numbers depends on the narrowing of the field of the subject matter itself and subsequently on the improvement of the sample basis of data. As concerns both of these problems, Fisher saw them as practical considerations toward augmenting accuracy. In particular, he calculated the accuracy of the ideal formula for a set of 36 commodities. Considering a set of commodities as a sample, this means that the statistical measurement has to reflect the accuracy not only of the smaller number of commodities included (36) but also of those excluded. "Thus are opened up two new lines of investigation with regard to the accuracy of index numbers, namely, the influence of (1) the assortment of samples, and (2) the number of samples. ... These two subjects are probably quite as important as the choice of formula" in Fisher's opinion. However, here he has mainly summarized the work done by Mitchell, Kelley, and Persons, stressing the fact that an "unwise choice" in both fields, or error in the original data, as a third kind of error possibility, "may result in much greater errors than the choice of formula itself."

Interestingly, Fisher held the view that whether the assortment of goods in a sample is fair largely depends on the purpose of the index number—whether general or rather special. But, in a way, it is "... intimately related to the process of fair weighting for which in fact it may be roughly used as a substitute." As regards the number of the price quotations used, he concluded that the minimal number is around 20, the optimal around 50. Some gradual improvement may be obtained by augmenting the quotations, but certainly not to over 200, where the extra trouble of additional work and expense is no longer worthwhile.

Fisher's analyses in connection with probability showed also that errors in the original data (that is, errors, within the sources of price quotations, committed by the collecting agency) do not play an essential role. The example chosen by him is convincing: on a 100 commodity basis with 10 quotations of which each is 10% too high, even if all fall into the same direction, i.e., cumulate, they account for no more than a 1% error in the whole set. If they mostly offset each other as deviations do, i.e., if they are randomly distributed, the resulting probable error will be only $\frac{1}{4}$%. This was one of the reasons why Fisher supported the idea that the increasingly specialized application of economic statistical index numbers could minimize problems arising from the utilization of probable error. But at the same time, he was quite convinced that the most important contribution of index numbers to macroeconomic calculations was still the measure of the purchasing power of money. As to future developments in the application of advanced probability methods, Fisher was rather skeptical and opposed to the use of least square methods, as proposed by A. A. Young, or assumptions of uncertainty, such as "these characteristics of the data may vary inversely as their deviation [or as its square] from any normal," as Edgeworth put it. Fisher retained the more sophisticated probability methods for the case "where the data are actually defective and uncertain."[48]

Within this new era of rapid progress in index-number calculations, brought about by technical perfection, the *International Statistical Encyclopedia* describes Gini's similar approaches of 1924 and 1937 (on the basis of similar criteria and several formulas) and the underlying investigations of Haberler of 1927. A new kind of chain index on the factor reversal basis was proposed by Divisia in 1925 and by Törnquist in 1937, and the *Encyclopedia* also attributed great importance to a survey and appraisal of recent developments in this field by Frisch, written in 1936.[49]

Insistence on the importance of the so-called normal sampling procedures in this period—concerning the selection of the consumers and the basic commodity items in relation to weighting or a comparison of prices—together with the assessment of the importance of changes in bases, qualities, and substitution possibilities coincided with a growing utilization of the so-called real probability sampling procedures, as illustrated by the activity of King in 1930. However, this new era, with the extended utilization of cluster and stratified sampling procedures, the estimation of sampling errors, and the variability of prices and commodities on the "a posteriori" probability basis, largely began only with the postwar period, according to the assessment of the *Encyclopaedia*.[50]

As a conclusion to this rather long and technical section, one may say that in this new era of probabilistic sampling for index number purposes, all of the most important achievements of econometrics, especially those of the statistical demand analysis, came to be fully utilized. Thus, the "a priori" probabilistic basis was retained, but only for the final induction. When we make use of adjusted (or nonadjusted) sample data, the "a posteriori" distribution represented by various, not normally distributed, empirical curves may be connected with an unknown theoretical distribution, still subject to some kind of assumption or hypothesis. Such a hypothesis could, as in every case of induction, only be determined on an "a priori" probabilistic basis because "it is either unknown or known only within very broad limits"—as Schultz had already pointed out in 1938 in the context of demand analysis.[51] Thus, hypothesis testing on the "a priori" probability basis is still part of any extended "a posteriori" probabilistic statistical sampling method—both of them are required in up-to-date index number theory and practical applications. Another technical innovation in this field may be seen in recent efforts to bring back the subjective value or marginalist element into index numbers, by constructing the so-called constant utility indices or by the construction of indifference price indices with upper and lower or even simultaneous limits. Nowadays the use of chain indices with the same method and, more recently, indices calculated with differential calculus is spreading. In the postwar period the problem of constructing linear indices was also satisfactorily solved. This concept made possible not only comparisons of averages for time points, but also, with the help of calculus, for continuous time periods.[52]

7

In concluding this survey of the important features of the development of statistical macroeconomic calculations and their relation to probability calculus, one is struck by the marked difference regarding the role of probability. During the development of the national income calculations, the deterministic trend had always been outstanding, due to the influence of economic policy aspirations. So it is no surprise that, even nowadays, despite the "econometric breakthrough," the economic statistical applications of probability in macroeconomic analysis rest on a deterministic basis "because of the great number of aspects included," as Frisch put it more recently, in 1964.[53] According to him, it is only on the basis of a first deterministic approach that one can attain a "grip of (statistical) probabilistic analyses," i.e. the real stochastic applications of economic programming. One is inclined to agree with this assessment of the development in econometrics, especially in view of the widespread use of national income calculations based on the standardized methods (either the social accounting system SAS in the Western world, or the material product system MPS in the socialist countries), and the increasing applications of input-output tables and analyses all over the world, as developed by Leontief. Compared with the rather scattered stochastic applications in this field, the real stochastic probabilistic approach has become almost universal through the recent development of index number calculations, based on the unchanged and solid foundations of the instrumental framework developed by Fisher.

Notes

1. Robert Aurel Horváth, "Probability and Epistemology in Theoretical Economics and Economic Statistics," in *Probability since 1800: Interdisciplinary Studies of Scientific Development*, ed. by Michael Heidelberger, Lorenz Krüger, and Rosemarie Rheinwald, (Bielefeld: B. Kleine Verlag, 1983), pp. 185–201.

2. Ragnar Anton Kittil Frisch, *Economic Planning Studies: A Collection of Essays*, selected, introduced and ed. by Frank Long (Dordrecht and Boston: Reidel, 1976), pp. 10ff.

3. De Bruyn Kops, "Revenu annuel de la nation" (Annual Revenue of the Nation), in *Congrès International de Statistique à la Haye, Septième Sèssion Première Partie: Programme* (La Haye: Nijhoff, 1869). pp. 137–144.

4. R. Dudley Baxter, "National Income and Taxation of the United Kingdom, I. National income, II. National taxation," *Congrès International de Statistique à la Haye, Compte-rendu des travaux de la Septième Sèssion, Second Partie*, ed. by M. von Baumhauer (La Haye: Nijhoff, 1870), pp. 528–532.

5. Georg von Mayr, "Jahreseinkommen der Nation" (Annual Revenue of the Nation) (note 3), p. 145.

6. Elek Fényes, "Párhuzam egyfelöl a magyar koronai birodalom, másfelöl az ausztriai, német, lengyel és cseh koronai országok közt, területi, népességi, hadi és pénzügyi tekintetben" (A Territorial, Demographic, Military, and Financial Comparison between the Empire of the Hungarian Crown and the Austrian, German, Polish and Czech Crown Countries), in *Statisztikai és Nemzetgazdasági Közlemények*, Vol. III (Pest: Hungarian Academy of Sciences, 1867), pp. 265–305.

7. Frigyes Fellner, "L'Evaluation de la richesse nationale" (Estimation of the National Wealth), in *Bulletin de l'Institute International de Statistique, 8e Sèssion*, Vol. XIII, Part 2 (Budapest: I.I.S., 1901), pp. 96–136.

8. Fellner, "L'Evaluation," p. 100 with references to Gustav von Rümelin, "Beiträge zur Ermittlung des Volksvermögens und Volkseinkommens," *Königreich Württemberg, Königliches statistisch-topographisches Bureau* (Contributions to the Estimate of National Wealth and National Income) (Stuttgart: W. Kohlhammer, 1863), and to Dr Schall, "Volksvermögen und Volkseinkommen" (National Wealth and National Income), ibid., Vol. 2 (1884).

9. Fellner, "L'Evaluation," pp. 105 and 108, with references to Alfred de Foville, *La fortune de la France* (The Wealth of France) (Nancy: 1883) and from the same author, "Richesse" (Wealth), in *Dictionnaire des Finances*, ed. by Léon Say (Paris: 1889), and *Report of the Tenth Census of the United States*, Vol. VII, *Report on Valuation, Taxation and Public Indebtedness in the United States* (Washington: Bureau of the Census, 1884).

10. Alfred Marshall, *Principles of Economics:· An Introductory Volume*, 8th ed. reprinted (London: Macmillan, 1930; 1st ed. 1890), pp. 18ff; 202ff.

11. Frigyes Fellner, "Die Schätzung des Volkseinkommens" (The Estimation of the National Income), *Bulletin de l'I.I.S., 9e Sèssion*, Vol. XIV, Part 3 (Berlin: I.I.S., 1905), pp. 109–120.

12. Fellner, "Die Schätzung," pp. 117, and 119, with references to the works of Rümelin and Schall (quoted in note 8), to Carl von Czoernig, *Statistisches Handbüchlein für die österreichische Monarchie* (Short Statistical Manual for the Austrian Monarchy), (Vienna: 1861), to Dudley Baxter (note 4), and to A. Soetbeer, *Umfang und Verteilung des Volkseinkommens im Preussischen Staate 1872–1878* (Volume and Distribution of the National Income in the Prussian State) (Leipzig: 1879).

13. Colin Grant Clark, *The Conditions of Economic Progress*, 2nd ed. (London: Macmillan, 1951; 1st ed. 1940; 3rd ed., largely rewritten, 1957.), Preface, p. VI, and, from the same author, *The National Income: 1924–1931, 2nd ed.* (London: Cass, 1965; 1st ed. 1932).

14. On Fellner's activity in this field see Robert Aurel Horváth, "A polgári nemzeti jövedelem számítás Magyarországon" (The Calculation of National Income in Prewar Hungary), *Statisztikai Szemle*, 4 (1956), 324–337.

15. John Maynard Keynes, *The Economic Consequences of the Peace* (London:Macmillan, 1920), and E. L. Hargreaves, "The Problem of Mark-Debts," in *London Essays in Economics: in Honour of Edwin Cannan* (London: Macmillan, 1927), pp. 155ff.

16. Josiah Charles Stamp, *Wealth and Taxable Capacity: The Newmarch Lectures for 1920–1921 on Current Statistical Problems in Wealth and Industry*, reprint (London: King, 1922).

17. Arthur Cecil Pigou, *Wealth and Welfare* (London: Macmillan, 1912) and, from the same author, *The economics of welfare*, 4th ed. reprinted (London: Macmillan, 1946 and 1960; 1st ed. 1920).

18. *A nemzeti jövedelemszámitás és az adóstatisztika problémái* (Problems of the National Income Calculation and the Statistics of Taxation) (Budapest: Hungarian Statistical Society, 1938), with reference to the contribution of Farkas Heller.

19. Horváth, "Probability and Epistemology," pp. 9ff. (note 1), with reference to Oskar Lange, *Introduction to Econometrics*, (New York and Warsaw: Pergamon Press, 1959), and, from the same author, *Essays in Economic Planning*, Indian Statistical Series, No. 4 (Bombay: Asia Publishing House, 1960).

20. John Maynard Keynes, *The General Theory of Employment, Interest and Money* (London: Macmillan, 1936), and, as regards the Keynesian schemes, Paul Marlor Sweezy, *The Theory of Capitalist Development: Principles of Marxian Political Economy*, 5th printing (New York: Monthly Review Press, 1964; 1st ed. London: Dennis Dobson Ltd., 1946).

21. *National Income and Expenditure of the United Kingdom*, Cmd. Nr. 6623 (London: H.M.S.O, 1945), and, Richard Stone, "The Use and Development of National Income and Expenditure Estimates," in *Lessons of the British War Economy* (Cambridge: Cambridge University Press, 1951).

22. Joan Violet Robinson, *The Accumulation of Capital* (London: Macmillan, 1956), pp. 389ff.

23. Simon Kuznets (Smith), *National Income: A Summary of Findings*, reprint (New York: Arno Press, 1975; 1st ed. 1946), and, from the same author, *National Income and Capital Formation, 1919–1935* (New York; Nat. Bur. of Ec. Res., 1937).

24. Stamp, *Wealth and Taxable Capacity* (note 16), p. 88. Adolphe Lambert Jacques Quetelet, *Lettres à S.A.R. le duc-régnant de Saxe-Cobourg et Gotha sur la théorie des probabilités, appliquée aux sciences morales et politiques* (letters to H.R.H. the Ruling Duke of Saxe-Coburg and Gotha Concerning the Theory of Probability, Applied to the Moral and Political Sciences) (Brussels: Hayez, 1846), pp. 70ff.

25. Joseph Alois Schumpeter, *History of Economic Analysis*, ed. from manuscript by Elisabeth Boody-Schumpeter, 6th ed. (London: Allen and Unwin, 1967; 1st ed. 1954), pp. 644ff.

26. Ernst Adolphe Théodor Laspeyres, "Hamburger Warenpreise und die californisch-australischen Goldentdeckungen seit 1848" (Commodity Prices in Hamburg and the Californian-Australian Discoveries of Gold since 1848), *Jahrbücher für Nationalökonomie und Statistik* (1864), 81ff., 209 ff., and, from the same author, "Die Berechnung einer mittleren Preissteigerung" (The Calculation of an Average Price Increase), ibid., 296ff.

27. H. Paasche, "Über die Preisentwicklung der letzten Jahre nach den Hamburger Börsenentwicklungen" (On Price Developments in the Last Years According to Hamburg Stock Market Movements), *Jahrbücher für Nationalökonomie und Statistik* (1871).

28. Moritz Wilhelm Drobisch, "Über Mittelgrössen und die Anwendbarkeit derselben auf die Berechnung des Steigens und Sinkens des Geldwertes" (On Central Values and Their Application to the Calculus of the Increase or Decrease in the Value of Money), *Jahrbücher für National ökonomie und Statistik* (1871).

29. Alfred Marshall, "Remedies for Fluctuations of General Prices," *Contemporary Review* (1887).

30. William Stanley Jevons, *Investigations in Currency and Finance*, reprint of the 1884 London ed. (New York: Kelley, 1964; 1st ed. 1863), and, from the same author, *The Variations of Prices and the Value of Currency since 1782* (London; 1865), *A Serious Fall in the Value of Gold Ascertained . . .* (London: 1863).

31. Harald Ludwig Westergaard and Hans C. Nybölle *Grundzüge der Theorie der Statistik* (Principles of Statistical Theory), 2nd rev. and enlarged ed. (Jena: G. Fischer, 1927; 1st ed. 1890), pp. 559ff.

32. Robert Aurel Horváth, "Adam Smith's Performance in the Field of Index-Numbers," in *Essays in Political Arithmetics and Smithianism, Acta Universitatis Szegediensis, Juridica et Politica*, 25 (1978) 67ff.

33. Quetelet, *Lettres* (note 24), pp. 71ff.—included also in his *Physique sociale ou sur l'homme et le développement de ses facultés* (Social Physics or on Man and the Development of his Faculties), 2d ed. (Brussels, Paris, St. Petersburg: Muquardt, 1869; 1st ed. 1835), Vol. I, p. 467.

34. Marshall, *Principles*, 8th ed. (note 10), pp. 18ff., with reference to Francis Ysidro Edgeworth, *Mathematical Psychics: An Essay on the Application of Mathematics to the Moral Sciences*, reprint of the London ed. of 1881 (New York and London: Kelley, 1967).

35. N. G. Pierson, "Further Considerations on Index-numbers," *Economic Journal* (March 1896).

36. Correa Moylan Walsh, *The Measurement of General Exchange Value* (New York: Macmillan, 1901), and, from the same author, *The Fundamental Problem in the Monetary Science* (London: Macmillan, 1903)—see also "Index-Numbers," in *Encyclopaedia of the Social Sciences*, 1st ed.

37. Francis Ysido Edgeworth, *Memorandum on the Best Method of Ascertaining and Measuring the Variations of the Money Standard* (London: Reports of the British Association, 1887, 1888, 1889), and, from the same author, "Index-numbers," in *Palgrave's Dictionnary of Political Economy*, ed. by Robert Harry Inglis Palgrave, reprint, Vols. 1–3 (New York: Kelley, 1963; 1st ed. 1894–1900), pp. 384ff.

38. Marshall, *Principles*, (note 10), pp. 317ff.

39. Francis Ysidro Edgeworth, "Miscellaneous Applications of the Calculus of Probability," *Journal of the Royal Statistical Soceity*, (1897–1898), 681ff., and, from the same author, *Papers Relating to Political Economy*, 1st ed. (London: Macmillan, 1925), Vols. 1–3.

40. Francis Ysidro Edgeworth, "A Defence of Index-Numbers," *Economic Journal*, 1 (1896), 356ff.

41. Francis Ysidro Edgeworth, "Applications of Probabilities to Economics," *Economic Journal*, (June 1910), 387ff.

42. Edgeworth, "Applications" (note 41), p. 390 and notes 4 and 5, with reference to D. Jules Dupuit, "De l'utilité et de sa mesure" (On Utility and Its Measure), in *Ecrits choisis*, republished by Mario de Bernardi (Turin: La Riforma Sociale, 1933; 1st ed. 1844), Vol. II, pp. 367 ff., and to Léon Clément Colson, *Cours d'économie politique à l'Ecole Polytechnique et de l'Ecole des Ponts et Chaussées* (Course of Political Economy), expanded and definitive ed. (Paris: Gauthiers-Villars, Alcan, 1907–1931), Books 1–6.

43. Schumpeter, *History* (note 25), pp. 1326ff.

44. Wesley Clair Mitchell, *Index-Numbers of Wholesale Prices in the U.S. and Foreign Countries*, Bulletins 173 and 284 of U.S. Bureau of Labour Statistics, and, from the same author, *The Making and Using of Index-Numbers*, reprint (New York: Kelley, 1965; 1st ed. 1938).

45. Irving Fisher, *The Making of Index-Numbers: A Study of their Varieties, Tests and Reliability*, 3d rev. ed. (New York: Kelley, 1967; 1st ed. 1922).

46. Irving Fisher, *Mathematical Investigations in the Theory of Value and Prices* (New York: Kelley, 1965; 1st ed. 1892), and, from the same author, *The Purchasing Power of Money: Its Determination and Relation to Credit, Interest and Crises*, reprint (New York: Kelley, 1971; 1st ed. 1911), *The Nature of Capital and Income*, reprint (New York: Kelley, 1965; 1st ed. 1906).

47. Fisher, *The Making* (note 45), pp. 229ff., 242ff., 330ff.—see also Appendix I (Notes to Text"), note to Chapter XI, §11 ("Does Skewness and Dispersion Matter?"), and Chart 65 on p. 409.

48. Fisher, *The Making* (note 45), Chapter XVII ("Summary and Outlook") §14 ("The Future Uses of Index-Numbers"), pp. 367ff., and Appendix I ("Note to Chapter III"), §12 ("Professor Edgeworth's and Professor Young's Probability Systems of Weighting Give Erratic Results"), pp. 387ff. See also Fisher's view on the possibility of probability theory use based on correlation by criticizing the method proposed by R. L. Kelley, note to Chapter XVI, §4 ("Probable Error by Professor Kelley's Method"), pp. 430ff.

49. "Index-Numbers, I. Theoretical Aspects," by E. Ruist, in *International Statistical Encyclopaedia*, ed. by W. H. Kruskal and J. M. Tanur, 2nd printing, (New York and London: The Free Press-Collier Macmillan, 1978), Vol. I, pp. 451ff., without mentioning the works of Gini, for which see under "Gini, Corrado," ibid., pp. 394ff., "Quelques considérations au sujet de la construction des nombres-indices des prix et des questions analogues" (Some Considerations on How to Construct Price Index-Numbers and Analogous Questions), *Metron*, 4 (1924/1925), 3ff., and, from Gini, "Methods of Eliminating the Influence of Several Groups of Factors." *Econometrica*, 5 (1937), 56ff. See also F. Divisia, "L'indice monétaire et la théorie de la monnaie" (The Monetary Index and the Theory of Money), *Revue d'Economie Politique*, 39 (1929), 842ff., L. Törnquist, "Finlands Banks Konsumtionsprisindex," *Nordisk Tidskrift för Teknisk Ökonomi*, 8 (1942), 79ff., Ragnar Anton Kittil Frisch, "Annual Survey of General Economic Theory: The Problem of Index-Numbers," *Econometrica*, 4 (1936), 1ff., and Gottfried von Haberler, *Der Sinn der Indexzahlen, Untersuchung über den Begriff des Preisniveaus und die Methoden zu seiner Messung* (Tübingen: Mohr, 1927).

50. "Index Numbers, III. Sampling," by P. J. McCarthy, in *International Statistical Encyclopaedia*, 2nd printing (note 49), Vol. 1, pp. 463 and 466, with a Postscript by the same author, with reference to Willford Isabel King, *Index Numbers Elucidated* (New York: Nat. Bur. Ec. Res., 1930).

51. Henry Schultz, *The Theory and Measurement of Demand*, reprint (Chicago: The University of Chicago Press, 1972; 1st ed. 1938), pp. 128ff., with a mathematical formulation in his note 37, and pp. 133ff., with the philosophical implications of the basic hypotheses.

52. "Index Numbers, I. Theoretical Aspects: The Economic Theory," in *International Statistical Encyclopaedia* (note 49), pp. 454 and 456, with references to A. A. Konus, "The Problem of the True Index of Cost of Living," *Econometrica*, 7 (1939), 10ff., Melville J. Ulmer, *The Economic Theory of Cost of Living Index Numbers* (New York: Columbia, 1950), H. Staehle, "A Development of the Economic Theory of Price Index Numbers," *Review of Economic Studies*, 2 (1934), 163ff., Henry Theil, "The Information Approach to Demand Analysis," *Econometrica*, 33 (1965), 67ff., Henri Theil, "Best Linear Index Numbers of Prices and Quantities," *Econometrica*, 28 (1960), 464ff., and T. Kloek and G. M. De Wit, "Best Linear and Best Linear Unbiased Index Numbers," *Econometrica*, 29 (1961), 602ff.

53. Frisch, *Economic Planning* (note 2), Preface to the Oslo-channel model, pp. 101ff.

8 Statistics without Probability and Haavelmo's Revolution in Econometrics

Mary S. Morgan

In the early part of this century economists began to use statistical methods to measure the laws and relationships of economic science. However, their use of statistics was accompanied by a rejection of probability theory because it was argued that economic data suffered from certain defects that made probability theory inapplicable. At the same time, the role of probability theory in statistical inference was also limited because economists had little use for inferential reasoning between theory and data. These early attitudes were changed by Haavelmo's introduction of the probability approach for the measurement and testing of economic theories in 1944. The revolutionary arguments employed by Haavelmo, and the logical basis of his new program for econometrics in probability theory, are both discussed in detail. Since Haavelmo's work, econometricians have regarded statistical methods and probability theory as one set of tools in applied economics. The comparison of the pre-Haavelmo era with the more recent period suggests that the real impact of his program was the provision of a formal framework for testing economic theories.

The "probabilistic revolution" occurred in econometrics with the publication of Trygve Haavelmo's "The Probability Approach in Econometrics" in 1944.[1] It may seem strange that this "revolution" should have been delayed until the midcentury since, from the early days of the twentieth century, applied economists had been using statistical methods to measure economic laws. However, their use of statistical methods was accompanied by an indifference to, or a positive rejection of, probability theory on the grounds that it was inapplicable to economic data. Herein lies the contradiction. Although the theoretical basis for statistical inference lies in probability theory, economists used statistical methods, yet rejected probability theory. An examination of this paradox is essential in order to understand the revolutionary aspects of Haavelmo's work in econometrics and why his arguments were successful in changing economists' ideas about the role of probability theory.

In the early part of the twentieth century applied economists believed that there were real and constant laws of economic behavior waiting to be uncovered by the economic scientist. The method of statistics seemed to offer a substitute for the experimental method of the physical sciences; this alternative way in which to uncover the relationships or laws of economic theory was labeled econometrics. Early econometrics took the form of two sorts of activity depending on the status of the theory concerned and type of law to be uncovered. Where a well-defined and generally held theory existed, the role of statistical methods was to measure the parameters or constants of the law. This measurement function was an important one, but one in which questions of inference were almost nonexistent. Because the

Financial support from the Volkswagen Foundation is gratefully acknowledged. I am also thankful to members of the research group and to Meghnad Desai and John Aldrich for their invaluable intellectual support.

theory was not in doubt, applied workers sought neither to verify nor to disprove; the measured law was taken to be the true law corresponding to that theory. In areas where the theoretical laws were in doubt, where theorists disagreed, or where empiricists reigned, the role of statistical methods was to act as midwife in bringing forth the true laws or phenomena from the data. Again the role of inference seemed to be limited—whatever lawlike relationships or phenomena emerged from the data were taken to be the correct representations of economic life. Sophisticated inference procedures—ways of comparing theoretical laws and empirical relationships, and the need to argue between these two levels—appeared to be unnecessary under either approach.

This description does not paint an entirely accurate picture of econometric work in the years prior to Haavelmo's paper—not because the picture of underlying beliefs is untrue but because applied results rarely came out neatly. Measured relationships did not correspond with theoretical laws for a variety of reasons that econometricians grappled with in the 1920s and 1930s. The introduction of formalized inference procedures as a way of dealing with some part of these correspondence problems entailed a recognition of the role of probability theory. But this recognition was delayed by another belief held by econometricians, namely, that economic data were not raw material to which probability theory could be applied. So there was no role for probability theory in inference (since the use of inference arguments was limited), and there was a rejection of the role of probability theory in measurement and in uncovering economic phenomena.

1 Statistics without Probability

Economists' perception of the role of probability theory in the 1910s to the 1930s was that it had a very narrow domain of application, which extended neither to the treatment of economic data nor to the activity of measuring and uncovering economic laws; still less was probability seen as an element in theory itself. The attitudes of economists in using statistical methods but rejecting probability can best be illustrated by taking examples of the two different types of use made of statistical methods. In one case—that of demand analysis—economists generally agreed on the theory, and the point of the statistical work was to measure the laws for specified goods. In the other case—that of business cycle analysis—there was much disagreement, not only over the theory of the cycle, but also over the empirical nature of the cycle, and the first aim of the statistical work was to uncover the phenomenon.

The early work on demand was designed to measure the elasticity of demand (the sensitivity of quantity demanded to small price changes) for particular commodities. Because of their strong belief in theory, econometricians expected that the measured law would be the "true" relationship. Since in practice the data showed deviations from an exact relationship, they adopted statistical ways of measuring the underlying "true" relationship. These measurement devices, such as the least squares method, were used without underpinnings from probability theory

—that is, without reference to underlying probability distributions or arguments. There was no obvious reason why probability theory should have entered, since the observations were usually for the aggregate of all consumers rather than the result of some sampling procedure. There was occasional use of standard errors of coefficients, or of the regression; these went along with the statistical methods, but use of these tools of inference was by no means a standard procedure.

The estimated single equations relating prices to quantities were held to represent the exact theoretical relationships that investigators believed were the "true" demand curves. Economists considered two explanations for the lack of exact fit of the data points to the estimated equation and therefore of the lack of correspondence with an exactly holding "true" demand curve. The first of these explanations—omitted factors—came from economic theory considerations, and the second—measurement errors—from statistical ideas.

Demand theory postulates a relationship between the price of a good and the quantity demanded, *given* that all other factors that influence the demand for the good remain constant. It soon became clear to early investigators that in practice these "ceteris paribus" clauses would not be fulfilled, since the data covered a number of time periods in which these other factors changed, causing disturbances in the observed demand relationship. They tried to make their econometric models match economic theory models by dealing with these changing factors. Initially econometricians adjusted the data to allow for these changes prior to measuring the demand elasticity; for example, they adjusted changes in consumption to allow for increases in population or in income. Later they included the most influential factors (such as income) in the equation to be estimated; minor disturbing factors were still omitted.

Perhaps the clearest representation of this view can be seen in the work of Henry Schultz (whose reputation was based on his work on demand) in 1928: "All ... statistical devices are to be valued according to their efficacy in enabling us to lay bare the true relationship between the phenomena under consideration. An ideal method would eliminate entirely all of the disturbing factors. We should then obtain perfect correlation between changes in the quantity demanded and corresponding changes in price." [2]

So, in the economic explanation, the omitted disturbing factors were the cause of the approximation in measurement of the "true" demand curve. However, these omitted factors were not explicitly modeled or discussed as a part of the econometric relationship. Instead, discussion was limited almost entirely to the necessity to apply data adjustment processes in an effort to exclude the disturbing influences prior to measuring the relationship.

The lack of correspondence between theoretical models and applied results that arose because of the inability to hold all disturbing factors constant led some econometricians to take a less sanguine view of their results. In the late 1920s Mordecai Ezekiel suggested that statistical laws have their own worth, but because of the changing conditions under which they are obtained they cannot be taken to represent fundamental laws of economic behavior:

... the results obtained by statistical determination of the relations are not fundamental "laws of nature" in the same sense as is the law of gravity. They are measures of the way that particular groups of men, in the aggregate, have reacted to specific economic conditions during a specified period in the past. If the study is elaborate enough, it may reveal the way in which the reaction has been changing during the period considered, and the direction and rate of change. But it does not tell how long the same reaction will continue to prevail, what new causes may arise to change the responses, or what the relations would be in the new situation. The theories of mathematical probability do not apply.[3]

Unfortunately, he did not spell out exactly why he thought probability theory was important in his argument.

In 1932 the economist Lionel Robbins satirized the econometric work on demand by describing the methods of one, Dr. Blank, researching the demand for herrings. Robbins was vehemently against econometric work, and it is interesting that his argument against econometrics echoed some of Ezekiel's reasoning:

Instead of observing the market for herrings for a few days, statistics of price changes and changes in supply and demand may be collected over a period of years and by judicious "doctoring" for seasonal movements, population change, and so on, be used to deduce a figure representing average elasticity over the period. . . . they have no claims to be regarded as "laws." . . . there is no reason to suppose that their having been so in the past is the result of the operation of homogeneous causes, nor that their changes in the future will be due to the causes which have operated in the past. . . . there is no justification for claiming for their results the status of the so-called "statistical" laws of the natural sciences.[4]

Robbins argued that since probability theory could not be applied to economic data because of changing conditions, logically statistical methods could not be used either: "The theory of probability on which modern mathematical statistics is based affords no justification for averaging where conditions are obviously not such as to warrant the belief that homogeneous causes of different kinds are operating."[5] Econometricians did not seem unduly worried about the logical foundations of the statistical methods that they used, but were aware that the conditions required for the application of probability theory were not fulfilled by economic data.

The presence of measurement errors in the observations provided the alternative (or in some cases, parallel) rationalization for the lack of exact fit in the estimated relationship. This statistical explanation found favor with a number of econometricians—perhaps because it threw the onus of failure onto the quality of the data rather than implying that the statistical "experiment" had failed to replicate the economic theory because of uncontrolled factors. At the same time the application of least squares as a method for dealing with measurement error had a respectable pedigree in other scientific fields.

There were only cursory references in the econometrics literature to an underlying probabilistic model for these errors, and again they were not explicitly modeled as part of the relationship. The agricultural economist Holbrook Working in 1925 went further toward linking his statistical techniques to sampling theory than most other economists in discussing the presence of measurement errors and disturbing variables: "If the form of the relationship is properly judged, a statistical determination should give the true theoretical relationship, subject to the fluctuations of sampling, whenever the significant effects of errors and of extraneous forces are reflected only in the dependent variables." [6] In fact, the theoretical discussion of measurement errors in econometrics developed in a specifically antiprobability (or antisampling) framework set by the Norwegian econometrician Ragnar Frisch.

Frisch was not against sampling theory as such, but against its unthinking application in economics: "Of course, this contains no reflection on the value of sampling theory in general. In problems of the kind encountered when the data are the result of experiments which the investigator can control, the sampling theory may render very valuable services. Witness the eminent works of R. A. Fisher and Wishart on problems of agricultural experimentation." [7]

Whereas the standard method of regression assumed errors only in the dependent variable, it seemed to Frisch (as it did to others) much more likely that all variables were measured with error. Frisch's work in 1934 on "confluence analysis" dealt with the compound problem of measurement errors in all variables and several relationships existing between the variables at the same time. Frisch's approach to this complicated problem has been described as statistical—this is how he described it:

... if the sampling aspect of the problem should be studied from a sufficiently general set of assumptions, I found that it would lead to such complicated mathematics that I doubted whether anything useful would come out of it. And, on the other hand, if the sampling aspect should be studied under simple assumptions, for instance, of not collinear and normally distributed basic variates, the essence of the confluence problem would not be laid bare.... I decided therefore first to attack the problem more from the experimental side, working out numerically—on actual economic data as well as on constructed examples—various other types of criteria which intuitively and heuristically may suggest themselves. [8]

In Frisch's model all variables were made up of a systematic part plus a measurement error, and an exact relationship was believed to hold between the systematic ("true") variables. Frisch had not specified the full distribution of these measurement errors, being interested only in their variances; he had assumed that the errors in different variables were uncorrelated with each other and with the systematic components. The solution to the problem of measurement errors, Frisch decided, lay in the choice of correct weights for each variable in the relationship, chosen in accordance with the relative size of the variance of their measurement errors. The correct choice of weights would enable the "true" demand curve or relationship of interest to be found. In the absence of knowledge about the relative size of errors it would be possible to place limits on the "true" relationship by taking the least squares measurements under different extreme assumptions.

Probability was not part of econometric demand analysis in the 1920s and 1930s, nor did the economic theory model of demand involve probability ideas. Econometric work *measured* the elasticity of demand using the least squares method. If this measured value seemed unreasonable, then economists usually assumed that the data were no good—doubt was rarely cast on the model.[9] These attitudes limited the role of statistical inference and appeared to obviate the need for probability theory. It is notable that probability theory was actually rejected rather than just ignored, as might be expected. This point is explored further in the context of business cycle analysis.

Business cycle analysis was very different in approach from that of demand analysis. To begin with, it seemed that each economist had his or her own theory about why there were irregular cycles in the level of economic activity, which were reflected in cycles in prices, output, employment, and so forth. In fact there was not even general agreement on exactly what constituted the business cycle or how long the cycle was.

The earliest statistical work on business cycles had been obsessed with the idea that the business cycle was periodic and exogenously caused by a periodic cycle in weather or related phenomena. The two most famous examples are the sunspot theory of W. S. Jevons and the Venus theory of H. L. Moore.[10] This applied work involved the decomposition of time series data in the frequency domain (using periodogram analysis, a forerunner of spectral analysis), plus use of the statistical methods of least squares and correlation techniques to try and isolate the business cycle, measure its length, and state its cause.

Another quantitative approach to business cycle analysis involved the comparative description of cycles—the best-known writers in this field were C. Juglar and W. C. Mitchell.[11] They were not committed to an exact cycle, and Mitchell held an evolutionary view of the business cycle as a phenomenon of the capitalist economy. This approach was perhaps the least statistical inasmuch as cycles were treated as individual events. Knowledge of their characteristics was to be obtained from individual comparison, while notions of statistical description, such as the reduction of cycle data to an average cycle, were rejected.

Alongside, but separate from, this work by Mitchell was a movement concerned with forecasting the business cycle from descriptive statistical information on past cycles (with little input from theory). The forecasting movement started in the United States but spread rapidly in the 1920s, and business cycle institutes (both academic and commercial) opened in Europe and in Russia. The methods consisted of statistical analysis of business cycle data (the use of least squares to fit trend lines, correlation analysis, etc.) to decompose each time series variable into a trend, seasonal and cyclical components. The leftover elements in the data were thought to be the result of accidents or random causes and of no intrinsic interest. Correlation analysis was used to determine the relationships between the cyclical elements of different variables, and thence to build up compound indices, called business barometers. Business forecasting institutes using these methods lost some credibility because of their failure to predict the great crash and depression of 1929–1932.

The use of statistical techniques in business cycle analysis was accompanied by clear denials of a positive role for probability theory and of the reasons why. For example, Persons—founder of the influential "Harvard Business Barometer"—in his 1923 address to the American Statistical Association stated, "The view that the mathematical theory of probability provides a method of statistical induction or aids in the specific problem of forecasting economic conditions, I believe, is wholly untenable."[12]

Persons's reason for rejecting the use of probability theory was that economic data are time related—that is, each observation is related to the previous observation. In elaborating this argument, Persons confused the notions of randomness and independence:

... the actual statistical data utilized as a basis for forecasting economic conditions, such as a given time series of statistics for a selected period in the past, cannot be considered a random sample except in an unreal, hypothetical sense; that is to say, unless assumptions be made concerning our material which cannot be retained in actual practice. Any past period that we select for study is ... not "random" with respect to the present.... If the theory of probability is to apply to our data, not merely the series but the individual items of the series must be a random selection.... Since the individual items are not independent, the probable errors of the constants of a time series, computed according to the usual formulas, do not have their usual mathematical meaning....

Granting, as he [the statistician] must, that consecutive items of a statistical time series are, in fact, related, he admits that the mathematical theory of probability is inapplicable.[13]

Though he rejected the mathematical theory of probability in business cycle forecasting, Persons believed that probability as a measure of rational belief had a role in social science:

It is obviously impossible to state, in terms of numerical probability a forecast or an inference based upon both qualitative and quantitative evidence; and even if all the evidence were quantitative, we have seen that it does not express a numerical measure of rational belief for the future. So when we say that "the conclusions of the social scientist are expressed in terms of probabilities" we mean merely that his conclusions do not have the certainty of those of the natural scientist. The probabilities of the economic statistician are not the numerical probabilities which arise from the application of the theorems of Bernoulli and Bayes; they are, rather, non-numerical statements of the conclusions of inductive arguments.[14]

More specific reasons for rejecting probability theory were given by Oscar Morgenstern. He was director of the Vienna Institute for Business Cycle Research, but became disillusioned with the attempts to forecast business cycles in 1928. He defined the problems as lack of homogeneity in underlying conditions, the non-independence of time series observations, and the limited availability of data. Marget wrote an eloquently argued reply to Morgenstern's criticisms, in which he agreed with Morgenstern that "whatever may be the case in the other sciences, the formal technique of probability analysis can only rarely, if ever, be applied to economic data with any hope of obtaining reasonably significant results."[15] Yet, this was the only point on which Marget found he could agree with Morgenstern, and even on this he sought to rob Morgenstern of any real victory: "One has only to try to recall concrete instances of formal attempts to employ the technique of probability analysis to the problem of business forecasting to be convinced that our author is, after all, fighting with a shadow."[16]

The statistical methods used in isolating the business cycle were similar to those used in demand. But in cycle analysis, the rejection of probability theory was more clearly articulated. Probability theory was believed to be inapplicable to economic data because economic data did not behave according to the laws of probability; data observations were not independent of each other (i.e. they were related through time), and the underlying conditions were not homogeneous throughout the time period. In practice, in demand work and in cycle analysis, econometricians rejected data from obviously nonhomogeneous time periods (such as the war years, 1914–1918) and removed some time-related elements in the data (e.g., the removal of trends or trending factors). But the ostensible reason for these data control procedures was to fulfil the ceteris paribus clauses of economic theory (in order to reveal the underlying "true" demand relationship or business cycle), not to fulfil the data demands of probability theory. Econometricians did not cite data control procedures to justify their use of statistical methods because they did not believe that they needed to do so. Nor did they, like Robbins, feel it was necessary to condemn the use of statistical methods because they did not see these methods as dependent on the rejected probability theory.

So, prior to the 1930s a strong current in econometric thought rejected the application of mathematical theories of probability to economic data. At the same time there was widespread use of statistical methods, which did not seem to be tainted with probabilistic ideas. The use of least squares as a measurement device without justification from probability theory had a long history to support its application by econometricians.

2 Signs of Change

During the 1930s there were several signs of the "revolution" to come. These stirrings were visible in the desire to develop better ways of measuring relationships and better methods of testing competing economic theories using probability ideas. Probability was even creeping into econometricians' theoretical models.

One of the most notable of Haavelmo's forerunners in developing the probability approach to *measurement* was Tjalling Koopmans, whose thesis of 1937 extended Frisch's work on measurement errors.[17] Koopman's main point of issue with Frisch was that he had concerned himself with only one type of error—the measurement error—to the neglect of the sampling error: "... if all variables contain an erratic component, an estimated regression equation is subject to two quite different kinds of error.... Only one of them is considered by Frisch, and is due to absence of knowledge on the ratios of the variances of the errors in the individual variables. The other one is the usual sampling error.... It arises from the fact that the errors in the variables, even if being uncorrelated, mutually and to the systematic components, in the parent distribution, will in general fail to be so in a sample." [18] Koopmans reset Frisch's problem into a probability scheme, and he thought that the loss in generality involved would be compensated for by a gain in rigor and definite results.

It was not at first clear what the sampling framework should be because of the lack of repeatable experiments or controlled variability in economics (compared with, for example, work on the design of agricultural experiments), and because of the time-related nature of economic data: ". . . variables are developing in time in cyclical oscillation, apparently to a large extent governed by some internal causal mechanism, and only besides that influence, more or less, according to the nature of the variable, by erratic shocks due to technical inventions, etc. At any rate, they are far from being random drawings from any distribution whatever." [19]

Koopmans proposed to adopt the standard econometric model of the period—a causal and exact model between the systematic ("true") parts of the variables: ". . . regression coefficients [are] conceived as a quantitative measure of a causal relationship. . . . It is assumed that there is a 'true regression equation' which would be exactly satisfied by the 'true values' of the variables." [20] He then explained his sampling framework as follows: "The observations constituting one sample, a repeated sample consists of a set of values which the variables would have assumed if in these years the systematic components had been the same and the erratic components had been other independent random drawings from the distribution they are supposed to have." [21] That is, only one sample of observations on each time series variable (one observation on each variable at each point in time) is actually observed; the other samples remain hypothetical.

Working in this framework, Koopmans showed how taking account of sampling error would affect Frisch's results on measurement error, but his treatment was theoretical, highly technical in nature, and probably understood (and used) by relatively few econometricians.

Koopmans in fact favored the "classical" estimation method (as propounded by R. A. Fisher) for use in econometrics. In Koopmans's interpretation, this method treated the dependent variable (x_1) as having both measurement error and omitted variable error, while the independent variables ($x_2, x_3, \ldots, x_k$) were thought to be without error. He suggested that there were several ways, compatible with the assumptions of probability theory, in which the problem of time-related data could be overcome:

A conspicuous advantage is that this specification does not imply any assumption as to the distribution of $x_2, \ldots, x_k \ldots$ these observational variables need not be a random sample drawn from any probability distribution, but may as well be the values assumed by variables which develop in time by a, possibly unknown, causal mechanism; or they may be, as an intermediate case between these extremes, drawings from a series of distributions ordered in time, the next of which depends on the values drawn in the preceding ones. . . . The generality of Fisher's specification is a point strongly in favour of its use in economic regression analysis. [22]

This specification was thought by Koopmans to be a particularly strong advantage in economics where the variables $(x_2, \ldots, x_k)$ might not be expected to have stable distributions.

The field of business cycle analysis saw developments that involved probability ideas both in *theory construction* and in testing models. First of all, during the 1930s statistical business cycle analysis was overtaken by the development of dynamic macroeconomic theory (which sought to explain how the variables of aggregate economic activity were related and how they moved over time). For a long time, accidental events from outside the economy had been thought to precipitate crises or turning points in the cycle. In 1933 Frisch introduced these "shocks" as an integral part of his macroeconometric model, which enabled him to bridge the gap between the economic theory and the observed data.[23] In his model, the shocks were responsible for maintaining irregular oscillations in the economic system, though the system itself had a cyclical pattern in the form of a damped sine wave. Frisch combined these elements to produce a macroeconomic model that incorporated random shocks as a necessary part of the model and seemed suitable for applied measurement and testing by econometricians.

The first econometrician ambitious enough to build a large-scale econometric model of an economy and its cycles was Jan Tinbergen. In his work for the League of Nations in 1937–1938 he was effectively the first to use statistical techniques to *test*, as well as to *measure*, the macroeconomic relationships of the business cycle.[24] His model incorporated in its structure the shocks suggested by Frisch.

In his first League of Nations report, Tinbergen distinguished measurement and verification as different tasks. He suggested that regression coefficients measure the strength of the influence of the independent variables on the dependent variable in a multiple regression, while the multiple correlation coefficient tells whether a theory is verified or not. He was aware that theory testing was difficult: "... no statistical test can prove a theory to be correct ... even if one theory appears to be in accordance with the facts, it is still possible that there is another theory, also in accordance with the facts, which is the 'true' one as may be shown by new facts or further theoretical investigations."[25]

Tinbergen discussed the need for both *statistical* and *economic criteria* to test his results, because he feared that sampling considerations and omitted variables might make the results unreliable. He used a wide variety of statistical methods and tests, including Fisher's method (the standard application of ordinary least squares), Frisch's method of confluence analysis, and Koopmans's additional tests. He also tested to see whether the residuals from his equations (the empirical equivalent to Frisch's shocks) were normally distributed.

Tinbergen's second volume for the League of Nations was a model of the U.S. economy made up of chains of causal relationships. In this volume he explained his ideas on testing theories, which involved a two-stage procedure: first the individual equations were estimated, and then, by a process of substitution and elimination of the variables through the links of the chains, one "final equation" representing the path of the economic system was obtained. Tinbergen proposed to test the theory by testing the results at each stage:

First, the explanation that a given theory provides for each of the variables of the economic system may be tested by the method of multiple correlation analysis, and secondly, it may be tested whether the system of numerical values found for the "direct causal relations" (or what comes nearest to them) really yields a cyclic movement when used in the final equation.

This may be clarified by indicating the two ways in which an unfavourable result for any theory may be found. First, it is possible that the explanation given for the fluctuations of any of the variables might prove to be poor; and, secondly, it might happen that, although these explanations were not too bad, the combination of the elementary equations would not lead to a cyclical movement.

Apart from these two ways in which a theory may fail, there is the third— already mentioned above—that the theory may prove to be incomplete—i.e., that it contains less relations than variables to be explained—or indeterminate, in that it does not indicate from what other variables each variable depends and in what way.[26]

The total model, an amalgam of many economic theories, ran to 48 equations— not necessarily all estimated or independent, but still a mammoth undertaking given the state of computing power in 1938. He used ordinary least squares with an underlying sampling framework, although neither the errors in the equations nor their distributions were explicitly specified. The main problem was with Tinbergen's testing procedure—there was no adequate inference framework for testing the many different economic theories inherent in his model and taking decisions about which theories to reject and which to explore further.

One other area of work gave early warning of the "revolution"; once again this was as a result of econometricians trying to build more adequate (realistic) models of economic behavior. In this case the problem was how to treat the uncertainty in an individual consumer's or firm's behavior resulting from unfulfilled plans and forecasts. For example, Tinbergen attempted to establish how far people planned ahead by modeling their plans and the corrections they made when the plans proved to be in error.[27] These ideas provided another channel for a natural application of probability in the theory models used by econometricians, rather than through purely statistical devices for measurement of economic relationships.

Economists thought of life as a largely deterministic exercise, with the unknown, unknowable, or immeasurable bits dealt with as something outside the concerns of economic theory. In this setting, probability theory could not be part of economic models until economists thought that economic life itself involved elements of chance. By the end of the 1930s there had been some movement toward the integration of "chance" into the theoretical models developed and used by econometricians both at the macroeconomic level (as in Frisch's model of the cycle) and at the microeconomic level (as in Tinbergen's models of unfulfilled plans). Despite these signs of change, economic theories were still largely formulated and stated as exact rather than as probabilistic models.

The role of probability theory in measurement and inference was also restricted— but not because of the lack of chance elements in economic theories. Its role had been restricted because probability theory was not thought to be applicable to

economic data and because econometricians were naive about inference. Because
of the lack of correspondence between applied results and economic theory and
because of a growing desire to test theories rather than be content with measurement, econometricians in the 1930s were gradually becoming more sophisticated
about matters of inference.

3 Haavelmo's "Probability Approach"

Trygve Haavelmo was born in Norway in 1911. He studied economics at the
University of Oslo and was a student of Frisch and later (1933–1938) his research
assistant. He traveled to the United States in 1939 and visited various institutions
while funded by fellowships from the American Scandinavian Foundation and the
Rockefeller Foundation. From 1942 he worked for Norwegian agencies in the
United States and was at the same time associated with the Cowles Commission
for Research in Economics (then based at the University of Chicago). He returned
to Norway in 1947 to become professor of economics at the University of Oslo, a
position he has retained ever since.

Haavelmo's paper, originally entitled "On the Theory and Measurement of
Economic Relations," was written in 1941 while he was visiting at Harvard. It was
privately circulated (in hectographed form) among a number of econometricians.
By the time the paper was published in 1944 it had already proved very influential.
It appeared, with minor changes and some additions, in *Econometrica* (the journal
of the The Econometric Society) under the title "The Probability Approach in
Econometrics." [28] It was this published version that earned Haavelmo a doctorate
at the University of Oslo in 1946.

Given the attitudes of econometricians of the time and of Frisch in particular
(who thought that probability theory had only limited application in economics),
it is not surprising that a slightly defensive tone is evident in Haavelmo's writing.
The paper also bore signs of the evangelicism of the newly converted. Haavelmo
by his own admission owed much to Frisch—many of his ideas in the paper (apart
from the probability program) were developed from work initiated by Frisch. But
Haavelmo was converted to the usefulness of probability ideas by the brilliant
theoretical statistician Abraham Wald, whom he credited as the source of his
understanding of statistical theory.

Haavelmo recognized that the bulk of econometricians thought probability
theory had nothing to offer, though they made use of statistical methods. He argued
that since probability theory was the body of theory behind statistical methods, it
was not legitimate to use the latter without adopting the former:

> The method of econometric research aims, essentially, at a conjunction of
> economic theory and actual measurements, using the theory and technique of
> statistical inference as a bridge pier. But the bridge itself was never completely
> built. So far, the common procedure has been, first to construct an economic
> theory involving exact functional relationships, then to compare this theory with
> some actual measurements, and, finally, "to judge" whether the correspondence

is "good" or "bad." Tools of statistical inference have been introduced, in some degree, to support such judgements, e.g., the calculation of a few standard errors and multiple-correlation coefficients. The application of such simple "statistics" has been considered legitimate, while, at the same time, the adoption of definite probability models has been deemed a crime in economic research, a violation of the very nature of economic data. That is to say, it has been considered legitimate to use some of the *tools* developed in statistical theory *without* accepting the very *foundation* upon which statistical theory is built. For *no tool developed in the theory of statistics has any meaning*—except, perhaps, for descriptive purposes—*without being referred to some stochastic scheme.*[29]

The problems of nonindependence of observations and of nonhomogeneous time periods (over which economic relationships were unlikely to remain stable) were the strong arguments of the antiprobability lobby. Haavelmo argued, to the contrary, that it was precisely the generality of probability theory that made it applicable to economic data:

The reluctance among economists to accept probability models as a basis for economic research has, it seems, been founded upon a very narrow concept of probability and random variables. Probability schemes, it is held, apply only to such phenomena as lottery drawings, or, at best, to those series of observations where each observation may be considered as an independent drawing from one and the same "population." From this point of view it has been argued, e.g., that most economic time series do not conform well to any probability model, because the successive observations are not "independent." But it is not necessary that the observations should be independent and that they should all follow the same one-dimensional probability law. It is sufficient to assume that the *whole set* of, say, n, observations may be considered as *one* observation of n variables (or a "sample point") following an n-dimensional *joint* probability law, the "existence" of which may be purely hypothetical. Then, one can test hypotheses regarding this joint probability law, and draw inference as to its possible form, by means of *one* sample point (in n dimensions).[30]

This reversal of the argument—that far from having a narrow domain of application, the probability approach has a very general domain—was the basis of Haavelmo's revolutionary scheme for econometrics.

By adopting probability theory, Haavelmo believed that economists would be providing themselves with a proper framework for conducting economic research and the rigorous testing of theories in place of their present vague notions: ". . . if we want to apply statistical inference to testing the hypotheses of economic theory, it *implies* such a formulation of economic theories that they represent *statistical* hypotheses, i.e., statements—perhaps very broad ones—regarding certain probability distributions. The belief that we can make use of statistical inference without this link can only be based upon lack of precision in formulating the problems."[31]

These were Haavelmo's major arguments as set out in the preface to his 1944 paper. The succeeding 115 pages of the paper involved a discussion of many issues in econometrics, in all of which he made use of probability ideas to provide an integrated treatment of the subject and practice of econometrics. Only the more important strands in Haavelmo's arguments, showing how and why the probability approach should be implemented, are discussed here.

3.1 Theory and Data

The first problem was that of comparing an economic theory to data. Haavelmo's argument (told through his own words) took the following line: "The facts will usually disagree, in some respects, with any *accurate* a priori statement we derive from a theoretical model." Therefore, ". . . it is practically impossible to maintain any theory that *implies* a non-trivial statement about certain facts, because sooner or later the facts will, usually, contradict any such statement." And so, "What we want are theories that, without involving us in direct logical contradictions, state that the observations will *as a rule* cluster in a limited subset of the set of all conceivable observations, while it is still consistent with the theory that an observation falls outside this subset 'now and then'. As far as is known, the scheme of probability and random variables is, at least for the time being, the only scheme suitable for formulating such theories." [32]

Haavelmo gave a simple example elsewhere that parallels his reasoning here. [33] He considered that, though they might not admit to it, econometricians already worked with an informal probability scheme. No economist, he said, would want to work with an economic theory that predicted that, next year, national income would be exactly X millions because it would almost certainly be contradicted by fact. Instead, applied economists prefer to work with the type of theory that predicts that the level of national income next year will be close to X millions. Haavelmo argued that using probability theory is a formal way of specifying such theories. Economic theories must therefore be formulated as probabilistic statements. But what did this mean, and how would it help to relate the theory to the data?

In order to compare theory and data, it is normal to specify the experimental conditions under which the theory is expected to hold. Haavelmo believed that the same was true for economics. Consequently, a theoretical economic model

. . . will have an economic meaning only when associated with a design of actual experiments that describes—and indicates how to measure—a system of "true" variables (or objects) $x_1, x_2, \ldots, x_n$ that are to be identified with the corresponding variables in the theory. . . .

The model thereby becomes *an a priori hypothesis* about real phenomena, stating that every system of values that we might observe of the "true" variables will be one that belongs to the set of value-systems that is admissible within the model. The idea behind this is, one could say, that Nature has a way of selecting joint value-systems of the "true" variables such that these systems are as if the selection had been made by the rule defining our theoretical model. Hypotheses in the above sense are thus the joint implications—and the only testable implications, as far as *observations* are concerned—of a theory *and* a design of experiments. [34]

There were of course several problems with this definition that a hypothesis equals theory plus a design of experiments.

In the first place, unfortunately, economists are rarely explicit about their experimental designs. Haavelmo recognized this nettle and grasped it firmly. He began by grouping experiments into two classes:

(1) experiments that *we should like to make* to see if certain real economic phenomena—when *artificially isolated* from "other influences"—would verify certain hypotheses, and (2) the stream of experiments that Nature is steadily turning out from her own enormous laboratory, and which we merely watch as passive observers. . . .

In the first case we can make the agreement or disagreement between theory and facts depend upon *two* things; the facts we choose to consider, as well as our theory about them. . . .

In the second case we can only try to adjust our theories to reality as it appears before us. And what is the meaning of a design of experiments in this case? It is this: We try to choose a theory and a design of experiments to go with it, in such a way that the resulting data *would be* those which we get by passive observation of reality. And to the extent that we succeed in doing so, we become master of reality—by passive agreement.

Now if we examine current economic theories, we see that a great many of them, in particular the more profound ones, require experiments of the first type mentioned above. On the other hand, the kind of economic data that we actually have belong mostly to the second type.[35]

Haavelmo reached to the heart of the fundamental problem of econometrics. Economists are not in a position to isolate, control, and manipulate economic conditions—they cannot undertake experiments. Instead, they have to make do with passive observations (those from Nature's experiments), which are influenced by a great many factors not accounted for by the theory.

Haavelmo pointed out that the problem of passive observation was not "a particular defect of economic time series. If we cannot clear the data of such 'other influences', we have to try to introduce these influences in the theory, in order to bring about more agreement between theory and facts."[36] (There was certainly nothing revolutionary about this last statement—it could even be interpreted as a justification of the ad hoc statistical practices and model manipulations of the early econometricians.) The real defect, if the theory was to be formulated as a probabilistic statement, seemed to be that "passive observation of reality" resulted in data that were time related and from changing circumstances. Haavelmo insisted that this was not a problem—probability theory could be applied to economic data.

Haavelmo argued that the qualities of independence and randomness can be associated not only with individual observations on individual variables but also with observations on a system of variables $x_1, \ldots, x_r$. For example, if this were a set of variables thought to influence members of a consuming population; each person's set of variables could then be represented by one point in r-dimensional space and the x would be subject to a joint probability law. A sample of size s taken from all consumers (giving a set of information of dimension r by s) could be regarded as a sample of size s on the r-dimensional population (i.e., on the system of values $x_1, \ldots, x_r$); alternatively the data could even be regarded as one observation of dimension r by s. In other words, economic data could be dealt with in a number of different ways in probability theory. Usually, random sampling would denote that the s different points in the r space were independently chosen and that the dependence within the system (i.e., the relationship between the r different x) was "given by Nature."

The second obvious point about Haavelmo's definition of a hypothesis as a theory plus a design of experiments (aside from the issue of experimental design) was that it did not seem to require the hypothesis to be formulated as a probabilistic statement. Yet this formulation was essential to Haavelmo's scheme. He used an example to explain how a hypothesis became a probabilistic statement in practice.

The case was the consumption (y) of a given commodity by individuals to be explained by a number of factors ($x_1, \ldots, x_n$):

$$y = f(x_1, \ldots, x_n).$$

Haavelmo defined the statistical population—all possible values of each y—as the "infinity of possible decisions which might be taken with respect to the value of y,"[37] which was also acceptable as an economic theory definition. Then the observed values of y (the data), described as "all the decisions taken by all the individuals who were present during one year,"[38] constitute one sample of data from the population, the decisions in the next year as a second sample, and so on. Economists could, if they wished, then select a subsample from the observed data on individuals using a random sampling procedure.

The economic factors in the example (the x) could be treated either as having a joint probability distribution or as fixed at their observed values. In a sample of individuals with the same set of x values, the y values would naturally differ. Haavelmo argued that an additional explanatory factor should be included—the shift factor (s)—to represent the factors specific to the individual:

$$y = f(x_1, \ldots, x_n) + s.$$

This factor was also subject to a probability law: "When we assume that s has, for each fixed set of values of the variables x, a certain probability distribution, we accept the *parameters* (or some more general properties) of these distributions as certain additional characteristics of the theoretical model itself. These parameters (or properties) describe the *structure* of the model just as much as do systematic influences of $x_1, x_2, \ldots, x_n$ upon y. Such elements are not merely some superficial additions 'for statistical purposes'."[39]

Haavelmo felt it proper to justify the stochastic scheme in his example: "It is on purpose that we have used as an illustration an example of individual economic behaviour, rather than an average market relation. For it seems rational to introduce the assumptions about the stochastical elements of our economic theories already in the 'laws' of behaviour for the single individuals, firms, etc., as a characteristic of their behaviour, and then derive the average market relations or relations for the whole society, from these individual 'laws'." [40] But, he stated that even those who believed in exact theoretical relationships and only allowed for the presence of measurement error should also accept stochastic schemes. This was because the exact equations of theory could only be satisfied approximately in real life. It was necessary to bridge the gap between the exact theory and the facts with stochastic measurement errors—that is, measurement errors following probability laws.

The example of consumption that Haavelmo used showed how the relationship between the population and the sample in probability theory provided a model for the correspondence between economic theory and passive economic data ("a sample selected by Nature"). A probabilistic formulation of the economic theory imposed a formal relationship between the nonexperimentally obtained data and the theory, which enabled the theory to be tested.

3.2 Testing Theories

Haavelmo's revolution was concerned with changing both attitudes and practice—from an unchallengeable belief in theory plus the use of data adjustment processes, which aimed to make the data correct for the given theory, to a new approach, which aimed to find the correct choice of model for the observed data by using statistical tests. The benefits of the probability approach lay in its ability to test theories and thus aid in the correct choice of model.

Haavelmo argued that a properly formulated stochastic model (with a design of experiment) is a hypothesis that states which set of values is admissible. Observations may sometimes fall outside this set of admissible values without leading to a rejection of the theory, since the probabilistic model "does not exclude any system of values of the variables, but merely gives different weights or probabilities to the various value-systems." [41] In other words, the hypothesis states which sets of values are highly likely to occur and which are highly unlikely. According to Haavelmo, it is this feature that provides the power to compare theories and demonstrates the true value of the probability approach: "For the purpose of *testing* the theory against some other alternative theories we might then agree to deem the hypothesis tested false whenever we observe a certain number of such 'almost impossible' value-systems." [42]

The problem of economic theory was essentially "... to construct hypothetical probability models from which it is possible, by random drawings, to reproduce samples of the type given by 'Nature'." [43] The problem of testing the theory against these samples was then one of statistical theory and technique. This brought Haavelmo round to his last major point—the advances in statistical testing theory particularly due to Neyman and Pearson. Following an outline of the Neyman-Pearson testing procedures, Haavelmo set out how the economist should formulate theories for testing purposes.

Haavelmo proposed that a theoretical model to be compared to data should involve a system of theoretical random variables corresponding to the observed variables. The observed data (nN values: $x_{1t}, x_{2t}, \ldots, x_{nt}, t = 1, \ldots, N$) should be considered a sample point in an nN-dimensional sample space with a certain joint probability distribution. He commented, "It is indeed difficult to conceive of any case which would be contradictory to this assumption. For the purpose of testing hypotheses it is not even necessary to assume that the sample could actually be repeated. We make hypothetical statements before we draw the sample, and we are only concerned with whether the sample rejects or does not reject an a priori hypothesis."[44] However, the theoretical model according to Haavelmo should consist not only of the nN random variables corresponding to the data but also of a set of "auxiliary random parameters": $(e_{1t}, e_{2t}, \ldots, e_{nt}, t = 1, \ldots, N)$, with a specified joint distribution. These auxiliary variables might be "counterparts to some real phenomenon," such as measurement errors. A set of constant parameters and restrictions completed Haavelmo's formulation of the econometric model. The assumptions made about the e and the x would, he thought, restrict the class of probability laws to which the model belongs. Despite these restrictions, different models might lead to the same set of probability laws and therefore, in testing, the models would be indistinguishable from the point of view of the observations.

In testing the theory against the data, a well-fitting model (a theory that fits the data well) might produce further useful restrictions upon the class of prior theoretical models. But, as Haavelmo pointed out, there is still no guarantee that it is the "true" or "correct" model. This is because many different probability schemes might be capable of producing the same observed data:

Since the assignment of a certain probability law to a system of observable variables is a trick of our own, invented for analytical purposes, and since the same observable results may be produced under a great variety of different probability schemes, the question arises as to which probability law should be chosen, in any given case, to represent the "true" mechanism under which the data considered are being produced. To make this a rational problem of statistical inference we have to start out by an axiom, postulating that every set of observable variables has associated with it one particular "true," but unknown, probability law.[45]

Choosing the "true" model remained in practical terms a difficult task.

Haavelmo's belief was plainly that economists should think of all economic variables as being governed by probability laws. On what foundations did these beliefs rest? Haavelmo considered discussions of the nature of probability and the foundations of the theory to be futile. He thought that the foundational approaches based on either frequency of occurrence, a priori confidence, or formal axioms all suffered from the same drawback—namely, that they all dealt with a concept of probability that was purely abstract. He grumbled that they were all required "... to satisfy some logical consistency requirements, and to have these fulfilled a price must be paid, which invariably consists in giving up the exact equivalence between the theoretical probabilities and whatever real phenomena we might consider. In this respect, probability schemes are not different from other theoretical schemes. The rigorous notions of probabilities and probability distributions 'exist' only in our rational mind, serving us only as a tool for deriving practical statements...."[46] Haavelmo went on, "When we state that a certain number of observable variables have a certain joint probability law we may consider this as a construction of a rational *mechanism*, capable of producing (or reproducing) the observable values of the variables considered. When we have observed a set of values ... we may, without any possibility of a contradiction, say that these ... values represent a sample point drawn from a universe obeying *some* unknown ... probability law. Whatever be the a priori statement we want to make about the values ... we can derive this statement from one of several ... probability laws." [47]

Whatever the basis for Haavelmo's beliefs, and it seems at times that he was closest to adopting a frequency interpretation of probability, the important thing for him was not the underlying foundations or the necessity of belief in probabilities in economics. It did not matter whether probabilities existed or not; the important message was that econometricians should adopt the probability approach because it was the best scientific practice available. Thus, statements that seemed to cast doubt on his own attitudes should be interpreted, rather, as his occasional concessions to those who did not share his views:

Purely empirical investigations have taught us that certain things in the real world happen only very rarely, they are "miracles" while others are "usual events." The probability calculus has developed out of a desire to have a formal logical apparatus for dealing with such phenomena of real life. The question is not whether probabilities *exist* or not, but whether—if we proceed *as if* they existed—we are able to make statements about real phenomena that are "correct for practical purposes." [48]

It did not matter to Haavelmo whether economists were doubtful about the existence of probabilities. It was much more important that they adopt a skeptical attitude toward economic theories, for, he warned, "Whatever be the 'explanations' [of economic phenomena] we prefer, it is not to be forgotten that they are all our own artificial inventions in a search for an understanding of real life; they are not hidden truths to be 'discovered'." [49] It was this concern that lay behind Haavelmo's attempt to redirect econometrics.

4   A New Consensus

The most immediate—and probably most important—effect of Haavelmo's paper was on the work of the Cowles Commission for Research in Economics. From the beginning of 1943 a new research program was initiated there, under the direction of Jacob Marschak, involving a group of talented young statisticians, mathematicians, and economists. Inspired by Haavelmo's methodological blueprint for econometrics, they aimed to continue the tradition of Tinbergen's pioneering work on business cycles but to improve on his work by implementing the working practices of the "Probability Approach." They saw the probability approach as providing very powerful, precise tools that, they believed, would solve a host of economic problems—in comparison with the previous econometric tools fashioned in the bronze age.

Having adopted Haavelmo's blueprint, the researchers at the Cowles Commission devoted a great deal of energy to developing the statistical technology to go with the framework laid down by Haavelmo. They attacked these problems with considerable technical expertise, statistical insight, and rigor. Applied work was not omitted from this program, although, almost inevitably given their idealism, its results were more disappointing than those of the theoretical work on statistical problems.[50] Despite this, the latter part of the 1940s and early 1950s saw an impressive succession of Cowles Commission Monographs and papers.[51]

Given the content of the Cowles Commission's research program, opposition to Haavelmo's ideas might have turned up disguised as opposition to the Cowles work. This turned out not to be the case—although there was strong opposition to the Cowles program. This opposition was most clearly stated in a debate sparked off by a review of a quantitative-historical study of business cycles by Burns and Mitchell of the National Bureau of Economic Research (NBER).[52] Koopmans's review in 1947, under the title "Measurement without Theory," attacked the Burns and Mitchell volume on a broad front: for their lack of an underlying behavioral economic model with specified structure, for their inadequate statistical treatment of the data, and for their almost complete lack of inference procedures. Koopmans argued for a more rigorous formulation of economic hypotheses in probabilistic terms so that the most advanced statistical estimation and inference procedures could be used. This attack led to a journal debate in 1949 in which Koopmans was inevitably identified with the econometric program of the Cowles Commission (of which he was by this time director). Vining, who replied on behalf of Burns and Mitchell, was equally inevitably seen as the spokesman for the quantitative/institutional research program of the NBER (where he was a research associate).[53]

It is intriguing that both Vining and Koopmans drew on Haavelmo's paper in support of their conflicting positions.[54] Koopmans had no need to cite Haavelmo; it was obvious—as Vining recognized—that Koopmans had taken Haavelmo's paper as his text. But, in his own reply, Vining also called upon Haavelmo's authority. He accused the Cowles Commission of measuring without economic theory, citing Haavelmo's emphasis on the need to undertake research in economic theory before hypotheses could be given a statistical formulation. He claimed that statistics had little or no role to play in the discovery of such economic hypotheses.

Vining did not reject probability theory as a way of investigating economic data, but argued for a broader interpretation, which would cover the more empirical approach practiced by the NBER:

Distributions of economic variates in as large groups as can be obtained should be studied and analyzed, and the older theories of the generation of frequency distributions should be brushed off, put to work, and further developed. That is to say, statistical economics is too narrow in scope if it includes just the estimation of postulated relations. Probability theory is fundamental as a guide to an understanding of the nature of the phenomena to be studied and not merely as a basis for a theory of the sampling behaviour of estimates of population parameters the characteristics of which have been postulated. In seeking for interesting hypotheses for our quantitative studies we might want to wander

beyond the classic Walrasian fields and poke around the equally classic fields
once cultivated by men such as Lexis, Bortkiewicz, Markov, and Kapteyn.[55]

Koopmans agreed, but preferred to start with some prior economic model: "I
believe that his term 'statistical theory in its broader meaning' is used in the same
sense in which the econometricians speak of 'model construction'. It is the model
itself, as a more or less suitable approximation to reality, which is of primary
interest. The problems of estimation of its parameters or testing of its features have
a derived interest only."[56] On the other hand, Koopmans was not prepared to
concede that statistical theory had nothing to do with hypothesis seeking: "It is
possible to take a formal view and argue that hypothesis-seeking and hypothesis-
testing differ only in how wide a set of alternatives is taken into consideration. . . .
To the extent that hypothesis-seeking is an activity that can be formalized by
(statistical) theory, there is little doubt that the concept of statistical efficiency will
remain relevant. . . . However, there remains scope for doubt whether all hypothesis-seeking activity can be described and formalized as a choice from a pre-
assigned range of alternatives."[57]

There seemed to be a considerable measure of agreement that probability theory
had an important role in economics, though this was partly hidden by the desire
of both parties to score points.

The debate spilled over into a review in 1951 by Hastay (from the NBER) of a
collection of the main theoretical results in econometrics from the Cowles Commission's work of the 1940s. Once again Haavelmo's name seemed an important
talisman:

Is it the most fruitful view of economic theory that which treats it in essential
analogy with mechanics and meteorology? Such is the philosophy of the econometric school, . . . a stochasticized Walrasian model is arbitrarily laid down as
the essence of economic reasoning, and the authors take as their text a principle
of Haavelmo that every testable economic theory should provide a precise
formulation of the joint probability distribution of all observable variables to
which it refers. It can be argued, however, that Haavelmo's principle is sounder
than the program for realizing it worked out in this book.[58]

Though the NBER did not adopt an econometric approach, it did use a variety of
statistical techniques, which perhaps accounts for its desire to claim Haavelmo.

In view of both institutions' need to attract and retain adequate research funding,
the argument could also be seen as an exercise in propaganda. The Cowles Commission had managed to obtain financial backing from the Rockefeller Foundation
in the 1940s, but the same foundation also supported the NBER and had made a
very substantial grant to them in 1947.[59]

Though Haavelmo's 1944 paper was dense and difficult, part of the program he
suggested had already been published in more accessible form. In 1943, Haavelmo
had defended Tinbergen's work on business cycles against Keynes's criticism and
had taken the opportunity to introduce the probability approach and its justification in a shortened version.[60] In another paper in the same year, Haavelmo had
published one of the chapters of the long paper, which marketed the probability

approach as a necessary adjunct to the simultaneous equations model.[61] Haavelmo showed economists that, if this familiar model (which implied that the relationships and values of the crucial variables of the economy were determined more or less simultaneously) was the correct model of the world, then they should adopt the probability approach in their applied work to avoid biased results when measuring the relationships. This paper was more influential than the main "Probability Approach"; it was easier to understand because the argument was presented through an example.

Another influential, but not easy, paper that came from the same stable as the Cowles Commission and Haavelmo's papers was that by Mann and Wald, also in 1943.[62] It dealt with maximum likelihood estimation of the typical econometric model used by the group—a linear stochastic difference equation (or system of equations). The mass effect of all the papers together suggested that econometrics had been successfully taken over by the probabilists. Certainly, there were no notable econometricians who publicly dissented from the probability approach in the late 1940s, though some disagreed with other parts of Haavelmo's program.

Econometricians from Moore, Schultz, and Frisch to Koopmans and Tinbergen had all thought of their models as being causal. For Herman Wold, simultaneously determined relationships involved the contradiction of simultaneous two-way causality. Wold favored a causal chain model of economic relationships in which relationships were not independent, but were determined individually and in turn according to their interrelationships. Consequently a group of economists led by Wold rejected the simultaneous equations model advanced by Haavelmo. Despite their rejection of the model form as an unacceptable model of reality, Haavelmo's probabilistic formulation and treatment of models was adopted with approval by Wold's group.[63]

Tintner was another econometrician who accepted the probability aspect of Haavelmo's work; but he believed that Haavelmo had not dealt adequately with the problem of measurement errors.[64] In fact, Haavelmo had allowed for the presence of such errors in all variables in his theoretical scheme, but, in the few examples he gave, had been forced to reduce their presence (though not to ignore them entirely) in order to make the models he dealt with tractable. On the other hand, Haavelmo's research program, as reinterpreted by the Cowles group, concentrated almost entirely on errors in the equation due to omitted variables, thus deserving Tintner's criticism.

By the late 1940s, Haavelmo's ideas for a probability approach were accepted in econometrics. But what difference did they make? A comparison of post-Haavelmo econometrics with the pre-Haavelmo era may help to pinpoint exactly what was revolutionary about Haavelmo's "Probability Approach."

Before Haavelmo's work, econometricians used statistical methods, but were either indifferent to or rejected any role for probability theory. Probability theory was rejected because it was believed that certain features of economic data (nonindependence of successive observations and nonhomogeneous conditions of observation) made the theory inapplicable. At the same time, econometricians perceived little need for the framework of probability theory as an aid to inference or in testing theories.

Haavelmo convinced econometricians of the value of the probability approach. He pointed out that the theory could be applied to economic data merely by redefining the sample space to accommodate the relationships between the observations. Economists were persauded by the example of the simultaneous equations model, where use of the probability approach produced the correct results, while the old-style statistical methods produced biased results. Having accepted the logic of the probability approach for measurement, there was no bar to the more gradual acceptance of its usefulness in the difficult task of inference and theory testing.

Haavelmo believed in a full-scale probability approach, where all the variables are thought of as being generated by some probabilistic process. Post-Haavelmo statistical methods and probability theory have not been seen as separate entities, but as one and the same bag of tools used by econometricians. Even when econometricians since the 1950s have been half-hearted or indifferent to the status of economic variables, they have continued to use a version of Haavelmo's program. In this sense, then, Haavelmo did cause a revolution in thinking—no subsequent econometrician thought, as the early econometricians clearly had done, that statistical methods could be used but that probability theory should be rejected.

It is easy enough to describe the probabilistic revolution in terms of the ideas. But what was the real difference in practice? In the first rash of textbooks in the early 1960s, the full-scale probability model proposed by Haavelmo (in which all variables were treated as random) was watered down so that only the dependent variable in the equation was treated as random and the other variables were treated as fixed.[65] In this diluted model, it was the addition to the economic equation of an error term representing omitted variables that imparted randomness to the dependent variable. This was very like the simple pre-1940s econometric model, except that now the random "error" variable was explicitly included in the equation and its distribution specified. Econometricians continued to use comparatively simple methods of estimation (such as least squares) appropriate to such models.

More recently (since the 1970s) this movement seems to have been reversed, with a return to Haavelmo's principle of specifying joint probability distributions of all the variables under study and to the use of maximum likelihood estimation techniques (due perhaps to cheap computing time). At the same time, the range of probability models specified for economic relationships has widened. Even in the somewhat grayer intervening period, the basis of econometrics in probability theory, though not obvious, was not questioned. Indeed, it was this intervening phase that saw the emergence of an alternative foundational approach in econometrics using Bayesian ideas.

Haavelmo's paper marked an important step forward in developing the theory-testing role of econometrics. This evaluation and testing aspect was important: in early econometrics, data could be rejected, but not theory. This was because applied results were judged according to a given theory; if the results were rejected, this usually implied a rejection of the data (e.g., because of measurement errors) and precluded a rejection of the theory. This problem became acute when researchers wanted to test or compare theories, as there was no obvious way of proceeding. Tinbergen had experienced considerable difficulty in executing the commission of the League of Nations to test the available theories of business cycles and sort out which were correct. Faced with many, apparently conflicting but possibly complementary theories in a verbal form, he had made one large econometric model, which incorporated them all. He had then compared his results with the individual theories, judging his results by criteria internal to the theory. He had some statistical tests, but they were insufficient to compare these theories using external (statistical) criteria.[66]

Haavelmo showed that if theoretical laws were thought challengeable—rather than unchallengeable truths—they could be treated as hypotheses about probability distributions, and the nonexperimentally obtained data could be considered as samples from these distributions. Moreover, such a framework might have a natural analog in economic theory. This formalization of the framework for testing economic theories using probability theory allowed applied economists to be more flexible about theory, since any chosen hypothesis might be incorrect and an alternative model correct.

The change in ideas since Haavelmo's work has been reflected in practice by a widening in the range of statistical tests applicable to economic data and a dramatic increase in their use. Now, econometricians commonly apply a battery of statistical tests to help evaluate a chosen theoretical model. The traditional role of econometrics in measuring the parameters of a given theory has not been upstaged, since Haavelmo's blueprint and the Neyman-Pearson testing methodology are as relevant to the measurement of one parameter of an economic theory as to a test of the whole theory. Haavelmo's probabilistic revolution changed the sort of question asked by applied economists when looking at their results. No longer would econometricians like Robbins's Dr. Blank ask themselves the naive question, "Have I found the 'true' demand curve for herrings?"

Notes

1. Trygve Haavelmo, "The Probability Approach in Econometrics," *Econometrica*, 12 (1944; Supplement), 1–118.

2. Henry Schultz, *Statistical Laws of Demand and Supply with Special Application to Sugar* (Chicago: University of Chicago Press, 1928), p. 33.

3. Mordecai Ezekiel, "Statistical Analyses and the 'Laws' of Price," *Quarterly Journal of Economics* 42 (1928), 199–227, on p. 223.

4. Lionel Robbins, *An Essay on the Nature and Significance of Economic Science* (London: Macmillan, 1932), p. 101.

5. Lionel Robbins, *Essay*, p. 102 (note 4).

6. Holbrook Working, "The Statistical Determination of Demand Curves," *Quarterly Journal of Economics*, 39 (1925), 503–543, on p. 539.

7. Ragnar Frisch, *Statistical Confluence Analysis by Means of Complete Regression Systems* (Oslo, University Institute of Economics, 1934), p. 6.

8. Ragnar Frisch, *Confluence Analysis*, p. 7 (note 7).

9. There is one example where the model was rejected because the results of the applied work struck contemporaries as outrageous—this is the case of Henry Ludwell Moore's positive

demand curve found in his *Economic Cycles—Their Law and Cause* (New York: Macmillan, 1914).

10. William Stanley Jevons, *Investigations in Currency and Finance* (London: Macmillan, 1884), and *Papers and Correspondence of W. S. Jevons*, Vol. VII, edited by R. D. Collison Black (London: Macmillan, 1981), contain most of Jevon's work on the sunspot theory. Henry Moore's work is to be found in his two books on business cycles: *Economic Cycles* (note 9) and *Generating Economic Cycles* (New York: Macmillan, 1923).

11. Clement Juglar's work was partly responsible for making cycles (as opposed to crises) the unit of analysis; *Des Crises Commerciales et de leur Retour Periodique en France, en Angleterre et aux Etats-Unis* (Paris: 1862; expanded 2nd ed. 1889). Wesley Clair Mitchell's extended research program is best represented in *Business Cycles*, Memoirs of the University of California, vol. 3 (Berkeley: University of California Press, 1913), and *Business Cycles. The Problem and its Setting* (New York: National Bureau of Economic Research, 1927).

12. Warren M. Persons, "Some Fundamental Concepts of Statistics," *Journal of the American Statistical Association*, 19 (1924), 1–8, on p. 6. His work on business cycle statistics can be seen in, for example, "Indices of Business Conditions," *Review of Economics and Statistics*, 1 (1919), 5–110.

13. W. M. Persons, "The Problem of Business Forecasting," in *The Problem of Business Forecasting* (London: Pollak Foundation for Economic Research Publications, No. 6, 1924), pp. 9–11, parentheses added.

14. Persons, "Problem," p. 15 (note 13). He admitted to influence from J. M. Keynes, *A Treatise on Probability* (London: Macmillan, 1921).

15. A. W. Marget, "Morgenstern on Economic Forecasting," *Journal of Political Economy*, 37 (1929), 312–39, on p. 315. Oscar Morgenstern, *Wirtschaftsprognose: Eine Untersuchung ihrer Voraussetzungen und Möglichkeiten* (Vienna: Julius Springer, 1928).

16. Marget, "Morgenstern," p. 316 (note 15).

17. Tjalling C. Koopmans, *Linear Regression Analysis of Economic Time Series* (Haarlem: Netherlands Economic Institute, 1937). There were some statisticians associated with econometrics who argued in favor of the use of probability theory during the 1930s, but they had little effect. One example is Hotelling; another is Yule, whose article "Why Do We Sometimes Get Nonsense Correlations between Time Series—A Study in Sampling and the Nature of Time Series," *Journal of the Royal Statistical Society*, 89 (1926), 1–64, frightened econometricians because he showed that correlation coefficients on time-related data were likely to be consistently overstated.

18. Koopmans, *Linear Regression*, p. 45 (note 17).

19. Koopmans, *Linear Regression*, p. 5 (note 17).

20. Koopmans, *Linear Regression*, p. 6 (note 17).

21. Koopmans, *Linear Regression*, p. 7 (note 17).

22. Koopmans, *Linear Regression*, p. 29–30 (note 17).

23. Ragnar Frisch, "Propagation Problems and Impulse Problems in Dynamic Economics," in *Economic Essays in Honour of Gustav Cassel* (London: Allen & Unwin Ltd., 1933). In formulating this model, Frisch drew on an analogy by Wicksell that likened the economy to a rocking horse being hit at random intervals with random amounts of force. He was also influenced by works on stochastic processes by George Udny Yule, "On a Method of Investigating Periodicities in Disturbed Series, with Special Application to Wolfer's Sunspot Numbers," *Philosophical Transactions of the Royal Society*, A22 (1927), 267–298, and by Eugen Slutsky, "The Summation of Random Causes as the Source of Cyclic Processes," *Problems of Economic Conditions*, 3 [Conjuncture Institute of Moscow, Russian with English summary, and in English in *Econometrica*, 5 (1937), 105–146].

24. Jan Tinbergen, *Statistical Testing of Business Cycle Theories*, Vols I & II (Geneva: League of Nations, 1939).

25. Jan Tinbergen, *Statistical Testing*, Vol. I, p. 12 (note 24).

26. Jan Tinbergen, *Statistical Testing*, Vol. II, pp. 18–19 (note 24).

27. Jan Tinbergen, "The Notion of Horizon and Expectancy in Dynamic Economics," *Econometrica*, 1 (1933), 247–264. Another example is Gerhard Tintner, "A Note on Economic Aspects of the Theory of Errors in Time Series," *Quarterly Journal of Economics*, 53 (1938), 141–149.

28. The original paper is Trygve Haavelmo, "On the Theory and Measurement of Economic Relations" (Mimeo, Cambridge, MA, 1941). The main change in the 1944 "Probability Approach" version (note 1) is the addition of the last section on forecasting.

29. Haavelmo, "Probability Approach," Preface, p. iii (note 1).

30. Haavelmo, "Probability Approach," Preface, p. iii.

31. Haavelmo, "Probability Approach," Preface, p. iv.

32. Haavelmo, "Probability Approach," pp. 1–2, 40.

33. The example is to be found in Trygve Haavelmo, "Statistical Testing of Business Cycle Theories," *Review of Economics and Statistics*, 25 (1943), 13–18, on p. 16.

34. Haavelmo, "Probability Approach," pp. 8–9 (note 1).

35. Haavelmo, "Probability Approach," pp. 14–15.

36. Haavelmo, "Probability Approach," p. 18.

37. Haavelmo, "Probability Approach," p. 51.

38. Haavelmo, "Probability Approach," p. 51.

39. Haavelmo, "Probability Approach," p. 51.

40. Haavelmo, "Probability Approach," pp. 51–52.

41. Haavelmo, "Probability Approach," p. 9.

42. Haavelmo, "Probability Approach," p. 9.

43. Haavelmo, "Probability Approach," p. 52.

44. Haavelmo, "Probability Approach," p. 70.

45. Haavelmo, "Probability Approach," p. 49.

46. Haavelmo, "Probability Approach," p. 48.

47. Haavelmo, "Probability Approach," p. 48.

48. Haavelmo, "Probability Approach," p. 43.

49. Haavelmo, "Probability Approach," p. 3. (Autonomously given elements could of course be constant—that is, fixed with probability equal to one.)

50. For one thing, applied work was held back by the very heavy computational burden imposed by the use of the maximum likelihood technique, which was part of their program.

51. These were Cowles Commission Monograph 10 [Tjalling C. Koopmans, ed., *Statistical Inference in Dynamic Economic Models* (New York: Wiley, 1950)], Monograph 11 [Lawrence R. Klein, *Economic Fluctuations in the United States, 1921–41* (New York: Wiley, 1950)], and Monograph 14 [William C. Hood and Tjalling C. Koopmans, Eds., *Studies in Econometric Method* (New York: Wiley, 1953)]. Other papers were published in *Econometrica* and elsewhere in the late 1940s by Cowles Commission members.

52. Arther F. Burns and Wesley C. Mitchell, *Measuring Business Cycles* (New York: National Bureau of Economic Research, 1946), and review by Tjalling C. Koopmans, "Measurement without Theory," *Review of Economics and Statistics*, 29 (1947), 161–172.

53. The exchange between the two sides is set forth in R. Vining and Tjalling C. Koopmans, "Methodological Issues in Quantitative Economics," *Review of Economics and Statistics*, 31 (1949), 77–94.

54. This perhaps indicates the extent to which the Cowles program was perceived to have diverged from the Haavelmo blueprint by some contemporaries.

55. Vining in Vining and Koopmans, "Methodological Issues," p. 85 (note 53).

56. Koopmans in Vining and Koopmans, "Methodological Issues," pp. 89–90 (note 53).

57. Koopmans in Vining and Koopmans, "Methodological Issues," p. 90 (note 53).

58. M. Hastay, Review of T. C. Koopmans (Ed.), *Statistical Inference, Journal of the American Statistical Association*, 46 (1951), 388–390, on pp. 388–389.

59. The grants awarded and taken up by the two bodies can be traced in the *President's Review and Annual Report* (New York: Rockefeller Foundation). This Foundation also supported a number of individual econometricians on fellowships.

60. T. Haavelmo, "Statistical Testing" (note 33), has already been mentioned. The original review and criticism by Keynes of J. Tinbergen's *Statistical Testing* is J. M. Keynes, "Professor Tinbergen's Method," *Economic Journal*, 49 (1939), 558–568. Following the review are J. Tinbergen's "A Reply" and J. M. Keynes's "Comment" in *Economic Journal*, 50 (1940), 141–156.

61. Trygve Haavelmo, "The Statistical Implications of a System of Simultaneous Equations," *Econometrica*, 11 (1943), 1–12. Koopmans also wrote an article along similar lines with the aim of popularizing Haavelmo's papers: T. C. Koopmans, "Statistical Estimation of Simultaneous Economic Relations," *Journal of the American Statistical Association*, 40 (1945), 448–466.

62. H. B. Mann and Abraham Wald, "On the Statistical Treatment of Linear Stochastic Difference Equations," *Econometrica*, 11 (1943), 173–220.

63. Examples are Herman Wold's "Review of T. C. Koopmans (Ed.), *Statistical Inference*," *Econometrica*, 19 (1951), 475–477, and H. Wold and R. Bentzel, "On Statistical Demand Analysis from the Viewpoint of Simultaneous Equations," *Skandinavisk Aktuarietidskrift*, 29 (1946), 95–114.

64. Gerhard Tintner, "Multiple Regression for Systems of Equations," *Econometrica*, 14 (1946), 5–16.

65. Haavelmo, "Probability Approach" (note 42), was used as a text by those learning econometrics in the late 1940s and early 1950s along with the highly technical Cowles Commission Monographs 10 and 14. There were early texts by Tinbergen and Klein, but the main rash of textbooks concentrating on techniques began in the early 1960s [one of the most widely used in the 1960s and 1970s was J. Johnston, *Econometric Methods* (New York: McGraw-Hill, 1963)].

66. Tinbergen's testing methodology, along with other matters dealt with earlier in this chapter, are explored further in M. S. Morgan, *The History of Econometric Ideas* (Cambridge: Cambridge University Press, 1987).

IV PHYSIOLOGY

9 Experimental Physiology and Statistical Inference: The Therapeutic Trial in Nineteenth-Century Germany

William Coleman

The experimental physiologist of the nineteenth century both feared and refused the familiar notion that life means spontaneity. His goal was to bring vital actions within the reach of mechanistic science, to make physiology deterministic. This ambition expressed itself in diverse ways, ranging from impassioned polemics to serious efforts to pursue rigorous laboratory investigation of organic functions. From the seemingly hard facts thus obtained, seemingly necessarily conclusions were then to be drawn. The deterministic ideal was not so easily attained, however; physiologists had overlooked the problem of inference, and particularly the problems posed by statistical inference.

These issues became a matter for public discussion and dispute toward midcentury. The physicist and mathematician Gustav Radicke undertook to evaluate the reasoning behind a series of innovative experimental inquiries, most dealing with purportedly effective therapeutic measures. In so doing, he laid bare serious argumentative difficulties and at the same time offered the physiologist-physician a singular novelty, a simple significance test that might render reasoning more assured and conclusions more persuasive. Radicke rejected the physiologist's "numerical method" (confined to arithmetical means and their comparison) and proposed a new method, one based upon simultaneous appreciation of the differences between means and between standard deviations in two experimental populations. His opponents did not understand this approach and reasserted the crucial importance of seeking strictly deterministic conclusions. Their views long prevailed; Radicke's proposal won little favor. The serious use of statistical methods in experimental physiology and medicine only began with the introduction of new techniques after 1900.

Effective experimentation demands clear circumscription of the question to be asked and domination of the conditions that bear upon the phenomena under inspection. The experimentalist appreciates the close correspondence between conceptual and laboratory tools and understands that the accuracy of experimental data is strictly conditioned by the apparatus employed. He also knows that measurements made with such instruments will vary, often considerably. He must nonetheless use these data in drawing conclusions from his investigation. In the inferential process he shifts from the tools of the laboratory bench to those of the mind, among which statistics is poised to assume an important role. In the mid-nineteenth century, however, when physiology and experimental medicine first

I am grateful for research support provided by the Stiftung Volkswagenwerk, Hanover, and the Zentrum für interdisziplinäre Forschung of the University of Bielefeld during the year 1982–1983. To my colleagues at Zif go thanks for much advice and criticism; I am especially indebted to John Beatty, Gerd Gigerenzer, and Zeno Swijtink for their close review of this chapter in its various drafts. Personal thanks of a very special order I offer Lorenz Krüger, Michael Heidelberger, Rosemarie Reinwald, and L. Jegerlehner.

made contact with statistical reasoning, neither the science nor this tool was prepared for a cooperative and fruitful engagement.

In this early setting the domain of physiology was broad and still problematic.[1] The "physiologist" was trained as a physician and occasionally practiced medicine. He found his professional home in the medical faculty of the university or its equivalent. His primary pedagogical function was to instruct students of medicine in the functioning of the human body. By 1850 experimental, that is, rigorous laboratory investigation of vital processes, not least physical and chemical processes, had begun in earnest in a small number of university faculties, particularly in the German-speaking lands. Experimentation was not yet, however, the rule, and its triumph came slowly and variously, a function of social needs, national scientific and educational styles, and personal inclinations.

Among other concerns, physiologists paid considerable attention to the discovery and evaluation of medicinal substances and therapeutic procedures. Their investigations included clinical trials using self-experimentation and experimental animals.[2] One observes in this work both the beginnings of a systematic quest for new or more effective drugs and the emergence of serious problems in experimental design; one encounters, too, the nascent hope of applying statistical methods to physiological reasoning.

The statistical issue is a complex one and may be restated as a question: given the results of a therapeutic trial, how might these data be used to draw a persuasive conclusion? This is, of course, not one but several questions, and these questions, together with the answers that were ventured in response to them, constitute even now, as they constituted in the previous century, a set of closely interrelated and often vexing problems.[2] Physiological-therapeutic experiments, for example, usually involved only a very small number of subjects. What techniques were available for comparing the differences between an experimental and a nonexperimental population? Were the experimental data truly suited to the task of providing a solid foundation for reasoning? Could one trust one's experimental design?

Generally, the statistician's concern was to insist that the physiologist-physician recognize the role that chance might play in populations of limited size—hence the constant injunction to increase experimental population size. Only rarely was an effort made to devise a test whereby the mean values of two necessarily very small populations could be compared. A test of this kind is basic to all experimental work, and the following discussion presents the major features of one such proposal, a rare species indeed, from the midnineteenth century. Decisive contributions followed only later, and are best seen in the work of W. S. Gosset and R. A. Fisher after 1900.

1 Physiological Determinism

Biostatistics little participated in the development of nineteenth-century physiology, finding itself frustrated by its own limitations and by the enemies its central, probabilistic claim called forth. The most alarmed and most formidable of these

enemies were on the leading edge of physiology. Impatient with a merely observational and comparative approach and wholly rejecting the speculative methods of
the preceding generation, younger physiologists in the 1840s sought their salvation
in the perceived rigor of a fundamental science, mechanics, and in the hope that
physiology would now develop roots in secure mechanistic soil. This step, it
appeared, would render physiology deterministic and therefore truly scientific.
Determinism in its mechanistic form had become the criterion of science—and it
posed a difficult hurdle for those who insisted that some conclusions could be
expressed in only in statistical form.

Discussion of statistics within the nineteenth-century physiological and medical
communities turned principally upon the merits or shortcomings of method. While
some thought statistics to be helpful or even necessary, others deemed it irrelevant,
deceptive, or false, and perhaps all three. Worst of all, the medical practitioner, the
supposed beneficiary of the progress made in experimental science, could not easily
appreciate the advantage of the new form of reasoning. It seemed scarcely possible
that mere calculation could replace or improve one's fund of hard-won experience.
The physician deplored the fact that the primary effect of statistical analysis
appeared to be to divert attention from the matter of greatest medical concern, the
individual case in all its particularities. Statistical reasoning seemed destined to
separate the physician and his client.

These issues in the articulation of physiological method were rarely addressed
during the nineteenth century. Observers in the 1850s duly noted the problem of
statistical inference, and the need for a usable and valid method or methods was
also recognized. A satisfactory solution, however, was not found. Some physiologists simply rejected the notion of statistical inference, virtually on first principles.
A few other investigators, however, found themselves caught up in an earnest
debate concerning the use of probabilistic reasoning in physiology; in fact, their
own scientific contributions were subjected to a numerical test that left little
standing of their efforts.

The mathematician Gustav Radicke posed this challenge. In 1858 he applied a
numerical test to the results of experimental work in physiology and of therapeutic
trials. Radicke sensed that physicians and physiologists would not be interested in
a new denunciation of the practice of drawing scientific conclusions from a mere
difference in the means of two observed or experimental populations. The latter
practice was the defining feature of the so-called numerical method. Radicke's task
was to provide a practicable and sound procedure, one to be used in frequently
encountered research situations. But, in effect, he proposed more; his procedure
demanded that some and perhaps many questions in physiology were to be
answered, if answered at all, in a probabilistic manner. If so, the physiologists'
deterministic ambitions and the physicians' responsibility for seeking an efficacious
course of action in the highly individualized domain of human disease and its treatment, a no less awkward and much more vital matter, would both be undermined.

Radicke met the experimentalists at an important turning point in the history of
medicine. In contrast to the extraordinary progress in diagnostics made during the
first third of the nineteenth century, the physician's ability to heal victims of the

diseases that he could now so splendidly circumscribe had advanced little or not at all. By the 1840s the medical profession gave dramatic vent to its frustrations, and there ensued a long period of therapeutic skepticism—indeed, on the part of some, of outright nihilism.[3] For those who did not capitulate to despair, one road to salvation seemed to be to seek new or renewed means of discovering new remedies and of testing their efficacy. This proved no easy task, and Radicke made very clear that more was at stake than the quest for new substances or even the discovery of the actual physiochemical mechanisms that were involved in therapeutic effectiveness. His criticism helped circumscribe an important problem in experimental medicine, and this was quickly recognized, even if the needed improvements could not be or were not implemented for another two generations.

My discussion begins with a description of responses to the challenge of therapeutic skepticism. Some observers thought statistical reasoning might offer a useful complement to experimental work; to others, statistics meant uncertainty, the failure of determinism and hence loss of scientific legitimacy. Radicke focused his attention on the needs, as he understood them, and applications of correct statistical reasoning. I therefore turn next to his consideration of the problem of inference, the character of his proposed test, and the application of the test to a series of therapeutic investigations. There follows consideration of varied reactions to his proposals and to Radicke's response to his critics.

2 Escape from Skepticism

The therapeutic issue had become a matter for methodological consideration largely due to the argument and conclusions of the energetic French clinician P. C. A. Louis. Louis's numerical method was introduced in order to evaluate the efficacy of various traditional healing measures.[4] The method consisted of collecting numerous comparable instances of therapeutic efficacy and nonefficacy, calculating the mean value for each group, and then directly comparing these means. A difference signified an effect, and a large difference a large effect. Although the method never moved beyond this elementary and untrustworthy procedure, it long continued in use in medicine.

There were other ways, however, to seek assurance in therapeutic matters. One such approach demanded rigorous investigation of the physiological and pathological condition of the organism. Rigor meant especially experiment, and experiment entailed not only suitable design of a given procedure and careful manipulation of relevant experimental conditions but also critical and logically sound reasoning upon the data generated by the experiment. This was the program of a new "physiological" or "rational" medicine.

Carl August Wunderlich, for example, was a major figure in this movement, and his *Archiv für physiologische Heilkunde* supported a veritable campaign for physiological, that is, scientific, healing. In today's medicine, he observed in 1842, some physicians "believe," others "know," and to still others it is "apparent" that a given therapeutic measure is effective. The tragedy, however, was that none of these physicians really *knew* anything of the sort; no one possessed truly secure grounds for his conviction.[5] "The time has come," Wunderlich proclaimed, "when one must seek *to establish . . . a positive science.* This science will seek its roots not in the authorities but in the foundations and empirical evidence that the phenomena permit us to comprehend. This science will also arm us against the illusions of [medical] practice and is sure to lead us to a rational and secure therapy."[6] Wunderlich's dreams were essentially the same as those put forward by Jacob Henle and Carl von Pfeufer, and somewhat later endorsed by Rudolf Virchow.

The varied programs of physiological medicine were designed to lead medicine out of the well-worked but fruitless fields of empiricism and speculation, the latter being the physician's inveterate vice and the former his eternal point de repère, and to introduce the ancient art to the potentially abundant harvest to be won from science. Wunderlich at first did not view statistics with great enthusiasm, beginning his career as a medical publicist with a categorical rejection of its procedures and conclusions. By the 1850s, however, the *Archiv für physiologische Heilkunde* contained numerous articles of therapeutic interest that used a form of the numerical method. It was here, also, that Radicke's analysis was published. Wunderlich himself had been quickly forced to admit that, if "mathematical certainty" is indeed attainable in the single instance, in most other medical matters one must remain content with only a "reckoning of probability," a *Probabilitätsrechnung* to be "conducted according to rigorous rules that take into consideration all available factors."[7] But the truth is that Wunderlich and like-minded physicians tolerated statistics, even in therapeutic investigations, because they had few choices. With the chemistry and physics of drug action still hopelessly obscure and unsystematic empiricism or imaginative theorizing carefully excluded, medical statistics simply offered another tool. Unfortunately, it seemed an incomplete tool and one subject to dangerous misuse.

Another group, the masters of the deterministic outlook, either paid little attention to therapeutic efficacy or even medical needs in general (thus Emil du Bois-Reymond) or insisted with unyielding fierceness on the priority of experimental mastery of physiological conditions and the fact that radical determinism excludes any notion of probabilism, even in the important realm of therapeutics (thus Claude Bernard). Du Bois-Reymond's mechanistic reductionism was intimately connected to his view of the role of experiment is physiology; it also informed his notion of the proper use of number in physiological inquiry. In essence, he asserted that the uncertainty of statistics excluded probabilistic reasoning from scientific inquiry. An experiment must be so designed, he announced, that it returns a result that is "clear, simple, certain, and constant in its effects," a result that can be reproduced at any time or place.[8] This claim rested, of course, upon the presumption of the absolute regularity of natural processes. It assumed, too, that the experimentalist could dominate all factors pertaining to a given event.

Mechanics thus became the proper language of physiology, the goal of the latter being to express active organic relationships in terms of a definite mathematical function. Du Bois-Reymond profoundly distrusted those numerical tables that presented, for example, the relation between stimuli and responses in an "abstract" and, it may be added, indeterminate lifeless manner.[9] The numerical table, the favored instrument of the early medical statisticians, portrayed the relationship between aggregates of data; the mathematical function, rigorous and unambiguous, a statement of the relationship between any given stimulus and any given response, offered the only suitable form of expression for truly scientific physiology.

The French physiologist, Bernard, also devoted great attention to methodological matters. He, too, rejected statistical reasoning.[10] He did so because statistics contradicted the deterministic stance the physiologist must assume if he wished to render his science a causal science. Bernard articulated a physiological but resolutely nonreductionist determinism. His view differed markedly from the mechanistic reductionism of du Bois-Reymond and the heated programmatics of other German physiologists. Nonetheless, these dissimilar determinisms had a common consequence; they expressed distrust of a form of reasoning that returned only probable answers.

By midnineteenth century the statistical approach to physiological and therapeutic problems had thus acquired both hesitant advocates and unyielding opponents. The utility of statistical argumentation remained uncertain amidst various confused efforts to escape therapeutic skepticism, and it was flatly denied by advocates of experimental, deterministic physiology. There was, too, a further problem. Perhaps the source of difficulty was not the legitimacy of reasoning statistically in physiology but the more technical matter of designing an appropriate, physiological statistical tool.

3 Radicke's Test

Radicke applied his test principally to the results of a healing procedure that had recently moved from the periphery to a central position in medical and popular consciousness. Radicke assayed especially the efficacy of the supposed therapeutic value of exposure to the air and waters of the sea. The seaside bath as a medical measure was an English invention of the late eighteenth century; the first oceanic *Kurort* in Germany was established in 1797, and other stations followed in due course. Inland salt baths also rose rapidly in popularity. By 1850 these methods were well established throughout central Europe.[11]

This form of the water cure, whose roots go back to classical antiquity, assumed all the trappings of a panacea. In this world, fraud and charlatanism seem to have been as common as simple ignorance or incompetence. Many physicians spurned the practice altogether; others, all too cognizant of the fact that a responsible physician had no right to reject offhand any reasonable prospect of a cure, realized that the proper course to follow was to evaluate the therapeutic efficacy of the so-called healing waters. Just here lay the problem—and opportunity. Radicke reviewed the usefulness of statistical reasoning in physiology by assessing a veritable deluge of numbers, these being the bounty of the *Badeärtze* themselves.

A novel feature of this work was the quantitative techniques used to establish the chemical constituents and volume of the urine. These techniques were a product of the same period and drew their inspiration from Justus von Liebig and his followers, to whom it seemed possible to infer internal bodily condition from externally sensible materials.[12] Metabolic processes were protean and their products were a sign of the abnormal as well as the normal. Urine analysis thus seemed a sensitive as well as significant indicator of generalized vital chemical operations, an indication that perhaps an important change had been induced by, for example, some therapeutic procedure.

Friedrich Wilhelm Boecker, *Kreisphysikus* in Bonn, brought these matters to Radicke's attention. Exploring the effect of a common medicinal substance, Boecker examined the effect of sasparilla on the nitrogenous and other constituents of the urine. An individual receiving a controlled diet was given a decoction of sasparilla for a period of twelve days, and the volume of urine passed daily was carefully measured. For a further twelve days that same individual, on the same diet, was given only distilled water, and the daily quantity of urine was again determined. The first series of researches gave the following figures (in cubic centimeters): 1,467, 1,744, 1,665, 1,220, 1,161, 1,369, 1,675, 2,199, 887, 1,634, 943, and 2,093 (mean = 1,499); the second series: 1,263, 1,740, 1,538, 1,526, 1,387, 1,422, 1,754, 1,320, 1,809, 2,139, 1,574, and 1,114 (mean = 1,549).[13] Much uncertainty surrounded the exactitude of these measurements, but this played little role in the ensuing discussion. The fundamental issue was not the quality of the experimental data but how inferences were drawn from those data.

Gustav Radicke (1810–1883), a physicist and mathematician, was in 1858 *ausserordentlicher Professor* of physics and meteorology at the University of Bonn.[14] He had published a major handbook of physical optics and a textbook of arithmetic and analysis, as well as a small number of papers dealing with both subjects. His plunge into the world of the therapeutic trial was not out of character; he understood physiology well. His general approach to the question was also not unusual. Writing for the practitioner and the academic physician, Radicke spoke of familiar matters. "It is well known," he observed, that the procedure

usually employed in ascertaining the influence a given agent, such as nutritive or medicinal substances, baths and the like, exercises upon the metabolism [of a body] is the following. One subjects an individual for a certain number of days to the influence of the agent in question, holding other aspects of the individual's manner of living as uniform as possible. The daily excretions are assayed, principally but not exclusively the urine, whose volume and the quantity of whose most important chemical substances are determined. The data obtained are then compared with the results gained in a second series of researches, which may precede or follow the first series and during which the same mode of life is observed, except for the exclusion from the second series of the agent whose influence is being investigated.[15]

These researches were designed to permit comparison of two trial series, one an experimental series and the other, to employ a term introduced only later, a control series. After noting the many problems associated with the design and execution of such experiments, Radicke stated his major contention:

The comparison is, however, usually limited to placing the arithmetical mean obtained from the numbers presented by the first series of experiments alongside the mean given by the numbers of the second series, and then to concluding in so far as the first mean is larger or smaller than the second, that the agent in question has either increased or diminished the excretion being observed. ... The defect of this procedure lies in the fact that those who employ it do not clearly comprehend the true significance and value of arithmetic means and have, as a consequence, attributed to them a value which in applications such as these they do not actually possess.[16]

The defect proved a challenge, and this challenge provoked a response. Radicke sought to introduce into physiology a new way to determine the meaning of differences.

He realized that the basis of the therapeutic trial was the isolation and differential treatment of two research groups. He understood, too, that conventional medical statistics began by calculating the arithmetical mean of the results of each set of trials. Each member of a pair of such means was then directly compared with the other. Now, what did a difference between these means signify, if anything? Was such use of arithmetical means alone sufficient, or was additional information necessary to ensure a useful comparison? And what criteria might be employed to discriminate between a "reliable" and an "unreliable" conclusion?

First of all, Radicke noted, the arithmetical mean offered only one element in the numerical circumscription of a research series. It was a "pure average" and, although easily obtained, was too commonly used alone to effect comparisons. The important point was that an arithmetical mean concealed much regarding the population it purported to describe. Two experimental populations could possess identical arithmetical means yet differ greatly in the distribution of the values of the individual elements whose sum contributed to the calculation of the two means. It was the fact of dispersion that Radicke particularly wanted to put before his readers and whose relevance to statistical reasoning determined his notion of how experimental physiological reasoning should proceed.

The physiologist, clinician, or therapeutic investigator needed a standard that simultaneously measured the central tendency and the dispersion of two research populations. The test using this standard had to be simple and easy to apply, and it would have to generate results that lent themselves to meaningful physiological interpretation.

Two steps were involved in applying Radicke's test. The first required estimation of the stability of the mean of a series of numbers (observations), the second the calculation of the likelihood that a difference between two such means was or was not due to chance. When dealing, for example, with a "complete" series of observations—really, a very long series of observations—Radicke held that his method of "successive means" (a procedure requiring the calculation of successive arithmetical means and the search for a predetermined decimal point about which these means stabilized) provided assurance that "*the most exact result*," namely, the value of the mean in question, had been obtained.[17]

In the second step, two such means were obtained, one taken from the experimental group and the other from controls, and for each a second statistic, called the mean error, was prepared. The latter provided the required index of dispersion and was calculated by taking the square root of the mean of the sum of the squares of the differences between each observed value and the arithmetical mean. This mean error (*mittlerer Fehler*), or, as it was also called, mean fluctuation (*mittlere Schwankung*), gave a measure of how far on average the observed values of a given series might be expected to depart from the mean of that series.[18] A comparison was then made on the basis of the following argument.

Each mean error offered a measure of the distribution of observed values around the computed mean value. It was an index of the uncertainty of that mean value and a warning not to accept the arithmetical mean as the unique, summary statement of the results of a given trial. The needs of a useful test were, therefore, twofold: a measure of the uncertainty of the principal values being compared (the two means) and terms that ensured that these means, if they were to be accepted as indicating an effect of, say, a physiologically active substance and not simply the play of chance factors of observation or manipulation, stood sufficiently far apart from one another. Radicke proposed, and this was his test, that if the difference of the two arithmetical means exceeds the sum of the two mean errors, one may conclude that the result obtained is probably correct, that is, an effect had probably been produced. If, however, the sum of the mean errors exceeds the mean difference, the experiment must be held to have generated no useful conclusions. "If, in a given experimental situation, we are to conclude that an effect has been produced, the difference between the means ... must be greater than double the largest errors of observation; that is, it must be *greater than the sum of the uncertainties* of these means."[19]

The above argument rested upon the assumption that a complete or long series of observations was commonly available in scientific research. Radicke knew, however, that this was almost never the case in physiology. Physiologists were often satisfied, or were forced to be satisfied, with very short series of data, ranging from perhaps eight to a dozen observations down to the individual case. This fact posed a serious problem for test procedure, a problem that was overcome with what was admitted to be a "makeshift."[20]

Radicke proposed that, when dealing with an "incomplete," that is, short series, of data, one should calculate successive means by moving both downward and upward. If greater fluctuations were not observed in the second half of the series, where their occurrence would suggest either divergence or continued instability, one might conclude that an "incomplete" series was, for practical purposes, "complete."[21] This conclusion then allowed one to proceed directly to application of the full test described above, the comparison of means and summed mean errors. Radicke's procedure was, of course, wholly ad hoc, did not successfully evade the small-sample problem, and was open to various and serious misinterpretation; one critic supposed that the entire operation had been devoted to obtaining, even manufacturing, more exact numbers, not to evaluating the possible meaning of those numbers (see below). In any case, Radicke's purpose was to provide experimentalists, especially physiological experimentalists, with a practicable test; he had no choice but to seek a rough-and-ready escape from the problem of short and often widely fluctuating runs of data. But how was this test applied? To what conclusions did it lead? Boecker's own researches on the excretory effect of sasparilla provide a good example of the use of the test. In question was the diuretic power of sasparilla, measured by the volume and chemical constitution of the urine excreted. Boecker's two series had given means of 1,499 cc (cubic centimeters) of urine voided (sasparilla administered) and 1,549 cc of urine voided (no sasparilla administered); the mean difference was thus 50. The mean error of the experimental trial was 386, that for the control 268; the sum of the mean errors was 654 (and not 674 as stated by Radicke). The sum of the mean errors thus far exceeded the mean difference and only a "negative" conclusion could be drawn, meaning, an effect had not been demonstrated.[22]

From earlier researches by Boecker, Radicke drew a further example. What effect did sugar exert upon the urine's constituents? Here Boecker's data series were very short (8 observations in two experimental series and 5 in the control series), but still the computations pointed to a positive result: experimental mean error = 0.15 mg (milligrams), control mean error = 0.68 mg, sum of mean errors = 0.83 mg; difference between the means of the two series (experimental: 0.93 mg, control: 2.06 mg) = 1.13 mg. It seemed, therefore, that "one might be permitted to conclude" that a "probable diminution" of phosphate excretion had resulted from the sugar diet.[23]

Radicke had much confidence in other physiologists whose work he examined. Friedrich Wilhelm Beneke, for example, a leading *Badeartz*, had attempted by self-experimentation to determine the effects of sea bathing and of sea air on urinary excretion.[24] Radicke exhibited the shambles that these researches in fact were and indicated that any conclusions drawn therefrom were without value. In this case it seemed impossible or, better, meaningless to attempt to apply the test. Beneke's experimental series were very short, usually offering four or fewer usable data, and consistent experimental conditions had not been maintained (diet often varied greatly; experiments were conducted during different seasons of the year and in different locations, Oldenburg and Wangerooge). Furthermore, observed values ranged widely within the series, and the chemical analysis being employed entailed (uncorrected) variations of some 10–20%.[25]

In handing Radicke Beneke's publication, Boecker had observed that, in his opinion, "not one of the results contained therein appeared to be well founded."[26] Radicke clearly agreed. He also sought test cases elsewhere. He explored, for example, an investigation of the connection between dietary sodium chloride and the elemental composition of urinary excretion. Wilhelm Kaupp, a medical student at Tübingen working with Carl Vierordt, had seized upon Liebig's new method for exhaustive chemical analysis of the urine and produced a stupendous fund of data, a veritable monument to the contemporary god, hard facts. Even if all these numbers were actually pertinent to physiological inference, they still allowed no conclusions to be drawn. Radicke applied his test to Kaupp's several experimental series and found that "the differences between two mean numbers is in no case greater than the sum of their uncertainties." Furthermore, experimental conditions had not been constant. Temperature had varied significantly, and it was impossible to allocate the relative roles of temperature and sodium chloride intake in influencing the chemical composition of the urine.[27] Questionable experimental design undercut the possibility of attempting to apply numerical reasoning to experimental results.

It is obvious that Radicke had found that contemporary therapeutic trials were being conducted without necessary attention to both experimental design and the needs of sound inference. He did not, as R. A. Fisher would do some seventy years later, indicate the crucial dependence of the latter upon the former.[28] Had a reader been looking, and there is no evidence that any was, this general lesson could have been easily drawn from the course of Radicke's discussion. But exploration of the requirements of sound experimental design was not his purpose, and he was content simply to record indirectly the manifold shortcomings he had discerned in contemporary experimental practice. There was perhaps novelty enough in the idea that physiological inference must proceed according to a set of binding numerical rules; this point, surely, had to be made before it would pay anyone to devise complex experimental procedures whose plan and rigor were intended to assure the production of data that could in fact be properly employed in one or another statistical test.

4 Vierordt and Beneke Reply

The physiological profession had not anticipated such criticism. Radicke knew a brief moment of celebrity, but, after a flurry of discussion lasting into the 1860s, his name as well as his test simply disappear from the literature. Those whom one might have expectd to develop the argument, the physician-mathematicians who in the following decades continued to ponder the application of statistics to medical investigation, appear to have paid Radicke no heed. It is not, therefore, the latter's influence that is of principal interest; rather, it is the value of his views in permitting a clearer specification of the problem-domain of medical statistics in the era following introduction of the numerical method and of the difficulties that faced the physiological-therapeutic investigator when he sought to draw conclusions from the perplexing results of his experimental studies.

Vierordt established the basic grounds for assaying Radicke's proposals. Vierordt's notion that physiology might justly place full faith in a special "logic of facts," even when faced by seemingly compelling mathematical arguments, pleased fellow investigators. He also encouraged discussion of medical statistics in the *Archiv für physiologische Heilkunde*, of which he was editor. Medical statistics, he argued, could take three forms.[29] First of all, it could mean that a conclusion based upon a large number of observations is likely to be more reliable than one founded upon a few, a useful if perhaps obvious insight. Second, medical statistics could be guided by the calculus of probabilities. Hereby one gained a notion of the limits of error of one's observations and of the limits, too, of the conclusions that might be drawn from these observations. While this criterion was "only a *purely formal one*," it was nonetheless "valuable and important." It was the third form of reasoning, however, that disclosed Vierordt's serious doubts regarding the competence of statistics in matters physiological.

Apply whatever methods you wish, Vierordt advised, but remember, "In addition to the formal consideration of the numbers of any investigation, there may be other and independent reasons, arising from the *nature of the object itself*, which may and even must be taken into consideration when drawing a conclusion." These "independent reasons" constitute the very essence of a physiologist's observations and experience. They are the physiologist's body of expectation, his anticipations and accumulated practical knowledge regarding the working of the organism. Ideally, of course, the formal or numerical approach allowed one to consider vital function in an "unprejudiced" manner. The problem, however, was that this ideal situation rarely obtained and, more important in Vierordt's eyes, the "independent" or "intrinsic reasons" with which a physiologist deals are "not amenable to numerical expression." [30]

Such "intrinsic reasons," Vierordt argued, must always temper conclusions based upon numbers alone. In some cases, detailed knowledge of the physiological conditions under which certain results were obtained (this technical knowledge was precisely what was meant by "intrinsic reasons") increased the certainty of an already secure "formal," that is, statistical conclusion. In other cases a weak statistical conclusion might be confirmed by an understanding of the intimate physiological operations involved. But there were cases, and Vierordt implied that such cases had been the object of Radicke's criticism, wherein a "*formally accurate* numerical result [taken on its own terms] may be in no way incorrect and yet, given the circumstances defined by [the] intrinsic reasons [of the case, that result will be] recognized immediately by experts as false." [31] In short, physiological understanding based upon varied and direct experience with the phenomena in question retained the power of contradicting the seeming constraints imposed by the "purely formal" approach of the mathematician.

Vierordt by no means sought to exclude statistics from physiology and medicine. When properly applied, its value was very great. Such applications, however, were uncommon, and it was important, too, to note that physiological insight possessed an approach uniquely its own, expressed in the notion of a logic of facts: "In addition to the purely formal and mathematically convincing logic of the calculus of probabilities, there is *in many cases* a logic of facts which, when applied in the proper manner and in the proper place (that is, to questions that are not too complex), carries with it, for the man who is acquainted with his subject, a small or even a very high degree of conviction." [32] It seemed to Vierordt that Radicke had ignored this fund of experience and thus focused too exclusively upon the rules of mathematical reasoning, even suggesting constraints upon the physiologist's accustomed way of drawing general conclusions.

Benecke, whose ox had been most gored, added a new cascade of confusions to these remarks. To mathematics itself he announced no opposition; the physiologist, he granted, can always use greater precision in his studies. But what kind of precision? The working capacity of a steam engine, Beneke proposed, himself boarding quite the wrong train, can be assayed "accurately enough"; for this purpose we obtain with no great difficulty "infallible numbers." The animal, however, is no such engine. At every moment "influences" work upon it and elicit varied responses.[33] Our task as physiologists is to capture these responses, and for this task our essential, indeed our unique, tool is experimental manipulation.

It followed, or so Beneke maintained, that it is more important "to estimate these influences correctly, that is, to provide a physiological test of the experiment, than it is to apply the keen analysis of the mathematician to the numbers that are obtained. It is, I believe, a mistake to expect to make the results obtained from previous experiments on the human organism much *more certain* by means of mathematical examination of the numbers obtained."[34] Obviously, Beneke had seized only one and the lesser of Radicke's points; statistical inference was in question, not enhanced experimental data.

Beneke either misunderstood Radicke's purpose or willfully misread it, for he insistently viewed his critic's words as bearing upon accuracy of observation and not upon inference based upon such observations. This gave him an admirable if misdirected argumentative opening, permitting him to agree that, of course, one should seek to render observation more accurate. This had nothing to do with the calculus of probabilities, yet it had everything to do with that "logic of the facts themselves," announced by Vierordt and eagerly extended by Beneke.

This special logic seemed to involve not only the question of the exactitude of the facts derived from an experiment but the more problematic question of whether such facts literally spoke for themselves. With regard to Radicke's evaluation of Boecker's experimental results, Beneke insisted that, whatever the numbers might announce, real trust could be placed only in the experience of an active physiologist. It was obvious to Beneke that "the first opinion as to the accuracy and utility of physiologicochemical investigations ... should continue to emanate from the physiologist or physician who will know how to estimate how far the way in which the investigation is conducted, the numbers obtained, the conclusions drawn and, still more, the ruling ideas and general cast of mind of the experimenter, are deserving of confidence and correspond to our previous knowledge."[35] Only after such professional opinion had been carefully formulated might the mathematician be "allowed to intervene."

Beneke was worried that in many experimental situations the sought-after "influence" could be very slight; in such cases, mathematical manipulation would mean that we should "probably seldom come to any conclusion at all." Furthermore,

The numbers in which the results of our researches are expressed carry with them intrinsically more or less confidence in proportion as they agree or disagree with general physiological or medical experience. If such agreement be wanting every experimenter and each critic will justly feel great doubt before the results obtained. They will insist upon repeating the experiments, and not even the most exact mathematical manipulation of the numbers put before them will given them a feeling of certainty. If, however, an agreement with well-known physiological facts is evident, there will be less doubt. ... I recognize with gratitude the instructiveness of Professor Radicke's labors on several points and have not perused them without advantage. Despite this, I maintain that the physiological tact (*Umsicht*) and knowledge of the experimenter seem to me more important than the mathematical [skills] that have been recommended. It is only the greatest possible repetition of observations on different individuals and under different circumstances that can render our results positive. At very least, I cannot consider it a misfortune if we retain provisionally our old method of determining mean values, provided that we do not place too great weight upon single instances, recognize the amount of uncertainty [in the data and means] that generally exists, and make the number of observations long and decided enough both for the question under investigation and for the experimenter himself.[36]

Again, Beneke had misconstrued Radicke's objective. The mention of "certainty" and "uncertainty," "mean values," and the like refer to control of the accuracy of the data themselves. Beneke simply did not appreciate that Radicke, who was well aware of this problem, had devoted his attention to a related but distinct matter, namely, articulation of a ready test by which to decide whether an effect had been produced by experimental intervention.

5   The Dubious Logic of Facts

This dispute reached well beyond disagreement over the proper approach to a therapeutic trial. The Radicke-Beneke contest is, in fact, best located within a more general philosophical debate in Germany.[37] The 1850s constituted a period of extreme intellectual conflict, a residuum of the uncertainties left by the collapse of the Hegelian synthesis and the retreat into empiricism that followed a generation of metaphysical excess in the medical and other sciences. The often harsh tone of these disputes and also much of their substance were set by the failed revolution of 1848–1849 and by renewed efforts to reestablish the dual authority of church and crown.

Within this framework the expression "logic of facts" assumed an important role. That facts might speak for themselves was a comforting thought indeed to the generation that followed the *Naturphilosophen* and other medical and scientific system builders. A generalization might err, yet the damage, the empiricist rigorist held, was not serious; the facts behind that generalization remained irrefutable contributions to the progress of science. "There exist," an informed observer declared, "certain and indubitable facts in nature, and these we are able to establish by means of experimental induction."[38] This was the era of nascent positivism, and members of the medical community were by no means the least enthusiastic in embracing this seeming escape from the whirlwind of rival doctrines.

But the fact-mongers in turn met their critics. Liebig, for example, declaimed that "empirical scientific inquiry in the usual [crude Baconian] sense simply does not exist. An experiment that proceeds without theory, that is, without a previously existing idea, stands in the same relation to scientific investigation as the clatter of a child's rattle stands to music."[39] He charged Bacon with blind and blinding empiricism; Bacon had banished active thought from scientific inquiry. Friedrich

Albert Lange, at the time engaged in preparing his celebrated *Geschichte des Materialismus*, rejected what he regarded as the excessive idealistic implications of Liebig's position, but also disallowed the minimalist claims of persons whom he aptly labeled "rigorists of exact scientific research." The latter seemed especially prominent in physiology, and this fact led Lange to proclaim the Radicke-Beneke exchange the "most notable example . . . in recent years in Germany" of an almost know-nothing empiricism. The rigorist's intrinsic flaw, Lange observed, lay in his "overestimation of immediate sense impressions" and excessive fear of even the most careful exercise of reason.[40]

Another philosophical observer, Friedrich Ueberweg, then at the outset of a distinguished career, entered the discussion at the request of his Bonn friend, Boecker. To the advocates of a special logic of facts Ueberweg addressed the reminder that the notion of logic is applicable only to a thinking subject.[41] Facts, whatever else they might be, are not a thinking subject. The human mind has no access to an immanent logic of nature, a logic that express itself directly through mute facts. A scientific contest is really based upon differing views of available facts, and these views vary depending upon how we each arrange our thoughts and compose our arguments. Logic is the set of rules designed to guide such a course of reasoning.

Ueberweg found it necessary to emphasize these elementary points because of the implied charge by the empiricists that adherence to facts was salutary. It prevented a plunge into subjective fancies, whereas logic was merely an elegant license for unrestrained speculation. He complained that the empiricist reduced the mind to utter passivity. The active intellect was allowed no greater role in scientific inquiry than were dreams. But natural science, Ueberweg insisted, can arise only from the interaction of sensory impressions and fully alert intellectual faculties. He did not deny the highly suggestive role of the bare fact, that is, the importance of one or several facts in stimulating what was essentially prescientific cogitation and even conclusion. But this was a matter of tact and one for psychologists to explore, not a matter that bore upon the articulation of rigorous scientific conclusions. Ueberweg respected the psychologists' notion of tact and granted that tact could often provide the first approximation to an answer or an insight into an otherwise impenetrable problem. What tact could not do, however, was assume the form of consequential rational development of the relation between observation and conclusion.

Now *Takt* was the core concept of the so-called logic of facts, and it was also and had long been one of the most highly esteemed qualities of the skilled physician. Tact described a particular form of psychological behavior. The philosopher-psychologist Friedrich Eduard Beneke observed that often a series of impressions presents itself to us in rapid succession.[42] These impressions are confused, and some are contradictory; we do not cope well with their rapid delivery. In such cases, only the last member of the series stands out in our minds, but its character is no doubt formed by our unconscious reaction to previous members. In ordinary usage we call this hidden mental activity tact. Tact, too, was an aesthetic response, a form of insight, perhaps, but only the starting point for further development.

In social intercourse, Ueberweg noted, tact leads us to form an estimate of character; in the arts it is the basis for instinctive sureness of performance; in medicine it constitutes the essential act of diagnosis. "One surrenders oneself to the impression worked by the observed facts and generalizes using professional standards as far as possible, yet nonetheless always [relies] more or less upon good luck. Not uncommonly, one reaches one's goal—and often enough one misses it, too. "This," urged Ueberweg, "is the phychological reality, one that is more yearned for than realized; it is that which one decorates with the proud name of logic of facts." [43]

The physiologist Friedrich Wilhelm Beneke had cast "physiological tact" and the special "knowledge of the experimenter" at his critic, Radicke. Ueberweg's discussion dealt with this challenge; he was respectful but uncompromising. Physicians, he knew, had need not only to make constant use of tact, of conclusions drawn from hastily observed symptoms, but were keen to use this professional skill to assert the distinctiveness of their practice and defend the autonomy of their profession. Medicine is, because it must be, individualistic, its critical tool is observation; its method, which is a virtual program, is inflexible empiricism. The physiologist such as Vierordt or Beneke, just like the clinician, found in tact, in the immediacy of experience, his reason, his argument, indeed his logic that set him apart from other investigators. It was to this argument that the empirically inclined physiologist turned in order to defend the cognitive rights of his discipline. Yet, as Ueberweg pointed out, tact could not be transformed into a new logic. There existed but one logic, and it was not a logic of facts. Ueberweg thus welcomed the use, whenever appropriate, of probabilistic reasoning. Dispensing with his customary politeness, he concluded by addressing a hard truth to the physiological and medical community. When it comes to matters of number, the physiologist must either master mathematics or appeal to the mathematician. Failing this, his effort is merely the "naturalism of a dilettante." It is the result, in fact, of willful ignorance. Citing Paul lecturing the Corinthians, Ueberweg let the physiologist know that, once a "precise method" had been obtained, failure to use it read one out of the scientific camp: "Da ich ein Mann ward ... that ich ab, was dem Kinde geziemte." [44]

6   Radicke Responds

In his reply to the physiologists' complaints, Radicke's concerns and argument remained largely unchanged. Due to "shyness of mathematical deduction," not to say ignorance or incompetence, Radicke noted of his opponents that they attended to numerous and often tangential matters but failed to address substantive criticism to the test itself.[45] The latter thus passed unchallenged, at least by members of the physiological community.

Here was an invitation to set forth anew the nature of the test. Radicke now attempted to devise and denominate a set of standardized levels of statistical meaningfulness, or, to use the term that has subsequently dominated usage, significance. He was particularly careful to insist that, since the objective was to make a comparison between two series of numbers, the principal task was to discover a method that would ensure that the comparison was made in a scientifically meaningful manner. This method did not require a knowledge of causes, physiological influences, or "intrinsic conditions," and tact had nothing to do with the matter. Skill and care in gathering experimental data were indeed important, but they were not part of the process of inference itself: "All that it is necessary to know in laying down a rule for the comparison of two series of observations is—what we may learn from the series themselves, viz., whether fluctuations exist in those series, and if so, how great those fluctuations are."[46] Statistical inference was an independent undertaking, a carefully constructed tool available to all the sciences.

Statistical inference was not, however, a matter of mathematics alone; it involved a "certain arbitrary element." A criterion, one necessarily imposed by the investigator, was needed for determining whether it was proper to infer that an effect had been produced. This criterion was to reflect the presumed canons of human expectation and conviction. In stating these significance criteria Radicke first indicated that in no case was it possible to attain "*absolute certainty*." The reason for this was that an investigator, a physiologist, for example, would never be able to control the totality of relevant instances, that is, work directly with all members of a population. All inference, therefore, involved a "degree of *probability*," and the physiologist's task was to determine just what this degree was and, the related step, decide whether as reasonable men we should accept such a degree of probability as a demonstration or refutation of the point in question. There followed the cardinal question in statistical inference: "What is the *minimum* mean difference, in comparison with the fluctuations, that justifies our concluding, with sufficient certainty, that the greater amount of the mean in series A [as compared with the mean of a control series, B] is due to the operation of the agency under investigation?"[47]

But the expression "sufficient certainty" was itself "undefinable"; it had not been produced by the numerical operations themselves. Radicke therefore imposed a set of conservative standards. Other writers, he granted, might favor other criteria. Radicke's remarks permit a scale of degrees of significance to be constructed. At the top of this scale is the wholly imaginary realm of "absolute certainty." Physiology, of course, faced only uncertainty, and for this was offered the basic measure stated in Radicke's test, namely, the relation between the difference between the means of two experimental groups and the sum of their mean errors. If the mean difference exceeded the summed mean errors, then a definite reference point, called "sufficient certainty," was attained. Radicke indicated that this meant that an investigator was warranted in postulating that the agent under investigation probably had had an effect. Suggestive but less persuasive instances could also arise. Here the results were stated to be "*conditionally reliable*." In this case, a mean difference less than the sum of the mean errors but greater than one-half that sum was allowed.[48] A rude scale of significance was thus constructed; numerical values had been attached to seemingly vague verbal expressions.

Radicke's verbalisms continued unsettled. He used certainty, uncertainty, probability, and reliability almost interchangeably. In truth, the operative words were not nouns but adjectives: absolute, sufficient, and conditional. One adjective that was not used, although the concept was crucially present, was "insufficient." Addressing a rebuke to yet another challenger, Louis Lehmann of Bad Oeynhausen, Radicke pointed out that the "meaning of my rules is *not* ... that the observations are to be thrown aside if they give no positive answer to the question, whether the agent produces any effect or not, but that, in such a case, the efficacy of the agent cannot be considered as sufficiently established. In this situation a potentiality will undoubtedly possess a certain degree of probability; and it would be well to catalogue the result under the head of 'probable, but as yet not satisfactorily established,' so as to keep the completion of certainty on the point open until similar results have been obtained from later researches." [49] Lehmann had argued that, in cases in which Radicke's test allowed no "determination" of the problem, meaning that the computed probability was very low, it was better simply to make use of the "individual" data themselves and ignore statistical computation. [50] This, of course, was but another appeal, somewhat restrained, to the supremacy of the logic of facts.

Beneke had earlier betrayed what this claim involved. Referring to Boecker's experiments on the influence of sugar on urine, Beneke noted that many agents can affect the volume of urine excreted. These effects could be quite large (or small), and the sum of the mean errors would therefore wax (or wane). In the former case, lesser degrees of likelihood on Radicke's test must follow. Does this then mean, Beneke asked, that these agents are "irrelevant" with regard to the excretion of urea? [51] Beneke obviously believed this was the intended conclusion. He failed to recognize that Radicke's purpose was to emphasize the necessity of dealing with inherently probabilistic conclusions. To Beneke and Lehmann and to deterministic physiologists in general, facts and conclusions to which facts led existed in really only two conditions, certain and not certain. The former constituted the acquisitions of science; the latter stood beyond the pale. Beneke, like Lehmann, was not prepared to read "insufficient" certainty as meaning, not proven yet perhaps deserving further scrutiny, but as signifying disproved, the data in question therefore to be "thrown aside" because they give "no positive answer to the question." Experimental physiologists were profoundly uncomfortable living in a land between clear-cut extremes, the strange realm of degrees of certainty and probabilistic conclusions. Experimental control over the spontaneous activities of the organism offered the deterministic physiologist his only assurance of solid scientific foundations. Certainty was his programmatic emblem, uncertainty a threat to cognitive and disciplinary identity.

Perhaps Radicke was responding to this concern when he proposed quite conservative criteria of significance. He realized conclusions drawn by physiologists and physicians might have very serious consequences. "*Large* [scientific] *superstructures are erected* upon physiological and pharmacological conclusions"; shaky foundations could lead only to precarious generalizations. In matters uncertain, excess caution is all the more important. Hence, he noted with irony, the mathematician sustained a skepticism that the physician had inaugurated, but that the new experimental physiologist and pharmacologist urgently hoped to escape. "I require a rather high degree of certainty," he announced. "I do this because the conclusions drawn in physiology and pharmacology may well be further extended and upon them further theories and perhaps new methods of healing may be founded. These new measures will be applied in science and practice all the more restrictively [that is, more freely] as one considers them well established. The problem is, one forgets that all of this is founded not upon *certain* but only upon more or less *probable* data." [52] No less than the physicians, who were naturally eager to exhibit the beneficial effects of familiar and, especially, alternative healing methods, notably hydrotherapy, Radicke recognized that experimental fact was only the investigator's point of entry. He understood, however, that, once past this point, statistical inference came into play and that caution, not audacity, was called for.

Radicke was, of course, an outsider looking in. Nonetheless, he belonged to a long series of mathematicians and mathematically informed physicians who attempted to familiarize the medical profession with the promise and demands of careful numerical reasoning. The basic theme of the medical numerists was the law of large numbers and its relevance, for example, to therapeutic reasoning. The special situation of experimental work, the fact that physiological and therapeutic experiments did not generate anything even approximating the requisite "large" numbers, was repeatedly overlooked amidst the exhortation.

As is well known, the character of numerical reasoning in medicine had been established by Jules Gavarret, responding to the provocation of Louis and his school. Gavarret accepted the claim that real certainty was possible in pathology and diagnostics. One might literally see, or believe that one had seen, the conjunction between symptom and lesion and boldly draw therefrom a causal conclusion. Therapeutics, however, was the disputed territory; here nothing was intuitively obvious. [53]

Gavarret reassesed Louis's therapeutic trials, insisting that the number of cases observed must be sufficiently large so that the effect observed could not be assigned to the variability inherent in any set of observations. He drew upon Denis Poisson's analysis to show that the odds were 212 in 213 (probability $P = .9953$) that the ratio of two values m/n, that is, the number of times an event will occur within the total series of two possible events (m and n), will fall within limits set by a pair of equations. He showed also how those limits grew steadily narrower as the number of events increased. [54] He then addressed this measure of likelihood to the arguments of the most celebrated product of the numerical method, namely, Louis's *Recherches sur les effets de la saignée.* Gavarret's point was to indicate that, even with, for example, 500 observations, the limits were quite wide, and that the results of the next or even next several trials could easily fall within them and, most important, that this was to be expected on the basis of chance. [55] Little of Louis's work could withstand so demanding a test.

In answering one of his own critics, Gavarret pointed out that

in therapeutics, the facts depend wholly upon *contingent* causes. In such a situation the isolated fact proves only one thing: the possibility of the phenomenon under consideration. One gives a drug to a sick individual and he is cured. This merely shows that a particular individual might be cured by a particular drug; it does not prove that he was cured by the agency of this drug and by nothing else. If, now, one tests this drug in a certain number of cases similar to the first and if one finds that successes greatly exceed failures, it is proper to suspect that this drug does indeed possess healing qualities. However, one does not yet know the *law* governing the action of the drug. To obtain this law, I have shown that it is indispensable to have several hundred such facts. One must have them, absolutely must have them, in order that the *numerical relations*, which are only a translation of the observed effects, represent with sufficient approximation the true influence that is being sought.

The risk, he added, was great, for, "in therapeutic matters, every law deduced from a small number of facts can differ so much from the true law one is seeking that it is of no value whatsoever."[56]

Gavarret's argument was widely remarked, at least in Germany, but it appears to have had little influence upon the conduct of therapeutic trials, and it obviously represents an approach to the problem of reasoning with numbers very different from that put forward by Radicke.[57] Radicke's test also had no great success, whatever be its strengths or shortcomings. I know of no reasoned mathematical assault on his proposals, and almost certainly none was put forward by men of the laboratory. Indeed, their inclinations tended strongly in another theoretical direction, that of asserting the deterministic principles upon which alone they believed a rigorous science might be erected.

7 Determinism Prevails

Statistical methods might or might not have a role in a deterministic program. Even the most insistent deterministic physiologist could admit the utility of statistics, provided he was not compelled to acknowledge that nature herself was indeterministic. His was, at most, the world of subjective probabilities. Even this perspective he found imprecise and unsatisfying; it was to be accepted only because scientific knowledge was incomplete. Statistics within such a framework possessed heuristic value and could be employed when deemed appropriate or when other methods failed; it was not, however, to venture other claims.

Friedrich Martius, an astute clinical investigator and keen critic of the statistical literature, had no real complaint with the application of statistical methods to biological problems. He understood, however, that when physiology spoke in probabilistic terms, it did so only because it could not yet express itself in deterministic language. The latter was the true language of science. Statistics, Martius declared, offered no access to causes; experiment, on the other hand, did reach down to the causal level. The chance event was, therefore, nothing but a present and, one hoped, remediable gap in our knowledge. "Each observed fact, each event, is in itself neither probabilistic nor due to chance. The fact itself that such an event generally recurs may be taken as a demonstration of the necessity of its recurrence. Probability and chance reside, therefore, not in things but merely in us; their basis lies really in our lack of knowledge of the phenomena to be assayed. To designate the occurrence of an event as being due to chance means, therefore, only that we do not know the causes that bring this event to pass."[58] This statement, another credo in the rich dogma of deterministic physiology, indicates the constitutive priority of causal science. Martius's view, however, is a modulated one and as such probably reflected the broad commonality of opinion among working physiologists. Radicke offered no public statement of his view of the matter, but the character of his analysis strongly suggests adherence to Laplacean determinism and no concern for the possibility of an objective probabilism.

Martius and other skilled experimentalists allowed themselves no real worry that a mere instrument might imperil the deterministic foundation of their science. Martius appreciated, and largely recapitulated, the distinction drawn by Gavarret between the deterministic accomplishments of pathology, anatomy, and diagnostics and the (large) residual uncertainty that troubled therapeutics. He was pleased to explain this difference by the rapid progress of inquiry in the former domains, in particular, the perfection of experimental inquiry, and content to regard the awkward situation in therapeutics as one that would in time yield to further experimentation, the one route to assured natural knowledge. This was common ground with other physiologists. It is best, therefore, to view the uncertain status of statistical reasoning in nineteenth-century physiology, and in biological inquiry in general, as a reflection of the widespread ambition to create, quite literally create, a truly scientific physiology, a goal to be attained only by thinking and acting deterministically, just as the other natural sciences were seen or presumed to be doing. From this perspective, statistical methods provided a useful addition to the investigator's armamentarium, a conceptual tool to accompany the host of other, physical instruments that populated the new laboratories. It was not a tool that would dominate its apparent master—Radicke scarcely believed his test capable of mechanizing inference and replacing individual judgment—but it was a tool both constructive and cautionary, a warning to as well as a weapon of the physiological investigator.

Nonetheless, a quarter-century before Martius expressed his views, Beneke in pained exasperation had declared that "the physiological tact and knowledge of the experimenter seem to me more important than the mathematical [skills] that have been recommended."[59] So much, then, for Radicke's suggestion: the physician-physiologist must keep his principles, practice, and subject matter inviolate. But physiologists can be enthusiastic learners, and still the critic may not be satisfied. A century after Martius's announcement, the shoe finds itself on the other foot. The change is extreme, indeed droll, for now the mathematician is obliged to remind the physiologist of his, the physiologist's, experimental object, namely, the living organism itself. Among the cautions addressed to the biologist by a recent manual of statistics stands this unexpected warning:

Don't Forget Nonstatistical Knowledge

Statisticians are often stunned by the overzealous use of some particular statistical tool or methodology on the part of an experimenter, and we offer the following caveat. Experimenters, when you are doing "statistics," do not forget what you know about your subject-matter field! Statistical techniques are most effective when combined with appropriate subject-matter knowledge. The methods are an important adjunct to, not a replacement for, the natural skill of the experimenter.[60]

This new passion for the use of statistics has good reason behind it. Numerical reasoning has become since the 1920s an important element in the training and practice of research biologists and physicians because it is recognized as an indispensable means of scientific inference. Modern biostatisticians possess, and this is surely the critical point, an array of techniques, from correlation and analysis of variance to small-sample significance tests, that was unknown to the nineteenth-century enthusiast and his critics.[61] In the 1850s the very notion of a test of experimental results to some seemed unnecessary (medical statistics should be, rather, the extension of reasoning on large numbers); to others it appeared unobtainable, absurd, or even impertinent. Had, however, a practicable test been available with which one could have dealt effectively with the small numbers of data customarily produced by the physiological experiment or therapeutic trial, the physiologist of the period might well have acknowledged its utility and perhaps also employed it in his inquiries. He would not for that reason, however, look any more favorably upon the possibility that physiology could do no more than arrive at probabilistic conclusions, as Radicke had suggested. The banner of determinism rallied the troops against that possibility, and in this campaign the perceived need for a certain, rigorous new science was not to be qualified by the implications of what came to be regarded, albeit always with some reluctance, as a valuable ancillary tool, statistics.[62]

Notes

1. A large literature deals with this theme: see especially J. E. Lesch, *Science and Medicine in France. The Emergence of Experimental Physiology, 1790–1855* (Cambridge: Harvard University Press, 1984); G. L. Geison, *Michael Foster and the Cambridge School of Physiology. The Scientific Enterprise in Late Victorian Society* (Princeton: Princeton University Press, 1978). See also H.-H. Eulner, *Die Entwicklung der medizinischen Spezialfächer an den Universitäten des deutschen Sprachgebietes* (Stuttgart: Ferdinand Enke, 1970), pp. 32–65.

2. See J. P. Bull, "The Historical Development of Clinical Therapeutic Trials," *Journal of Chronic Diseases*, 10 (1959), 218–248; A. B. Hill, "The Philosophy of the Clinical Trial," in *Statistical Methods in Clinical and Preventive Medicine* (Edinburgh: E. and S. Livingstone, 1962), pp. 3–43; T. C. Chalmers, "The Clinical Trial," *Health and Society*, 59 (1931), 324–339; A. M. Lilienfeld, "Ceteris paribus: The Evolution of the Clinical Trial," *Bulletin of the History of Medicine*, 56 (1982), 1–18.

3. Julius Petersen, *Hauptmomente in der geschichtlichen Entwicklung der medicinischen Therapie* (Copenhagen: A. F. Høst. 1877; reprint, 1966), pp. 189–212; E. H. Ackerknecht, *Therapeutics from the Primitives to the Twentieth Century* (New York: Hafner, 1973), pp. 97–100, 106–110.

4. See Jacques Piquemal, "Succès et décadence de la méthode numérique en France à l'époque de Pierre-Charles Louis," *Médecine en France*, 250 (1974), 11–24; E.-R. Müllener, "Pierre-Charles-Alexandre Louis' (1787–1872) Genfer Schüler und die 'méthode numérique,'"

Gesnerus, 24 (1967), 46–74; Ulrich Tröhler, "Quantification in British Medicine and Surgery, 1750–1830, with Special Reference to Its Introduction into Therapeutics," Ph.D. dissertation, University of London, 1978.

5. C. A. Wunderlich, quoted in Petersen, *Hauptmomente der medicinischen Therapie*, p. 232 (note 3).

6. C. A. Wunderlich, quoted in Petersen, *"Hauptmomente,"* pp. 225–226 (note 3).

7. C. A. Wunderlich, "Das Verhältniss der physiologischen Medizin zur ärztlichen Praxis," *Archiv für physiologische Heilkunde*, 4 (1844), 13. See K. E. Rothschuh, "Deutsche Biedermeiermedizin. Epoche zwischen Romantik und Naturalismus," *Gesnerus*, 25 (1968), 167–187.

8. My discussion follows Brigitte Lohff, "Emil Du Bois-Reymonds Theorie des Experiments," in *Naturwissen und Erkenntnis im 19. Jahrhundert: Emil Du Bois-Reymond*, ed. Gunther Mann (Hildesheim: Gerstenberg, 1981) pp. 117–128; passage cited, p. 117.

9. B. Lohff, "Du Bois-Reymonds Theorie," p. 122.

10. William Coleman, "Neither Empiricism Nor Probability: the Experimental Approach," in *Probability since 1800: Interdisciplinary Studies of Scientific Development*, Report Wissenschaftsforschung 25, ed. Michael Heidelberger et al. (Bielefeld: B. Kleine, 1983), pp. 275–286.

11. Johannes Steudel, "Aus der Geschichte der Balneologie," *Deutsche medizinische Wochenschrift*, 79 (1954), 497–500; "Therapeutische und soziologische Funktion der Mineralbäder im 19. Jahrhundert," in *Der Arzt und der Kranke in der Gesellschaft des 19. Jahrhunderts*, ed. W. Artelt and W. Ruegg (Stuttgart: Ferdinand Enke, 1967), pp. 82–97; Heinz Müller-Dietz, "Frühe balneologische Forschungen an der deutschen Ostseeküste," *Medizinhistorisches Journal*, 2 (1967), 239–247.

12. F. L. Holmes, "Introduction," Justus von Liebig, *Animal Chemistry or Organic Chemistry in Its Application to Physiology and Pathology* (Cambridge: John Owen, 1842; reprint, 1964), pp. vii–cxvi, esp. xc–cix.

13. Gustav Radicke, "Die Bedeutung und Werth arithmetischer Mittel mit besonderer Beziehung auf die neueren physiologischen Versuche zur Bestimmung des Einflusses gegebener Momente auf den Stoffwechsel, und Regeln zur exacten Beurtheilung dieses Einflusses," *Archiv für physiologische Heilkunde*, N.F. 2 (1858), 145–219; data cited, pp. 190, 192; English trans. (see below), pp. 228–230. Radicke's paper was translated as *On the Importance and Value of Arithmetic Means*, etc. (London: New Sydenham Society, 1861). This translation includes also English versions of some commentaries on Radicke's proposals. I have quoted these translations but occasionally have silently introduced minor emendations and have made a few major changes. I cite both the German original and the English translation, whenever the latter is available. Data from Boecker cited by Radicke, "Bedeutung und Werth," pp. 192–193, 230; "Arithmetical Means," pp. 153, 193. See F. W. Boecker, "Die Sasparile, physiologisch, historisch, und kritisch untersucht," *Journal für Pharmakodynamik, Toxikologie und Therapie*, 2 (1858), 1–154, plus tables.

14. See J. C. Poggendorf, *Biographisch-literarisches Handwörterbuch zur Geschichte der exacten Naturwissenschaften*, Vols. Iff. (Leipzig: J. A. Barth, 1863ff.), II, pp. 557–558, III (2), p. 1082.

15. G. Radicke, "Bedeutung und Werth," p. 146 (English, p. 186) (note 13).

16. G. Radicke, "Bedeutung und Werth," p. 146 (English, p. 186).

17. G. Radicke, "Bedeutung und Werth," p. 160 (English, p. 199).

18. G. Radicke, "Bedeutung und Werth," pp. 160–164 (English, pp. 199–202). This value is today called the standard deviation.

19. G. Radicke, "Bedeutung und Werth," pp. 165–175 (English, pp. 204–212). Passage cited, pp. 166–167 (English, p. 205) (note 13).

20. G. Radicke, "Bedeutung und Werth," p. 182 (English, p. 219).

21. G. Radicke, "Bedeutung und Werth," pp. 175–182 (English, pp. 212–219).

22. G. Radicke, "Bedeutung und Werth," pp. 190–192 (English, pp. 228–230).

23. G. Radicke, "Bedeutung und Werth," pp. 194–196 (English, pp. 232–235).

24. F. W. Beneke, *Ueber die Wirkung des Nordsee-Bades. Eine physiologisch-chemische Untersuchung* (Göttingen: Vandenhoeck und Ruprecht, 1855).

25. G. Radicke, "Bedeutung und Werth," pp. 183–190 (English, pp. 220–228) (note 13).

26. G. Radicke, "Bedeutung und Werth," p. 189 (English, p. 227).

27. G. Radicke, "Bedeutung und Werth," pp. 196–208 (English, pp. 235–248). Passage cited, pp. 198–199 (English, p. 238). See Wilhelm Kaupp, "Beiträge zur Physiologie des Harnes," *Archiv für physiologische Heilkunde*, 14 (1855), 384–424; 15 (1855), 125–164.

28. See W. G. Cochran, "Early Development of Techniques in Comparative Experimentation," in *On the History of Statistics and Probability*, ed. D. B. Owen (New York: Dekker, 1976), pp. 3–25; J. F. Box, *R. A. Fisher. The Life of a Scientist* (New York: John Wiley, 1978), pp. 140–166.

29. Karl Vierordt, "Bemerkungen über medicinische Statistik," *Archiv für physiologische Heilkunde*, N.F. 2 (1858), 220–227 (English, pp. 251–256). Passage cited below, p. 221 (English, p. 252).

30. K. Vierordt, "Bemerkungen," pp. 221–222 (English, pp. 252–253) (note 29).

31. K. Vierordt, "Bemerkungen," p. 223 (English, p. 253).

32. K. Vierordt, "Bemerkungen," p. 223 (English, p. 254).

33. F. W. Beneke, "Entgegnung auf den Aufsatz des Herrn Prof. Radicke: 'Ueber den Werth und die Bedeutung arithmetischer Mittel'," *Archiv für physiologische Heilkunde*, N.F. 2 (1858), 550–567; passage cited, p. 552 (English, p. 259).

34. F. W. Beneke, "Entgegnungen," p. 552 (note 33).

35. F. W. Beneke, "Entgegnungen," pp. 553–554 (English, pp. 260–261).

36. F. W. Beneke, "Entgegnungen," pp. 560–561 (English, pp. 267–268).

37. See Frederick Gregory, *Scientific Materialism in Nineteenth Century Germany* (Dordrecht: Reidel, 1977), pp. 1–48; T. E. Willey, *Back to Kant. The Revival of Kantianism in German Social and Historical Thought, 1860–1914* (Detroit: Wayne State University Press, 1978), pp. 13–39; D. G. Charlton, *Positivist Thought in France during the Second Empire 1852–1870* (Oxford: Clarendon, 1959), pp. 5–23, 72–85.

38. Friedrich Martius, "Die Numerische Methode (Statistik und Wahrscheinlichkeitsrechnung) mit besonderer Berücksichtigung ihrer Anwendung auf die Medizin," *Virchows Archiv für pathologische Anatomie und Physiologie und für klinische Medizin*, 83 (1881), 351.

39. Justus von Liebig, "Ein Philosoph und ein Naturforscher über Francis Bacon von Verulam," *Reden und Abhandlungen* (Leipzig: Winter, 1874), p. 249.

40. F. A. Lange, *Geschichte des Materialismus und Kritik seiner Bedeutung in der Gegenwart* (Iserlohn: J. Baedeker, 1866), pp. 352–353, 354.

41. Friederich Ueberweg, "Ueber die sogenannte 'Logik der Tatsachen' in naturwissenschaftlicher und besonders in pharmakodynamischer Forschung," *Virchows Archiv für pathologische Anatomie*, 16 (1859), 400–407.

42. F. E. Beneke, *Lehrbuch der Psychologie als Naturwissenschaft*, 2nd ed., (Berlin: E. S. Mittler, 1845; reprint, 1964), pp. 147–148.

43. Ueberweg, "Ueber die sogenannte 'Logik der Tatsachen'," p. 405 (note 41).

44. F. Ueberweg, "Ueber die sogenannte 'Logik der Tatsachen'," p. 405.

45. Radicke, "Notiz über die Herleitung physiologischer und pharmakodynamischer Wahrheiten aus coordinirten Beobachtungsreihen. Entgegnung auf Aeusserungen, die auf Missverständnissen des vom Verfasser im Roserschen Archiv veröffentlichen Aufsatzes: 'Ueber arithmetische Mittel' beruhen," *Untersuchungen zur Naturlehre* (Moleschott), 6 (1859), 307–314; passage cited, p. 308 (English, p. 269).

46. G. Radicke, "Notiz," p. 308 (English, p. 270) (note 45).

47. G. Radicke, "Notiz," pp. 309–310 (English, pp. 271–272).

48. G. Radicke, "Notiz," pp. 310–313 (English, pp. 272–275).

49. G. Radicke, "Notiz," pp. 312–313 (English, pp. 273–274).

50. G. Radicke, "Notiz," p. 312 (English, p. 272); see L. Lehmann, "Zur Würdigung der physiologischen Wirkung der Sitzbäder," *Untersuchungen zur Naturlehre* (Moleschott), 6 (1859), 186–187.

51. F. W. Beneke, "Entgegnung auf den Aufsatz des Herrn Prof. Radicke" (note 33), p. 555 (English, p. 262).

52. Radicke, "Notiz," p. 311 (English, p. 273) (note 45).

53. Jules Gavarret, "De l'application de la statistique à l'examen critique auquel M. le docteur Valleix a soumis, dans le numéro de mai 1840 des *Archives générales de médecine*, l'ouvrage de M. Gavarret intitulé Principes généraux de statistique médicale ou developpement des règles qui doivent présider à son emploi," *L'expérience*, 2 juilliet 1840, 12 pp.

54. J. Gavarret, *Les principes de statistique médicale ou developpement des règles qui doivent présider à son emploi* (Paris: Bechet jeune, 1840), pp. 264–271.

55. J. Gavarret, *Principes*, pp. 153–162, 141–142 (note 54).

56. J. Gavarret, "De l'application de la statistique à la médicine," p. 9 (pagination of offprint).

57. Gavarret's work was translated as *Allgemeine Grundsätze der medicinischen Statistik*, trans. S. Landmann (Erlangen: Ferdinand Enke, 1844). The following works represent a long generation's commentary on Gavarret's principles: G. Schweig, "Auseinandersetzung der statistischen Methode in besonderem Hinblick auf das medicinische Bedürfniss," *Auchiv für physiologische Heilkunde*, 13 (1853), pp. 305–355; Friedrich Oesterlen, *Handbuch der medicinischen Statistik* (Tübingen: H. Laupp, 1865), pp. 59–74; Adolf Fick, "Ueber die Anwendung der Wahrscheinlichkeitsrechnung," in *Medizinische Physik*, 2nd ed. (Braunschweig: Vieweg, 1866), pp. 430–447; Willers Jessen, "Zur analytischen Statistik," *Zeitschrift für Biologie*, 3 (1867), 128–136; Julius Hirschberg, *Die mathematische Grundlagen der medizinischen Statistik, elementar dargestellt* (Leipzig: Veit, 1874); Carl Liebermeister, "Ueber Wahrscheinlichkeitsrechnung in Anwendung auf therapeutische Statistik," Volkmanns *Sammlung klinischer Vorträge*, No. 110: Innere Medizin, No. 39, pp. 935–962 (Leipzig: 1877); Martius "Die numerische Methode" (note 38). See also Martius, "Die Principien der wissenschaftlichen Forschung in der Therapie," Volkmanns *Sammlung klinischer Vorträge*. No. 139: Innere Medizin, No. 47, pp. 1169–1188 (Leipzig: 1887); O. B. Sheynin, "On the History of Medical Statistics," *Archives for the History of Exact Sciences*, 26 (1982), pp. 241–286.

58. Martius, "Die numerische Methode," p. 360 (note 38).

59. Beneke, "Entgegnung auf den Aufsatz des Hern Prof. Radicke," p. 561 (English, p. 268) (note 33).

60. G. E. P. Box et al., *Statistics for Experimenters. An Introduction to Design, Data Analysis and Model Building* (New York: John Wiley, 1978), p. 15.

61. All is not yet well, however, in the medical use of statistics; recent studies reveal that, in a selection of major clinical journals, errors (in both computation and interpretation)

appeared in some 40–50% of all articles employing statistical procedures. See S. A. Glantz, *Primer of Biostatistics* (New York: McGraw-Hill, 1981), pp. 6–9, and esp. the references, p. 9n.

62. Physiological determinism long continued to exclude the prospect that chance constitutes "an autonomous aspect of the world subject to regular mathematics": Ian Hacking, "Nineteenth-Century Cracks in the Concept of Determinism," in *Probability and Conceptual Change in Scientific Thought*, Report Wissenschaftsforschung 22, ed. Michael Heidelberger and Lorenz Krüger (Bielefeld: B. Kleine, 1982), p. 27.

V EVOLUTIONARY BIOLOGY

The Probabilistic Revolution in Evolutionary Biology—an Overview
John Beatty

Prior to whatever it is that we call the "Darwinian revolution" (which began around Darwin's time and is hardly over yet), chance accounts of natural phenomena were invoked mainly as alternatives to teleological accounts. More specifically, they were posed as strictly *mechanistic* alternatives to teleological explanations, which sounds, at least from a contemporary point of view, like a funny notion of "chance," since it has nothing to do with statistics or probability.

Prominent among the antichance, teleological approaches to nature in the nineteenth century was the "natural theological" approach to organic form, so famously defended by the Reverend William Paley and the authors of the nine *Bridgewater Treatises*. It is all very well, proponents of this approach maintained, to attribute wens and warts to chance. But eyes, hands, and other complicated structures serving adaptively appropriate uses must be attributed to the Creator, who foresaw the needs of organisms and provided for them.

There were other important antichance, teleological approaches to organic form as well, including the tradition exemplified (more or less) by Karl Ernst von Baer. According to this perspective, the purposeful directing agencies were not divine, but rather inhered in the organisms themselves. So, for instance, species-specific "types" were invoked to organize and integrate chemical reactions in order to ensure adaptively appropriate ends. Von Baer later blamed the natural theological approach version of teleology for the "teleophobia" of the mid-to-late nineteenth century. Whatever the differences of the two approaches to teleology, though, they were united against the understanding of appropriateness and adaptation in the organic world as a series of coincidences of blind mechanical laws.

The gradual acceptance of Darwinism spelled the end of strictly teleological explanations in biology. In his alternative, strictly mechanical account of form, Darwin invoked chance in much the same sense as teleologists prior to him had understood that notion: that is, as resulting in adaptiveness, if at all, only coincidentally. Darwin did not, however, construe organic form as *merely* the result of chance, as teleologists had supposed mechanists *must* construe it.

According to Darwin's account, chance entered into only the first of two steps leading to adaptation. First, variations arise among the offspring of an organism that, by chance, confer adaptive benefits on their possessors. Second, possessors of those variations, *on account of their possession of those variations*, contribute per capita more offspring to the next generation than do possessors of alternative traits. As a result, the adaptively beneficial traits increase in frequency from generation to generation. The second step is (according to Darwin) not a matter of chance, so the process of adaptation is only partly a matter of chance.

In saying that adaptively beneficial variations arise (if not increase in frequency) "by chance," Darwin meant only that their occurrence is not actually occasioned by the adaptive purpose they serve for their possessors—no vital or divine agency summons them to that end. The variations have mechanical causes (which Darwin did not pretend to understand), but those causes act blindly with regard to their consequences.

The chapter by Jonathan Hodge in this part treats the development of Darwin's notion of chance variation at length. The chapter by Bernd-Olaf Küppers makes very clear that Darwinian evolution does not rely entirely on chance. Küppers discusses, among other things, the conceptual difference between saying that the origin of life was a matter of chance and that the origin of life was a matter of evolution by natural selection.

Darwin's notion of chance variation was very little, if at all, influenced by developments in probability theory and statistics. By the turn of the century, however, two approaches to the study of variation within species had emerged, the "biometrical" and the "Mendelian" schools, both of which brought developments in probability theory and statistics to bear on the subject. This broad similarity between the two approaches aside, the two schools found very different aspects of probability theory and statistics useful for their purposes.

The biometricians focused on means and distributions of variation in successive generations. They were not the first to do so. Already in the mid-nineteenth century, the statistician Adolphe Quetelet had used the so-called "law of errors" (described by the normal curve) to describe variation in humans. The law of errors had originally been used to describe such things as the distribution of repeated measurements of a particular object. In applying the law to humans, Quetelet understood human variation as something very akin to measurement error or, more generally, replication error. What was being replicated in this case was the "average man," the timeless essence of humankind (or, alternatively, the timeless essence of a particular race).

The biometricians did not attach such essentialistic connotations to the means and distributions of variations with which they characterized species and races. In the first place, for biometricians like Francis Galton and Karl Pearson, offspring deviations from the mean were not errors in the representation of species or racial prototypes, but were rather reflections of parental (and more remote ancestral) deviations. More specifically, they understood the average deviation of offspring from the mean in terms of the average deviation of their parents from the mean (or in terms of the weighted average deviations of their ancestors). Reflecting very strongly the influence of Darwin, the biometricians (Pearson and the rest more so than Galton) also differed from Quetelet in attaching no notion of eternal stability to species and racial means. Pearson spoke of evolution in terms of *changes* in species and racial means. With all such talk of shifting racial and species means, it is not clear what a timeless racial or species essence would amount to.

The Mendelians applied very basic combinatorial methods of analysis to problems concerning the resemblance between parents and offspring. In the simplest case, a difference in traits was explained in terms of one pair of hereditary factors, *A* and *a*. On this scheme, every organism is supposed to have a pair of factors, one of which it inherits from its male parent (via the male gamete), and the other from the female parent (via the female gamete). *AA* organisms produce only *A* gametes; *aa* organisms produce only *a* gametes. *Aa* organisms produce, in the long run, 50% *A* gametes and 50% *a* gametes and thus, in the long run, contribute *A* to one-half of their offspring and *a* to the other half. Purely combinatorial reasoning, based on this mechanism, can be used to explain/predict the frequency distributions of genetic outcomes of any cross; and given a particular connection between the traits in question and the various possible pairings of hereditary factors, the same combinatorial reasoning can be used to predict frequency distributions of trait outcomes of any cross. Thus, $Aa \times Aa$ crosses lead, in the long run, to the familiar $AA : 2Aa : aa$ distribution of offspring; and assuming that AA and Aa cause the manifestation of one and the same trait, while aa causes the manifestation of another, the same cross leads, in the long run, to the famous 3 : 1 trait distribution. (Evolution from a Mendelian point of view will be discussed shortly.)

For all their own emphasis on probabilistic reasoning, Mendelians like Charles Davenport scorned the biometricians' approach as being overly statistical. The Mendelian's approach at least located the mechanical source of the stochasticity with which they dealt—namely, in the process of gamete formation. The biometricians, on the other hand, were little interested in the mechanical sources of their character distributions and correlations. Other more technical biological issues also kept the two schools at odds during the first decades of the twentieth century. By the thirties, the two approaches had been reconciled (the work of R. A. Fisher was significant in this regard). To this day, the two approaches (with modifications) exist side by side. Some biologists study the inheritance and evolution of traits that clearly abide by the 3 : 1 ratio, or some combinatorial variant thereof. Other biologists ("quantitative" geneticists) study continuously distributed traits whose patterns cannot (in practice) be so neatly resolved into Mendelian ratios.

Interestingly, in spite of the fact that the biometricians and Mendelians brought probability theory and statistics to bear on the study of variation, while Darwin did not, the notion of "chance" variation or "chance" mutation still carries the connotation that Darwin attached to it. As one evolutionary biologist explains the notion in a current textbook, "Mutation is random in that *the chance that a specific mutation will occur is not affected by how useful that mutation would be*" (Douglas Futuyma, *Evolutionary Biology*, 1979, p. 249).

The peculiarly Darwinian notion of "chance variation" or "chance mutation" has also outlasted attempts on the parts of those like Max Delbrück and Erwin Schrödinger, who have tried to reduce the problem of mutation to a problem of quantum mechanics, and have tried thereby to attach to that problem the problems of indeterminacy characteristic of quantum mechanics. It is certainly not the case that the biometricians', the Mendelians', or the physicists' special interests in chance aspects of variation have proved fruitless programs of research. Quite the opposite is the case. It is just that the notion of chance variation, as originally conceived by Darwin, continues to play a conceptually very important role in discussions of evolutionary change.

Somewhat more sophisticated notions of chance entered evolutionary biology in the thirties, shortly after the reconciliation of biometry and Mendelism. In considering, among other things, the consequences of Mendelian heredity for evolutionary change, evolutionary theorists of the likes of Fisher and Sewall Wright came to see the need for a stochastic theory of evolution, one that would take into account not only evolution by natural selection, but also something that Wright called evolution by "random drift."

Evolution by random drift involves chance variation, in the Darwinian sense, but it also involves chance changes in the frequencies of variation, something that Darwin did not take into account. One of the sources of these chance changes is Mendelian inheritance. As a result of the fact that *Aa* organisms can contribute, *in the short run*, more *A*s than *a*s (or vice versa) to their offspring, gene and trait frequencies can change from generation to generation by chance—they can "drift randomly."

While Fisher and Wright (and others) agreed that evolutionary theory would have to be formulated stochastically in order to account for these sorts of fluctuations in frequency, they (and others) still found plenty to disagree about. The stochastic theory of evolution that we call the "synthetic theory" (partly to signify that it is a synthesis of Darwinian and Mendelian theories) tells one, for instance, that the smaller the population and the smaller the selection pressures acting on that population, the greater the extent to which gene frequencies drift randomly. But the theory does not tell one whether population sizes are generally, or in any particular case, large or small, nor whether selection pressures are generally, or in any particular case, large or small. So the theory does not tell one just how important random drift is. And this has been a much disputed issue—right up to the present day.

The chapter by Hodge deals with conceptual issues concerning the difference between evolution by natural selection and evolution by random drift. John Turner and I then discuss the history of the controversies between proponents of the importance of natural selection and proponents of the importance of random drift.

10 Natural Selection as a Causal, Empirical, and Probabilistic Theory

M. J. S. Hodge

Darwin's conforming of his theory to the old vera causa *ideal shows that the theory of natural selection is probabilistic not because it introduces a probabilistic law or principle, but because it invokes a probabilistic cause, natural selection, definable as nonfortuitous differential reproduction of hereditary variants.*

Chance features twice in this causal process. The generation of hereditary variants may be a matter of chance; but their subsequent populational fate is not; for their physical property differences are sources of causal bias giving them different chances of survival and reproduction. This distinguishes selection from any process of drift through fortuitous differential reproduction in the accumulation of random or indiscriminate errors of sampling. To confirm the theory of natural selection empirically is to confirm that this probabilistic causal process exists, is competent, and has been responsible for evolution. Such hypotheses are both falsifiable and verifiable, in principle, if not in practice.

Natural selection has been accepted and developed by biologists with very diverse attitudes toward chance and chances. But the theory and its acceptance have always involved probabilistic causal judgments that cannot be reduced to correlational ones. So, the theory has contributed to a probabilistic shift within the development of causal science, not to any probabilistic rebellion in favour of science without causes.

1

This chapter proposes a framework for integrating biologists' and philosophers' analyses of natural selection as a causal, probabilistic, and empirical theory of evolution. Throughout, the argument will be that the probabilistic character of the theory, whether in Darwin's day or ours, can only be properly understood when its distinctively causal and empirical character is kept in view.

Until fairly recently, perhaps only a decade and a half ago, philosophical commentary on natural selection rather rarely interested biologists, who were understandably impatient, for example, with endless variations on the old tautology complaint. Equally, biologists' disagreements over genetic drift, the classical and balance theories, group selection, and so on attracted little attention from philosophers.

I am much indebted here to discussions with John Beatty, David Hull, Larry Laudan, Rachel Laudan, Ernst Mayr, Bernard Norton, Michael Ruse, Sam Schweber, and John Turner. For the excellent interdisciplinary opportunities provided by two Bielefeld conferences, I am very grateful indeed. The present chapter incorporates material from an earlier piece, where some historical points were treated somewhat more fully: "Law, Cause, Chance, Adaptation and Species in Darwinian Theory in the 1830's, with a Postscript on the 1930's," in M. Heidelberger, L. Krüger, and R. Rheinwald, eds., *Probability since 1800. Interdisciplinary Studies of Scientific Development* (Bielefeld: Universität Bielefeld, 1983), pp. 287–330.

Happily, this phase is now past, as anyone will know who reads in such journals as *American Naturalist, Annual Review of Ecology and Systematics, Biology and Philosophy, Journal of Theoretical Biology, Paleobiology, Philosophy of Science, Studies in History and Philosophy of Science, Synthese,* and *Systematic Zoology.* For there one finds the technical resources provided by both biology and philosophy often combined and applied to problems of common concern to the two disciplines.[1]

This chapter aims to contribute to this most welcome trend. But it does not set out to do so directly. Rather the hope is to clarify the probabilistic character of the theory of natural selection, as it concerns biologists and philosophers alike, by beginning from a point of departure that lies within the discipline of neither party.

This point of departure is an historical one, namely, Darwin's original understanding of the theory of natural selection as a causal, empirical, and so explanatory theory. There will be, however, no concern with history for its own sake, nor any attempt to settle current disputes by invoking venerable authority. Instead, it will be argued that Darwin's conception of the character of the theory is still appropriate today, because it conforms to what is common to the best explications of the theory given of late by biologists and philosophers. Accordingly, the proposal will be that those developments in this century—most notably in Mendelian population genetics and molecular biology—that have made the current versions of the theory no less causal, no less empirical, and no less probabilistic have also made Darwin's original conception of what kind of theory it is more, and not less, instructive.

Although this is a historical proposal, albeit an overtly normative one, it will not be defended by offering a narrative analysis of developments from Darwin to Dobzhansky and beyond. For it will be proposed that we can abstract and generalize, from the century and a quarter since 1859, and insist that answering certain enduring clusters of questions in certain ways has always been characteristic of any thoroughgoing commitment to natural selection as a theory of evolution.[2] The principal thesis will be, accordingly, a very simple, even simple-minded, one, namely, that in trying to understand the theory of natural selection, whether in the original Darwinian or in any subsequent neo-Darwinian context, it is always best to follow Darwin's own strategy and concentrate on distinguishing some four clusters of questions:

1. *The definition question:* What is natural selection? How is this process (or agency or force or whatever) to be defined?

2. *Existence, that is, occurrence and prevalence, questions:* Does it exist, is it going on anywhere? How widespread, how prevalent is it? Among what units and at what levels—organisms, colonies, species and so on—is it occurring?

3. *Competence, that is, consequence and adequacy questions:* What sorts, sizes, and speeds of change does it suffice presently to produce? What are its possible and actual consequences?

4. *Responsibility, that is, past achievement questions:* What has it done? For how much of past evolution has it been responsible? What is to be explained as resulting from it?

Obviously, such question clusters might be distinguished in other ways. But this way will serve our purpose here. For it will be argued throughout this chapter that any correct answers to the definition question will have as a corollary that all the other three clusters concern empirical questions, albeit very diverse questions, about probabilistic causal processes, questions that can be given answers that are testable in principle if not in practice.

2

Before proceeding to argue for this thesis, however, mention should be made of three analytic resources that will be deployed in due course.

First, I have not assumed the privileged correctness of any particular account of testability, whether Carnapian, Popperian, Duhemian, or whatever. But I have assumed what I take to be accepted by most accounts of theory testing, namely, (a) the elementary logical point that "denying the consequent" is valid or deductively correct while "affirming the consequent" is not, so that a statement's truth is not validly inferred from the truth of its consequences, while its falsity is validly inferred from their falsity; (b) the familiar epistemological point that theories can often only be made to have testable predictions as deductive consequences by conjoining them with additional auxiliary hypotheses, so that the elementary logical contrast does not in itself sanction any methodological imperative whereby predictive errors should always be instantly accepted as conclusively falsifying the theory rather than the auxiliaries; (c) the obvious methodological point that, in judging the respective merits of two or more theories confronted with some observational reports accepted as facts, we have to evaluate, among other things, the different sets of auxiliary hypotheses that need to be conjoined with those respective theories if they are to predict those facts.[3]

Second, I sometimes distinguish between something being true as a matter of definition or meaning and something being true as a matter of fact or experience. Now, thanks especially to the teachings of Quine (whose own position has shifted more than is usually appreciated), all such distinctions are often held to be difficult to defend.[4] However, my uses of them do respect these difficulties. For it is not assumed here that something is definitionally rather than factually true independently of all contextual considerations; with changes in the theoretical context, especially, one might well want to revise earlier judgments as to what is best taken as definitional and what as factual. But that possibility only confirms that one can often clarify the assumptions constituting the theoretical framework current at any time by seeing how they make it reasonable to draw the line here rather than there. To drop all such distinctions would thus be to forgo needlessly a very useful source of light on those assumptions themselves.

Third, in conformity with Salmon's developments of Reichenbach's views, I take a causal process to be a physical process, one wherein energy is transformed and transmitted, and one whose later temporal stages can be altered by interfering with earlier stages.[5] Thus, by contrast, an abstract, mathematical "process," whereby successive numerical values are "generated" by an algorithmic "operation," is not a causal process because not physical, while the successive positions of a shadow passing over the ground do not form a causal process because no alterations of later positions can be wrought by actions at earlier ones. Among the successive stages of a causal process there hold relations of causal relevance. And, in conformity with recent explications of what it is to be a causal factor, Giere's, for instance, I accept that relations of statistical relevance cannot in and of themselves constitute causal relevance.[6] Accordingly, I assume that insofar as a demand for explanation is construed as a demand for an identification of what is causally relevant to what is being explained, then neither mere mathematical representations nor purely statistical description can by themselves meet this demand.

Finally, I should emphasize that I have not brought to this analysis of natural selection theory any one philosophical account of what a scientific theory is: the "received view," for instance, dominant in the heyday of logical empiricism, or its more recent rival, the "semantic view." Both of these have been shown to clarify some leading features of Darwin's own and later evolutionary theorizing.[7] But no one such view is uniquely helpful in exhibiting the most instructive continuities in the roles of probabilistic notions in natural selection theory as those roles have developed over the last century and a quarter.

3

On going back now to our starting point in Darwin's work, it will be evident that for my purposes here it is not necessary to read into his writings any philosophical resources developed only in our own times. On the contrary, if our understanding of natural selection theory is eventually to benefit from those resources, we need to begin by taking Darwin on his own terms. We need an analysis of the problems he saw natural selection as solving, an analysis that brings out why he deliberately gave his argument for natural selection a very distinctive structure.

The structure is most easily discerned by comparing *On the Origin of Species* (1859) with its predecessors, the manuscript *Sketch* of 1842 and *Essay* of 1844. Then it is apparent that Darwin knowingly conformed his argument to the *vera causa* ideal for a scientific theory.[8]

The phrase *vera causa* meant not the true cause but a true cause, that is, a real cause, one known to exist, and not a purely hypothetical cause, merely conjectured to exist. So, the *vera causa* ideal, as Darwin sought to conform to it, required that any cause introduced in a scientific theory should be not merely adequate to produce the facts it is to explain on the supposition that it exists. For the existence of the cause is not to be accepted on the grounds of this adequacy. Its existence should be known from direct independent evidence, from observational acquaintance with its active presence in nature, and so from facts other than those it is to explain.

This *vera causa* ideal had been first given this formulation in the Scots moral and natural philosopher Thomas Reid's teachings on the import of Newton's first rule of philosophizing, the one that specified that no causes are to be admitted except such as are both true and sufficient to explain the phenomena.[9]

Reid, going beyond Newton's own understanding of this principle, used his novel explication of it to argue for the conclusive epistemic superiority of the Newtonian gravitational force over the Cartesian ethereal vortices as explanatory of the planetary orbits. That force with its determinate law, unlike those vortices, was a well-evidenced cause for the orbits up there, because the orbits themselves were not the sole evidence for its existence. It was a real and true cause, not a hypothetical and conjectural one; for it was known to exist from our direct and familiar experience of swinging pendulums and falling stones down here on earth.

It was the teachings of Charles Lyell, in the three volumes (1830, 1832, and 1833) of his *Principles of Geology*, that mediated between this Reidian *vera causa* tradition, in the epistemology of physics, and Darwin's understanding of what evidential demands would have to be met in solving the problem of the origin of species.

All the causes of change on the earth's surface were presumed, in Lyell's system, to persist undiminished in intensity, and so in efficacy, into the present, the human period, and on into the future. Now, as at all times, habitable dry land is being destroyed by subsidence and erosion in some regions, while it is being produced by sediment consolidation, lava eruption, and elevationary earthquake action in others. Likewise, there is a continual, one-by-one extinction and creation of animal and plant species, a constant exchange of new species for old, adequate to bring about a succession of faunas and floras in the long run.[10]

Lyell's most explicit rationale for this presumption of the persistence of all such causes of change into the present and future was the ideal of explanation by *verae causae*, causes known to exist from direct observational evidence independent of the facts they are to be invoked to explain. Accordingly, his system was to exemplify, no less explicitly, an epistemological analogy. In geology, only causes active in the present, human period are accessible in principle, although often not in practice, to direct observation. So, in this science, the brief human present is to the far vaster prehuman past as the terrestrial is to the celestial in Newtonian physical astronomy.

We have here, then, in this *vera causa* ideal the source for Darwin's structuring his argumentation in the *Origin* as he did. For, to conform to this ideal, the argumentation on behalf of a causal theory had to make three distinct evidential cases: for the existence of the cause, for its adequacy for facts such as those to be explained by it, and, finally, for its responsibility for those facts. Thus the first two cases, in establishing that the cause exists and can produce effects of the appropriate sort and size, would constitute argumentation showing what the theory is, what cause it introduces, that it is a true theory, one of a true cause, and an adequate theory, one of a true cause capable of such effects, and so no mere conjectured hypothesis, while the third would constitute its verification, as the true theory for the particular facts it is to explain.

As anyone who has consulted the *Origin* will recall, the one long argument of that book does indeed make three distinguishable evidential cases for three theses about natural selection: first, that it is a really existing process, one presently at work in nature; second, that, as it now exists, it is adequate for the adaptive formation of species and their adaptive diversification, in the very long run, into genera, families, and so on; third, that it has probably been the main agency in the production in the past of the species now extant and of those extinct ones commemorated in the rocks.

Thus the first case evidences both a tendency in wild species, as in domesticated ones, to vary heritably in changed conditions, and a struggle for life wherein variant individuals are surviving and so reproducing differentially, as when domesticated species are bred selectively by man. The second argues for the ability of this natural selective breeding—so much more sensitive and sustained, precise, and prolonged than man's as it is—to produce and diversify species in eons of time and changing conditions. Finally, the third adduces many facts of various kinds—geological, embryological, and so on—about extant and extinct species; and it is argued that these facts can best be explained—most intelligibly connected by referring them to unifying laws, that is—on the supposition that this cause was mainly responsible for producing and diversifying those species in irregularly ramified lines of descent diverging adaptively from more or less remote common ancestral stocks.

4

We have, then, to acknowledge that Darwin, in the *Origin*, was bringing to a problem in biology evidential and explanatory ideals first explicated, principally if not solely, in legitimating Newtonian celestial mechanics. But to acknowledge this is to raise at once an issue of immediate bearing on our efforts to understand the probabilistic character of Darwinian theory. How could Darwin have thought that such an ideal, originally legitimating the subsumption of the solar system under the laws of a deterministic classical mechanics, was at all appropriate to his arguments on behalf of a probabilistic causal process, natural selection as a cause for the adaptive formation and diversification of species? And, more generally, even if Darwin saw no inconsistency here, has it not turned out that Darwinian theory has developed since so as to make quite inappropriate his notion of the evidential and explanatory challenges raised by any commitment to natural selection as a theory of evolution?

These are questions that could be resolved by constructing and contrasting caricatures of "Laplacean" and "Darwinian" science, so as to secure a quick verdict, namely, that in its concern with what is "historical," "unique," "statistical," and so on, Darwinian science is utterly unlike any Laplacean program and must, therefore, have arisen in a repudiation of everything we associate with the French mathematician and physicist.

Nor surprisingly, to anyone who has studied how Darwin came to natural selection, while he was filling his *Notebooks* B–E and M–N in the years from July 1837 to July 1839, two decades before publishing the theory, this response is much too quick.[11] These decisive sources reveal a far more complicated history than this response can allow. Moreover, to see why these sources should do so, is to be in a better position to decide how far such a response can be vindicated by what has happened to natural selection theory since Darwin. In understanding how Darwin's notebook theorizing is related to the Laplacean heritage, or indeed to any prior precedents, it is indispensable to recognize that although Darwin had broken fundamentally with his mentor Lyell over the organic world, before *Notebook* B was opened in July 1837, he had not departed, nor would he ever, from Lyell's main teachings on the physical world of land, sea, and climate change.

On the physical side, then, Darwin remains in conformity with the Laplacean heritage. Lyell himself had revised but not repudiated the Huttonian theory of the earth, as it was upheld by John Playfair. A principal interpreter of Laplacean science to the British, Playfair, in his *Illustrations of the Huttonian Theory of the Earth* (1802), drew explicit analogies between Laplace's and Lagrange's conclusion as to stability in the solar system and the leading Huttonian thesis of a permanent stable balance in the actions of the igneous and aqueous agencies modifying the earth's surface.[12] John Herschel supported Lyell's neo-Huttonian position, in his *Preliminary Discourse in the Study of Natural Philosophy* (1830), and Darwin allied himself with Herschel in the late 1830s in efforts to construct a theory of crustal elevation and subsidence on the assumption that for the whole earth the forces arising from subterranean heating and cooling were balanced, and so subject only to reversible local fluctuations in the long run of the geological past, back at least to the time when the oldest known fossiliferous rocks were laid down.

Such commitments to stability and reversibility in the system of physical causation at work on the earth's surface were upheld in the face of explicit dissent from those, notably Sedgwick and Whewell, who thought these analogies between celestial mechanics and geology profoundly erroneous. As Lyell saw it, however, to give up this stability and reversibility was to forgo the possibility of meeting the *vera causa* ideal in geology. If it is to be possible to have direct evidence of the real existence of a cause deployed in explaining some past effects recorded in the rocks, then it must be at work still today. And if its adequacy for such past effects is to be properly evidenced, then we should be able to presume that it could, indeed must eventually, reproduce effects of that character and magnitude in the indefinitely long run of the future. So, this reproductive adequacy requirement presupposed limited variability in the conditions of working of these persistent causes. For, a terrestrial world without that limitation would not be one where the same causes continued to produce similar effects because working in similar conditions. Lyell held, therefore, that such a terrestrial world would not be one safe for analogical inferences from the short human period to the vast, past, prehuman periods, the inferences that allowed one to bring those periods within a science of geology conforming to the *vera causa* ideal essential to all inductive science.

For the inorganic world of land, sea, and climate changes, Darwin was always to uphold this teaching as he found it proposed by Lyell and seconded by Herschel. So we have to ask how far, in developing his own quite novel account of the organic world to complement this Lyellian one for the inorganic world, he was in intent and in effect departing from the presuppositions made by such presumptions of stability and reversibility.

5

Here, we need to consider Darwin's theorizing over the year and a half before he arrived at the theory of natural selection late in 1838. For, when he eventually constructed this theory, he drew on views about chance and chances and about *verae causae* that he had already been working with explicitly.

From summer 1837 on, his theorizing was dominated by two theses: first, that all structural change, and so all adaptive change, is ultimately due to the effects of changing conditions on the impressionable immature organization possessed by the offspring of sexual as distinct from asexual generation; second, that the outcome, in the long run, of the changes wrought by changing conditions is an irregularly branched tree of life, wherein species are multiplied when lines split without ending and become extinct when they end before splitting. So, sex is the immediate means and the tree of life is the eventual result of adaptation to an earth's surface everywhere changing à la Lyell.[13]

Direct experience indicates, according to Darwin, that sexual generation leads to adaptive heritable variation in altered conditions; it may then be a *vera causa* for unlimited adaptive diversification with unlimited time and changing circumstances. Equally, common ancestry is a known cause for similarities among relatives, especially similarities not creditable to common adaptations to common ends. So, in the tree of life, the resemblances among the species of some supraspecific group, a genus, family, or class, may be explained as due to a common inheritance from common ancestors, while their differences will be largely due to adaptive divergences.

Darwin was led to reflect on chance and chances in the propagation of the tree of life by the very considerations that led him to prefer branching descent over special, independent creations of species.[14]

Lyell had had each species created independently of any other. The character of a new species was not determined by the structure and instincts of any older species already in the area. Rather, its character and so its supraspecific type are entirely determined by the conditions where it is being created together with those conditions it is destined to meet subsequently on later spreading into other areas. Here, then, conditions determine character not only through present needs but also prospectively, in that the species, thanks to divine prevision, is provided in advance for future contingencies.

It was from this view that Darwin dissented directly. A new species, he argued, has the characters distinctive of its genus thanks to heredity, to inheritance from older congeneric species already present. And it owes the structures and instincts distinguishing it from its congeners not to any provisionary adaptation to conditions not yet encountered, but to those conditions encountered during and so determining its divergence from those ancestral stocks.

To see how this alternative to provisionary character determination by conditions led Darwin to reflect on chance and chances, consider how he used his new tree of life to reinterpret and extend various demographic analogies expounded by Lyell. For Lyell had developed an analogy between the births and deaths of human individuals and the coming and going of species as quasi individuals. The intermittent recording in any region of species births, lives, and deaths, by the fossilization process, was thus compared with the periodic visits of census commissioners to one place in the nation.[15]

Darwin continued such demographic analogies in his *Notebook* B, in elaborating his branching tree representation of species multiplications and extinctions. He dwelt, especially, on the quantitative implications of high average degrees of relatedness, many species, that is, descending from a few ancestral species, "father" species in his phrasing. With high degrees of relatedness and a constant total species population, the chances are small of any individual species having living descendant species a long time from now; and the causes determining which ones will succeed in doing so may be impossible to analyze, just as with two flourishing human families, today, where many causes, such as hereditary disease and dislike of marriage, may eventually determine that one family rather than another has living successors.

These same genealogical, demographic analogies are elaborated further in considering the wanderings and colonizations whereby species settle in fresh sites, such as new land emerging from the ocean. Here Darwin again opposes prevision and provision; for he argues that, in any generic or familial group, those species that colonize some area do so because they happen to be fitted for it by structures and habits already acquired in adapting to conditions previously encountered. One would expect, then, Darwin reflects, that by the law of chances, larger groups of species would supply more successful colonists than others.

It is likewise with the structural diversifications whereby an ancestral stock may have one or more aberrant species among its descendants, a ground-dwelling species, say, arising in a group, such as the woodpeckers, that was originally and is still predominantly arboreal. Here, too, Darwin opposes providential plans in favor of adaptive opportunities and successes that accord with circumstantial contingencies and so numerical chances; such an aberrant species has been formed when some one species among an ancestral group has succeeded in adapting to the aberrant way of life, successful aberrances happening, therefore, more often in larger than smaller groups of species.

Pursuing these quasi-demographic and quasi-genealogical concerns, Darwin considers two ways whereby the adaptation in a father species may be related to the adaptations in descendants. It could be, he reflects, that in adapting to its own circumstances the father species ensures the adaptation of the offspring species to its circumstances. Or, over the long run, it could be that a father species is adaptively influenced by a succession of changing circumstances and produces numerous varieties, among which the best adapted alone are preserved and diverge to become species.

So far, then, in his *Notebook* B, Darwin is considering adaptive variety and species formation, and supraspecific group proliferation, without relating these changes to the individual maturations that are distinctive of sexual reproduction, and that ultimately make adaptive variation possible. In *Notebook* C he does take up this challenge, and he starts with two possibilities.

On the first possibility, adaptive species formation would trace to the production of "chance offspring" characterized by some slight peculiarity and by exceptional vigor, so that, among the males, they would be eventually more successful than others in winning mates in competitive combat and in passing on their peculiarity. By contrast, on the second possibility, the first step is not a chance prenatal innovation in an offspring, but a postnatal change in habits in some parents—as when some jaguars are tempted to swim after fish prey, on their region turning swampy, the changes in structure thereby acquired then being transmitted to their offspring.[16]

Now, it is this second possibility that Darwin develops most fully in the next six months, from spring to autumn 1838.[17] And in developing it, he continues his opposition to prevision and provision in adaptive species formation. For, although the structural variations acquired in habit changes are not chance variations or chance productions, they are initiated by chance encounters with new conditions, and not in planned anticipations of changed circumstances.

Darwin makes these corollaries of his habit theorizing most explicit in his *Notebook* M, begun in July 1838, where he reflects on the human will and its apparent freedom.[18] For in higher animals, new habits may arise from willed responses to whatever conditions or circumstances are being encountered. So, if willings are uncaused and unlawful, then structural changes tracing to habit changes may be too. To avoid this conclusion, Darwin takes a resolutely deterministic line. Free will is to mind, he says, as chance is to matter; that is, in both there is a misleading appearance, but only an appearance, of a lack of lawfully determining causation. Accordingly, Darwin insists that all responses to whatever is encountered are determined by existing hereditary organization and prior education and so on.

This determinism is complemented by a no less explicit materialism whereby everything mental arises from determinate material causation. In particular, any mental traits can be transmitted without conscious awareness in the material organization passing from one generation to another. By being subsumed within the corporeal, mental changes, such as instinct changes, are brought within the general account given for all adaptive change.

Darwin's reluctance, at this time, to credit adaptive change to chance variation is not then due to any misgivings about chance productions in general and as such. For his determinism makes chance variations the products of lawful, albeit hidden, causation. And here he was in conformity with the customary view that chance congenital variants, polydactylous offspring, for example, were not uncaused but, in accord with the commonplace ignorance interpretation of chance, caused by unseen prenatal conditions. For Darwin at this time, the most evident drawback to chance variations as a contribution to adaptive species formation in the wild was their rarity. Breeders, he emphasized, could deliberately pair rare, chance, congenital variants together and so make and perpetuate a race distinguished by that peculiarity. But, in the wild, free crossing ensured that such rare, chance congenital variation is counteracted. This contrast between species, as formed adaptively in the wild, and the products of the selective breeder's art was, for

Darwin, reinforced by the reflection that the artificial, selected varieties were monstrous, not adaptive, being fitted not to natural ends or conditions but human purposes.

Nor did Darwin's thinking about chance variation in relation to species formation change on his assimilation of Malthus late in September 1838. For Darwin's immediate response was to see Malthusian superfecundity as complementing crossing in contributing to the adaptiveness of structural change in the wild. Crossing ensured that structure changes were adapted to the slow, permanent changes affecting a whole country in the long run of physical change studied by geologists. For, with crossing, any adaptive variation elicited by local fluctuating changes in conditions is eliminated through the blending in offspring of parental characteristics. Complementing this means of retention and elimination, the Malthusian crush of population ensures, further, that only variations that were adaptive for the whole life of the individual, from conception to adulthood, would be retained and so, in the course of many generations, embedded permanently in the hereditary constitution as structure is adapted to permanent long-run changes in conditions.

Such, then, was the drift of Darwin's thinking about chance and chances, in relation to sexual generation and the tree of life, before he came, most likely in late November 1838, to his positive analogy between artificial and natural selection as both means of adaptation, and so before he came to that deployment of probabilistic notions that is so characteristic of the theory of natural selection as known in the *Origin*.

Plainly, before the selection analogy was arrived at and before it transformed his thinking in decisive ways, he had already the conceptions of chance and chances that he would integrate in that analogy. Those conceptions had always been such as to rest on two assumptions for which there were ample precedents: first, that, in an individual, the future consequences of its properties may be determined by conditions that can arise independently of the conditions that produced those properties in it earlier in its life; second, that, in a population, physical property differences may be causally adequate to move outcomes away from what frequency differences, rarity and commonness, alone would determine.

To confirm that these assumptions were, in themselves, not innovative, consider how Darwin's conceptions of chance and chances relate to the old distinctions of chance, necessity, and design.

In accord with the ignorance interpretation of chance, Darwin is not ultimately working with a trichotomy of chance, necessity, and design, only a dichotomy between design and the rest. For, while he will judge a chance production unplanned, unintended, or undesigned, he will not call it uncaused. Although not known to be necessitated by discernible causal antecedents, it is presumed to be so, no less than is one of known necessitation.

As with chance, so with chances: only ignorance requires us to distinguish them from causes. Chances are what are determined by numbers, by quantitative differences, when as far as we know all else is qualitatively, physically, causally equal. When all we know of two families is that one is larger than the other, then we give it the better chance of having descendants centuries hence. As soon as we learn that it is less healthy, that bet is off, for we now know a cause sufficient to shift the chances away from where they would be were the numerical difference all that nature was working with in determining what future will be produced from the present.

A distinction between accidental and necessary adaptations can then be upheld, by Darwin, as a distinction among productions that are, as all productions are, causally necessitated. For in an accidental adaptation, its productive causes are independent of its adaptive consequences, in that the conditions producing it are not those causing it to have consequences such as to make it count as adaptive. The fetal conditions producing extra length in some puppy's legs may be arising independently of the presence of the hares that make such legs post-natally advantageous. By contrast, a variant is a necessary adaptation if it is produced by the very conditions that make it advantageous, as in the thicker fur grown by puppies moved to a cold climate or the webbed feet developed by the swimming jaguars. Unlike necessary adaptations, then, accidental adaptations are chance productions, because they are produced by hidden causes acting early enough in life to be independent of the conditions encountered later, the conditions determining their advantageous consequences for survival chances.

In his conceptions of chance and chances, Darwin was not innovative; his leading distinctions relate in traditional ways to venerable contrasts between the accidental, the necessary, and the designed. We need next to see how, nevertheless, natural selection made a novel deployment of these conceptions.

6

As it emerged in late 1838 and early 1839, in *Notebook* E, Darwin's theory of natural selection made a dual deployment of chance and chances. Heritable variants sometimes arise "by chance," and among them will be some that have a "better chance" than the normal individuals of surviving to reproduce. It is thus a matter of chance as to what variations are arising in the conditions the species is now living in, but it is not a matter of chance as to which are most successful in surviving to reproduce.

Darwin did not have this dual deployment of chance and chances until he was explicitly drawing analogies between the selective breeding practiced by man and that going on in nature. To appreciate his own understanding of the place of these probabilistic notions in his new theory, it is necessary to consider his original rationale for developing these analogies as he did.

He seems to have come to these analogies, late in November 1838, in reflecting on sporting dog varieties, such as greyhounds; in these cases, man had not only made and maintained varieties adapted to his own purposes but also, in doing so, had given them structures and instincts that would be adaptive in the wild, in preying on hares, for example.[19] Accordingly, Darwin was soon reflecting that the outward structural form distinctive of greyhounds could be produced by selective breeding away from all hunting and hares. So, he reasoned, the superior adaptive power of the selective breeding going on naturally in the wild was not due to variation arising there differently from the case of domestication; rather, this superiority was due to the greater persistence and precision of the selection that would arise in a species making its living in the wild by hunting. So, those conditions of life would not be necessary for the elicitation of the requisite variation, but they would be determining the selective retention of it required for such an adaptation to be produced.

Going further, Darwin soon concluded that the conditions a structure was adaptive for would sometimes not be sufficient. To take an example he would elaborate later, he could not see how in an area the presence of woolly animals would affect the growth of plant seeds so that they became hooked and fitted for attaching themselves to the animals.

What Darwin's new theory did, therefore, was not to make changes in conditions less determining of adaptive change, but to make them less directly so. He was content to drop the thesis that conditions always had the power to determine adaptive change directly by working heritable effects upon growth and maturation, because the analogy with the breeder's art convinced him that adequate determination would come from the way different conditions determined chances of survival and reproduction among chance variants.

Darwin was used to considering the conjunctive chances of two rare chance variants coming together to breed and so to perpetuate their peculiarity.[20] He had emphasized that, in nature, their rarity made the chances very slight, there being no inherent tendency of like variants to pair together.[21] So, on going over to selective breeding of chance variants in nature, Darwin was from the first concerned with the consequences of their rarity. He had long argued that, with the reproductive isolation of a few individuals, following migration to an island, say, the conservative effect of crossing could be circumvented as it was in the breeder's assortative matings. Selective breeding in nature was, then, for Darwin a cause that worked slowly because of the initial rarity of chance variants, and was effective despite the counteractive effect arising from the crossing of these new variants with commoner older ones.

In his new appeal to chance variation, Darwin was thus concerned not only with the lack of direct adaptive determination by conditions of variant growth in individuals but also with the determination, by frequency alone, of the populational fate of such variants, as long as no causal interactions were biasing their chances of survival and reproduction. So, natural selection was, from its very inception, a theory as to when and why frequency alone was not solely determining, because causes for bias in those chances were present.

As an agency working causally to bias population outcomes away from where frequencies alone would otherwise have them, nature as a selective breeder, in Darwin, may remind us of the demon in Maxwell.[22] However, the resemblance must not be allowed to mislead us as to the contrasting rationales motivating the two theorists' essentially different proposals. Maxwell was concerned to dramatize how utterly improbable in nature is anything like the outcome secured by the demon; for under all natural conditions there will be no such quasi-purposive interference as the demon exerts. By contrast, Darwin was out to establish that a quasi-designing form of selective breeding is an inevitable consequence of the struggle for existence and superfecundity, tendencies so ubiquitous and reliable as not to be construed as interferences at all.

These distinctive features of the theory were manifested vividly in Darwin's own earliest metaphors: of thousands of trials, that is, individual variations and struggles for survival over many generations, and of a grain in the balance, that is, of a consistent slight causal bias effective in a very long run in conformity with the law of large numbers.[23] Darwin did not need to have been a youthful afficionado of the gaming table, and of chemical paraphernalia, to find in loaded dice and tipped scales illustrations of the probabilistic causation constituting the process of change through the natural means of selection.

Darwin took the reiterated process of chance variants tried in the wild for adaptation, and so for causal bias in their chances of survival and reproduction, to be a quasi-designing process. At this time, he held ontogeny to recapitulate phylogeny; so he was also prepared to say that a growing mollusk was able to make a hinge for itself because it has in its heredity the innate, unconscious equivalent of a human craftsman's skill in making hinges, the skill gained by a long sequence of conscious trials, rejections of failure, and retention of successes.[24]

One can say, then, that Darwin gave up having variation arise as "necessary" adaptations, as necessary effects of conditions, in favor of having it arise "accidentally" or "by chance," when and only when he came to see that its fate was under the quasi-designing control of a natural selection analogous to the skilled practice of the breeder's quasi-designing art.

Darwin's insistence on the analogy, ever thereafter, was no accident of expository tactics, but an essential component in the original construction of the theory. What he did give up later, before the *Origin* (1859), was his early thesis that the maturation distinctive of the products of sexual generation was what made possible all hereditary adaptive variation. He went over to the view that the production of the male and female elements, ovules and pollen in plants, eggs and sperm in animals, was a disruptible budding process that could provide chance variation with which natural selection could work perfectly well. So, instead of the variation that made adaptation possible arising in a proper, distinctive, functioning of sexual reproduction, it now arose in disruptions of the replicative function common to all generation, sexual or asexual.

There was thus no inconsistency between Darwin's commitment to the old ideal of known causes and his acceptance of the venerable ignorance interpretation of chance. It was only in the interpretation of the production of chance variation that Darwin invoked this ignorance; and in doing so he did not preclude arguing that the process of selection of that variation counted as a *vera causa*, a cause known— from the struggle for life—to be existent in nature, and known—from the breeders' results—to be efficacious in producing permanent adaptive change in structures and instincts.

7

We are now in a position to see more clearly how Darwin's theorizing, with its twofold invocation of chance and chances, stands in relation to the legacy of the physical sciences as found in Laplacean exemplars.

For a start, we can see the disadvantages in one tempting historiographic view of the probabilistic elements in Darwin's theorizing: the view that insofar as Darwin retained deterministic, nomological commitments to reversible, stable causation, he was still in the thrall of those Laplacean physics precedents, while, by contrast, insofar as he took seriously chance occurrences, accidental changes, statistical trends, circumstantial contingencies, and the like, he was being liberated from this thralldom by those influences on him that came from the sciences of man and society, from Fergusonian conjectural history, Hartleyan associationist psychology, Queteletian societal arithmetic, Malthusian demographic theodicy, and the like.

This view is attractive because social theory obviously reflects social practice, so these influences could provide the mediation whereby Darwin's science was conditioned by his society, in accord with presumptions long made by many sociologies of knowledge.

The trouble with this view is that it tends to reduce all the diverse biological concepts of the early nineteenth century to so many echoes, emulations, borrowings, projections, extrapolations, analogies, and metaphors from physics, on the one hand, and social science on the other. And such reductions are very difficult indeed to square with all the traces we have of Darwin's deep indebtedness to rich traditions in physiology and natural history concerning vital forces, generation, geography, and so on. Moreover, it is surely arbitrary, as well as half-hearted, for a sociology of knowledge to presuppose that scientific theories are conditioned by their social context principally insofar as they are indebted to overtly social theories. More consistent and confident presuppositions for a whole-hearted sociological historiography would not require reading biological concepts and contexts out of the intellectual narrative, but would take them, and the physical science sources, too, as all suitable subjects for social conditioning, social construction, or social relations analyses.

To see one way whereby biological considerations led Darwin's theorizing away from the precedents set by the physics of the day, we may consider how heredity, for Darwin, could ensure long-run irreversibility. In Lyell, species losses, extinctions, are not reversible, in that no individual species returns once extinct; but any supraspecific group losses are reversible, for such a group may come to be represented again on the earth after a period when it has been temporarily missing. For that group, a genus or family or whatever, will reappear when conditions have changed, so as to be once more fitting for the creation of species with the structure and instincts characteristic of that group.

Darwin added heredity to adaptation in explaining the spatial and temporal representation of supraspecific groups on the constantly changing Lyellian earth's surface. In doing so, he gave a genealogical interpretation of classifactory groups that denied the possibility of any reversibility such as Lyell upheld. For Darwin, each supraspecific group has a single father, a monophyletic ancestry; and the assumption that a particular species never returns is extended to wider and wider groups, through the assumption that no group could ever arise again from a different future father species. This contingent impossibility was, for Darwin, confirmed by considering the cumulative changes in the lines leading from the simplest, remotest ancestral animals to the highest: birds and mammals. Here, he argued that heredity worked so that the oldest characters were most deeply embedded in the constitution, and so least susceptible to loss or alteration later. New characters would thus be added to these older ones and not substituted for them. The irreversibility of splitting and branching divergences likewise depends, for Darwin, on this same power of hereditary constitutional embedding, for inability or disinclination to interbreed depended on the accumulation and embedding of constitutional changes in two races descending from a common stock.

Heredity, although a principle of conservation, for Darwin, is therefore not at all closely analogous to the nearest equivalent principle in Newtonian physics: inertial mass. Nor, then, is the adaptive transformation and multiplication of species closely analogous to the outcome of successive actions of impressed forces on a body of constant mass. Selection, working with and complementing the powers of heredity, as Darwin judged it to do, was a force or cause with no close precedent in the exact sciences of the day.

However, in breaking with those precedents Darwin did not need, in his theory of natural selection, any new conceptions of chance and chances. As his heredity, adaptation, and selection theorizing developed, his twofold deployment of chance and chances could be made without breaking with available conceptions of these two.

Darwin could see natural selection as a *vera causa* because of his understanding of the relations between the short run and the long, especially as those relations were illumined by the analogy between artificial and natural selection. In the short run, in the hands of expert human practitioners, selective breeding was a causal, quasi-designing process with an approximately predictable outcome, in any instance, given a prior knowledge of the materials and the objectives. Superfecundity and the susceptibility of heredity to disruption in altered conditions entailed that an analogous process existed in nature, while comparison of that natural analogue with the achievement of the breeders indicated that, if they could produce marked racial differentiation in the short run, nature could produce distinct and diversely adapted species in her longer run.

There was the implication that the changes wrought by any selective breeding of hereditary variants would become less and less predictable, as one moved away from the shortest of artificial runs to the longest natural ones, but that loss of predictability was entirely tolerable, because Darwin was not seeking a theory as to why organisms of one particular group would tend to have descendants of another particular group: why fishes tend to have mammal descendants, or mastodons elephant descendants. There was for him no tendency in such descents, as so described. They had only happened once and were not expected to be repeated.

The ramifying, diversifying, complexifying tendencies for which Darwin sought adequate causation were general and so could be introduced with referential anonymity. His tree diagram, in the *Origin*, is labeled abstractly with letters and numbers, not proper names, so that it can represent generalizations about types of trends. To fail to appreciate the implications of this generality and abstraction is to misunderstand the entire Darwinian enterprise in the *Origin* and ever since. For their most general explanatory purposes, the generalizations Darwin and his latter-day successors have needed to establish have had to indicate, for example, when adaptive divergence rather than convergence was more, rather than less, likely to occur, or when extinctions rather than species splittings are to be expected.[25] And, in this generality and abstractness of its predictive and explanatory functions, Darwin's natural selection was in conformity with such exemplary *verae causae* as the gravitational force.

There was a break with the gravitational theory precedent, however, in that there was no law that was to natural selection as the inverse square law with proportionality to mass products was to that force. For gravitation, this law enabled the exact consequences of that force to be deduced for certain simple, suitably specified cases of one mass in motion around another in an otherwise empty universe. Natural selection has no equivalent law because its very existence requires, causally, processes of reproduction, heredity, and variation; and while these processes may be and were presumed to be conforming to laws of their own, they cannot exist and conform to those laws in an empty universe void of complex interactions between what is changed and the conditions determining how it is changed.

These sources for the lack of any equivalent for the gravitational law do not entail that selective breeding is on all occasions as likely to produce one outcome as another. Indeed, Darwin does not have to deny that, if on some occasion exactly the same heredity and variation were to be subject to exactly the same selection as on some earlier occasion, then the outcome would exactly resemble the earlier outcome. However, it is not a possibility that will ever be approximately instantiated. What Darwin needs, for his explanatory purposes, is the presumption that the departures from the impossible sequence specified by that assumption are not capricious, but are occurring because of causation similar in kind, although different in degree, from those producing the controllable, approximately predictable results of the animal and plant breeders.

That the existence of natural selection requires the causal processes of heredity and variation shows that there is no Anselmian demonstration possible for its existence from its essence, no demonstration in the style of Anselm's ontological proof for God's existence, which argued that His essence, as completely perfect, must necessarily include the perfection of actually existing. In the *Origin*, Darwin introduced the term natural selection, definitionally, after he had argued for the existence of the process in nature. And so he proposed natural selection as an appropriate name because the process was analogous to artificial selection. But this analogical definitional procedure presupposed no Anselmian aim of making selection exist in nature by virtue of its very essence as stated in its definition. On the contrary, Darwin argued for the appropriateness of the name because of the character of the process—arising from heredity, variation, and the struggle for existence—as already shown to exist in nature. He did not argue for its existence on the ground that its definition was such as to make that name appropriate.

We have seen how Darwin deployed traditional conceptions of chance and chances in a *vera causa* solution for his often unprecedented problem situation. And in doing so we are brought to see that it was intrinsic to his whole enterprise that he take up in certain distinctive ways various clusters of questions about the definition, the existence, the competence, and the responsibility of natural selection as a cause of evolution. We do well to work with such question clusters even after leaving Darwin behind.

8

On considering the first of these four clusters of questions, it will be evident that there are many ways to construe the definitional task it sets us. But that is only to be expected. Biologists call natural selection various things: an agency, a process, a factor, a cause, a force, and so on. Philosophers, meanwhile, always insist that definitions themselves come in several genres, although they rarely agree on what those genres are. The possible permutations are, therefore, plentiful for anyone proffering a definition of natural selection.

For our purposes here, however, one construal of this definitional task makes the best point of departure, for it introduces us directly to the issues of chance and causality raised by Darwin's and later versions of natural selection as a theory of evolution. On this construal, we confine ourselves, for a start, to intrapopulational selection of heritable traits distinguishing individual organisms. We take this selection to be a process, and we do what is always instructive for any process, namely, define it by stating the conditions necessary and sufficient for the process to occur. Such a definition allows us to integrate current textbook expository practices with longstanding themes about fundamental principles.[26]

In textbook presentations of definitions for natural selection, it is instructive that everything goes along standard lines only up to a certain stage, and then different authors tend to go in one of two directions. Thus, as to necessary conditions, there is agreement that these include variation, heritability of variation, and differential reproduction of heritable variation. So we quickly reach differential reproduction of hereditary variants as indispensable to the definition, because necessary for the process. Moreover, there is almost always explicit recognition that some further condition is necessary, so that once this is given there will be a set of necessary conditions that are jointly sufficient. The need for the further condition is apparent, because in a finite population of hereditary variants, even without selection, there will be differential reproduction in genetic drift, that is, in the accumulation of any successive indiscriminate or random sampling errors in the same direction. And such drift must not be allowed to count as natural selection.

However, in proposing a further condition to distinguish selection from drift, some framers of definitions lay down that the differential reproduction in selection be "consistent" or "systematic" or "nonrandom," all terms with no peculiarly biological content and drawn often from the terminology of statistics, while other authors insist that the differential reproduction must be due to differences in "fitness" or "adaptation," terms characteristic of the biologists', even formerly indeed the theologians', lexicon, terms with an apparent teleological import.

Any diversity of definitional proposals reminds us that it is rarely easy to decide on what grounds one should be preferred over another. At a minimum, however, we have surely to take two sorts of considerations into account. First, there are judgments already being made, independently of particular definitional analyses, as to which real or imaginary processes count or would count as cases of natural selection. So, we need a definition that respects these judgments without merely reflecting them uncritically. Second, we should be guided by what has motivated the development of the theory of natural selection, and by what has influenced the way this concept is related to others, just as, to take another example, with the concept of mass in physics. For, with mass, awareness of the different presuppositions about space, time, motion, matter, and force, in Cartesian, Newtonian, and Einsteinian physics, has motived decisions as to how to define mass itself and associated concepts such as weight, length, and so on.

Bringing these considerations to bear on natural selection, we are led, I shall argue, to one preferred resolution of the disagreements over what is needed, in defining selection, beyond differential reproduction of heritable variants. For we have to avoid two errors. On the one hand, we may be tempted to have a purely formal or mathematical restriction on differential reproduction in selection as distinct from drift: choosing a term such as "nonrandom" and then seeking for that term a purely formal or mathematical explication. On the other hand, we may be tempted to be finalistically biological rather than formalistically mathematical, and to require that the differential reproduction be due to differences in fitness or adaptation, with these terms explicated by reference to standards of design.

These two moves are errors, I submit, because their formalistic and finalistic quests lead us away from what is manifestly desirable: an explicit definitional insistence on causation itself, on, that is, its physical ingredients rather than on mathematical representations or teleological interpretations of its inputs and outputs. For, obviously enough, differential reproduction in selection is distinguished from any in drift by its causation; by contrast with drift, it is occurring because the physical property differences constituting the hereditary variation that is being differentially reproduced are not merely correlated with differences in reproduction—they are causally relevant to them.

When such causal relevance is present, we may call the differential reproduction nonfortuitous, a term better suited to biologists' conceptions than nonrandom precisely on account of its connotation of causation rather than mere correlation. So, intrapopulational selection may be defined as what is occurring when and only when there is the nonfortuitous differential reproduction of hereditary variants.

9

To bring out the advantages in such a physicalist, causalist explication of the concept of natural selection, we may start with those corollaries of it that can be clarified through elementary imaginary exemplars; then we can proceed to the issues raised by more complicated cases of intrapopulational selection, and by the extension of the explication to other levels, such as the interspecific. Elementary, imaginary scenarios are not to be dismissed as conceptually uninstructive merely because they are unrealistic. The interactions that make up any process that counts as a cyclone are bafflingly complex, obviously; but to acknowledge this is quite consistent with holding that the concept of a cyclone is such as to be usefully explicated through its illustrative exemplification in elementary, imaginary scenarios.

Consider two small populations of butterflies. In both, the only variation is in color: half are red, half are green, and this difference is inherited. Both populations are living and breeding in green environments and only die from predation by birds. However, one population is preyed on by birds that are color-blind, the other by birds that are color-sighted. Now consider four particular runs of breeding and predation over several generations of the butterflies. In each population, it is found that there is one run in which the proportion of red butterflies goes down from a half to about a quarter, and one run where it rises to about three-quarters.

Presumptively, only one of these four runs involves selection rather than drift, because only in one population is the heritable color difference causally relevant to survival and so to differences in reproduction. Where the predators are color-blind, increases in red or in green are equally to be expected. In the other population the increase in red is presumably an unusual outcome, as the increase in green is not. However, even in this population, we have to admit the possibility that in a particular run the red butterflies may sometimes have been picked off at a higher rate by the color-sighted birds through the bad luck of landing up near the birds, perhaps at night, rather than as a result of being spotted on account of their color.

So, to presume that the differential reproduction in such a cause is nonfortuitous, rather than fortuitous, is to presume that in this environment this physical property difference is causally relevant to reproductive success.

In making this contrast between selection and drift, we maintain an obvious analogy with paradigm cases of discriminate versus indiscriminate samplings, when those samplings are considered as physical, causal processes. If a person, without looking, is picking balls out of a bag, some of them red, some white, and differing in no other property, there is indiscriminate sampling with respect to color differences, whereas if he is looking, and going for red ones deliberately, the sampling is discriminate. Even with indiscriminate sampling no ball jumps spontaneously out of the bag into the sampler's hand, out of the population and into the sample. So, likewise with the birds and the butterflies: no spontaneous uncaused deaths are occurring whether predatory sampling is indiscriminate or otherwise. What the balls, picked by the indiscriminate sampler from the bag, have in common is that they were in the right place at the right time to come to hand. And likewise, in indiscriminate predatory sampling, a disproportionate number of red or green butterflies may happen to be in the right place at the right time—or wrong place from the prey's viewpoint.

All these samplings, indiscriminate or otherwise, are physical, causal processes with energy transformed and transmitted. So, it would be fallacious to think that because drift is a corollary of a mathematical property of the population, its finite size, while selection is a consequence of physical property differences among the individuals, it follows that drift is somehow a mathematical rather than a physical process. For selection in a finite population is sampling error, too, a discriminate sampling error. So, here, drift and selection are not to be contrasted as sampling with and without error, but as causally discriminate rather than causally indiscriminate erroneous sampling. An explanation that invokes drift invokes causation no less than a selection explanation does, but it invokes indiscriminate causation and so no causes of discrimination.

To explicate selection by contrast with drift allows for the indispensable distinction between selection for a property or trait and selection of a property or trait.[27] A property that is selected for is one that is itself causally relevant to its own differential reproduction. However, there will be selection of any properties that are correlated with one that is causally relevant, even though they are themselves not relevant causally. Thus if size differences are correlated with color differences and there is selection for color differences, then there will be selection of size differences, even if these size differences are not themselves of causal relevance, being only statistically relevant to this differential reproduction thanks to their correlation with those for which there is causal discrimination and so selection.

Defining selection as nonfortuitous differential reproduction is consistent with accepting that the instantiation of fortuitousness is description relative, so that, in our drift scenario, the deaths of the red butterflies were not fortuitous events qua deaths of butterflies, but were qua deaths of red butterflies in a population also including green ones preyed on by color-blind predators in green surroundings. For, even with such description relativity, it is still an empirical, causal matter whether a particular process of differential reproduction instantiates some description under which the process is not fortuitous.

Again, explicating selection by contrast with drift allows for—indeed encourages—the admission that in real life the ecology and genetics of butterflies and their predators may include some changes that are equally plausibly categorized as selection or drift.[28] Once one considers such familiar complications as linked genes, correlated responses to selection, patchy environments, frequency dependent effects, habitat preferences, and so on, it is possible to think of scenarios for which it is impossible to draw a sharp line between fortuitous and nonfortuitous differential reproduction. But, here, as always, it is to be emphasized that such impossibilities do not nullify the rationales for making the conceptual distinction.

To see why the distinction is indispensable, consider what sorts of causal theories of evolution are left once we have set aside orthogenetic and saltationist views that deny gradual change under the control of environmental conditions. To be schematic, we could draw up four options: two selectionist positions and two neutralist positions, where neutralism is the view that ascribes evolution to the populational fixation by drift of mutations that are adaptively neutral. For there would be Lamarckian selectionism and Weismannian selectionism, and likewise two versions of neutralism. Thus, in Lamarckian selectionism neither individual generation of hereditary variation nor its populational fate would be a matter of chance in relation to conditions. By contrast, in Lamarckian neutralism the generation but not the fate of the variants would be a matter of chance.

Now, in the last half-century, Lamarckian views have declined greatly, the relation between phenotype and genotype implied by the molecular biology of protein synthesis making them very implausible in most eyes. So, to seek to define selection in contrast to drift, as one of two options to go with Weismannist presuppositions about hereditary variation, is now the more appropriate.

10

Such a definition is implicit in Wright's familiar classification of factors of evolution. He starts from the point that in any selection, by contrast with drift, the direction and amount of the gene frequency change increment may be determinate in principle; and he concludes that selection is a "wastebasket category" including all causes of directed change in gene frequency not involving mutation or introduction of hereditary material from outside. Biologically speaking, he stresses, it will include factors as diverse as differential viability, dispersal beyond the breeding range, fertility differences, and so on.[29]

One advantage of this classificatory characterization of all selection is that we can circumvent the difficulties inherent in drawing lines between natural versus artificial selection or between natural versus sexual selection. This is an advantage, because insofar as all selection is nonfortuitous, drawing such lines is not desirable definitionally.

Another line, drawn within the category of selection, is inappropriate if our definitional proposal is acceptable; and that is Lewontin's suggested contrast between "tautological selection" and "functional selection."[30] In cases of the first, he says, we do not know why some genotype difference is subject to selection, but we think that it is and hence have to presume, tautologically as it were, that there must be some cause for its being so, while in the second the selection is arising from the known relation of a trait, such as color, to a known function, such as concealment from predation. Now, the very use of the unfortunate term tautological here may carry an unwelcome suggestion, albert presumably unintended by Lewontin, namely, that if the causal workings constituting some selection are not manifest, then we are somehow reduced to asserting even the existence of this selection, Anselm style, for reasons to do with the way selection as such is defined. But by any reasonable definition this will not be so. For a process to be nonfortuitous rather than fortuitous, in the requisite sense, is for it to be so quite independently of how far we have succeeded in observing the causal bases for the causal relevance. There have been many cases—some chromosomal polymorphisms in *Drosophila* being once a famous one—where it was accepted that some genotypic differences are causally relevant to survival and reproduction differences, and so subject to selection rather than merely to drift; and yet it remained unknown as to how and why these genotypic differences are making for this reproductive success difference.

To conclude that there is causal relevance does not require knowing, much less observing, how it works. Many experimental routines with caged populations of *Drosophila* establish that, under some range of conditions, one genotype or karyotype will improve its frequency far too consistently for drift to be a credible explanation; and yet the causal basis for this consistent superiority in reproductive performance is undisclosed.

Consider next how this definitional explication of selection can be extended to levels other than the intraspecific. Here, one needs to keep constantly in mind the conditions that must be met if the different contrasts of Lamarckian versus Weismannian and selectionist versus neutralist are to work. Failure to keep them in mind has led to confusions. Thus, for instance, it is sometimes suggested that evolution in some prebiotic molecular systems is somehow Lamarckian. But in the absence of any distinction between analogues of germ plasm and somatic tissue, or phenotype and genotype, the contrast between Lamarckian and Weismannian modes of change loses all force. Again, we have been offered various games as models for selectional processes whereby life might originate. But on inspection, it often turns out that the games are equally good models for neutralist drift scenarios, in that there is differential reproduction but it is left open as to whether it is fortuitous or not.[31]

Going the other way, to higher levels of organization than the species, there are proposals current as to "species selection." The proposals appear to presuppose a quasi-Weismannist rather than quasi-Lamarckian view of the relation between changes in conditions and the generation of interspecific differences; and the term "species selection" indicates an implicit contrast with a species level analogue of drift. But the proposals would often be clearer than they are if the quasi-Lamarckian and quasi-neutralist analogues had been argued against more explicitly.[32]

Hull has suggested that any selectionist proposal for any level can be clarified by insisting that it distinguish between replicators and interactors, examples of these being, respectively, genes and organisms at the familiar level of intraspecific selection. He emphasizes, what is more, that causal interactions, as a basis for consistent differential reproduction, are required, definitionally, if the process is to be properly called selectional. However, even Hull does not always sustain successfully the contrast between all selectional processes and any drift processes. Thus he defines a replicator as "an entity that passes on its structure directly in replication" and an interactor as "an entity that directly interacts as a cohesive whole with its environment in such a way that replication is differential." And he defines the selection process as one "in which the differential extinction and proliferation of interactors cause the differential perpetuation of the replicators that produced them."[33] Hull makes it clear, then, that he is as resolved as anyone to distinguish, causally, selectional from drift processes. However, the definition may not do this as it stands. After all, in our color-blind predation scenario there was differential extinction and proliferation of interactors (organisms) causing differential perpetuation of the replicators (genes) that produced them. So, once again, we need to insist that the differential perpetuation be caused nonfortuitously, that is, as a result of the causal relevance for reproductive success differences of those particular physical property differences.

For, consider how a difference can fail to be causally relevant, how it may be causally neutral. A genotype difference may make no phenotypic difference, or it may make a phenotype difference that the environment cannot detect and causally discriminate. By contrast, then, we should count as selection any differential reproduction of any genotypic variation that is occurring because of the causal effects of that genotypic variation. This will include, therefore, Lewontin's "tautological selection." All that some gene may do for its possessors is to enhance their enzymatic heat stability, and it may do this independently of any particular environmental conditions; but if its doing that makes a difference causally to reproductive survival, then its frequency must be credited to selection, not drift.

It will now be evident why the mathematics associated with the theory of natural selection may mislead us as to how to demarcate selection definitionally from drift. When the so-called coefficient of selection is introduced in textbooks, it is introduced merely as a coefficient of differential reproduction. So the manner of its introduction makes it equally applicable to differential reproduction in drift, provided that the drift has already occurred. Indeed, we can imagine a teacher of theoretical population genetics setting an instructive exam question. Here the student is told how some consistent differential reproduction in each generation has happened by luck, by drift, in some small population, and he or she is asked to use the mathematics introduced for selection to work out the resultant change in genotype frequencies. In an infinite population, granted, there is no random sampling error, and so any differential reproduction is selective, not drifting; and if the coefficients of selection are constant and all other sources of uncertainties are suppressed *ex hypothesi*, then the selection equations can be deterministic, not stochastic. It is only under such conditions that the so-called selection coefficient is appropriately termed; otherwise, it can mislead us in our efforts to understand selection as nonfortuitous by contrast with drift. It is on grounds of metaphysical presuppositions and empirical findings about causation, not for purely mathematical reasons alone, that we accept that deterministic equations become better representations of selective differential reproductions as population size gets larger.

Just as the implications of the mathematics of selection must be supplemented with physicalist and causalist notions as to their applicability, if they are not to mislead us in explicating the concept of selection, so with considerations of adaptation and fitness; these too will mislead us unless supplemented similarly.

11

There is no canonical explication of adaptation as it is understood in evolutionary biology. But there seems widespread agreement that adaptation is relative, in that an organism or an organ is not judged to be adapted unqualifiedly, but to be adapted or to be an adaptation, for example, to low rainfall or a herbivorous diet or for heat regulation or defense. Moreover, it seems accepted that such judgments often presuppose decisions as to how to distinguish one feature from another in the organism, and also decisions about the problems, resources, and constraints involved in its way of life. Here, the application of what have been called engineering criteria of structural and functional efficiency are often thought to have heuristic value.

On all such proposals for the explication of adaptation, we would have two ways of relating adaptation to selection. We might define adaptation narrowly enough so that some variation not counting as adaptive could nevertheless be subject to selection. Thus we might decide that a heritable prolongation of the period of fertility is not an adaptive difference even though it has selectional consequences. On the other hand, we might make having selectional consequences definitionally part of what is meant by adaptation. Either way, however, what is indispensable to the concept of selection remains the same to this extent: whether some adaptive differences are judged to be selectional differences as a matter of fact or whether all are to be so as a matter of definition, a selectional difference may still be, by definition, only one that is causally relevant to reproductive success. Putting the point another way: we should define selection by contrast with fortuitous differential reproduction. And if we do that, we may—although I see no strong reason to do so—define adaptation in terms of selection so defined. And even if we did this, selectional explanations would not lose their force when given for structures judged to be adaptations. It would be an empirical question whether a particular structure was produced by selection or not. And that question could not be justifiably settled merely by declaring in advance of inquiry into its history that it is adaptive; because for that judgment to be justified one would have had to make that inquiry. So, whatever empirical or definitional relations we decide do hold between adaptation and selection, the appropriateness of a physicalist, causalist explication of selection is not compromised.[34]

Turning now to fitness, we should note that many authors use this term more or less synonymously with adaptedness. But the concept of fitness that is of concern here is the one standard in population genetics, where the fitness of a genotype is defined in terms of the expected contribution of offspring made to the next generation by individuals of that genotype.

That fitness so defined is an expected quantity makes it appropriate to see it as a reproductive expectancy analogous to a life expectancy. Now, expectancies are in themselves not causal and so without explanatory content. If Jones has outlived Smith we cannot explain this difference by establishing that he earlier had the higher life expectancy and arguing that this duly caused him to live longer. It is physical differences that can make a causal difference and provide explanations; perhaps Smith smoked while Jones jogged. But expectancies estimated from physical differences can neither constitute causes nor provide explanations. For these reasons, it is misleading to conceive of fitness differences as causally mediating between the causes of a difference in reproductive performance and the difference in performance itself. We should not suppose that physical property differences somehow make organisms differ in fitness and that those fitness differences then somehow make them reproduce differentially. The contrasts between an expectancy and a performance or outcome must never be overlooked. According to the reference class to which an individual is assigned, he may have a high life expectancy or a low one. A smoking jogger has one qua smoker, another qua jogger, and a third—not algorithmically computable from the other two—qua smoking jogger. Likewise, then, for the genotype of an individual organism when referred to its various allelic classes, it has many reproductive expectancies. But the organism eventually makes only one reproductive contribution performance, and that is free of all reference class relativity.[35]

Because the relationship between fitness differences and reproductive performance differences is not causal and explanatory, we do not need one common but mistaken stratagem in defending natural selection against the tautology charge. It is often thought that a successful defense depends on showing that fitness differences are contingently and fallibly rather than definitionally and infallibly related to performance differences. And it is often thought that to show this requires establishing criteria of fitness independent of reproductive success.[36] But this whole line of defense is mistaken because no such defense is appropriate. All that is needed if selection is to be saved from charges of tautology is that it be an empirical question as to whether any differential reproductive performance be nonfortuitously or fortuitously caused. Suppose in some case that the expectancy judgments are turning out false. As the tautology objector insists, they can always be revised retrospectively so as to be made true retroactively. But where the tautology objector and his opponent are both mistaken is in thinking that this point about fitnesses is decisive for the empirical rather than tautologous status of selection theory. What fitness estimates are being made ahead of time, and what revisions of them are made later, does not determine whether the physical differences the estimates are based on are or are not causally relevant to the reproductive performance differences. We now know that smoking is causally relevant to lung cancer, and to know this is to know that it is so whatever odds the life insurance companies are offering the tobacco companies' best customers.

In mathematical deductions of the consequences of a fitness having a given value, understood as an expected value, no epistemic gap, so to speak, between expectation and outcome ever arises. But this does not mean that the presence of such an epistemic gap is all that distinguishes real selection processes in nature from mathematical representations of them on paper. When we ask what it is that allows us to put some empirical content with the otherwise empty formalisms of the mathematics of selection theory, the answer should not refer to the riskiness as predictions of prospective estimates of fitness differences, but to the empirical status of our conclusions about causally relevant physical property differences.

Such an emphasis on these property differences can accommodate frequency dependent effects, as when a color variant in a prey population is at an advantage in survival and reproduction only as long as it is rare. In specifying the environmental conditions wherein a physical property difference has causal relevance to survival and reproduction, we may have to specify what the populational proportions of the variants are. But to do this is not to substitute mathematical for physical factors, for the frequency differences only make a causal difference because they have different physical consequences for such processes as predator discrimination learning.[37]

Considerations of fitness, then, its probabilistic and frequency dependent aspects included, bring us back to causation rather than away from it in any explication of what selection is, in the sense of how it is to be definitionally demarcated.

These considerations also confirm that it is fundamentally mistaken to answer the question as to what selection is by proposing that it is a law of nature, like a gas law, or, alternatively, that it is a framework principle, perhaps like the principle of the uniformity of nature.[38] In answering the question as to what natural selection is, we need only to insist that there is the term "natural selection," that the term's meaning can be given in a definition, that a concept is thereby explicated, and that when that is done natural selection is, definitionally, a kind of causal process, and so neither a law nor a principle. It is these reasons for this conclusion, as much as the conclusion itself, that are useful when we turn to those further question clusters concerning the existence and the competence and responsibility of natural selection for evolution.

12

The existence, competence, and responsibility questions are not independent of one another in every way. If selection does not really exist, it cannot be competent, nor therefore responsible for evolution. Again, if selection is judged to exist and to be competent for the job and is also thought to be the only conceivable cause competent, then it must be thought responsible. Thus Dawkins has argued along these lines for this last conclusion. His argument is not Anselmian, that is, solely from the essence of a cause, as expressed in a definition, to its existence; rather, as he acknowledges, it is in the Paleyan manner of the old arguments from and to design. For he argues that selection must have been responsible for evolution because no other cause could have done it. And his reasoning explicitly depends not only on premises concerning what selection is, definitionally, deemed to be, but also on premises concerning what anything adapted and complex enough to count as organisms will be like, whether here on earth or anywhere else where they have been produced.[39]

What I ask to be conceded is thus merely that some answers to these questions leave the answers to the others open. If selection is judged to exist but not to be competent, then that rules out responsibility; but if it is judged to exist, that leaves competence as an open question, and if it is judged competent, that leaves responsibility open. For, by competence is meant competence for the sorts and sizes of change that the whole past course of evolution includes; and by responsibility is meant responsibility for these past changes. Obviously, if any selection exists in nature, it is having some consequences and so is competent and responsible for those consequences, but this does not imply competence and responsibility for evolution in the senses needed here.

It should be stressed that in this century, as in the last, the disagreements have often been over matters of more or less rather than all or none. Neutralists, today, agree with their selectionist opponents that selection exists; but they think that there is less of it. They agree that it can produce and has produced some major changes. They agree with selectionists that manifestly adaptive structures such as vertebrate limbs should be credited to evolution by selection. They doubt, however, that selection can bring about changes as quickly as drift sometimes can; they also think drift responsible for much past evolution of molecular structures. Equally, for their part, selectionists admit some drift as existing and even as competent and responsible for some evolutionary changes.[40]

The controversies between selectionists and neutralists have thrown into a clear light the relations among various answers to existence, competence, and responsibility questions. The controversies themselves, and the acceptance on both sides that empirical findings bear decisively if not conclusively on their resolution, undermine entirely any notion that natural selection as a theory of evolution is somehow forced on us a priori because it is merely tautological in the truth-functional sense of a logic text. If any such version of the tautological view were correct, selectionists and neutralists should have sought a verdict on their dispute using only a lexicon of definitions and a table of truth functions. As has rightly been observed, if selectionism is tautologous, then neutralism is contradictory.[41] Any historical and philosophical analysis of natural selection as a theory of evolution must make plain why these recent controversies have not been about whether to prefer a tautology over a contradiction.

We may start by considering the existential issue. Here we should distinguish modest claims from sweeping ones, specific claims from general ones, and claims for one level from those for others. Thus someone might hold not merely that there is some selection going on, but that for most heritable traits there exist organism-environment interactions making those traits causally relevant to reproductive success. And such a general claim for intrapopulational selection might be matched by equivalent ones for interspecific selection, for example.

The generality of such a claim allows for it to be supported empirically in many ways. One might point to particular instances where selectional causation is established, and then argue that these instances are typical.[42] Or, more in the manner of Darwin, one might argue that, from what is known generally concerning hereditary variation and concerning life, breeding, and death in natural conditions, it is probable that there is plenty of selection going on. And both lines have been taken, as can be seen from any recent survey of selectionist work.

The quantificational form of any claim may have implications for its testability: its verifiability or its falsifiability. And two points, familiar especially from Popper's teachings, should be introduced here.

The first, an uncontroversial corollary of what entailment includes, is a quite general one about falsification and verification. It is that one can falsify any statement deductively by falsifying what it entails, that is, by falsifying what can be deduced from it, but not verify it deductively by verifying what it entails, while, conversely, one can verify a statement deductively by verifying what entails it, but not falsify it by falsifying what entails it.

The second point, far from uncontroversial in its entirety, comes when Popper applies this first one to law statements. Law statements, he holds, have unrestricted universal form; the law that all metals expand when heated is universal and unrestricted in that it is about all metals and about them wherever and whenever they are, here today or on Mars a million years hence. Now, Popper insists that we cannot learn the truth of such a law statement through deductively valid inferences from any statements that we can verify. For we can verify only statements about the existence of finite numbers of instances of such a law, as we only have experiential access to these instances. However, we may have experiential access to a counterinstance, and so can learn the falsity of the law through a valid deductive inference from a statement of the existence of a counterinstance.[43]

For law statements there is thus, Popper insists, an asymmetry; they are deductively falsifiable but not deductively verifiable. Moreover, for unrestricted existentially qualified statements the reverse holds. The unrestricted statement that there exist centaurs, somewhere at some time, is deductively verifiable by finding some centaurs at a particular time and place, but not falsified by not finding some at any particular time and place. And human experience is always limited, Popper holds, to finite findings at particular times and places.

For our purposes what is decisive, as Popper himself emphasizes, is that for statements that are not unrestricted these asymmetries do not hold. A restricted statement of universal form may be both falsifiable and verifiable, as the statement is that all the finite number of snails in my finite garden during this finite summer day are hermaphrodite. This can be falsified or verified by checking its finite instances. Equally, the statement that there exist now some fish in some particular pond is an existentially qualified statement that, thanks to its restrictedness, is both falsifiable and verifiable through finite observational findings.

We see, then, that if universal or existential statements are restricted in these ways, how easily falsifiable or verifiable they are depends on their content, it being much easier to verify or falsify a claim that there is an elephant in the room now than a claim that there is a gravity wave in the room now, or the claim that there was an elephant in Scotland in Roman times.

Now, as Popper seems to appreciate, testing natural selection as a theory of evolution involves testing claims that are restricted.[44] Consider even the very general selectionist claim that for most heritable traits there exist some organism-environment interactions making them causally relevant to reproductive success. This claim is of mixed quantificational form; the logic of *most* (very different from that of *all*, obviously) and of *exists* are both in play. But also we have to appreciate that there are several restrictions implicit in the content of the claim. Organisms cannot interact now with past or future environments, only with present ones; they cannot interact instantaneously with distant present ones. That the interactions are causal means that they are subject to restrictions that hold for all causation; further, as organism-environment interactions they are implicitly restricted even more.

Nothing I have said suggests that testing existence claims on behalf of natural selection is possible in practice, much less easy. The selectionist-neutralist controversy has highlighted the difficulties. But testability in principle cannot be denied.

13

There are many diverse issues involved in the competence question cluster. We have to distinguish between real competence, that is, the abilities of natural selection as it is found to exist in nature, and the hypothetical competence of selection as specified in abstract idealized models or imaginary scenarios. There is the obvious distinction between the short-run abilities of selection as experimentally accessible to us and its long-run powers. Again, very diverse questions can be asked about the ability of selection to cause equilibrium states or to cause change, to produce a particular sort of structure, such as the python's jaw, or to produce very general trends in adaptive diversification.

What is most instructive about the answering of all such questions is that nothing is thought to be true of all the workings of natural selection—except the definitional truth that they all include the conditions necessary and sufficient for selection to occur. In that sense there is no law of natural selection, no law of action for this cause.

One should not be surprised by this lack of law. For we would not expect all the cases of causal bias in the differential reproduction of hereditary variants to have anything in common that could be summarized in a statement of law, any more than we would expect all the cases of causal bias in any other sort of process to do so. A famous attempt to legislate for natural selection, Fisher's so-called fundamental theorem, only confirms this point. The proportionality it asserts between genetic variance in fitness and rate of increase in fitness only holds under some conditions, and so is not equivalent to the proportionalities asserted for gravitational attraction by Newton. As for the comparison Fisher made between his theorem and the second law of thermodynamics, this comparison is weakened on the same account. The reasons for his theorem not holding when it does not are not matched by any equivalent limitation on the applicability of the second law of thermodynamics.[45]

It is tempting to say that there is no nondefinitional general statement of law to be made about all the workings of natural selection, because what its consequences are depends on what materials it is working with and what conditions it is working in. But this would be misleading in that it would suggest that natural selection is somehow an agency distinguishable from the hereditary materials and their interaction with their environmental conditions. And, of course, that is just what natural selection consists of—the processes of interaction—and that is why there can be no such general statement of law.

This lack of law does not make it impossible in principle to have well-confirmed generalizations about the workings of selection in specified ranges of conditions and with genetic systems of specified properties. And the contribution here of mathematical theory and experimental inquiry seems to be much as it is in most physical sciences.

Mathematics can provide the means for consequence extractions. And in this role it leaves open the status of the premises, assumptions, or axioms whose consequences it extracts. These premises can be completely conjectural or they can be statements of more or less well-confirmed generalizations about what is found in nature. In either case, the issue of responsibility for past evolution may be left untouched. Fisher knowingly invoked all the phrases associated with the old *vera causa* ideal when he explained, in the preface to his *Genetical Theory of Natural Selection*, what his book was intended to do. For it was to develop, mathematically, the generally theory of natural selection without reference to the responsibility issue. Indeed, he emphasized in correspondence that most of his book would stand if the world turned out to be created a few thousand years ago as in the Bible account.[46]

Many writers have dwelt on the deductive structures that can be given to the theory of natural selection. But what is meant and not meant by this has to be kept in mind. One may deduce a consistent differential reproduction in a postulated system of suitably variable, reproductive entities. In that sense a formal modeling of natural selection can be exhibited as a theorem derivable from an appropriate set of axioms.[47] Again, one can postulate that the conditions of Hardy-Weinberg equilibrium are satisfied by some Mendelian system, and one can then add supplementary specifications as to a coefficient of consistent differential reproduction, and so, much as in classical rational mechanics, deduce the equations for the resultant change.[48]

The fact that deductive techniques of consequence derivation have such application to formal representations of selective processes does not mean that in the defining of natural selection, and in the asking and answering of existence, competence, and responsibility questions about it, biologists are matching closely what is done by physicists working with a theory, such as Newtonian gravitation theory, that has a force law intrinsic to its very formulation. However, framing and testing generalizations about the competence of natural selection is not made impossible by the lack of such a force law analogue. A causal process can have testable causal consequences, and the theory of such a process can have derivable logical consequences, notwithstanding that lack.

Most natural selection theorists have, of course, wanted to go beyond questions about consequences, causal or logical, to the responsibility issue. They have sought to establish whether or not natural selection really did it, or at least most of it.

To establish this, more is required than merely making it plausible that selection could have caused evolution. Reconciling the supposition that it did with the fossil record or biogeographical data is not enough. Beyond that, from Darwin on, upholders of the theory have sought to show the explanatory superiority of that supposition over rival theories, in relation to such facts, and thereby to confirm that supposition itself. Insistence on some such link between explanation and confirmation has been a commonplace for centuries, and is not distinctive of biology, much less evolutionary biology. Nor is there a source of difficulty here that is sometimes thought to be peculiarly damaging to Darwinian explanations. For most structural or functional characters, it is easy to construct, imaginatively, some selectionist scenario for its production as an adaptation. Such scenarios have been deprecatingly dubbed "Just So" stories in recollection of Kipling's tale as to how, for instance, the leopard got his spots.[49] But, as the deprecators have appreciated, there is little that is surprising or distinctive about natural selection in the way it lends itself to such exercises of imaginative ingenuity. To have any theory supply a reasonably confirmed rather than merely a conjectured explanation, it has to be supplemented with auxiliary assumptions, if only assumptions as to initial conditions, and these assumptions must themselves be independently confirmed from their own evidence.

In the standard cases, such as industrial melanism in moths or sickle cell anemia in man, the requisite auxiliary assumptions about the genetics of the variation and ecology of the interactions with the environment are evidentially pretty well secured. Securing such assumptions is never easy, and it is conspicuous that in very few cases indeed has this been done. There is a vast way to go if biologists are ever to have for the evolution of reptiles from their fish ancestors any selectionist explanation that is on a par, evidentially, with what they have for the melanism and anemia cases. But moving further toward such a goal is possible in principle, if often impossibly difficult in practice; and this shows that responsibility theses on behalf of natural selection are empirical theses, testable in principle.

14

In taking natural selection as a theory of evolution conforming to the old *vera causa* ideal, we are led, therefore, to appreciate the tremendous complexities inherent in any sophisticated articulation and thoroughgoing acceptance of the theory. Simple-minded though it may be, the approach suggested by that old *vera causa* ideal can show us clearly why such acceptance cannot be reduced to anything at all simple, such as "belief in natural selection or the survival of the fittest." The answers to the existential, competence, and responsibility questions involved in accepting the theory have always been highly complex answers, conceptually and evidentially, for reasons arising from the distinctive probabilistic, causal, and empirical character of the theory itself, as it has descended historically from Darwin's own exposition. Inevitably, many writers on the theory, even today, may think their analytical obligations to the theory can be largely met by formulating some single propositional expression of a "law" or "principle" of natural selection, exhibiting its "structure" and then arguing for its "status" as "falsifiable" or "nontautologous" or whatever. But such exercises canot be counted as seriously engaging the theory as it has been developed, conceptually and evidentially, over the last century and a half. Equally, anyone working in some domain beyond evolutionary biology, whether linguistics, epistemology, or immunology, who wishes to represent his theorizing as Darwinian or selectionist, in some serious sense, should always be asking whether there are in his proposals any significant and not merely superficial analogues to the definitional, existential, competence, and responsibility issues that biologists have always had to face in making up their minds about natural selection.

This last declaration may seem to presume that there has always been a consensus among selectionists in biology over all fundamental matters. But, on the contrary, one virtue of approaching the theory from the old *vera causa* ideal is that we can see how the theory could be embraced and developed by people with strikingly contrasting metaphysical and ideological orientations.

Consider for a moment what diverse metaphysical and ideological attitudes toward chance itself are found within the Darwinian tradition. Darwin's own attitude was ultimately negative, in that he never saw the chanciness of chance variation as itself making a positive contribution to adaptation or progress. For him the decisive contrast was between the chanciness in the generation of chance variation and the determinacy in its populational fate insofar as that was directed by selection. Here, then, were echoes of all those schemes, such as Plato's in the *Timaeus*, with Manichaean affinities. Variation, as chancy, is wild, chaos, error, so many failures of orderly replication, adding up to nothing when left to themselves, while selection is the cosmogonic, quasi-designing shaper of this otherwise helpless material.

Among dominant figures in our century, Fisher can be seen as largely perpetuating similar attitudes. Moved by Christian sympathies and thermodynamical preoccupations as well as eugenic zeal, he gave us a view of the living world as subject not only, as the inanimate world is, to the degenerative tendency entailed by the second law of thermodynamics but also to another, counterdegenerative agency, natural selection, so that this natural selection, although formally comparable, according to Fisher, with the second law, is, cosmologically, to be contrasted with it, as he emphasizes quite explicitly. Thus, for Fisher, gene mutations, as errors in gene replication, and drift, as erroneous genetic wanderings in populations, can never add up to anything by themselves except bigger errors and wider wanderings. So, on the variation side, where there rule such disordering tendencies as the living world shares with the inanimate world, small is beautiful; for small variations can be most precisely shaped by selection and made into what they cannot make themselves. To have small mutations in large populations is thus the optimal combination, for then drift is minimal and variation is plentiful, while even small selection pressures can control the outcome. Thus, too, was Fisher attracted to indeterminism, as required by free will and creativity in man. For, to exercise the possibility of free will and creativity is to master and direct, as natural selection does, what is, insofar as it is indeterminate, merely a material contribution to any adaptive progress, such as human eugenic advances.[50]

By contrast, Wright, drawing on his Wundtian panpsychism and his shorthorn cattle breeding strategies, has seen chance processes as able, by virtue of their very chanciness, to contribute to the overcoming of the limitations arising from the determinacy of selection. For Wright, the stochastic exploration, by drift, of the field of variational possibilities can lead to a population moving to new adaptive peaks through selectional forces that would not otherwise have been brought into play.[51] So, a population structure that entails such drift is in the long run more optimal than one that does not. One can even think of such a population structure as a quasi-designed error machine.

Likewise, sexual reproduction itself, when it includes meiosis, with its randomized combinatorial consequences, can be conceived of as a well-designed mechanism for making replicative mistakes at the individual level. So, from such reflections, a much more positive attitude toward chance can arise, one that allows for a teleology of chance. For, if the mechanisms, such as meiosis, that now supply natural selection with its material, have themselves evolved by earlier selection, then the benefits of chanciness in variation, when changing through natural selection in an unpredictably changing world, are decisive for any understanding of how that evolution took place.[52]

Finally, an approach to natural selection through the old *vera causa* ideal does not have to deny that the theory has often been championed and elaborated by people, such as Pearson, who have followed Comtean and other positivist precedents in repudiating the very notion that science can and should construct causal rather than correlational theories and laws. The contribution of Pearson and other biometrical correlationists included new quantitative analyses of variation and differential reproduction. However, such measurement and computation analyses could proceed purely correlationally, avoiding issues of causal discrimination and relevance, only as long as questions about adaptation and population size in nature were not confronted. For, when biologists of the Pearsonian persuasion, as Weldon was, sought to instantiate empirically their claim for the effectiveness of selection in changing structures gradually and adaptively, they appealed to the sensitivity and intricacy of organism-environment interactions, as providing adequate causes, and so explanations, for the preservation of particular structures in natural populations; and in making that appeal they never claimed to reduce all their causal judgments to correlational ones.[53] It should now be plain why such a reductive claim was not made. Statistics may be a science of probabilities and correlation may be a statistical measure; but it does not follow that natural selection insofar as it is probabilistic is correlational rather than causal. A historical view of the theory confirms what philosophical examinations and scientific expositions suggest; the theory arose as a probabilistic contribution to causal science, not as a rebellious rejection of causation in favor of science without causes.[54]

A proper appreciation of Darwin's place in the probabilistic revolution can be enhanced, therefore, by recalling our own historical location. From Comte, in Darwin's day, to Hempel, in our own, empiricism in its more positivistic forms has construed the main questions about evidence and explanation in science as questions about universal statements of law rather than existential claims for causation. Darwin worked in an older, *vera causa*, tradition of empiricism than the Comtean one. We have had now two decades of philosophers priding themselves on seeing the limitations in any positivistic form of empiricism. So, perhaps, the present is an appropriate time to develop further the original interpretation of natural selection as a causal and empirical theory.

Notes

1. Two books by Michael Ruse introduce many of the issues and much of the literature: *Is Science Sexist? And Other Problems in the Biomedical Sciences* (Dordrecht and Boston: D. Reidel, 1981) and *Darwinism Defended. A Guide to the Evolution Controversies* (Reading, MA: Addison-Wesley, 1982). An invaluable collection of papers is now available: Elliott Sober, ed., *Conceptual Issues in Evolutionary Biology. An Anthology* (Cambridge, MA: The MIT Press, 1984). See also Sober's *The Nature of Selection* (Cambridge, MA: MIT Press, 1984), a masterly analysis of natural selection theory.

2. Peter Bowler brings out well the usefulness of analyzing enduring clusters of questions regarding evolution in his excellent study, *The Eclipse of Darwinism: Anti-Darwinian Evolution Theories in the Decades around 1900* (Baltimore: Johns Hopkins University Press, 1983). In doing so he acknowledges direct debts to Stephen Gould.

3. For a lucid introduction, see R. Giere, *Understanding Scientific Reasoning* (New York: Holt, Rinehart and Winston, 1979).

4. S. Haack, *Philosophy of Logics* (Cambridge: Cambridge University Press, 1978), introduces the complexities in Quine's position.

5. See, especially, W. Salmon, "Causality: Production and Propagation," *PSA 1980: Proceedings of the Biennial Meetings of the Philosophy of Science Association* (Philosophy of Science Association: East Lansing, Michigan, 1981) vol. 2, pp. 49–69, and earlier papers cited there.

6. Giere, *Understanding Scientific Reasoning* and "Causal Systems and Statistical Hypotheses," in L. J. Cohen, ed., *Applications of Inductive Logic* (Oxford: Oxford University Press, 1980). Giere's views are discussed in relation to natural selection theory in an important trio of articles: E. Sober, "Frequency-Dependent Causation," *The Journal of Philosophy*, 79 (1982), 247–253; John Collier, "Frequency-Dependent Causation: A defense of Giere," *Philosophy of Science*, 50 (1983), 618–625; and R. Giere, "Causal Models with Frequency Dependence," *The Journal of Philosophy*, 81 (1984), 384–391. The contrast between mathematical and causal considerations is insisted on in E. Sober and R. Lewontin, "Artifact, Cause, and Genic Selection," *Philosophy of Science*, 49 (1982), 147–176, an article reprinted in Sober, ed., *Conceptual Issues in Evolutionary Biology* (note 1).

7. Ruse prefers the received view, for example, in the essay "The Structure of Evolutionary Theory" in his collection *Is Science Sexist?* (note 1). J. Beatty argues for the semantic view in "What's Wrong with the Received View of Evolutionary Theory?" *PSA 1980: Proceedings of the Biennial Meetings of the Philosophy of Science Association* (Philosophy of Science Association: East Lansing, Michigan, 1981) vol. 2, pp. 397–426.

8. M. J. S. Hodge, "The Structure and Strategy of Darwin's 'Long Argument,'" *The British Journal for the History of Science*, 10 (1977), 237–246.

9. L. Laudan, "Thomas Reid and the Newtonian Turn of British Methodological Thought," in his *Science and Hypothesis: Historical Essays on Methodology* (Dordrecht and Boston: D. Reidel, 1981), pp. 86–110.

10. R. Laudan, "The Role of Methodology in Lyell's Science," *Studies in History and Philosophy of Science*, 13 (1982), 215–249; M. J. S. Hodge, "Darwin and the Laws of the Animate Part of the Terrestrial System (1835–1837): On the Lyellian Origins of His Zoonomical Explanatory Program," *Studies in History of Biology*, 6 (1982), 1–106.

11. *Notebooks* B–E were published in G. De Beer, M. J. Rowlands, and B. M. Skramovsky, "Darwin's Notebooks on the Transmutation of Species," *Bulletin of the British Museum (Natural History). Historical Series*, 2 (1960), 27–200, and 3 (1967), 129–176; and *Notebooks* M–N were published by Paul Barrett in H. E. Gruber, *Darwin on Man* (New York: E. P. Dutton, 1974).

12. On matters concerning reversibility in this period, see S. G. Brush, *The Kind of Motion We Call Heat: A History of the Kinetic Theory of Gases in the 19th Century* (New York: North-Holland, 1976) and "Irreversibility and Indeterminism: Fourier to Heisenberg," *The Journal of the History of Ideas*, 37 (1976), 603–630.

13. For a more detailed account of Darwin's pursuit of these theories, see M. J. S. Hodge and D. Kohn, "The Immediate Origins of Natural Selection," in D. Kohn, ed., *The Darwinian Heritage. A Centennial Retrospect* (Princeton: Princeton University Press, 1986).

14. Darwin's early thinking on chance and chances has been discussed in S. S. Schweber, "The Origin of the *Origin* revisited," *The Journal of the History of Biology*, 10 (1977), 229–316, and in E. Manier, *The Young Darwin and His Cultural Circle* (Dordrecht and Boston: D.

268   M. J. S. Hodge

Reidel, 1978), pp. 117–122. For a comprehensive survey of notions of chance and chances in the biology of this period, see O. B. Sheynin, "On the History of the Statistical Method in Biology," *Archive for the History of Exact Sciences*, 22 (1980), 323–371.

15. M. J. S. Rudwick, "Charles Lyell's Dream of a Statistical Paleontology," *Paleontology*, 21 (1978), 225–244, is an invaluable study.

16. *Notebook* C, MS pp. 61–63 (note 11).

17. See, further, Hodge and Kohn, "The Immediate Origins of Natural Selection" (note 13).

18. See *Notebook* M in Gruber, *Darwin on Man* (note 11).

19. For the details, again, see Hodge and Kohn, "The Immediate Origins of Natural Selection" (note 13).

20. See his *Notebook* D (note 11).

21. Compare, for example, the discussion of congenital variation in Darwin's *Notebook* C (note 11), with the view of chance varieties in J. C. Prichard, *Researches into the Physical History of Mankind*, 2nd ed., 2 Vols. (London: T. Hughes, 1826), Vol. 2, p. 548. Whether Darwin had read Prichard by this time is not clear.

22. S. S. Schweber, "Aspects of Probabilistic Thought in Great Britain: Darwin and Maxwell," in Heidelberger, Krüger, and Rheinwald, eds., *Probability since 1800: Interdisciplinary Studies of Scientific Development* (Bielefeld: Universität Bielefeld, 1983), pp. 41–97.

23. Hodge and Kohn, "The Immediate Origins of Natural Selection" (note 13), gives a fuller account of these metaphors and analogies.

24. See, for example, Darwin's notes on this, probably from March, 1839, published in Gruber, *Darwin on Man*, p. 420 (note 11).

25. On the themes of this paragraph, see D. Hull, *Philosophy of Biological Science* (Englewood Cliffs, NJ: Prentice Hall, 1974). The view that evolutionary theory is not predictive is effectively countered in two articles by M. B. Williams: "Falsifiable Predictions of Evolutionary Theory," *Philosophy of Science*, 40 (1974), 518–537, and "The Importance of Prediction Testing in Evolutionary Biology," *Erkenntnis*, 17 (1982), 291–306.

26. See, for example, F. J. Ayala and J. A. Kiger, *Modern Genetics* (Menlo Park, CA: Benjamin-Cummings, 1980), pp. 657–658, and D. J. Futuyma, *Evolutionary Biology* (Sunderland, MA: Sinauer, 1979), pp. 300–301, for careful attention to the definitional issue. For complementary emphasis on physical properties and property differences, see Sober and Lewontin, "Artifact, Cause, and Genic Selection" (note 6).

27. On the importance of this distinction, see E. Sober, "Force and Disposition in Evolutionary Theory," in C. Hookway, ed., *Minds, Machines and Evolution. Philosophical Studies* (Cambridge: Cambridge University Press, 1984), pp. 43–62.

28. See the incisive analysis of discriminate and indiscriminate samplings in selection and drift in John Beatty, "Chance and Natural Selection," *Philosophy of Science*, 51 (1984), 183–211.

29. S. Wright, "Classification of the Factors of Evolution," *Cold Spring Harbor Symposium on Quantitative Biology*, 20 (1955), 16–24.

30. R. Lewontin, "Testing the Theory of Natural Selection," *Nature*, 236 (1972), 181–182.

31. See Beatty's critique, in his "Chance and Natural Selection" (note 28), of a game proposed by Eigen and Winkler.

32. J. Maynard Smith raises such questions about such proposals in "Current Controversies in Evolutionary Biology," in M. Grene, ed., *Dimensions of Darwinism. Themes and Counter-Themes in Twentieth-Century Evolutionary Theory* (Cambridge: Cambridge University Press, 1983), pp. 273–286.

33. D. Hull, "Individuality and Selection," *Annual Review of Ecology and Systematics*, 11 (1980), 311–332—see, especially, pp. 317–318.

34. There is now a sizable literature devoted to adaptation. Two influential contributions by biologists are G. C. Williams, *Adaptation and Natural Selection. A Critique of Some Current Evolutionary Thought* (Princeton: Princeton University Press, 1966), and R. Lewontin, "Adaptation," *Scientific American*, 239 (1978), 212–230. See, further, R. N. Brandon, "Adaptation and Evolutionary Theory," *Studies in History and Philosophy of Science*, 9 (1978), 181–206. Brandon's article is in Sober, ed., *Conceptual Issues in Evolutionary Biology* (note 1), together with other related discussions of adaptation, including one by Lewontin. Many recent proposals are discussed in valuable analyses by R. Burian, "Adaptation," in M. Grene, ed., *Dimensions of Darwinism* (note 32), pp. 241–272, and by C. B. Krimbas, "On Adaptation, Neo-Darwinian Tautology and Population Fitness," *Evolutionary Biology*, 17 (1984), 1–57.

35. The literature on fitness also grows apace. In relating my suggestions to others, it would be best to start, perhaps, with R. Brandon and J. Beatty, "Discussion: The Propensity Interpretation of 'Fitness'—No Interpretation Is No Substitute," *Philosophy of Science*, 51 (1984), 342–357. There is much in the propensity interpretation, with its application of the propensity view of probability associated with Popper, that I can agree with; but a confusion surely remains precisely where we need to be unequivocal: are fitnesses as propensities causal and so explanatory or are they not? The difficulties inherent in population-genetic fitness definitions and measurements are discussed in R. Lewontin, *The Genetic Basis of Evolutionary Change* (New York: Columbia University Press, 1974); R. Dawkins gives an instructive attempt to relate population-geneticists' and ecologists' notions of fitness in his book *The Extended Phenotype* (San Francisco:Freeman, 1982), Chapter 10. For a fuller discussion of fitness and the reference class relativity of probabilities, see S. Mills and J. Beatty, "The Propensity Interpretation of Fitness," *Philosophy of Science*, 46 (1979), 263–286, reprinted in E. Sober, ed., *Conceptual Issues in Evolutionary Biology* (note 1). See also A. Rosenberg, "Fitness," *Journal of Philosophy*, 80 (1983), 457–473, and E. Sober, "Fact, Fiction and Fitness: A Reply to Rosenberg," *Journal of Philosophy*, 81 (1984), 372–383.

36. See, for example, S. J. Gould, *Ever Since Darwin* (Harmondsworth: Penguin Books, 1978), pp. 39–48, reprinted in E. Sober, ed., *Conceptual Issues in Evolutionary Biology* (note 1).

37. See Collier, "Frequency-Dependent Causation: A Defense of Giere" (note 5).

38. I can agree, then, with many points made by H. C. Byerly against E. S. Reed's proposal that natural selection is a law of nature; but I cannot accept his own suggestion that it is a framework principle: E. S. Reed, "The Lawfulness of Natural Selection," *The American Naturalist*, 118 (1981), 61–71; and H. C. Byerly, "Natural Selection as a Law: Principles and Processes," *The American Naturalist*, 120 (1983), 739–745. I am in full agreement with Beatty when he insists that there is no law or principle of natural selection. See his discussion in "What's Wrong with the Received View of Evolutionary Theory" (note 7); see also on this question, in the same book, R. N. Brandon, "A Structural Description of Evolutionary Theory."

39. R. Dawkins, "Universal Darwinism," in D. S. Bendall, ed., *Evolution from Molecules to Men* (Cambridge: Cambridge University Press, 1983), pp. 403–425.

40. Futuyma, *Evolutionary Biology* (note 26), provides a good introduction here. See, also, R. Lewontin, The Genetic Basis of Evolutionary Change (note 35).

41. M. Ruse, *Darwinism Defended*, pp. 140–141 (note 1).

42. R. Lewontin, "Testing the Theory of Natural Selection" (note 30).

43. For these themes in Popper, the best introductions are his own books, *The Logic of Scientific Discovery* (London: Hutchinson, 1959) and *Conjectures and Refutations* (New York: Harper and Row, 1968).

44. Popper's views on natural selection are discussed in Ruse, *Is Science Sexist?* (note 1).

45. Fisher's original account is in his *Genetical Theory of Natural Selection* (Oxford: Clarendon Press, 1930), a revised edition of which was published by Dover in 1958. On the conditions

for the theorem holding, see J. R. G. Turner, "Changes in Mean Fitness under Natural Selection," in K. Kojima, ed., *Mathematical Topics in Populations Genetics* (New York: Springer-Verlag, 1970), pp. 32–78; G. R. Price, "Fisher's 'Fundamental Theorem' Made Clear," *Annals of Human Genetics (London)*, 36 (1972), 129–140; and R. C. Olby, "La Théorie Génétique de la Selection Naturelle Vue par un Historien," *Revue de Synthèse: Actes du Colloque R. A. Fisher et L'Histoire de la Génétique des Populations*, 103–104 (1981), 251–289.

46. Thus Fisher to J. S. Huxley, 6 May 1930: "... if I had had so large an aim as to write an important book on Evolution, I should have had to attempt an account of very much work about which I am not really qualified to give a useful opinion. As it is there is surprisingly little in the whole book that would not stand if the world had been created in 4004 B.C., and my primary job is to try to give an account of what Natural Selection *must* be doing, even if it had never done anything of much account until now." J. H. Bennett, ed., *Natural Selection, Heredity, and Eugenics. Including Selected Correspondence of R. A. Fisher with Leonard Darwin and Others* (Oxford: Clarendon Press, 1983), p. 222.

47. M. B. Williams, "Deducing the Consequences of Evolution: A Mathematical Model," *The Journal of Theoretical Biology*, 29 (1970), 343–385, and "The Logical Status of the Theory of Natural Selection and other Evolutionary Controversies," in M. Bunge, ed., *The Methodological Unity of Science* (Dordrecht: D. Reidel, 1973), pp. 84–101. Hull, in his *Philosophy of Biological Science* (note 25), brings out the advantages and limitations of Williams's approach. See also Sober, "Fact, Fiction and Fitness" (note 35).

48. M. Ruse, "The Structure of Evolutionary Theory," in his *Is Science Sexist?* (note 1).

49. S. J. Gould and R. C. Lewontin, "The Spandrels of San Marco and the Panglossian Paradigm: A Critique of the Adaptationist Programme," *Proceedings of the Royal Society of London, Series B*, 205 (1977), 581–598.

50. Fisher's views on drift and related issues are discussed authoritatively, in this volume, in J. R. G. Turner, "Random Genetic Drift, R. A. Fisher, and the Oxford School of Ecological Genetics." For Fisher on determinism, free will, and associated subjects, see the correspondence published in J. H. Bennett, *Natural Selection, Heredity, and Eugenics* (note 46), and two published papers: "Indeterminism and Natural Selection," *Philosophy of Science*, 1 (1934), 99–117; "The Renaissance of Darwinism," *The Listener*, 37 (1947), 1001. Both are in the *Collected papers of R. A. Fisher*, edited by J. H. Bennett (Adelaide: University of Adelaide, 1971–1974). See further J. R. G. Turner, "Fisher's Evolutionary Faith and the Challenge of Mimicry," *Oxford Surveys in Evolutionary Biology* 2 (1985): 159–196.

51. The classic sources are S. Wright, "Evolution in Mendelian Populations," *Genetics*, 16 (1931), 97–159, and "The Roles of Mutation, Inbreeding, Crossbreeding and Selection in Evolution," *Proceedings of the VIth Congress of Genetics*, 1 (1932), 356–366. Wright's metaphysical views are instructively related to his biology in his "Biology and the Philosophy of Science," *The Monist*, 48 (1964), 265–90. W. B. Provine has a biography of Wright in preparation.

52. Here, issues are relevant that go back to H. J. Muller and, beyond him, to Weismann, and are now pursued in, for example, J. Maynard Smith, *The Evolution of Sex* (Cambridge: Cambridge University Press, 1978), and G. C. Williams, *Sex and Evolution* (Princeton: Princeton University Press, 1975).

53. B. J. Norton, "The Biometric Defense of Darminism," *Journal of the History of Biology*, 6 (1973), 283–316.

54. (Added in proof) On causation, probability, and explanation, see now W. C. Salmon, *Scientific Explanation and the Causal Structure of the World* (Princeton: Princeton University Press, 1984). For all of these topics the reader should now see Sober's *The Nature of Selection* (note 1).

11 Dobzhansky and Drift: Facts, Values, and Chance in Evolutionary Biology

John Beatty

In the thirties and early forties, evolutionary biologists attributed considerable significance to stochastic evolutionary changes. That is, while they attributed many evolutionary changes to natural selection, they attributed many others to chance instead. So-called "evolution by random drift" was, however, invoked less and less frequently in the period that followed (from the early forties to the early seventies). During this period, natural selection was the increasingly favored agent of evolutionary change. Empirical grounds cited in support of the shift consisted of a number of selectionist reinterpretations of evolutionary changes originally attributed to random drift.

The case of Theodosius Dobzhansky, one of the (if not the) most important figures in this shift, is considered here. Dobzhansky's case suggests, first, that the shift in question was different, in interesting ways, from the manner in which it is usually characterized, and second, that the empirical grounds usually cited for the shift actually played only a part (and by no means the most important part) in the change in thinking.

With regard to these two points, it is important to recognize that Dobzhansky always *attributed a major role to selection. He never considered random drift an important "alternative" to selection. The most he ever accorded to drift was a role that "complemented" the role of selection. In this sense, Dobzhansky was never* simply *a proponent of the importance of drift. Moreover, Dobzhansky was never* simply *a selectionist. Although, over time, he accorded selection a greater and greater role relative to drift, he never favored selection* simpliciter. *Rather, he* favored *particular forms* of selection—*forms that had some of the same "desirable" (in the value-laden sense) consequences as the combined drift-selection mechanism that he favored early on. Throughout his career, he played down the significance of* other forms *of selection—forms that had "less desirable" consequences. Value considerations thus played an important part in the shift in Dobzhansky's thinking about the relative importances of random drift and natural selection.*

1 Introduction

At this point, some evolutionists will protest that we are caricaturing their view of adaptation. After all, do they not admit genetic drift ... and a variety of reasons for non-adaptive evolution? They do, to be sure, but we make a different point. In natural history, all possible things happen sometimes; you generally do not support your favoured phenomenon by declaring rivals impossible in theory. Rather, you acknowledge the rival, but circumscribe its domain of action so

I have been helped and encouraged by many people in this and related projects. I am especially grateful to Richard Burian, James Crow, Richard Lewontin, Ernst Mayr, Diane Paul, William Provine, and all my friends in the ZiF Group.

narrowly that it cannot have any importance in the affairs of nature. Then, you often congratulate yourself for being such an ecumenical chap. We maintain that alternatives to selection for best overall design have generally been relegated to unimportance by this mode of argument. Have we not all heard the cathechism about genetic drift: it can only be important in populations so small that they are likely to become extinct before playing any sustained evolutionary role?[1]

Natural historians are reluctant to rule out anything *altogether*. Even the staunchest rivals in natural history are willing to "give or take" a few real exceptions to their positions, and to admit a few actual instances where their opponents are in the right. As evolutionary biologists Stephen Gould and Richard Lewontin note, "In natural history, you generally do not support your favoured phenomenon by declaring rivals impossible in theory." Most natural historians are, at least in this sense, "pluralists" with regard to most issues.

The snag, as Gould and Lewontin proceed to point out, is that pluralism does not preclude polarization. Natural historians may be tolerant of alternative outlooks; nevertheless, they assign different degrees of "importance" to the alternatives, and polarize among themselves as to which is the "most" important, and which the "least."

Exemplary issues in this regard concern the plurality of so-called "modes," "mechanisms," "factors," "agents," or "forces" of evolutionary change. Ever since Darwin, evolutionists have haggled over the relative importance of the various evolutionary mechanisms. Darwin and Moritz Wagner argued about the relative importance of *migration* and *selection* as agents of evolutionary change.[2] In the early twentieth century, William Bateson and W. F. R. Weldon argued about the relative importance of *mutation* and *selection* as agents of evolutionary change.[3] There have even been disputes *among* the proponents of the importance of selection as to what *kind* of selection is most important. There was, for instance, a long and complicated controversy between Theodosius Dobzhansky and H. J. Muller as to the predominance of *selection in favor of heterozygotes* versus *selection in favor of homozygotes*.[4] Representative of this interest of evolutionary biologists in "ranking" evolutionary agents is the title of a book written by the Hagedoorns in 1921: *The Relative Value of the Processes Causing Evolution*.[5]

These were not (are not) all-or-none issues. The disputants defended (defend) the importance of their favorite evolutionary agents without ruling the others entirely out of question. Darwin did not altogether deny the evolutionary significance of migration, nor did Wagner completely overlook the importance of selection. Bateson certainly did not ignore selection, nor did Weldon ignore mutation. Dobzhansky always admitted case of selection in favor of homozygotes, and Muller always admitted cases of selection in favor of heterozygotes. But their pluralism did not preclude their polarization.

The story that I am about to tell also has to do with pluralism, polarization, and the relative evolutionary importances of the various modes of evolutionary change, the issue in question being the one alluded to by Gould and Lewontin in a tongue-in-cheek manner in the opening quotation—namely, the issue of the relative evolutionary importance of random drift versus natural selection. For the time being, the following very brief discussion of the differences between evolution by random drift and evolution by natural selection will have to do.[6] Most simply put, evolution by random drift is a "matter of chance," in a sense in which evolution by natural selection is not. That is, we attribute the increase in frequency of a particular trait to natural selection when the possessors of that trait, *because of the possession of that trait*, leave a greater average number of offspring than possessors of alternative traits. We attribute the same change to random drift, on the other hand, when it is simply a *matter of chance* that possessors of the trait in question leave a greater average number of offspring than possessors of alternative traits (or otherwise a matter of chance that the relative frequency of the trait in question should increase). Sometimes, for instance, when the survival and reproductive abilities of possessors of different traits are not significantly different, possessors of one trait may just *by chance* leave more offspring on the average than possessors of alternative traits.

So much, for now, for the main conceptual differences between evolution by random drift and evolution by natural selection. As for the disputes concerning their relative importance, proponents of the importance of evolution by random drift do not deny the importance of evolution by natural selection, nor do proponents of the importance of selection completely deny the importance of drift, but their pluralism has not precluded their polarization.[7]

According to Gould, the architects of modern evolutionary theory—the authors of what we call the "synthetic theory of evolution"—were originally genuinely pluralistic with regard to the importances of the various agents of evolution (including drift and selection). But during the forties, fifties, and sixties, they are their students came to emphasize more and more the importance of natural selection, to the extent that the other agents, like random drift, became fairly unimportant. Gould has thus far documented this change—which he refers to as the "hardening of the synthesis"—as it occurred in multiple-edition works of two architects of the synthesis, Theodosius Dobzhansky and George Gaylord Simpson.[8] William Provine has documented the same sort of change in the writings of another architect of the synthesis, Sewall Wright.[9] Provine also considers briefly the same sort of change of heart in the works of Ernst Mayr, yet another important spokesman for the evolutionary synthesis.[10] Gould and, following him, Provine, have shown fairly convincingly that the attitudes of these evolutionists toward alternative evolutionary mechanisms—random drift in particular—became increasingly skeptical as they "hardened" in favor of the importance of selection.

But Gould, for one, finds it easier to document the hardening of the synthesis than to explain it. Of course, as an advocate of the original, more genuinely pluralistic pluralism, it is understandable that he finds the polarization that has since occurred unjustifiable and, at least in that sense, unaccountable. On the other hand, Gould is also a sharp and sensitive historian interested in the hardening of the synthesis as a historical phenomenon. He is right, after all; it *is* hard to understand. It is hard to see how the arguments usually advanced by those who seek selectionist accounts of everything, those to whom Gould refers as "panselectionists," stand up to the force of the pluralist appeal in natural history. The arguments in question concern the ability of selectionists to account for particular patterns of variation in nature that were originally attributed to random drift. The arguments are based, in other words, on the ability of selectionists to come up with causally significant differences in survival and reproductive capabilities where such differences were previously overlooked. Six or seven such "successes" were commonly invoked by panselectionists during the hardening of the synthesis in order to make their case for the super-importance of selection. But it is not clear how these few cases, of *all* the particular evolutionary changes that had yet (and have yet) to be adequately investigated, could have and should have decided the issue that polarized panselectionists and the increasingly outnumbered advocates of the importance of drift.

In order to get at "what more" was involved in the change, I shall investigate the case of Dobzhansky's apparent change of attitude. Of course, Dobzhansky is just one of many evolutionary biologists whose attitudes are at issue. But he is by no means *simply* one. As architects of the synthesis and historians alike have recognized, Dobzhansky's formulation of the synthesis, in his multiedition *Genetics and the Origin of Species*, was in many respects the most influential.[11] Dobzhansky made the significance of random drift clear to the community of evolutionists in the first edition of *Genetics and the Origin of Species*. As Gould documents, the significance of drift became less and less clear in later editions. Accordingly, I would add, the significance of drift became less and less clear to the community of evolutionary biologists for whom Dobzhansky's book was the bible.

But why did Dobzhansky change his mind in this regard? Certainly part of the reason, as Gould notes in passing, has to do with the course of Dobzhansky's own empirical research.[13] Much of Dobzhansky's early research consisted in documenting the existence of what originally appeared to him to be selectively insignificant variations. Differences in the frequencies of these variations in populations of the same species were originally attributed by him to evolution by random drift. But he later gathered rather indisputable evidence that the differences in the frequencies of the variations he studied were better understood in terms of selection (more specifically, in terms of a *special form* of selection that I shall discuss later). Dobzhansky's ultimate account of those variations is included in the short list of selectionist accounts—accounts of phenomena previously attributed to random drift—that panselectionists invoked in favor of their position.

But how could Dobzhansky have changed his mind concerning the general importance of drift versus selection on the basis of such a small sample of evolutionary events? Dobzhansky himself never overestimated the significance of the selectionist successes. In reporting them, he was usually careful to point out that they did not have much bearing on the all-importance of selection. This, of course, makes his own change of heart all the more enigmatic. There seems to have been no good reason, even on Dobzhansky's own admission, to have changed in this regard. It is this sort of absence of reason, I take it, that leads Gould to end his analysis of Dobzhansky's role in the hardening of the synthesis on a note of incredulity: "I do not fully understand why this hardening occurred, but I regard it as an important topic for historical research since its result so dominated the research program of evolutionary biology for many years." [13]

In what follows, I shall try to fill the apparent gap in reasoning—to show how and why the importance of selection became increasingly compelling from Dobzhansky's point of view. I shall argue that the shift was *reasoned*, leaving aside the issue of whether it was, by one or another philosophical criterion, *entirely reasonable*. [14] To this end, I must first correct, or qualify, some assumptions concerning the event to be explained. In the first place, it is important to recognize that Dobzhansky *always* attributed a major role to selection. He never considered drift an important "alternative" to selection. The most he ever accorded to drift was a role that "complemented" the role of selection. In this sense, Dobzhansky was never *simply* a proponent of the importance of drift. Moreover, Dobzhansky was never *simply* a selectionist. Although, over time, he accorded selection a greater and greater role relative to drift, he never favored selection *simpliciter*. Rather, he favored *particular forms* of selection—forms that had some of the same evolutionary consequences as the combined drift-selection mechanism that he favored early on. Throughout his career, he played down the significance of *other forms* of selection—forms that had evolutionary consequences more at odds with his early drift-selection model. I do not mean to belittle the shift in Dobzhansky's attitude toward the evolutionary significance of random drift. But we cannot hope to understand that shift until we are clear as to the *extent* of the shift.

After discussing the respects in which Dobzhansky did (and did not) "harden" with regard to the evolutionary significance of selection, I shall offer an explanation of the change—one that accords a major role to value considerations. Part of the reason that random drift was originally of such interest to Dobzhansky was that it had evolutionary consequences that were very "desirable," in a value-laden sense. This rosy assessment of the status quo was threatened by proponents of the importance of one particular form of selection, a form of selection whose prevalence would have had consequences that were very undesirable from Dobzhansky's point of view. In response to proponents of the importance of this undesirable form of selection, Dobzhansky began to champion the importance of a different form of selection that had the same evolutionarily desirable consequences as random drift, but that better guaranteed those consequences. Dobzhansky thus turned from the importance of drift to the importance of *one particular form of selection*, largely in response to a perceived threat posed by the importance of *one other form of selection*.

It is crucial to keep separate the two forms of selection in question when discussing the hardening of the synthesis. At least as far as Dobzhansky was concerned, the hardening of the synthesis was not just a matter of promoting selection, but a matter of promoting one form of selection in particular. Relative to the form of selection whose importance he minimized, the form of selection he championed was actually very similar to random drift. The latter two modes of evolution had importantly similar consequences, much valued by Dobzhansky. What the selectionist mode offered over the drift mode was a stronger guarantee that those desirable consequences would obtain.

2 The Problematic of Dobzhansky's Research and the Appeal of Chance

Dobzhansky, on the other hand—but I don't have to tell you about
Dobzhansky! Suffice it to say that you have always and characteristically
stressed the intrinsic value of variability, the flexibility of life . . . .[15]

There is a point about Dobzhansky as a natural historian that is perhaps best made
by relating the following scene involving another great natural historian, Ernst
Mayr. Perturbed by a letter he had just received, Mayr disclosed its seemingly
innocuous request. "This fellow wants some information on 'the Greylag goose.'
But there is no such thing as *the* Greylag goose! There are only Greylag *geese!*" A
nitpicking point? Not from the point of view of many evolutionary biologists and
historians of evolutionary biology, who consider this distinction one of the central
conceptual developments in modern biology. That is, it is generally acknowledged
that, from an evolutionary point of view, there is something very wrong with
"typological" views of species, according to which members of a species are all
alike, all instances of *the* specific type. Such typological views of species were better
suited to the antievolutionary perspectives of past natural historians. For them,
the perpetuation of a species consisted precisely in the production of ever more
organisms of the same type. Deviations from the type were errors—the result of
interfering circumstances that disturbed the natural tendency of organisms of a
particular type to produce offspring of the same type.[16]

From an evolutionary perspective, the occurrence of intraspecific variation is
quite natural—such variation is the material of evolutionary change. Members of
a species are not all alike—they cannot be if that species is to evolve. To speak of
the evolution of Greylag geese, for instance, is to deny that there is anything like
the typical Greylag goose.

No one, including Darwin, has emphasized the significance of intraspecific
variation more than Mayr. Nor is Mayr any greater an advocate of the significance
of intraspecific variation than Dobzhansky. Dobzhansky and Mayr together proposed and defended new definitions of the term "species" that were designed to
replace older definitions of "species" as uniform types.[17] Both were inspired in
this task, in large part, by their extensive backgrounds in field natural history.
Mayr had observed extensive intraspecific variation in South Sea island birds.
Dobzhansky had observed the same in ladybug species (of the family *Coccinellidae*),
and in fruitfly species (of the genus *Drosophila*).

What especially interested Dobzhansky were those variations whose frequencies
differed not only *within* but *among* geographically isolated populations of the same
species. These were, for instance, the sorts of data that he emphasized in an early
summary of his *Coccinellidae* studies.[18] Figures 1 and 2, taken from that summary,
show the sorts of intraspecific variations he was observing, and the extent to which
their frequencies differed from population to population within the same species.
The species in question is the geographically widespread, Asiatic species *Harmonia
axyridis*. The variations in question are color patterns. Figure 1 shows seven such
patterns found within *Harmonia axyridis*. As shown in figure 2, the frequencies of

Figure 1
Elythra variations within *Harmonia axyridis* (drawn to relative scale—see text for explanation): A, variety *succinea*; B, variety *frigida*; C, variety *novemdecimsignata*; D, variety *axyridis*; E, variety *spectabilis*; F, variety *conspicua*; G, variety *aulica*.

these patterns differ drastically among the geographically isolated populations of the species. Rather than referring to each *type of variation* as a separate race or subspecies, as was all too often the practice at the time, Dobzhansky referred to each *population* as a race, thereby extending to races the same variability that the Darwinian revolution extended to species.[19]

Dobzhansky did not work much on ladybugs after 1927, the year in which he moved from Russia to the United States to study genetics with T. H. Morgan. Like the other members of Morgan's "fly group," Dobzhansky was expected to work on *Drosophila*, the organisms that had already yielded so much information about heredity.[20]

Dobzhansky undertook *Drosophila* research without hesitation. But in the following recollection of his commitment to *Drosophila* research under Morgan, Dobzhansky reveals, interestingly enough, that a *third* group of organisms, besides ladybugs and fruitflies, was actually of *most* interest to him. The recollection thus foretells part of our story to come. "But within these limits [i.e., working on a *Drosophila* project], he [Morgan] wanted everybody to choose his own way, and I think I am not being conceited if I say that I was at that time perfectly capable of choosing my own way. I knew what I wanted to do. I was interested in problems

Population location	succ	frig	19-sig	axyr	spec	cons	aul
Altai Mts.			0.05	99.95			
Yeniseisk Province		0.9		99.1			
Irkutsk Province			15.1	84.9			
Transbaikalia (western part)		4.9	45.9	49.2			
Amur Province	7.3	29.3	41.5				
Maritime Province (Khabarovsk)	18.6	18.1	38.7	0.2	13.2	10.7	0.3
Maritime Province (Vladivostok)	16.9	31.1	37.6	0.8	6.0	6.8	0.8
Manchuria	12.9	32.8	34.0		11.2	8.6	0.5
Korea	28.1	26.6	26.6		6.2	12.5	
Japan		16.4	3.0	4.5	16.4	59.7	
China (Chi-Li, Shan-Si, Shan-Tung)	36.8	12.5	27.0		12.5	10.5	0.7
China (Kan-Su, Sze-Chuan)	3.7	26.0	40.7		11.1	14.8	3.7

Figure 2
Proportions of different elythra patterns in different populations of *Harmonia axyridis* (see text for explanation).

of evolution, evolution particularly as it applies to the case of man, although I did work, and I was planning to continue to work, on *Drosophila*."[21] As this quote suggests, and as we shall see more clearly later, Dobzhansky's investigations of *Drosophila* did not merely serve the purpose of understanding *Drosophila* genetics and evolution, but also the purpose of understanding the genetics and evolution of *Homo sapiens*.[22] But more on mankind later.

As far as intraspecific variation was concerned, *Drosophila* at first seemed to be very different from *Coccinellidae*. "Outwardly" (in a sense to be explained), *Drosophila* species seemed to substantiate older typological views of species. *Drosophila* workers had been quite taken by the extent to which members of any one *Drosophila* species resembled each other. One need only remember the *excitement* of Morgan upon discovering a mutant in his laboratory stocks, thus providing him with an alternative trait whose pattern of inheritance he could study.[23] Following the discovery of the first mutant, variations were found in ever greater numbers in Morgan's laboratory *Drosophila* stocks. But it was held by many that these variations were laboratory artifacts. *Drosophila in the wild* seemed to have few if any intraspecific variations, hence little if any material for the evolution of *Drosophila* species.

The solution to the apparent invariability of *Drosophila* species came from reflecting on the fact that the mutations that appeared in laboratory stocks were recessive mutations—i.e., mutations that resulted in observable phenotypic differences only when they occurred in double doses in the homozygous state, not when they occurred in single doses in the heterozygous state. In small laboratory stocks, in which there was significant inbreeding, these recessive mutations tended to occur in the homozygous state more frequently than they did in wild populations in which there was less inbreeding. It was for this reason that variations could be observed more frequently in the lab than in the wild. Wild populations of *Drosophila* had plenty of recessive variations after all, though mainly in the heterozygous state in which they resulted in no observable phenotypic differences. As the Russian natural historian Sergei Chetverikov suggested in 1926, "In nature the process of mutation proceeds in precisely the same way as it does in the laboratory, or among domestic animals and plants. Only a series of special conditions hampers its observation in the natural state." [24] Indeed, Chetverikov argued, since recessive mutations are so rarely *apparent* (i.e., so rarely phenotypically expressed) in wild populations, and since in their "hidden" state they are protected from selection, they can accumulate to significant frequencies within wild populations. As he put it, a wild population "soaks up" recessive variations "like a sponge": "All these mutations, originating within a 'normal' species, pass, as a result of crossing, into the heterozygous state, and are thus swallowed up, absorbed by the species, remaining in it in the form of isolated individuals. As a result, we arrive at the conclusion that a species, like a sponge, soaks up heterozygous mutations, while remaining from first to last externally (phenotypically) homogeneous." [25]

Chetverikov was able to substantiate his hypothesis about intraspecific genetic variation in *Drosophila* in the wild, basically by inbreeding flies collected in the field. [26] Several other Russian evolutionary geneticists, including H. and N. W. Timofeef-Ressovsky and Nicolai Dubinin, further substantiated Chetverikov's expectations with similar studies. [27] Dobzhansky was one of the very few in Europe or America who could follow the work of the Russian investigators, and was, accordingly, one of the first outside Russia to enter upon this program of research.

Coupled with these investigations were studies of genetic variations of a rather different sort—more *observable* genetic variations—namely, variations in chromosome structure. In the early thirties, Dobzhansky, together with another of Morgan's coworkers, Alfred Sturtevant, began to investigate structural variations in the microscopically visible chromosomes of fruitfly salivary glands. [28] The sorts of chromosomal variations that Dobzhansky studied, so-called chromosomal "inversions," are rearrangements of the genes along a chromosome—specifically, 180° turn-arounds of chromosomal segments (see figure 3).

These variations in gene arragement are not to be confused with variations in gene substance (so-called "point" mutations). But like variations in gene substance, variations in chromosome structure may serve as materials for evolutionary change. This is, in part, because variations in gene arrangement, like variations in gene substance, may cause phenotypic variations. This is also, in part, because different chromosome structures may be accompanied by differences in gene substance. That Dobzhansky considered the evolutionary role of variations in chromosome

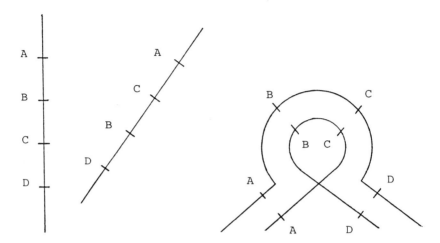

Figure 3
Chromosome inversions: (left) chromosome pair with an inverted segment; (right)
observable loop formed when the two chromosomes to the left pair together.

structure to be importantly similar to the evolutionary role of variations in gene
substance is evident enough from the organizational structure of the first edition
of *Genetics and the Origin of Species*.[29] Chapters 2 and 3 are on "Gene Mutations"
and "Mutation as a Basis for Racial and Specific Differences." Accordingly,
chapter 4 is on "Chromosomal Changes," including also an extensive discussion
of chromosomal changes as a basis for racial and specific differences.

Dobzhansky's field investigations of *Drosophila* revealed a considerable amount
of gene variation and variation in chromosome structure, hence a good deal of
material for the evolution of *Drosophila* species. As in the case of *Coccinellidae*
species, he found significant variation within and between the populations of
Drosophila species. Figure 4, for instance, shows the different frequencies of four
inversion types in eleven populations of *Drosophila pseudoobscura*.

Thus, *Drosophila* in the wild was not so different from *Drosophila* in the lab, nor
so different from *Coccinellidae*, as far as intraspecific variation was concerned. As
Dobzhansky introduced a summary of early *Drosophila* research (his and others'),

Gene mutations and chromosomal changes are currently supposed to be the
mainsprings of the evolutionary process. A rather imposing body of knowledge
has accumulated regarding these phenomena as observed under laboratory
conditions; relatively little is known about their occurrence and behaviour in
natural populations. In no other field is this disproportion more glaring than in
Drosophila genetics. In the minds of some biologists a false impression has been
created, as though a very high mutability under experimental conditions coexists
in *Drosophila* with a striking constancy under natural ones. Mutations have been
branded "monstrosities," "domestication products" and the like, and their

Population location	Arrowhead	Chiricahua	Standard	Mammoth
Lida (Mt. Magruder)	76.80 ± 1.80	6.00 ± 1.10	16.80 ± 1.59	0.40
Mt. Whitney	69.57 ± 4.57	8.70 ± 2.73	21.73 ± 4.10	
Coso	72.27 ± 1.89	14.45 ± 1.48	13.28 ± 1.43	
Cottonwood	51.20 ± 2.13	9.60 ± 1.25	38.80 ± 2.08	0.40
Grapevine (Funeral)	50.86 ± 2.23	18.70 ± 1.73	30.43 ± 2.04	
Panamint (Telescope)	67.41 ± 2.11	18.75 ± 1.75	13.83 ± 1.56	
Awavaz	62.20 ± 3.61	19.51 ± 2.95	18.29 ± 2.88	
Kingston	64.08 ± 2.25	5.34 ± 1.06	30.58 ± 2.17	
Charleston	68.75 ± 1.96	19.14 ± 1.66	12.11 ± 1.38	
Sheep Range	88.11 ± 1.53	1.98 ± 0.64	9.90 ± 1.42	
Providence	82.00 ± 1.50	10.00 ± 1.17	8.00 ± 1.06	

Figure 4
Proportions of different gene arrangements of the third chromosome in different populations of *Drosophila pseudoobscura* (see text for explanation).

existence in free-living populations was doubted or even denied outright.

In a remarkable paper published in 1926, Chetverikov has outlined a programme of studies on the genetics of natural populations of *Drosophila*. Owing to the well-known fact that most mutations are recessive and deleterious for viability, they may exist in natural populations without manifesting themselves except on rare occasions. Hence, a mere inspection of wild individuals is inadequate to detect mutations; genetic methods must be devised to reveal the concealed genetic variability. Only a quantitative study of this variability, taking into consideration factors of geography, ecology, subdividion of populations in semi-independent colonies, etc., can furnish a true picture of population structure.[30]

And as he concluded the same summary, "The gene mutations and the chromosomal changes obtained in *Drosophila* under laboratory conditions have their counterparts also in free living populations."[31]

Dobzhansky later described the significance of his *Drosophila* findings with regard to the organisms of his *ultimate* interest (here again we are looking ahead to part of our story to come):

Now we are reasonably sure that *Drosophila* populations are in this respect not particularly different from populations of other sexually reproducing outbred organisms, including man.

We are dealing with a phenomenon of fairly general significance and of considerable interest.

It was consequently about 1937 or '38—that was in my life as a researcher crucial—I became convinced that here is a chance of discovering something both of general biological interest, and something which would have bearing and relevance to man and to the human condition. We were playing with flies, but the existence of this concealed genetic variability, or of the genetic load, is a necessary consequence of the occurrence of mutation process, most mutations being harmful and natural selection not being absolutely effective. Natural selection cannot eliminate all these harmful mutations at once. They persist for a greater or lesser number of generations in the populations of flies and of man.[32]

Dobzhansky was not only concerned with documenting the extensiveness of interpopulation and intraspecific variation, though. He was also concerned with understanding how the forces of evolution act upon that variation to bring about evolutionary change. Actually, the most accurate construal of the question that most concerned him was, "How is extensive intraspecific variation *maintained*?" It was, for reasons that I hope will become clear in what follows, a matter of great significance to Dobzhansky that intraspecific variation was not "used up" in the course of evolution.

In particular, it was important to him that intraspecific variation was not used up in the course of evolution *by natural selection*. The most common conception of natural selection at the time was that it acted to reduce variation by "weeding out" the less fit variants. From Dobzhansky's point of view, it would have been a shame indeed if evolution had proceeded thus. It would have been a shame, that is, if the process of adaptation to an environment resulted in the depletion of the store of variations upon which adaptation to new, changed environments depended. As Dobzhansky expressed this, the problematic of all his reasearches, in a rather long passage in the first edition of *Genetics and the Origin of Species*,

... the accumulation of germinal changes in the population genotypes is, in the long run, a necessity if the species is to preserve its evolutionary plasticity. The process of adaptation can be understood only as a continuous series of conflicts between the organism and its environment. The environment is in a state of a constant flux, and its changes, whether slow or catastrophic, make the genotypes of the past generations no longer fit for survival. The ensuing contradiction can be resolved either through the extinction of the species, or through a genotypical reorganization. A genotypical change means, however, the occurrence of a mutation or of mutations. But nature has not been kind enough to endow the organism with the ability to react purposefully to the needs of the changing environment by producing only beneficial mutations where and when needed. Mutations are random changes. Hence the necessity for the species to possess at all times a store of concealed, potential variability. This store will presumably contain variants which under no conditions will be useful, other variants which might be useful under a set of circumstances which may never be realized in practice, and still other variants which were neutral or harmful at the time when they were produced but which will prove useful later on.

It has already been pointed out ... that mutational changes that are unfavorable under a given set of conditions may be desirable in a changed environment. Mutations that decrease viability when taken separately may have the opposite effect when combined. It should be kept in mind that selection deals not with separate mutations and separate genes, but with gene constellations, genotypes, and the phenotypes produced by them (Wright, 1931, 1932). A species perfectly adapted to its environment may be destroyed by a change in the latter if no hereditary variability is available in the hour of need. Evolutionary plasticity can be purchased only at the ruthlessly dear price of continuously sacrificing some individuals to death from unfavorable mutations. Bemoaning this imperfection of nature has, however, no place in a scientific treatment of this subject.[33]

Dobzhansky was thus concerned to understand how, *in the face of selection*, intraspecific variation could be maintained.

He had suggested, early on, in the context of his ladybug research, that interpopulation-intraspecific variation might be maintained by different selection pressures facing different populations. He suggested, in other words, that the coloration patterns that proved adaptive in one area/environment did not prove adaptive in another area/environment. And indeed there were correlations between the latitudes at which the various populations of *Harmonia axyridis* were located and the amounts of pigmentation of their elytra.[34]

I am not sure whether Dobzhansky ever changed his mind about the adaptive differences between populations of *Harmonia axyridis*—i.e., whether he ever gave up the idea that those differences were due to different selection pressures. But by 1937, when the first edition of *Genetics and the Origin of Species* was published, selection was certainly not the *only* means at Dobzhansky's disposal of accounting for interpopulation differences. The relative importance of natural selection, as compared to other evolutionary agents, was very much an open question in his mind at that time, as the following, open-minded account of the (then ongoing) dispute about the importance of selection reveals.

Whether the theory of natural selection explains not only adaptation but evolution as well is quite another matter.... No agreement on this issue has been reached as yet. Fisher (1936), who is probably one of the most extreme among the modern selectionists, has expressed his opinion very concisely as follows: "For these two theories (Lamarckism and selectionism) evolution is progressive adaptation and consists of nothing else. The production of differences recognizable by systematists [supposedly nonadaptive differences—see further] is a secondary by-product, produced incidentally in the process of becoming better adapted." And further: "For rational systems of evolution, that is, for theories which make at least the most familiar facts intelligible to the reason, we must turn to those that make progressive adaptation the driving force of the process." A good contrast to this is provided by the conclusions reached by Robson and Richards (1936): "We do not believe that natural selection can be disregarded as a possible factor in evolution. Nevertheless, there is so little positive evidence in its favor ... that we have no right to assign to it the main causative role in evolution." And: "There are many things about living organisms that are much more difficult to explain than some of their supposed 'adaptations'."[35]

G. C. Robson and O. W. Richards, to whom Dobzhansky referred in the above passage, were among a number of eminent naturalists who questioned the adaptive significance of interpopulation and interspecific differences. J. T. Gulick,

David Starr Jordan, Henry Fairfield Osborn, Henry Edward Crampton, Vernon
Kellogg, Alfred Kinsey, and Francis B. Sumner were also among the prominent
naturalists who questioned whether evolution *by natural selection* was responsible
for differentiating related populations and related species.[36] These naturalists
simply could not find any adaptive significance behind most interpopulation and
interspecific differences. As Robson and Richards summed up their own survey of
interspecific differences,

A survey of the characters which differentiate species (and to a lesser extent
genera) reveals that in the vast majority of cases the specific characters have no
known adaptive significance. A few special cases where such a significance has
been suggested are [here] considered in detail (pp. 283–290). Most of these
examples still require confirmation. As we have frequently insisted, without some
sort of direct evidence for selection such examples prove very little. It may be
conceded that in a number of instances structures apparently useless may in the
future be found to play an important part in the life of the species; further, many
'useless' characters may be correlated with less obvious features which are of real
use, but, even allowing for this, the number of apparently useless specific
characters is so large that any theory which merely *assumes* that they are
indirectly adaptive is bound to be more a matter of prediliction than of scientific
reasoning.[37]

Attempting to account for these nonadaptive differences, a number of naturalists
proposed various versions of the role of *chance* in evolution.[38] Certainly the most
influential analysis of the role of chance as an evolutionary agent was that of
Wright, although it is at least debatable whether its early influence derived as much
from Wright's version of it as from Dobzhansky's elaboration of it in the first
edition of *Genetics and the Origin of Species*.[39] Dobzhansky added a bit to Wright's
theory in the form of explication, and left out a lot of the mathematics, but was
otherwise true to Wright's analysis. (At least, he was true to Wright in the first
edition of *Genetics and the Origin of Species*, though there is some question as to
whether he remained true in later editions.[40] Of course, Wright's own version also
changed somewhat over time.[41])

Wright conceived of species as being subdivided into smaller, more or less
reproductively isolated populations within which random shifts—what he referred
to as "random drifts"—of gene frequency were inevitable. In presenting Wright's
views, Dobzhansky relied on what has since become the most popular means of
explaining why random drift in small populations is inevitable. This approach,
which he in turn borrowed from Nicolai Dubinin and D. D. Romaschov, is the
classic means of modeling chance processes—namely, the blind drawing of balls
from an urn.[42]

The balls in this case are alternative genes, i.e., different alleles at a locus. The
alternative genes are different colors, but they are otherwise indistinguishable by
a blindfolded sampling agent. One urn of balls represents one generation of alternative genes—a finite number, characterized by particular gene frequencies. The
frequencies of the next generation of genes are determined by a blind drawing of
balls from the urn. This second generation of genes fills a new urn, blind drawings

from which determine the frequencies of genes in the third generation. And so on. The frequencies of genes will inevitably differ from urn to urn—generation to generation—as a result of the fact that frequencies of otherwise indistinguishable balls sampled by blind drawings will inevitably not be representative of the frequencies in the urns from which the samples were drawn. (The probability of drawing a representative sample from a population of a given finite size is easy to calculate—the smaller the population, the smaller that probability.)

Imagine now that we start with two urns, representing two separate populations, each with the same initial gene frequencies. For each of the two urns, there is a second-generation urn, a third-generation urn, and so on. If the genes are again indistinguishable by the blindfolded sampling agent, then not only will the frequencies of genes differ from generation to generation in each set of urns, but also the frequencies of genes in the *same* generation of the two sets will inevitably differ—e.g., the frequencies of genes in the two tenth-generation urns will inevitably differ, as will the frequencies of genes in the two eleventh-generation urns, the two twelfth-generation urns, etc.

Within each set of urns, the time inevitably comes when one or another of the alternative genes attains (and thereafter maintains) a frequency of 100%. The smaller the population, the smaller the time it takes. Moreover, just as the gene frequencies change independently in different populations, different genes may become fixed in different populations. *The important point* is that, in a species divided into a number of isolated, independently evolving populations, a gene may, by chance, reach a low frequency, or even a frequency of 0%, in one population, and yet, by chance, reach a high frequency, and even a frequency of 100%, in another. So although random drift inevitably decreases variation within any one population, it promotes the maintenance of variation *between* populations, and hence *within* species.

For purposes of introductory explication, Dobzhansky (like most pedagogues since) supposed that the alternative genes in question conferred no differences in fitness upon their possessors. But as Wright had shown, and as Dobzhansky went on to explain, in small enough populations, the frequencies of genes associated with small fitness differences were also subject to random drift. As Dobzhansky put it, "These events [random drift of alternative gene frequencies] may take place even in spite of natural selection, genes which produce slight adverse effects may reach fixation and more advantageous genes may be lost."[43]

Wright originally believed that random drift in isolated populations might be part of the solution to the puzzle presented by nonadaptive differences between populations and species.[44] But he also suggested early on that random drift could play an important role in the *adaptive* evolution of species, by helping to maintain interpopulation (hence intraspecific) variation that might ultimately serve as material for the evolution of species by natural selection.[45] As he stated early on, random drift in isolated populations of a species would result in "a rapid differentiation of local strains, in itself non-adaptive, but permitting selective increase or decrease of the numbers in different strains and thus leading to relatively rapid adaptive advance of the species as a whole."[46] This eventually became

Wright's official stance on the role of drift in the evolution of species. As Dobzhansky relayed the message, "In an isolated self-sufficient population it [random drift] leads toward a genetic uniformity, a loss of variance, and consequently to a restriction of the adaptive potencies. In a species subdivided into numerous semi-isolated colonies, the same process leads toward a greater differentiation of the species population as a whole, which may mean an increase instead of a decrease of the potentialities for adaptation." [47]

In light of Dobzhansky's concern, discussed above, that the process of adaptation might, by using up available variability, preclude further adaptation, it is not difficult to imagine how comforting he found Wright's analysis of the variation-maintaining effects of random drift in appropriately structured species.

In his initial reports on intraspecific variation in *Drosophila*, Dobzhansky invoked random drift to account for interpopulation variation, with respect to both gene frequencies and inversion frequencies. Absence of any detectable correlations between the environments inhabited by different populations and the frequencies of genes and inversions within those populations licensed, in Dobzhansky's mind, the conclusion that the interpopulation differences were due to random drift rather than to different selection pressures. Granted, he had not "proved" that such correlations did not exist, and that different selection pressures had not caused the differentiation after all. But differentiation through random drift was, he insisted, a simpler account of the data. As he urged in the context of a summary of alleged instances of nonadaptive racial differences, "The difficulty of proving that a given trait has not and never could have had an adaptive significance is admittedly great; nevertheless, the facts at hand are explicable without stretching any logical point, on the assumption that racial differentiation is due to mutations and to random variations of the gene frequencies in isolated populations." [48]

Dobzhansky later changed his mind about the importance of drift, though it is not very easy to say why. Given his early hesitation to invoke selection in cases where it was not absolutely called for, we (like Gould) would expect quite *strong* selectionist support in favor of his change of heart. But there was really very *little* evidence to justify his change of attitude—there were really very few selectionist successes relevant to the shift.

But before we say anything more about those cases of selection, and about Dobzhansky's change of heart, it is worth considering in a bit more detail how he originally conceived of the relative evolutionary importances of drift and selection. At the population level—i.e., as far as the evolution of the isolated populations of a species was concerned—drift and selection were *alternative* accounts. Dobzhansky accounted for interpopulation differences via drift *instead* of via different selection pressures. At the species level, however—i.e., as far as the evolution of whole species was concerned—drift and selection were *complementary* accounts, each with a different role to play in the explanation. Drift provided or maintained the intraspecific variation that ultimately served as the material for the evolution of the species by natural selection. Indeed, at the species level, drift was complementary in a *subsidiary* sort of way to natural selection, since it played a contributory role in the evolution of species by natural selection, while natural selection did not, in turn, play a contributory role in the evolution of species by random drift. As Wright, the source of Dobzhansky's views, now responds to those who consider him preeminently a proponent of the importance of drift, "I emphasize here that while I have attributed great importance to random drift in small local populations as providing material for natural selection among interaction systems, I have never attributed importance to nonadaptive differentiation of species."[49]

Like Wright, Dobzhansky never accorded to random drift anything quite like the status he accorded to natural selection, namely, the status of being the *ultimate* agent in the evolution of species. Nevertheless, at least early on, Dobzhansky attributed to drift a substantial subsidiary role in the course of the evolution of species by natural selection. Later, he attributed much less of a role to drift. In the meantime, he had come across rather incontrovertible evidence that natural selection (rather than random drift, as he had previously supposed) controlled the population frequencies of the genes and inversions he had been studying.

In 1939, Dobzhansky first detected seasonal fluctuations in population frequencies of the chromosomal inversions he was studying. In the years that followed, he collected more and more evidence that frequencies of the various inversion types rose and dropped regularly with changes of season.[50]

It appeared then, that the inversion frequencies were not fluctuating randomly, but were under the control of natural selection. Indeed, Dobzhansky later obtained strong evidence that possessors of different inversion types had different survival abilities in different seasons. Moreover, the selection pressures in question seemed to be quite large. For, the sizes of the populations studied were small, and in small populations (as was discussed above), chance fluctuations will mask the effects of any but large selection pressures.[51]

Dobzhansky communicated the results to Wright, at first with some surprise: "This is something neither of us expected, and I must confess that when a few years ago Ake Gustafsson suggested that the changes in the old San Gabriel population may be seasonal, both myself and Sturtevant laughed at him ...."[52] A couple of weeks later, he reported with less surprise, but lingering ambivalence, "I believe there is, unfortunately or fortunately, no longer any doubt that the changes are cyclic."[53] As time passed, the results became more and more "agreeable," in a sense to be discussed.

Finding one case in which selection rather than drift was occurring was not in itself, however, enough to cause Dobzhansky to reappraise the importance of random drift. Recall Dobzhansky's concern to understand how variation could be maintained—not "used up"—in the course of evolution by natural selection. It worried him that natural selection of the fitter variants might deplete a species of its variations, thus rendering it incapable of further evolutionary change. Consider also how the evolution of a *Drosophila* species in response to seasonally fluctuating selection pressures creates just such a problem. Given the short generation time of *Drosophila*, and given strong enough seasonal selection pressures for organisms of a particular genetic type, selection could, in the course of successive seasons, purge a *Drosophila* population of the genetic types best suited for later seasons. It was Dobzhansky's discovery of a *variation-maintaining* (versus a variation-reducing) form of selection that really caused him to reappraise the importance of drift. Only a variation-maintaining form of selection provided him with a satisfactory alternative to drift.

Actually, over the years, Dobzhansky considered a number of different variation-maintaining forms of selection. First one, then another was tried out, until finally in the early fifties he settled on an account to which he adhered until his death in 1975. These mechanisms mark, as a whole, a shift in interest toward natural selection. But there is also another shift discernable *among* the various selective mechanisms entertained by Dobzhansky. The latter shift is as important a part of our story as the former. So it is worth considering briefly several of those mechanisms in the order in which they were proposed.

All the mechanisms Dobzhansky proposed depended on the assumption, in support of which he gathered substantial evidence, that the different inversion types under investigation were accompanied by differences in gene content as well.[54] He attributed the differential success (in different seasons) of possessors of different inversion types not to the different gene *rearrangements*, but to the differences in gene *content* associated with the different arrangements. Dobzhansky's attempts to give a selective account of variation in chromosome structure were thus, at one and the same time, also attempts to give a selective account of variations in gene content.

All the mechanisms proposed by Dobzhansky depended also on a second assumption that was already well established at the time—an assumption concerning gene exchange between chromosomes during cell replication. During cell division, when the two homologous chromosomes of each pair come together, genes at the same locus may be exchanged. It was widely held (and is still held) that when the two homologous chromosomes of a pair differ by an inversion, gene exchange *in the area of the inversion difference* is unlikely. Physiological reasons concerning the difficulty of close pairing between chromosomes with inversion differences, and hence the difficulty of exchanging genes between them, were cited.[55]

In the case of the first selection mechanism proposed by Dobzhansky, genes improving the fitness of their possessors in different seasons became somehow (through some series of historical accidents) associated with different inversion types. Since different inversion types do not exchange genes, the associations in question were maintained. Thus, different inversion types regularly attained their highest frequencies in different seasons.

But that did not explain how, in the course of a season—and especially in the course of an "extreme" season—the frequency of the inversion containing the most advantageous gene did not rise to somewhere near 100%, resulting in the complete elimination of the intrapopulation variation. That did not explain how, in other words, a population retained inversion types containing genes that were not the most advantageous in the season at hand, but that would be advantageous in following seasons. Dobzhansky explained all that as follows. Every inversion type has, for the same sorts of reasons that Chetverikov offered, at least one recessive mutation associated with it that is lethal, or that is at least deleterious when doubled up in the homozygous state. For any inversion type to reach 100% frequency within a population, each organism in the population would have to have a double does of the type in question. But organisms with double doses of any inversion type are severely selected against. The most advantageous inversion in any season thus confers its advantages only when paired with a different inversion type—i.e., when in the heterozygous state. Consequently, the most advantageous inversion never increases in frequency to 100%. Alternative inversions, and hence the alternative genes they contain, are thus retained in the population, even during seasons when one particular gene, associated with one particular inversion, confers the greatest survival and reproductive benefits on its possessors. As Dobzhansky summarized this variation-maintaining form of selection,

A mutant gene which is advantageous at a certain season arises in a chromosome with, for example, the Standard arrangement [the "Standard" inversion type]. The descendants of this chromosome may attain a considerable frequency within the colony, and if the gene in question lies in part of the chromosome in which crossing over is rare in inversion heterozygotes, the association may persist for a long time. In another colony the same mutant gene, however, may arise in a chromosome with a different gene arrangement. Thus, in some colonies a gene favorable at a given season may be associated with chromosomes with the Standard arrangement, in other colonies with Arrowhead [another inversion type], and in still others no such association may exist. The fact that most chromosomes in natural populations of *Drosophila pseudoobscura* contain deleterious recessives (Dobzhansky, Holz, and Spassky, 1942) will, of course, counteract too high an increase in frequency of the descendants of an individual chromosome, even if that chromosome contains a mutant which is *per se* favorable. The optimum condition is probably the presence in a population of a variety of chromosomes with different gene arrangements and different gene contents.[56]

Note that, in the case of this first selection mechanism, there is, for any given environment, an optimal gene for every locus. What prevents these optimal genes from prevailing over all other alternatives in the environments for which they are optimal—i.e., what circumvents Dobzhansky's worst fears—is just that the optimal gene at every locus is inevitably linked to a lethal recessive at another locus, which means that possessors of the optimal gene are at an extreme disadvantage when the optimal gene, and hence the lethal recessive to which it is linked, is doubled up in the homozygous state. So homozygotes for the "otherwise" optimal gene at a locus are actually at a disadvantage to heterozygotes at that locus, though (and this is the important point) not because of their heterozygosity with respect to that locus, but rather because of their associated heterozygosity at the loci occupied by the lethal recessives.

Looking ahead for just a moment, in the cases of the second and especially the third selection mechanisms defended by Dobzhansky, heterozygotes at most loci are at an advantage to homozygotes at those loci *precisely* because of their heterozygosity at those loci. There are, in other words, no optimal genes at these loci—no genes that, other things being equal, will prevail over all other alternatives. Genes work best in combinations with alternatives, so it does not make sense to talk about any one optimal gene at a locus.

The second mechanism involved a rather different set of basic assumptions than did the first. Two of those assumptions were especially important. First, Dobzhansky assumed that inversions were a way of keeping together on one chromosome those genes that worked well together. Moreover, he added, a gene complex held together by an inversion on one chromosome often worked best in combination with an *alternative* complex on the homologous chromosome. Within any population, the alternative gene complexes held together by inversions were "coadapted" to one another, in the sense that those inversion types that worked well together in the heterozygous state were retained in moderate frequencies, while those inversion types that did not work well in combination with the rest were selected against. As Dobzhansky summarized this form of selection,

The chromosomes with different gene arrangements carry different complexes of genes (arising ultimately through mutation). The gene complexes in the chromosomes found in the population of any one locality have been, through long continued natural selection, mutually adjusted, or "coadapted," so that the inversion heterozygotes possess high adaptive values. But the genes in chromosomes with the same or with different gene arrangements vary also from locality to locality. The gene complexes in different localities are not coadapted by natural selection, since heterozygotes for such foreign gene complexes are seldom or never formed in nature. Heterosis [heterozygote advantage] is, therefore, an outcome of a historic process of adaptation to the environment.[57]

Finally in the third selection account, Dobzhansky placed the greatest weight on heterozygous advantage, and extended his account beyond genetic variations associated with inversion variations to genetic variations in general. Evidence gathered by his coworkers M. A. Vetukhiv and D. Brncic suggested that heterozygous advantage was not restricted to inversion heterozygotes, but applied to individual heterozygous loci, independently of their association with heterozygous inversions. Vetukhiv's and Brncic's evidence also suggested that heterozygous advantage was not a matter of the coadaptation of alternative genes and gene complexes within a population. Rather, their evidence suggested, heterozygosity *in and of itself* was more advantageous than homozygosity.[58]

Around the same time that Vetukhiv's and Brncic's results were published, Dobzhansky's colleague and close friend I. M. Lerner proposed a theory of "genetic homostasis," which also had some impact upon Dobzhansky's thinking.[59] According to this theory (at least according to Dobzhansky's understanding of it, which seems to differ from Lerner's own version), heterozygotes were often fitter than homozygotes by virtue of the dual capacities that result from having two alternative genes for a characteristic rather than two genes of the same kind. This flexibility would supposedly prove important in fluctuating environments, where specialized homozygotes would fare poorly relative to more versatile heterozygotes. Heterozygosity thus "buffered" fitness in variable environments, and in that sense regulated fitness "homeostatically." As Dobzhansky quoted Lerner in this regard, "In his important book on *Genetic Homeostasis*, Lerner (1954) concludes 'that heterozygosity has a dual function in the life of Mendelian populations. On the one hand, it provides a mechanism for maintaining genetic reserves and potential plasticity [for the evolution of populations and species], and on the other, it permits a large proportion of individuals to exhibit combinations of phenotypic properties near the optimum. Underlying both processes is the superior buffering ability of heterozygotes as compared with homozygotes'."[60]

Combining the empirical considerations of Vetukhiv and Brncic and his version of the theoretical considerations of Lerner, Dobzhansky placed more weight than ever on heterozygosity per se as the reason for the preservation of intraspecific variation. Variation not only contributed to the adaptive flexibility of species, but first and foremost (in a causal sense) to individual organisms as well. Dobzhansky found further support for this notion in field studies and experiments of his own.[61]

It is worth pausing now to consider how much Dobzhansky's later selection accounts (especially the last one) differ from his early drift account of intraspecific variation. What is the same, in all accounts, is the natural preservation of intraspecific variation. In the drift account, the separate populations of a species may become uniform for particular variations, but the chance that they will all become uniform for the same variation is small, so *interpopulation*, or intraspecific, variation is bound to be preserved. In the selection accounts defended by Dobzhansky, not only is intraspecific variation preserved, but *intrapopulation* (emphasis on "intra-") is as well. And the reason is that, in the case of these mechanisms, *intraorganismic* variation is advantageous—i.e., heterozygous organisms are at an advantage to homozygous organisms. Note that the three selection mechanisms, inasmuch as they preserve *intrapopulation* variation, as well as *interpopulation* variation, better guarantee the preservation of intraspecific variation, since they do not rely for their variation-maintaining ability on the assumption that the species is subdivided into small populations.

The selection mechanisms also differ *among* themselves in interesting respects. In the first selection account, heterozygotes at a locus are at an advantage to homozygotes at that locus, but only as a result of the "mitigating" chromosomal circumstances explained above—not on account of their heterozygoisty per se. Were it not for these mitigating circumstances, optimal genes would ultimately prevail over all other alternatives at all loci. In the second and especially third selection accounts, on the other hand, heterozygotes at most loci are at an advantage to homozygotes at those loci precisely because of their heterozygosity at those loci. Genes work best in combination with alternatives, so it does not make sense to talk about optimal genes that would prevail over all alternatives, other things being equal. The third selection mechanism places the most weight on heterozygosity per se as the reason why variation is maintained in populations and hence in species. For in the case of the third mechanism, unlike the first and second mechanisms, heterozygous advantage is not limited only to those heterozygous loci that are associated with inversion heterozygotes, but extends to most all individual loci.

On all the positions entertained by Dobzhansky, intraspecific variation was good for the species, though it was not maintained *on that account*. That it was good for the species was rather a happy result of the evolution of species by random drift (he once thought) or by special forms of natural selection (he later thought).

In time, Dobzhansky found better guarantees for intraspecific variation, in the form of selection for variation at the organismic level (heterozygous advantage). Moreover, the various conceptions of heterozygous advantage that he entertained placed increasingly more importance on heterozygosity per se. Variation was thus, for Dobzhansky, not only an ideal that he hoped to find in nature (at the species level), but also (at the organismic level) the means upon which he relied more and more to "secure" that end.

During the forties and fifties, while Dobzhansky was trying out various selectionist accounts of chromosomal inversion frequencies, other investigators were succeeding in providing selectionist accounts of other characteristics previously thought to have no adaptive significance. Of particular importance were the selectionist accounts offered of color patterns in the snail genus *Cepaea* and of blood groups in humans.

Arguments for the importance of random drift in both these cases had been made in the early forties, around the same time that Dobzhansky had been pursuing drift accounts of his *Drosophila* inversions. Charles Diver, for one, argued that the frequencies of the various banding and color patterns in species of *Cepaea* were governed mainly by drift.[62] Wright himself had suggested that blood group frequencies within human populations were governed mainly by drift. Differences between alternative blood groups were, Wright claimed at the time, of no adaptive significance "as far as known."[63]

Diver's and similarly minded views on *Cepaea* were criticized severely by A. J. Cain and Phillip Sheppard in the early fifties.[64] They had found that the seemingly innocuous differences in color (yellow versus brown) and banding pattern (0–5 bands) of these snails were correlated with differences among the environments in which populations of *Cepaea* species lived. In particular, they found that the color and banding patterns most frequent in an environment were those that provided more camouflage in that kind of environment, thus better protecting the snails from their bird predators. Interestingly enough, they proposed a variation-maintaining form of selection, involving heterozygous advantage, to account for the fact that particular color and banding patterns reached high frequencies in particular environments *without reaching frequencies of 100%* in populations in those environments.[65]

The change of attitude that took place with regard to the insignificance of blood-group variations was succinctly expressed by the human geneticist William Boyd in the mid-fifties: "Realization of the preeminent role of natural selection in evolution has been slow. Until recently it was the fashion ... to state that characteristics suitable for the classification of Man into races should be non-adaptive, meaning not influenced by selection, and the present author ... was maintaining in 1940 (Boyd, 1940) that the blood groups were suitable for racial classification partly because they were nonadaptive. That point of view has been completely abandoned ...."[66] Among other things, workers found that possessors of different blood groups were susceptible in different degrees to different diseases. For instance, O-group individuals were shown to be more susceptible to peptic ulcers than either A- or B-group individuals, while A-group individuals were shown to be more susceptible to stomach cancer.[67] No straightforward correlations between blood group frequencies and environments were established in the time period in question, but findings that different blood groups had different effects on survival led many investigators to expect that such correlations would be found. Though there was also little evidence in favor of one or another form of selection governing blood-group frequencies, there was initially considerable faith that some form involving heterozygous advantage must be responsible for maintaining blood-group variability.[68]

In retrospect, Dobzhansky's selective account of chromosomal inversions in *Drosophila* pretty well stood the test of time. But there has been considerable controversy concerning the selective accounts of color patterns in *Cepaea* and blood groups in humans. Maxine Lamotte has argued forcefully that selection alone is not responsible for *Cepaea* color patterns—that drift is also largely responsible.[69] And Luigi Cavalli-Sforza has successfully accounted for differences in blood-group frequencies among different populations in northern Italy in terms of drift alone.[70] Though, of course, what has happened in the meantime does not really concern us here.

Combined with Dobzhansky's success in accounting for *Drosophila* inversions in terms of selection, the apparent successes of selectionists in accounting for color patterns in *Cepaea* and blood groups in humans seemed to many of them to constitute overwhelming support for their efforts. It was particularly important to proponents of the importance of selection in the fifties and sixties to trot out these successes as evidence of the shortsightedness of proponents of the importance of drift, as though the issue of the relative evolutionary importance of drift versus selection rested on these few cases alone.[71]

Whether so few successes could *reasonably* have added up to a significant case against the importance of drift, the fact is that these cases *were* thought by selectionists to be devastating to Wright's position (and hence to Dobzhansky's early position) on the importance of drift. In 1950, Dobzhansky reported to Wright the victorious mood of the British selectionists in particular.

That you and your works have been annihilated in Oxford is well established by testimony of several high authorities. You seem to be the only person who has any doubts about it. (1) E. B. Ford announced it in a triumphant letter to Cain, a very capable (really capable) young Oxford instructor (or whatever they call it in Oxford), who works on snails and who has spent the last winter with Mayr and myself. . . . (2) Julian Huxley in the first draft of his speech which he delivered in Columbus last September has stated that the genetic drift has recently been shown to be an imaginary phenomenon. . . . (3) Darlington has spent a few days with us living in our house and he inquired whether or not I am satisfied that all your stuff is dead. He had no fixed opinion himself, but this is clearly the consensus of those who know in England or at least in the Cambridge-Oxford circles.[72]

Dobzhansky reported this news to Wright tongue-in-cheek. And yet he did not go on to offer Wright much consolation or support in the rest of the letter. Rather, he proceeded directly to report further findings of heterozygous advantage with regard to his inversion types. Following that, he concluded with this final remark: "My principal work for this winter is preparation of a 3rd edition of 'Genetics and the Origin of Species'. The damned thing must be rewritten almost entirely—partly because so much new work has been done partly because I have changed my own opinions on some things."

Included among the revisions in the third edition of *Genetics and the Origin of Species* were new views on the relative evolutionary importances of random drift and natural selection.[73] Consider, for instance, how Dobzhansky rewrote the passage quoted above (p. 283) from the first edition of that book. In the third edition, he no longer referred to Robson's and Richard's doubts about the all-importance of selection as a "good contrast" to Fisher's enthusiastic pan-selectionism, but only as a "contrast." Moreover, instead of just acknowledging the difference of opinion that existed between evolutionists like Fisher, on the one hand, and Robson and Richards, on the other (as he had in the first edition), Dobzhansky added an assessment of the merits of the two positions—one that was very critical of the latter. Having first quoted Fisher on the all-importance of natural selection, then Robson and Richards on the limited importance of the same, Dobzhansky reported that "the development of population genetics in the last two decades has considerably strengthened the theory of natural selection. It is fair to say that, among the two opinions just cited, the first is believed by a majority of modern evolutionists to be much nearer the truth than the second."[74]

This is a fairly straightforward expression of what Gould has described as the "hardening of the evolutionary synthesis." Dobzhansky acknowledged the hardening of the synthesis even more explicitly in 1970 in his *Genetics of the Evolutionary Process*, which was something like a fourth edition of *Genetics and the Origin of Species*:

In the nineteen thirties, Wright's random drift idea appeared to offer a ready explanation of adaptively neutral changes, although, as pointed out at the beginning of this chapter, Wright himself stressed the interactions of deterministic [selective] and stochastic processes, and not drift alone. For about a quarter of a century thereafter, a hyperselectionism became fashionable. All differences between populations and species were assumed to be products of natural selection. Even so judicious an author as Mayr (1963) wrote, "Selective neutrality can be excluded almost automatically wherever polymorphism or character clines (gradients) are found in natural populations," and "For these reasons, it appears probable that random fixation is of negligible evolutionary importance."[75]

Thus far, I have stressed the empirical grounds for the hardening of the synthesis—what grounds there were. It is hard to imagine what sort of evidence, in addition to the evidence of selection that Dobzhansky gathered in the case of *Drosophila*, and in addition to the evidence mustered in support of the other selectionist successes discussed above, could have been invoked in behalf of the hardening of the synthesis. Gould, for one, doubts that this amount of evidence justified the change of attitude, especially given that proponents of the importance of drift—good pluralists that they were—never claimed that drift was *all-important*. As Gould argues,

Empirical aspects certainly influenced the hardening. Once the pluralistic version had reemphasized classical Darwinism as a respectable alternative, the search for actual measures of selection and adaptation in nature intensified and succeeded. The British panselectionist school, headed by Ford and Cain (see Ford, 1963, for example), presented many examples from butterflies and snails. More importantly, Dobzhansky, the key figure in the transition, discovered that his favorite example of potential nonadaptation needed to be reinterpreted in selectionist terms. In 1937, he attributed differences in inversion frequencies within natural populations of *Drosophila* to genetic drift, but he then discovered (see Dobzhansky, 1951) that these frequencies fluctuate in a regular and repeatable way from season to season, and decided (with evident justice) that they must be adaptive. Still, we surely cannot attribute such a major change as the hardening of the synthesis entirely to induction from a few empirical cases. After all, no pluralist had doubted that selection regulated many examples, so the elegant display of a few should not have established a generality.[76]

With regard to Dobzhansky's own change of attitude, Gould recognizes that Dobzhansky's empirical findings must have been "part" of the story, but could only have been *part*.[77]

It is indeed questionable whether selectionist successes like those discussed here justified Dobzhansky's or anyone else's role in the hardening of the synthesis. Dobzhansky himself was well aware that the case for the all-importance of selection had yet to be made. He always distanced himself from those whom he considered guilty of "hyperselectionist" excesses, like the members of the British school of evolutionary geneticists whose views he reported in the letter to Wright quoted above (p. 33).[78] He also warned others against placing too much weight on the importance of the selectionist successes. For instance, in reporting the conclusions of Cain and Sheppard in the case of *Cepaea*, he reasoned,

Cain and Sheppard (1950) find ... that at least some of these microgeographic variations (of *Cepaea hortensis*) are correlated with certain features of the local environment. Thus, unbanded reddish snails are common to populations of beechwoods, while banded ones predominate in the hedgerows. Cain and Sheppard rightly argue that such correlations indicate that the variations observed by them fall in the class of adaptive polymorphism, possibly balanced polymorphism .... The local differentiation is then due to natural selection and not to genetic drift. It should, however, be noted that not all variable characteristics in the snail show such correlations, and, hence, the participation of the genetic drift in the microgeographic differentiation is by no means excluded in the species which Cain and Sheppard have studied.[79]

Dobzhansky's hesitation to push the selectionist successes of the forties, fifties, and sixties too far must please Gould to some extent, but must also frustrate him considerably. Dobzhansky did, after all, change his mind about the importance of drift. But *why* did he change his mind about the importance of drift, if not just on account of empirical findings?

Before I turn to further reasons for Dobzhansky's change of heart, let me just summarize briefly the respects in which Dobzhansky did and did not harden against drift and in favor of selection. As was explained above, Dobzhansky never touted drift as a strict alternative to natural selection, as far as the evolution of species was concerned. Rather, following Wright, he conceived of random drift and natural selection as complementary mechanisms of the evolution of species. Random drift of gene frequencies within populations contributed to the maintenance of interpopulation—and hence intraspecific—variation, upon which natural selection ultimately acted in the process of adaptively modifying species.

It would have been a shame, from Dobzhansky's point of view, if selection of the fittest variations resulted in a depletion of intraspecific variation, restricting the possibility of *further* evolution by natural selection. There had to be some way of maintaining intraspecific variation, in the face of the selection of the fitter variations, if species were to remain responsive to selection pressures. Drift played that role. Thus, the position away from which Dobzhansky shifted his position was not *simply* a prodrift position, but rather a position according to which drift and selection *together* brought about evolutionary change, with drift actually playing the more subsidiary role of "serving" natural selection.

Moreover, Dobzhansky's later perspective was not *simply* selectionist. Rather, he championed *particular* forms of selection, variation-maintaining forms, which solved the same sorts of problems that drift had earlier been invoked to solve. Dobzhansky's role in the hardening of the synthesis cannot be understood without first recognizing that he was never *simply* a drifter, nor ever *simply* a selectionist.

3   The Appeal of Selection

Commonly, the social theorist [read "theorist"] is trying to reduce the tension between a social event [read "event"] or process that he takes to be real and some value which this has violated. Much of the theory work is initiated by a dissonance between an imputed reality and certain values, or by the indeterminate value of an imputed reality. Theory-making, then, is an effort to cope with a threat; it is an effort to cope with a threat to something in which the theorist himself is deeply and personally implicated and which he holds dear.[80]

Throughout the shift in his attitude toward the relative importance of natural selection versus random drift, Dobzhansky never wavered in his opposition to the importance of *one particular form* of selection. Thus, it is quite misleading to say of him that he was, later in his career, *simply* a slectionist—at least as misleading as to say of him that he was, earlier in his career, *simply* a proponent of the importance of drift.

It is worth considering the form of selection that Dobzhansky opposed, however, not just in order to make the point that the later Dobzhansky is mischaracterized as a selectionist *simpliciter*, but also to clarify further the shift in his thinking about the relative evolutionary importances of drift and selection. It is important to recognize that at the same time he was moving away from his prior interest in drift, he was also distancing himself more and more from a form of selection whose importance he had always found "objectionable" in a sense to be explained. Relative to this "foil" position, there are some clear respects in which Dobzhansky never changed his mind, even in the course of the hardening of the synthesis. The effect of taking this foil position into consideration is to minimize somewhat the differences between Dobzhansky's earlier and later positions. The earlier and later positions that Dobzhansky entertained actually look much more alike when viewed in contrast to the particular view of evolution that he so strongly opposed. His long-standing objections to this foil position "constrained," in this sense, the other changes in theoretical outlook that he underwent.

The form of selection whose importance Dobzhansky seems *always* to have questioned was a variation-reducing form championed by his archrival, the geneticist H. J. Muller. Best known for his work in classical genetics, specifically for his work on x-ray induced mutations (for which he was awarded the Nobel Prize), Muller was also known in evolutionary circles for his very traditional view of natural selection.

The controversy in which Dobzhansky and Muller were rivals persisted throughout the fifties and sixties, until Muller's death in 1967. During this time, they did their best to split fellow geneticists and evolutionary biologists along the dividing lines that they had imposed. By the mid-fifties, their positions and assembled schools had names. Dobzhansky identified himself with the "balance" view of genetic variation and evolution, and identified Muller with the traditional-sounding "classical" view.[81]

On the classical view, selection is supposed to be constantly reducing intra-population and intraspecific variation. There is supposed to be an optimal state for every character, and likewise an optimal gene or set of genes for every character (i.e., an optimal allele at every locus). Organisms that have a homozygous pair of the optimal gene or set of genes are supposedly fitter than organisms that have any other homozygous or heterozygous pair. Once the optimal genes have been introduced into a population by mutation, selection supposedly favors them to the exclusion of all other alternatives. In this sense, selection is supposed to result primarily in the elimination of variation.

Muller was convinced, on the basis of his experiences with x-ray induced mutations, and on the basis of theoretical considerations, that extant species are already quite close to being perfectly adapted.[82] The optimal genes and sets of genes for most characters of most species are already the norm. Presently, then, mutational changes are most always deleterious—a fact apparently well confirmed by his own work on the disadvantageous effects of x-ray induced mutations. Muller believed that selection currently serves only the task of weeding out variations that continue to arise. Any variations present in a population are just variations yet to be eliminated.

How different were Dobzhansky's various accounts of evolution by natural selection. In none of them was intraspecific variation belittled as simply variation yet to be eliminated. In all of them, variation was *intrinsic*.

The first selection mechanism that Dobzhansky entertained came closest to the sort of mechanism championed by Muller. In the case of both, there are optimal

	Situations predicted by extreme models based on	
	Homozygote superiority	Heterozygote superiority
Selection's goal—the "ideal" genotype	ABCDEFGH	$A_1B_9C_2D_7E_3F_5G_4$
	ABCDEFGH	$A_7B_5C_6D_2E_1F_4G_3$
Genotype of an average individual under ordinary conditions	ABCDeFGH	$A_1B_9C_7D_6E_4F_5G_8$
	aBCDEFGH	$A_1B_6C_5D_7E_4F_9G_{23}$
Genotype of an individual that is homozygous for a chromosome of the sort commonly found in populations	ABCDeFGH	$A_1B_9C_7D_6E_4F_5G_8$
	ABCDeFGH	$A_1B_9C_7D_6E_4F_5G_8$
Genotype of an individual similar to the one above but now with a new mutation (') in the heterozygous condition	ABCDeFGH	$A_1B_9C_7D_6E_4F_5G_8$
	Ab'CDeFGH	$A_1B'\,C_7D_6E_4F_5G_8$

Figure 5
The classical and balance positions (see text for explanation).

genes for all loci. But in the case of Dobzhansky's first mechanism, there are additional factors (involving chromosomal "linkages" between otherwise optimal genes and lethal recessives) that prevent the optimal genes from completely prevailing, and that thus result in the preservation of a good deal of variation at each locus.

Far more different from Muller's position was Dobzhansky's third and final view of evolution by natural selection—the view he referred to as the "balance" view. In the balance view, there are no "optimal" genes—genes work best in heterozygous combinations. Thus heterozygotes at a locus are fitter than homozygotes at the same locus, precisely because of their heterozygosity at that locus. The result of this sort of heterozygous advantage is a balance of the frequencies of genetic variations present at each locus (no single variation ever prevails as long as heterozygous combinations do better than homozygous combinations).

Figure 5 is the first diagrammatic illustration of the classical and balance views.[83] Here we see "ideal" and "average" chromosome pairs, according to the "extreme" classical and balance models. The ideal classical chromosome pair is homozygous at all loci; the ideal balance chromosome pair is heterozygous at each locus. The average classical chromosome pair, however, inevitably has some heterozygous loci due to mutations that selection has yet to eliminate. The average balance chromosome pair inevitably has some homozygous loci due to inbreeding within a finite population. (For our purposes, the bottom two rows of the figure can be ignored.) Notably, in neither model does the variation at a locus have anything to do with drift.

Dobzhansky and Muller, as I said, did their best to split fellow geneticists and evolutionary biologists along the dividing lines of their dispute. And they succeeded in large measure. I think it is quite fair to say that this was considered the most important dispute within evolutionary biology in the sixties. That a dispute among *selectionists*—about which form of selection is the most important agent of evolution—should have captured the interests of evolutionary biologists during this period is, of course, a pretty good indication that interest in the selection-drift controversy had waned considerably by this time. With the classical-balance controversy, selection took center stage in evolutionary biology. This did not escape the attention of the few proponents of the importance of drift at the time. As one of them wrote in a review of a book about the classical-balance controversy, "This is an important and balanced statement of the *selectionist* point of view in population genetics."[84]

Whether interest in drift had already died by the time the classical-balance controversy really began to rage in the fifties, and/or whether the raging of the classical-balance controversy distracted evolutionists away from their prior interests in drift, I am not certain. It may also be the case, as William Provine has suggested to me, that the drift-selection controversy had pretty much "run its course," in the sense that it had proved fairly irresolvable, by the mid-fifties.[85] In this case, the classical-balance controversy would not *initially* have been responsible for the waning interest in the drift-selection controversy, but it may nonetheless have contributed to a prolonged neglect of the drift-selection issues. At any rate, as the classical-balance controversy grew in intensity, the drift-selection controversy subsided.

As philosophers of science have recently urged, we are unlikely to understand why scientists take the positions they do without considering the *rival* positions from among which they may choose.[86] Scientific positions are not *intrinsically* worthy of pursuit, nor intrinsically unworthy of pursuit. All scientific positions suffer from anomalies of one sort or another. The available rival scientific positions at any one time offer different strengths and different weaknesses. Some offer *fewer* and/or *less significant* problems than their rivals, and perhaps solve *more* problems, or at least solve problems at a *greater rate*. Taking a position in science involves *comparing* positions to see what each has to offer over the others. There is no scientific position whose "acceptability" or "pursuability" is entirely intrinsic.

For just these sorts of reasons, the classical theory is an important "foil" position in our story. Dobzhansky took his early prodrift position *against* the classical theory (though not necessarily against Muller's articulation of it). He also took his later probalance position *against* the classical theory (this time against Muller's version in particular). Of course, he adopted his later probalance position also over his early prodrift position. We have already considered, in part, what the probalance position offered over the prodrift position, although perhaps what it offered in that respect seems so slight that it is still incomprehensible why Dobzhansky would have moved to that pole. Actually, I think that one of the most important, "missing" parts of that story concerns Dobzhansky's opposition to the classical theory. The shift in Dobzhansky's views on the relative importance of drift and selection is somewhat more understandable in light of his long-standing opposition to the classical position.

What the prodrift and probalance positions *both* offered over the proclassical may already be obvious. The classical theory all but denied the existence of the variation that Dobzhansky and othe naturalists had observed in the field, explaining it away as variation on the decline, variation yet to be eliminated. Suffice it to say, however, that the field data that Dobzhansky collected did not spell doom for Muller's theory, and that Muller and the other proponents of the classical view had considerable empirical and theoretical arguments against Dobzhansky's position. The closest thing to *compelling* evidence against Muller's position was not gathered until 1966, the year before Muller's death.[87]

Perhaps, though, the most telling advantage of the prodrift and probalance positions over the proclassical, as far as Dobzhansky was concerned, was a matter of *value* rather than a matter of fact—i.e., a matter of the way Dobzhansky *wanted* the world to be, rather than simply a matter of the way it was.

How did Dobzhansky want the world to be? He wanted it to be full of variation. In particular, he wanted the human species to be full of variation. Remember that Dobzhansky was ultimately interested not in the condition of ladybugs and fruit-flies, but in the human condition. His *Coccinellidae* and *Drosophila* studies were means to this end—means not only of understanding but also of protecting (in ways that we shall consider shortly) the human estate.

In all the accounts of evolution that Dobzhansky considered seriously, intra-specific variation was a virtue, and a *naturally occurring* virtue at that. Recall his belief that intraspecific variation rendered species more adaptable to changing circumstances. As for *Homo sapiens* in particular, Dobzhansky believed that human variation might play a part in enabling the species to respond adaptively to environmental upheavals of various sorts:

Let us suppose that the human species inhabits an environment at least as uniform as the ones used for laboratory animals. It is just conceivable that a genetic endowment might ultimately become selected which would be the best possible one for that particular environment. Any further genetic change could then only be harmful. The reality is otherwise. Not only do people follow different ways of life, engage in different occupations, have different duties and interests, but human environments change rapidly and most rapidly of all in technologically advanced societies . . . .

Environmental instability presents challenges to the organism—both to an individual and to a population or a species. To maintain itself in harmony with a changing environment, the organism must be not only adapted but also adaptable (Thoday, 1953, 1955, Waddington, 1957). A species should not only possess genetic variety but also be able to generate variety. It may then respond to changing environments by genetic changes.[88]

Assuming that humans, like ladybug and fruitfly species, had sufficient intraspecific genetic variation to ensure their adaptability, Dobzhansky judged the future of the human species to be quite rosy. His world was thus a happy one.

The classical position on evolution by natural selection represented a threat to all this. It did not threaten the view that intraspecific variation was necessary for the adaptive flexibility of species, but it did threaten the view that species actually possessed the intraspecific variation necessary for such adaptive flexibility. By denying the natural prevalence of intraspecific variation, Muller seriously threatened the "rosiness" of Dobzhansky's worldview.

A worse threat yet was posed by the variation-reducing eugenic directives associated with the classical position. Muller himself had long defended variation-reducing eugenic policies on the supposition that variations are generally fitness decreasing and are eventually eliminated by natural selection. According to Muller, the protections of civilization against natural selection had resulted in the accumulation of ever-increasing loads of variation. He further believed that unless loads were artificially eliminated, they would accumulate to such an extent that available therapies would be insufficient to save the human species from extinction.[89]

No matter that variation-reducing eugenicists like Muller were sincere in their concerns; Dobzhansky did not want their policies to be taken seriously. His particular prodrift and proselection positions provided him with platforms from which to attack those policies as misguided. Already in the first edition of *Genetics and the Origin of Species*, Dobzhansky lashed out at the "eugenical Jeremiahs" who viewed the variation present in human populations as a liability to the species. Having just explained Chetverikov's views on how species naturally soak up variations "like a sponge," he turned to the eugenicists' fears of variaton:

It is not an easy matter to evaluate the significance of the accumulation of germinal changes in the population genotypes. Judged superficially, a progressive saturation of the germ plasm of a species with mutant genes a majority of which are deleterious in their effects is a destructive process, a sort of deterioration of the genotype which threatens the very existence of the species and can finally lead only to its extinction. The eugenical Jeremiahs keep constantly before our eyes the nightmare of human populations accumulating recessive genes that produce pathological effects when homozygous. These prophets of doom seem to be unaware of the fact that wild species in the state of nature fare in this respect no better than man does with all the artificiality of his surroundings, and yet life has not come to an end on this planet. The eschatological cries proclaiming the failure of natural selection to operate in human populations have more to do with political beliefs than with scientific findings.[90]

Immediately following this discussion is the passage quoted above (pp. 282–283) where Dobzhansky explained the benefits of intraspecific variation—i.e., in promoting the adaptive flexibility of species. The connection between the passages is clear. The eugenical Jeremiahs would have purposefully depleted the intraspecific variation that rendered species adaptively flexible—a dangerous act, based on what Dobzhansky believed was a misconception of the evolutionary process. As was already explained, the mechanism of evolutionary change that Dobzhansky himself was pursuing at the time—the mechanism in which random drift played a significant role—was one in which the variability necessary for species adaptability was preserved.

Later, in his book *Mankind Evolving*, Dobzhansky took on Muller specifically. By this time (1962), Dobzhansky had settled on his final, balance theory of evolution by natural selection, the theory that best guaranteed intrapopulation and intraspecific variation. Following a section on Muller's "fire and brimstone" prophecies concerning the fate of the human species, Dobzhansky proceeded to point out, much as he had done twenty-five years earlier in the first edition of *Genetics and the Origin of Species*, that "Drosophila flies are doing nicely [thank you] in their natural habitats, despite the fact that they bear enormous genetic loads."[91] Humans too were doing quite nicely, thank you, in spite of their "loads" of variation: "The adaptive norm of the human species consists of persons burdened with genetic loads. Nor is there anything new in this situation—all of human evolution occurred in populations that carried heavy genetic loads."[92]

The question is, he proceeded, whether these loads are detrimental, as proponents of the classical view maintained, or a virtue—something that "natural selection might somehow turn to an advantage"—as proponents of the balance view believed.[93] He reminded the reader that the scientific basis of Muller's eugenics policies was a matter of some dispute. He argued, in particular, that although the classical theory of evolution supported Muller's eugenics proposals, the rival balance theory did not.

If one wanted to engineer the ideal human species, the balance view (as opposed to the classical view) would dictate the *promotion* of variability. That is, if the balance view were correct, "Eugenics must be ... more dextrous, for instead of making everybody alike, possessing some optimal genotype, it will have to engineer a gene pool of the human population that would [by promoting variability] maximize the frequency of the fit and minimize that of the unfit."[94] Intraspecific variation was, Dobzhansky believed, crucial to the well-being of mankind.

By this time, Dobzhansky saw virtues in variation at the *organismic* level as well as at the *species* and *population* level. Recall that, in the balance account, heterozygotes were supposedly fitter than homozygotes, partly because of the fact that possessors of two different genes for a trait had dual capabilities with respect to that trait—i.e., because heterozygosity contributed to adaptability at the organismic level as well as the specific level. Variation-reducing eugenics programs were thus, at the time, *doubly* threatening to Dobzhansky. They threatened not only the long-run well-being of the *species*, but also the more immediate well-being of the *members of the species*.

Although Dobzhansky did not have many established cases of heterozygote superiority in humans to use in his case against Muller's eugenics proposals, he did employ theoretical considerations, like those just raised, in this regard.[95] All in all, the balance view, according to which not only intraspecific but also intraorganismic variation were naturally maintained, provided Dobzhansky with his strongest case against variation-reducing eugenics directives.

Summing up, the various theories of evolution entertained by Dobzhansky cannot be considered independently of the classical position, *over which*, and *in reaction to which*, he pursued his alternatives. The views of evolution that he entertained served him as platforms from which to respond to the threats posed by the classical position and the eugenics proposals associated with it. In shifting from his prodrift to his particular proselection positions, Dobzhansky assumed a *stronger* position against the classical theory, and in shifting from his earliest proselection position to his ultimate probalance position, he assumed a *still stronger* position against the classical position. His positions were not just more and more different from the classical position. They offered stronger and stronger guarantees of the intraspecific variation that Dobzhansky cherished, and they offered stronger and stronger rebuttals to proposals to eliminate that variation.

Dobzhansky's shift from a prodrift to a proselection position thus represented more than just a second thought concerning the importance of chance in evolution. It represented increasing opposition to a "foil" position—a position that threatened Dobzhansky's happy picture of the status quo—moreover, a position upon which were based policy proposals that threatened to change the status quo for the worse.

Of course, in addition to value considerations that played a role in his thinking, Dobzhansky also had empirical support for the various positions he took; he did not just *hope* that his positions were true. Those empirical considerations were covered in preceding sections. Dobzhansky's positions would have been weak platforms from which to attack Muller if they had not had empirical support. The point of emphasizing the empirical support might be different, though. One might argue that Dobzhansky took the positions he did on purely empirical grounds, *only then* coming to "cherish" the positions he took as good ways for the world to be, and *only then* coming to see Muller's positions as more and more threatening with respect to values.

Perhaps that is the correct account of Dobzhansky's earliest attempts to explain intraspecific variation. Perhaps it was only after he found lots of variation in ladybugs, and then also in fruitflies, and after he could account for that variation in one way or another, that he came to believe that the world was really good in this respect. But once he was originally assured of the value of intraspecific variation, I am arguing, further changes were constrained—not determined, to be sure—but constrained. The facts further constrained the changes. To insist instead that his values were entirely determined by his empirical findings is to construe as purely coincidental the *direction* in which the facts led over the course of Dobzhansky's life, namely, further and further in the direction of supporting the picture of the world that he originally found comforting.

In closing this section, let me just add one further indication that values and threatened values played a role in Dobzhansky's shift of opinion with regard to the relative importances of drift and selection. In the course of that shift, Dobzhansky not only found more and more threatening the variation-reducing eugenic directives associated with the classical position, but he also seems to have found it increasingly necessary to make his *own* positions palatable to those who might be threatened by them.

He was especially concerned to waylay fears that the naturalness of variation (and especially his promotion of the preservation of variation) might be misused by opponents of equality of opportunity. His case for the extensiveness and naturalness of intraspecific variation, if extended to cover the mental and physical capabilities that limit peoples' roles and places in society, might, he realized, be misused as a case against granting all people the same access to those roles and places. His *promotion* of the preservation of variation might, he also realized, be further misconstrued as support for constraints on opportunity.

This concern was expressed by Dobzhansky's student, Bruce Wallace, in a fairly early review of the classical-balance controversy, in which he suggested (whether seriously or not, I do not know) that the classical position "is a moral system in that, under ideal conditions, every individual is his neighbor's equal." The balance position, he suggested, is "morally deficient" in this regard.[96]

Whether or not Wallace took this issue seriously, Dobzhansky certainly did.[97] To ensure that his positions did not offend proponents of equality of opportunity, Dobzhansky developed, over the years, an argument to the effect that equality of opportunity was not only reconcilable with variations in capability among humans but was in fact the *best way to take advantage of those differences.* The argument ran as follows. There are no castes, classes, or any other subgroupings of society within which there is uniformity with respect to capabilities. Making role, place, and occupation decisions on the basis of any such subgroupings would thus have the effect of placing at least some people in positions for which they were not the most capable, as well as having the effect of placing some of the people most qualified for certain positions in quite different positions. Equality of opportunity, on the other hand, would at least *allow* people to take up the positions for which they were best suited. A society that opted for equality of opportunity would accordingly have the best chance of having its various roles, places, and occupations filled by the people most capable of assuming them. Equality of opportunity is, therefore, the most efficient way to manage a society whose members differ in their capabilities. Dobzhansky even went so far as to add that the extensiveness of variation constituted the *strongest* grounds for adopting equality of opportunity: "Equality is necessary if a society wishes to maximize the benefits of genetic diversity among its members. With anything approaching full equality, every trade, craft, occupation, and profession will concentrate within itself those who are genetically most fit for these roles."[98]

Dobzhansky's concern to make his own positions nonthreatening valuewise constituted the "flip-side" of his concern to attack positions that threatened, directly or indirectly, his values. I have discussed the former side of the coin only for the purpose of further calling attention to the latter side, and hence to reiterate that Dobzhansky's shift of positions concerning the major modes of evolution was constrained in part by what he found objectionable about the classical view of evolution—hence, in part by value considerations.

Postscript

Events of the late sixties and seventies that bear upon the story I have told are worth relating, if only briefly. These events involve allegiances that I do not understand very well, but hope to understand someday. Even though I do not fully understand them now, though, they are so interesting in connection with what I have said so far that they really deserve some attention.

I have argued that Dobzhansky shifted from a drift position to a balance position, in part on account of his opposition to the classical position advocated by Muller. This thesis—specifically, the notion that the balance position represented a stronger alternative to the classical position than did the drift position—would seem to receive at least indirect support from recent events, specifically the rejuvenation of the drift position in the late sixties and seventies, and its affiliation with Muller's classical position. The rejuvenated drift position has even been labeled the "neoclassical" position in contrast to the balance position.[99]

As I mentioned earlier, the closest thing to compelling evidence in the classical-balance dispute was not gathered until 1966, the year before Muller's death. This evidence consisted of fairly direct measurements of the prevalence of genetic variation, via gel-electrophoretic analyses of the immediate "products" of the genes at a locus. The gel-electrophoretic techniques in question are capable of revealing double-product (i.e., heterozygous loci), as well as single-product (i.e., homozygous loci). Those techniques are also capable of revealing differences between organisms with respect to products of genes at the same loci. The amount of variation actually revealed by those studies seemed (and still seems) much greater than that allowed by Muller's classical position. But while having to concede in some respects, defenders of the classical position have nonetheless interpreted the evidence differently from defenders of the balance position.

Defenders of the old classical position have acknowledged substantial variation, but have denied that the variation is retained by balancing selection. They attribute the variation instead to drift. They continue to insist that the most prevalent form of selection is classical selection, but also admit a very significant role for random drift. The combination of classical selection and drift distinguishes their position from the balance position, which is, in and of itself, mute on the importance of drift, and which, understandably enough, seems to be considered an antidrift position on account of its silence in this regard.

This is, *in brief*, the picture presented by the most accepted and influential review of the classical-balance controversy: the gel-electrophoretic data occasioned a modification in the classical position, giving rise to the neoclassical position, with its emphasis on classical selection at some loci and drift at others, a position that supposedly bears a very real "historical continuity ... with the classical position."[100]

The classical heritage of the neoclassical position has not been without its detractors, however. Dobzhansky himself was among them. Labeling the resurrected prodrift position "neoclassical," Dobzhansky contended, "is in my opinion nonsense.... The 'classical' model assumed that most individuals of a species are homozygous for almost all their genes, and are alike among themselves .... The last to defend this was the late H. J. Muller, and to my knowledge, nobody is now supporting that point of view."[101]

This lack of agreement concerning the connections between the classical position and the rejuvenated drift position of the late sixties and seventies raises interesting questions about one of the main theses of this paper—namely, the thesis that Dobzhansky's shift away from a drift position was occasioned in large part by his increasing opposition to the classical position. Suggested connections between the classical position and the new prodrift position at least indirectly support the view that the balance position later advocated by Dobzhansky really was a stronger alternative to the classical position than was his old drift position (although there is, of course, much work to be done in elucidating the nature of that support).

What does Dobzhansky's denial of the classical-drift connection mean as far as the previous arguments of this paper are concerned? That is hard to say. By calling into question the suggested connections between the classical and new drift positions, Dobzhansky may only have wanted to emphasize that it was *he* who was an "old drifter," not Muller, and *he* who even later attributed some small role to drift, not Muller. Of course, even if Dobzhansky was right to minimize the connections between the classical and new drift positions, and right to emphasize instead the connections between the balance and his old drift positions, that would still not rule out the possibility that his shift away from his old drift position was motivated in large part by his desire to take a stronger stand against the classical position. It is just that the subsequent alliance of proponents of the classical and new drift positions would not give us any special hindsight on Dobzhansky's shift in that case.

Could it, though, have been just a *coincidence* that Dobzhansky gave up a drift position in response to the classical view, while proponents of the classical view adopted a drift position in response to the balance view? Perhaps. Perhaps not. But that is grist for another mill.

Notes

1. S. J. Gould and R. C. Lewontin, "The Spandrels of San Marco and the Panglossian Paradigm: A Critique of the Adaptationist Programme," *Proceedings of the Royal Society of London*, B205 (1979), 581–598, on pp. 585–586.

2. F. J. Sulloway, "Geographical Isolation in Darwin's Thinking: The Vicissitudes of a Crucial Idea," *Studies in the History of Biology*, 3 (1979), 23–65.

3. W. B. Provine, *The Origins of Theoretical Population Genetics* (Chicago: Chicago University Press, 1971).

4. This controversy will be discussed later. See also R. C. Lewontin, *The Genetic Basis of Evolutionary Change* (New York: Columbia University Press, 1974).

5. A. L. and A. C. Hagedoorn, *The Relative Value of the Processes Causing Evolution* (The Hague: Nijhoff, 1921).

6. For more thorough analyses of the differences between drift and selection, see J. Beatty, "Chance and Natural Selection," *Philosophy of Science*, 51 (1984), 183–211; E. Sober, *The Nature of Selection* (Cambridge: MIT Press, 1984), pp. 103–134; and M. J. S. Hodge, "Natural Selection as a Causal, Empirical, and Probabilistic Theory," in this volume.

7. See J. Beatty, "Chance and Natural Selection" (note 6), for a review of this controversy.

8. S. J. Gould, "G. G. Simpson, Paleontology, and the Modern Synthesis," in E. Mayr and W. B. Provine eds., *The Evolutionary Synthesis* (Cambridge: Harvard University Press,

1980); Gould, "Introduction," to the Columbia Classics in Evolution Series reprint of the first edition of T. Dobzhansky's *Genetics and the Origin of Species* (New York: Columbia University Press, 1982); Gould, "The Hardening of the Synthesis," in M. Grene, ed., *Dimension of Darwinism* (Cambridge: Cambridge University Press, 1983).

9. W. B. Provine, "The Development of Wright's Theory of Evolution: Systematics, Adaptation, and Drift," in M. Grene, ed., *Dimensions of Darwinism*; Provine, *Sewall Wright and Evolutionary Biology* (Chicago: University of Chicago Press, 1986).

10. Provine, *Sewall Wright: Geneticist and Evolutionist*.

11. T. Dobzhansky, *Genetics and the Origin of Species* (New York: Columbia University Press, 1937, 1941, 1951). For assessments of the importance of *Genetics and the Origin of Species*, see Gould, "Introduction," p. xx (note 8); R. C. Lewontin, "Introduction: The Scientific Work of Theodosius Dobzhansky," in Lewontin et al., eds., *Dobzhansky's Genetics of Natural Populations*, Series I–LXIII (New York: Columbia University Press, 1981), p. 98; E. Mayr, *The Growth of Biological Thought* (Cambridge: Harvard University Press, 1982); W. B. Provine, "Origins of the Genetics of Natural Populations Series," in Lewontin et al., eds., *Dobzhansky's Genetics of Natural Populations*, Series I–LXIII; and Provine, *Sewall Wright: Geneticist and Evolutionist* (note 9).

12. Gould, "Introduction," p. xxxviii (note 8).

13. Gould, "Introduction," p. xxxviii.

14. With regard to the latter issue, see J. Beatty, "Pluralism and Panselectionism," in P. Asquith and P. Kitcher, eds., *PSA 1984*, Vol. 2, Proceedings of the Biennial Meetings of the Philosophy of Science Association (East Lansing, Michigan: Philosophy of Science Association, 1985).

15. J. L. King to T. Dobzhansky, May 21, 1970, Theodosius Dobzhansky Papers at the American Philosophical Society Library in Philadelphia.

16. Mayr, *Growth of Biological Thought*; E. Sober, "Evolution, Population Thinking, and Essentialism," *Philosophy of Science*, 47 (1980), 350–383.

17. T. Dobzhansky, "A Critique of the Species Concept in Biology," *Philosophy of Science*, 2 (1935), 344–355; Dobzhansky, *Genetics and the Origin of Species* (1937); E. Mayr, "Speciation Phenomena in Birds," *American Naturalist*, 74 (1940), 249–278; Mayr, *Systematics and the Origin of Species* (New York: Columbia University Press, 1942), pp. 102–146.

18. T. Dobzhansky, "Geographical Variation in Lady-Beetles," *American Naturalist*, 67 (1933), 97–126.

19. Lewontin, "Introduction: The Scientific Work of Theodosius Dobzhansky," pp. 95–96 (note 11).

20. T. Dobzhansky, *The Reminiscences of Theodosius Dobzhansky* (transcript of interviews conducted by B. Land for the Oral History Research Office of Columbia University in 1962 and 1963), p. 244.

21. Dobzhansky, *The Reminiscences of Theodosius Dobzhansky*, p. 244.

22. Provine, "Dobzhansky's Genetics of Natural Populations Series," p. 11 (note 11).

23. G. E. Allen, *Thomas Hunt Morgan: The Man and His Science* (Princeton: Princeton University Press, 1978), pp. 148–153.

24. S. S. Chetverikov, "On Certain Aspects of the Evolutionary Process from the Standpoint of Modern Genetics," originally published in Russian in *Zhurnal Eksperimental'noi Biologii*, A2 (1926), 3–54, translated in *American Philosophical Society Proceedings*, 105 (1961), 167–195, on p. 191.

25. Chetverikov, "On Certain Aspects of the Evolutionary Process from the Standpoint of Modern Genetics," p. 178.

26. S. S. Chetverikov, "On the Genetic Constitution of Wild Populations," originally published in German in *Proceedings of the Fifth International Congress of Genetics*, 1927, pp. 1499–1500, translated in D. L. Jameson, ed., *Evolutionary Genetics* (Stroudsburg, PA: Dowden, Hutchinson, and Ross, 1977).

27. H. and N. W. Timofeef-Ressovsky, "Genetische Analyse einer freilebenden *Drosophila melanogaster*-Population," *Roux's Archiv für Entwicklungsmechanik*, 109 (1927), 70–109; N. P. Dubinin et al., "Experimental Study of the Ecogenotypes of *Drosophila melanogaster*" (in Russian), *Biologichesky Zhurnal*, 3 (1934), 166–216; See especially M. Adams, "The Founding of Population Genetics: Contributions of the Chetverikov School, 1924–1934," *Journal of the History of Biology*, 1 (1968), 23–39; Adams, "Towards a Synthesis: Population Concepts in Russian Evolutionary Thought, 1925–1935," *Journal of the History of Biology*, 3 (1970), 107–129; and Adams, "Sergei Chetverikov, the Kol'tsov Institute, and the Evolutionary Synthesis," in Mayr and Provine, eds., *The Evolutionary Synthesis* (note 8).

28. See Provine, "Dobzhansky's Genetics of Natural Populations Series," with regard to the ill-fated Dobzhansky-Sturtevant collaboration (note 11).

29. Dobzhansky, *Genetics and the Origin of Species* (1937) (note 11).

30. T. Dobzhansky, "Experimental Studies on Genetics of Free-Living Populations of *Drosophila*," *Biological Review*, 14 (1939), 339–368, on p. 340.

31. Dobzhansky, "Experimental Studies on Genetics of Free-Living Populations of *Drosophila*," p. 366.

32. Dobzhansky, *The Reminiscences of Theodosius Dobzhansky*, pp. 413–414 (note 20).

33. Dobzhansky, *Genetics and the Origin of Species* (1937), pp. 126–127 (note 11).

34. Dobzhansky, "Geographical Variation in Lady-Beetles," pp. 113–115 (note 18).

35. Dobzhansky, *Genetics and the Origin of Species* (1937), pp. 150–151 (note 11).

36. See Provine, "The Development of Wright's Theory of Evolution: Systematics, Adaptation, and Drift" (note 9).

37. G. C. Robson and O. W. Richards, *The Variation of Animals in Nature* (London: Longmans and Green, 1936), pp. 314–315.

38. J. T. Gulick, *Evolution, Racial and Habitudinal* (Washington: Carnegie Institute, 1905); C. Elton, "Periodic Fluctuations of the Numbers of Animals: Their Causes and Effects," *Journal of Experimental Biology*, 2 (1924), 119–163; Elton, *Animal Ecology and Evolution* (Oxford: Oxford University Press, 1930); Hagedoorn and Hagedoorn, *The Relative Value of the Processes Causing Evolution* (note 5); G. C. Robson, *The Species Problem* (London: Oliver and Boyd, 1928). For reviews of this literature, see Robson and Richards, *Variation of Animals in Nature*, and Provine "The Development of Wright's Theory of Evolution: Systematics, Adaptation, and Drift" (note 9).

39. S. Wright, "Evolution in Mendelian Populations," *Genetics*, 16 (1931), 97–159; Wright, "The Roles of Mutation, Inbreeding, Crossbreeding, and Selection in Evolution," *Proceedings of the Sixth International Congress of Genetics*, 1 (1932), 356–366; Wright, "The Statistical Consequences of Mendelian Heredity in Relation to Speciation," in J. S. Huxley, ed., *The New Systematics* (Oxford: Oxford University Press, 1940); and Wright, "On the Roles of Directed and Random Changes in Gene Frequency in the Genetics of Populations," *Evolution*, 2 (1948), 279–294.

40. See Gould, "Introduction," and Gould, "The Hardening of the Synthesis" (note 8).

41. See below.

42. N. P. Dubinin and D. D. Romaschoff, "The Genetic Structure of Populations and their Evolution" (in Russian), *Biologichesky Zhurnal*, 1 (1932), 52–95; Dobzhansky, *Genetics and the Origin of Species* (1937), p. 129.

43. Dobzhansky, *Genetics and the Origin of Species* (1937), p. 131 (note 8).

44. See Provine, "The Development of Wright's Theory of Evolution: Systematics, Adaptation, and Drift" (note 9).

45. Provine argues that the role of drift in the adaptive evolution of species is something that Wright emphasized very little early on, but more and more over time. See Provine, "The Development of Wright's Theory of Evolution: Systematics, Adaptation, and Drift," and Provine, *Sewall Wright and Evolutionary Biology* (note 9). Whether or not Wright emphasized the adaptive effects of drift "little" early on, though, he emphasized it *enough* early on to influence Dobzhansky in this regard. See further.

46. Wright, "The Genetical Theory of Natural Selection: A Review," *Journal of Heredity*, 21 (1930), 349–356, on p. 355.

47. Dobzhansky, *Genetics and the Origin of Species* (1937), pp. 185–186 (note 8).

48. Dobzhansky, *Genetics and the Origin of Species* (1937), p. 136.

49. S. Wright, "The Shifting Balance Theory and Macroevolution," *Annual Review of Genetics*, 16 (1982), 1–19, on p. 12. But see Provine "The Development of Wright's Theory of Evolution: Systematics, Adaptation, and Drift" (note 9).

50. See his first report of these changes in T. Dobzhansky, "IX. Temporal Changes in the Composition of *Drosophila pseudoobscura*," *Genetics*, 28 (1939), 162–186.

51. See, for example, S. Wright and T. Dobzhansky, "XII. Experimental Reproduction of some of the Changes Caused by Natural Selection in Certain Populations of *Drosophila pseudoobscura*," *Genetics*, 31 (1946), 125–156. See also Provine, *Sewall Wright and Evolutionary Biology*, on the details of Dobzhansky's discovery (note 9).

52. T. Dobzhansky to S. Wright, May 4, 1941, Sewall Wright Papers at the American Philosophical Library in Philadelphia.

53. Dobzhansky to Wright, May 21, 1941, Wright Papers.

54. See, for example, T. Dobzhansky and M. L. Queal, "II. Genic Variation in Populations of *Drosophila pseudoobscura* Inhabiting Isolated Mountain Ranges," *Genetics*, 23 (1938), 463–383, and Dobzhansky, "IV. Mexican and Guatemalan Populations of *Drosophila pseudoobscura*," *Genetics*, 24 (1939), 391–412.

55. Dobzhansky, *Genetics and the Origin of Species*, pp. 73–117 (note 11).

56. T. Dobzhansky, "IX. Temporal Changes in the Composition of Populations of *Drosophila pseudoobscura*" (note 50).

57. Dobzhansky, *Genetics and the Origin of Species* (1951), pp. 122–123 (note 11).

58. M. A. Vetukhiv, "Viability of Hybrids between Local Populations of *Drosophila pseudoobscura*," *Proceedings of the National Academy of Science*, 39 (1953), 30–40; D. Brncic, "Heterosis and the Integration of the Genotype in Geographic Populations of *Drosophila pseudoobscura*," *Genetics*, 39 (1954), 77–88.

59. I. M. Lerner, *Genetic Homeostasis* (Edinburgh: Oliver and Boyd, 1954).

60. T. Dobzhansky and H. Levene, "XXIV. Developmental Homeostasis in Natural Populations of *Drosophila pseudoobscura*," *Genetics*, 40 (1955), 797–808, on p. 797.

61. For example, Dobzhansky and Levene, "XXIV. Developmental Homeostasis in Natural Populations of *Drosophila pseudoobscura*."

62. C. Diver, "The Problem of Closely Related Snails Living in the Same Area," in Huxley, ed., *The New Systematics* (note 39).

63. Wright, "The Statistical Consequences of Mendelian Heredity in Relation to Speciation," p. 179 (note 39).

64. A. J. Clain and P. M. Sheppard, "Selection in the Polymorphic Land Snail *Cepaea nemoralis*," *Heredity*, 4 (1950), 275–294; Sheppard, "Fluctuations in the Selective Value of Certain Phenotypes in the Polymorphic Land Snail *Cepaea nemoralis* (L.)," *Heredity*, 5 (1951), 125–134; Sheppard, "Natural Selection in Two Colonies of the Polymorphic Land Snail *Cepaea nemoralis*," *Heredity*, 6 (1952), 233–238.

65. A. J. Cain and P. M. Sheppard, "Natural Selection in *Cepaea*," *Genetics*, 39 (1954), 89–116.

66. W. C. Boyd, "Detection of Selective Advantages of Heterozygotes in Man," *American Journal of Physical Anthropology*, n.s. 13 (1940), 37–52, on p. 37.

67. With regard to the former finding, see C. A. Clarke et al., "The Relationship of the ABO Blood Groups to Duodenal Ulcer and Gastric Ulceration," *British Medical Journal*, 1 (1955), 643–646. With regard to the latter finding, see I. Aird et al., "A Relationship between Cancer of the Stomach and the ABO Blood Groups," *British Medical Journal*, 1 (1953), 799–801.

68. For example, E. B. Ford, *Genetics for Medical Students* (London: Chapman and Hall, 1942), and Boyd, "Detection of Selective Advantages of Heterozygotes in Man" (note 66).

69. M. Lamotte, "Polymorphism of Natural Populations of *Cepaea nemoralis*," *Cold Spring Harbor Symposia on Quantitative Biology*, 24 (1959), 65–84.

70. For example, L. L. Cavalli-Sforza, "Genetic Drift in an Italian Population," *Scientific American*, 223(2) (1969), 26–33.

71. See, for example, E. Mayr, *Animal Species and Evolution* (Cambridge: Harvard University Press, 1963), pp. 204–214, and consider the entire structure of E. B. Ford, *Ecological Genetics* (London: Methuen, 1964). On the historical importance of the selectionist successes in question, see also Provine, *Sewall Wright and Evolutionary Biology* (note 9), and J. R. G. Turner, "Random Genetic Drift, R. A. Fisher, and the Oxford School of Ecological Genetics," in this volume.

72. T. Dobzhansky to S. Wright, November 9, 1950, Wright Papers (note 52).

73. See also Gould, "Introduction" (note 8).

74. Dobzhansky, *Genetics and the Origin of Species* (1951), p. 77 (note 11).

75. T. Dobzhansky, *Genetics of the Evolutionary Process* (New York: Columbia University Press, 1970), p. 262.

76. Gould, "The Hardening of the Synthesis," p. 89 (note 8).

77. Gould, "Introduction," p. xxviii (note 8).

78. For the reference to hyperselectionists, see T. Dobzhansky to J. L. King, June 8, 1970, Dobzhansky Papers (note 15).

79. Dobzhansky, *Genetics and the Origin of Species* (1951), pp. 170–171 (note 11). See also Dobzhansky, *Genetics of the Evolutionary Process*, p. 279 (note 75).

80. A. Gouldner, *The Coming Crisis of Western Sociology* (New York: Basic, 1970).

81. T. Dobzhansky, "A Review of Some Fundamental Concepts and Problems of Population Genetics," *Cold Spring Harbor Symposia in Quantitative Biology*, 20 (1955), 1–15.

82. See especially H. J. Muller, "Evidence of the Precision of Genetic Adaptation," *The Harvey Lectures* (1947–1948), 43 (1950), 165–229, and Muller, "Our Load of Mutations," *American Journal of Human Genetics*, 2 (1950), 111–176.

83. B. Wallace, "The Average Effect of Radiation-Induced Mutations on Viability in *Drosophila melanogaster*," *Evolution*, 12 (1958), 532–556, on p. 536.

84. J. L. King, "The Genetic Basis of Evolutionary Change: A Review," *Annals of Human Genetics*, 38 (1975), 507–510, on p. 507.

85. W. B. Provine, personal communication.

86. I. Lakatos, "Falsification and the Methodology of Scientific Research Programmes," in I. Lakatos and A. Musgrave, eds., *Criticism and the Growth of Knowledge* (Cambridge: Cambridge University Press, 1970); L. Laudan, *Progress and Its Problems* (Berkeley: University of California Press, 1977).

87. J. L. Hubby and R. C. Lewontin, "A Molecular Approach to the Study of Genic Heterozygosity in Natural Populations. I. The Number of Alleles at Different Loci in *Drosophila pseudoobscura*," *Genetics*, 54 (1966), 577–594, and Lewontin and Hubby, "A Molecular Approach to the Study of Genic Heterozygosity in Natural Populations. II. Amount of Variation and Degree of Heterozygosity in Natural Populations of *Drosophila pseudoobscura*," *Genetics*, 54 (1966), 595–609. An excellent treatment of the empirical and theoretical strengths and weaknesses of the classical and balance positions is Lewontin, *The Genetic Basis of Evolutionary Change* (note 4).

88. T. Dobzhansky, *Mankind Evolving* (New Haven: Yale University Press, 1962), p. 289.

89. Muller, "Evidence of the Precision of Genetic Adaptation"; and Muller, "Our Load of Mutations" (note 82).

90. Dobzhansky, *Genetics and the Origin of Species* (1937), p. 126 (note 11).

91. Dobzhansky, *Mankind Evolving*, p. 295 (note 88).

92. Dobzhansky, *Mankind Evolving*, pp. 295–296.

93. Dobzhansky, *Mankind Evolving*, p. 296.

94. Dobzhansky, *Mankind Evolving*, p. 127.

95. Dobzhansky, *Mankind Evolving*, pp. 296–298.

96. B. Wallace, "Some of the Problems Accompanying an Increase of Mutation Rates in Mendelian Populations," in *Effect of Radiation on Society* (Geneva: World Health Organization, 1957), pp. 58–59.

97. Wallace did not remember having raised this issue himself—personal communication.

98. T. Dobzhansky, *Genetic Diversity and Human Equality* (New York: Basic Books, 1973), pp. 44–45, and Dobzhansky and E. Boesigner, *Human Culture: A Moment in Evolution* (New York: Columbia University Press, 1983), p. 149.

99. Lewontin, *The Genetic Basis of Evolutionary Change* (note 4).

100 Lewontin, *The Genetic Basis of Evolutionary Change*, p. 197.

101. T. Dobzhansky to T. Jukes, May 22, 1975, Dobzhansky Papers (note 15).

12 Random Genetic Drift, R. A. Fisher, and the Oxford School of Ecological Genetics

John R. G. Turner

Random drift is a stochastic process in populations which disorders their genetic structure, just as mutation disorders the genetic structure of cells. Both mutation and random drift must be overridden by natural selection if adapted organisms are to exist. Evolutionists have found it difficult to reconcile this statement with the fact that selection must operate on entities generated by these same stochastic processes.

Fisher achieved a synthesis in which mutation was awarded a minimal role by being required to deliver variation to the selective machinery in the most finely divided state that was imaginable. His opposition to Sewall Wright's proposals that random drift was a creative force when allied to selection was a natural outgrowth of this philosophy. Fisher maintained that as a general proposition he believed in the essentially stochastic nature of causation, and the indeterminacy of the future. His opposition to the creative role of stochastic processes in evolution arose therefore not from a belief in determinism, but from a belief in the sole power of natural selection to create order and adaptation.

The Oxford School of Ecological Genetics continued Fisher's opposition to random drift and allied it to the view that differences between races and species could not be described as nonadaptive; their empirical demonstrations of natural selection in cases where drift had been invoked by other workers gave them good grounds for believing that this view was correct. Whether or not they were right, they performed the signal service of gathering strong empirical evidence for the action of natural selection in natural populations, and thus underpinning the central tenet of Darwinism.

1 Randomness and Creativity in Evolution

1.1

Random genetic drift is the stochastic change of gene frequency which occurs from generation to generation within populations. It results simply from the fact that the genes which form one generation are a finite sample drawn randonly from the genes in the previous generation. It is a cause of evolutionary change, but at first sight, only change of a degenerative, random kind, following the second law of thermodynamics, increasing the entropy in the living world. It disorganizes the genetic material at the *population* level, as mutation disorganizes it at the *molecular* level, and therefore works in the direction opposite to natural selection, which

This chapter would not have been written without the substantial encouragement and help of Jonathan Hodge; it contains more of his good ideas than I am willing to admit. I am also much indebted to William Kimler and Robert Olby for enlightening discussions on the subject. Hodge and Kimler read and made very valuable comments on the draft.

produces adaptive changes in organisms, leading to greater degrees of organisation, and as creationists are fond of pointing out, flouting the second law.

Random drift is mathematically much "tidier" than natural selection, and easier to handle. Theorists tend to feel comfortable with it. It tends to make empirical evolutionary biologists on the other hand distinctly uncomfortable, and there is a strong tradition of insisting that it has no significant part to play in evolution, being normally overridden by the organizing effects of natural selection. As organisms are manifestly adapted, by no means perfectly, but certainly above the random level of existence exhibited by rocks, this must indeed be true. But a stochastic process can have other results besides disordering the order created by a selective process. It may simply create its own random patterns among entities that are not being selected —that is, it can cause neutral evolution. It can also, in itself, be a creative force. Both these roles have been proposed for random genetic drift in evolution, the neutral effect notably by Motoo Kimura,[1] and the creative effect by Sewall Wright.[2] Both theories have been strongly opposed by an English school of evolutionary biology centered in Oxford.[3] Relegation of random drift to the role of a trivial "fluctuation" in evolution has come to characterize the distinctive English national style in population genetics (the adjective is English, not British). The origins of such a national school of thought must be of some general historical interest; this paper is a brief history of this opposition.

It would be easy enough to imagine that this English viewpoint represented in effect, if not in intent, a rearguard action in favor of determinism against the newer and more flexible stochastic philosophy. This interpretation[4] becomes all the more appealing when we consider that the prime mover was R. A. Fisher, a eugenist, and therefore perhaps a believer in that now favorite strawman among the radical interpreters of scientific history—"genetic determinism." Fisher's views could be seen as an attempt to keep evolution in some kind of right-wing, determinist, Calvinist mould. The immediately obvious paradox is that this leads to the conclusion that Fisher, the premier statistician of his generation, was opposed to stochastic and statistical ways of thinking! At the best, that can be only part of the truth. I shall argue that the question rested on deeper foundations—not the problem of stochasticity versus determinism, but of order versus chaos.

1.2

Evolutionary biologists are as concerned with creation as any biblical fundamentalist. Organisms, which adapt to their environment rather than merely suffering it, require a special kind of explanation. We do not hear much from the creationist about the special creation of immutable elements, and even though he views the stars as individually created, their apparently random scatter in the sky has never been much of an argument for design; as one man in holy orders put it, stars arranged in geometrical figures would have edified mankind to a far greater degree.

Only two passable explanations have ever been offered for the existence of adapted organisms: the creative working of mind, whether of an external creator or of the internal will of the organism, and natural selection. Natural selection was at once the pivotal theory of Darwinism and the point on which it was most prone to scientific and popular attack. It is a difficult concept to understand, as anyone knows who has tried to teach students to formulate it elegantly and correctly; the structure of European languages, which dictates the way we think (or is dictated by it), is resolutely set against describing natural selection without the most labored of circumlocutions. If one talks to educated nonbiologists about evolution, one as often as not finds that they adhere to some vaguely Lamarckian system.

Natural selection is cruel. The individual feels himself ground, in the course of both evolution and his own development, between the upper and nether millstones of the environment and the genes. Where is his free will? Where is his power over his own destiny? Natural selection uses an unconscionable amount of destruction to produce a little construction. Not for nothing has Iago, a far from plain rogue, whose destructiveness defies even his own comprehension in Shakespeare, come by the end of the nineteenth century to be a Darwinian rationalist: "I believe in a cruel God, who made me in his image. . . . In the being of a germ or some vile Particle I had my birth. For that I am human, I am a villain—and in myself I know the vice of the primordial mud."[5] Lamarckism is a more comforting doctrine. Natural selection has a bad press. Darwinists are rightly defensive about it.

But the statement "Without selection was nothing made" has two meanings: natural selection might be the sole creative process or it might merely be a necessary process among others. If it is not to be like pure mind, creating the adaptation, or adaptations, of organisms out of a void, then it must select from the products of the very processes of randomness and disorder that it seems to defy. Perhaps indeed all creation is like this, and the difference between Titian and Jackson Pollock is not the nature of the process employed but its location: Titian did in his mind what Jack the Dripper did on the canvas. The problem for Darwinists has been to admit the stochastic, disordering processes into the theory without seeming to give too much ground on the cardinal principal of natural selection. This has been a continued source of debate throughout this century; I believe it has involved the incorporation into evolutionary theory of four nonadaptive processes, three of them—mutation, random drift, and speciation—fairly clearly describable in stochastic terms, and the fourth—the mechanism that maintains genetic variation in the population—a more subtle affair involving natural selection in a non-adaptive mode.[6]

1.3

The problem with creating by selection, that there must first be something from which to select, had been correctly stated by Lucretius and by Darwin. Both of them had inadequate explanations for the generation of the variation that was needed, a random fecundity of the earth itself—"Du temps que la Nature en sa verve puissante/Concevait chaque jour des enfants monstrueux,"[7]—or the direct action of the changing environment inducing the variation in organisms. This lacuna in Darwin's theory tended to be concealed by the use of a much wider definition of the term "natural selection" than is in vogue today, to encompass the whole of the process postulated by Darwin, to include heredity with variation and the ensuing evolutionary change, as well as the now customary meaning of differential survival;[8] in this way "natural selection" could be seen as the sole creative force. Wallace described variation as "a constant and necessary property of all organisms," much as one might describe mass as a necessary property of matter.[9] This description meets the criticism that the variation has not been explained. Only things that are contingent require explanation; necessity is its own justification. In this way natural selection was seen as molding adaptations as it were out of some plastic substance, the continuous, quantitative variation that was to be observed, as Wallace had been at pains to show with a series of histograms of metric characters in birds,[10] as a normal feature of the organism.

But a materialist explanation of evolution demanded an explanation of the origin of variation. The way in which the discovery of Mendelian inheritance and mutation supplied the missing information is now a well understood story.[11] The supply of variation on which selection worked was produced not in a continuous form, but by quantum changes, mutations of genes. It commenced, not as a property of whole populations, which would then be moulded by selection, but as an event in a single individual, which was then spread through the population by natural selection; the problem of how the gene, commencing as a mutant form in that single individual, could eventually spread through the whole population, was solved in the mathematical appendix (by H. T. J. Norton) to Punnett's book *Mimicry in Butterflies*[12] in 1915. "Natural selection," said Punnett, "plays the part of a conservative, not of a formative agent."[13]

What then was the creative force? If natural selection merely conserved beneficial mutations, and rejected the harmful ones (at the same time doing nothing at all with those that were neutral in their effects on survival), then the creative force of evolution must be the process of mutation itself. Natural selection made nothing.

If an organism is like a house, the mutationist view of Punnett and of his mentor Bateson is that houses are constructed from prefabricated windows, wall panels, even whole roofs and chimneys; a late nineteenth-century Darwinist would assume that the selective construction began with bricks, mortar, and timber, perhaps even with clay, lime, and three trunks. In the mutationist book, mutation could produce whole adaptations, such as the elaborate mimetic patterns of butterflies, and selection had only to preserve them and spread them through the population.

The Modern Synthesis, the generation of the theoretical system now known as neo-Darwinism, can be viewed as the realization that creation requires both processes: "We cannot regard mutation as a cause likely by itself to cause large changes in a species. But I am not suggesting for a moment that selection alone can have any effect at all. The material on which selection acts must be supplied by mutation. Neither of these processes alone can furnish a basis for prolonged evolution."[14] This leaves open the question of how prefabricated are the units produced by gene mutation: Does the whole pattern of a butterfly alter? Does the whole mammalian middle ear appear from the alteration of a single gene, or a single alteration affecting a number of genes, or does it have to build up from a large number of individually small differences? That question is still under debate.

2   R. A. Fisher's Synthesis

2.1

The penchant of Punnett for crediting mutation with large and spectacular changes, producing perfected adaptations, arose in part from the dismissal by the mutationist school of the continuous variation that had been the mainstay of the Darwinists. This could be written off as mere "fluctuation," variation of no evolutionary significance because it was not inherited. R. A. Fisher's paper on the correlation between relatives[15] finally established the sophisticated mathematical expression of the growing recognition that the laws of inheritance exhibited by such continuously varying characters could be explained by the segregation of a large number of Mendelian genes, each of small effect. Even Punnett was being converted: in the 1919 revision of his book,[16] he added a simplified account—simplified no doubt for his own benefit as well as the reader's, for he never claimed to be a mathematician—of the inheritance of size in poultry, based on a model of four such small Mendelian genes.

The issue on which Punnett would not give way to the Darwinian approach was evolution. Even in the final edition of his book in 1927,[17] he insisted that continuous variation was nonheritable, and constituted no more than a fluctuation about the general trend of evolution; only large mutational changes were of importance. It was a view that would be taken up, with more sophisticated arguments to explain how perfected adaptations could be produced at the behest of a random process like mutation, by Richard Goldschmidt.[18] Fisher, whose interests in evolution and in its application to human populations had for long been equal to his interests in statistics and genetics, set out to make the final reconciliation: if the continuous variation beloved of the Darwinists could be shown to be compatible with the findings of experimental genetics, then genetics and Darwinian evolutionary theory could also be synthesized. Fisher therefore generated a system which would produce the evolutionary processes postulated by the Darwinians, but on a basis of Mendelian genetics; he had established many of the basic ideas as early as 1922.[19] His approach differed markedly from that of Haldane, the other British synthetist, who followed up more directly the lead given by Norton's calculations on the way natural selection would alter the incidence of genes of rather large effect in populations; Fisher's aim was to restore the smooth and gradual picture of evolution, to give natural selection the major constructing role, and mutation only the role of feeding it the most finely divided raw material.

Fisher's scheme of evolution is comprehensive.[20] He establishes that with Mendelian inheritance the variation upon which selection can work is dissipated neither by blending (as Darwin had supposed) nor to any significant degree by the random loss of genes from the population. With those two points established, evolution is imagined as a flux of new genes passing through the species under the driving power of natural selection. As the environment is constantly changing, the organism[21] never achieves maximal adaptation, but can be imagined as in perpetual pursuit of an *ignis fatuus*, the point of maximum fitness. The speed with which the flux of new genes is driven through the population can be described globally, summing the effects of all the genes, by the Fundamental Theorem of Natural Selection:[22] The change in the mean fitness of the organism is equal to the genetic variance in fitness at any one time.[23] Thus the mean fitness of the organism constantly increases, at a rate that is determined by its capacity to produce genetic variation that affects fitness, but the increase is just as surely offset by decreases produced by the changing environment. In particular, the evolutionary arms race with other evolving species ensures that all will evolve in response to each other, so that all are in what is now known as a "Red Queen" race, running as fast as they can to stay in the same place.

Appropriately, this law of evolution, in which fitness or adaptedness is perpetually on the increase, could be seen as a photonegative of the second law of thermodynamics, the ordering process of natural selection being set against the disordering of the stochastic processes in the universe in laws that showed strong mathematical similarities to one another.[24] Fisher showed that the effects of the random drift of gene frequencies about their steady progress through the species, and hence the perturbation produced in the increase of fitness, would be "very small compared to the average rate of progress," and to some extent cushioned by the statistical summation of the effects of many genes, as the behavior of a gas was the summation of the stochastic movement of many molecules.[25]

A rational starting point, perhaps the actual starting point, of Fisher's theory, can be found in his correspondence with Major Leonard Darwin, in which, as Fisher's letters are not at this date preserved, we must see his opinions through those of Charles Darwin's son.[26] Leonard Darwin and Fisher had been taxed by the problem that arises from the nature of mutations, as it was being revealed in the experimental studies of Morgan, Muller, and others,[27] which has often been urged against the synthetic theory of evolution: if gene mutations are so very often, perhaps universally harmful, how can the progressive and adaptive process of evolution arise from them?

As to big mutations, I have no doubt they are generally harmful. But are not they rare and soon stamped out? If so, they are of no great importance in evolution. As to small mutations, these are what I believe evolution mainly relies on, and it seems to me difficult to prove that they are more often harmful than not. . . . Perhaps there may be such a thing as an organism which is as perfectly adapted to its environment as selection can make it. In that case, *ex hypothesi*, every mutation must be harmful. . . .[28]

From this last proposition, it is a short step to consider that if an organism were fairly, but imperfectly, adapted, then a very few mutations might be beneficial.

If we imagined the peak of fitness which the species engaged in vain pursuit as a point in a three-dimensional space, it was easy enough to show that large changes in the organism were rather more likely to move it away from the peak than toward it (if you are standing blindfold on the side of a mountain, spin round at random, and take a jump, the curvature of the mountain around the summit means that you are more likely to go somewhat downhill than to climb higher). Only small changes (analogous to shuffling one's feet) have an approximately fifty-fifty chance of improving the organism—the difference becomes more impressive if the problem is conceived in n rather than three dimensions. Therefore with possible statistical exceptions, only mutations of very small effect—so small as to be indistinguishable from continuous, quantitative variants—were likely to be of importance in evolution. Fisher's argument here defeats the mutationist belief that mutation itself could produce creative novelties; creativity would all be in what selection did with the supply of continuous variation.[29]

However, this argument depends crucially on the organism being in pursuit of a single adaptive peak, with no other peaks in the neighborhood. If there are two such peaks within striking distance of each other, than a large mutation may achieve what gradual alteration cannot, the transfer of the organism to the region of a new adaptive peak. Fisher therefore tended to play down the possibility of interaction between genes, as it is this that can cause multiple peaks to appear in the landscape. The chief kind of interaction that he allowed was *specific modification*, in which genes at one site alter the expression of genes at another in such a way that, if one plots the adaptive surface, there is still only one peak. Fisher placed a great deal of emphasis on this kind of interaction, for it permitted him to solve two problems: why the wild-type gene was usually dominant even though most dominant mutations were highly deleterious (so how could a dominant mutation have come to establish itself in the first place?—answer: the dominance of wild-type over its mutation is enhanced by modification during the course of evolution), and why mimicry in butterflies gave such clear evidence of the occurrence of single, large mutations generating new adaptive types (answer; the apparent single large change had in fact been generated by this same process of modification in gradual stages).[30]

But Fisher also sought an explanation for sexual reproduction, and the concomitant processes leading to genetic recombination, one of the hardest problems of Darwinian biology. He pointed out that when two genes interacted in such a way that, let us say, AB and ab were advantageous, and Ab and aB disadvantageous, then selection would favor tighter linkage between the two genes. As the whole genome had not "congealed" (to use the modern term—that is, lost recombination and then sexual reproduction), there must be some other force opposing this tendency, and Fisher found it in the need to pick out advantageous combinations within the constant flux of new mutations.[31] He felt that this consideration might give one some conception of the size of the flux. The theory was quite probably right, and is at least one of the most plausible explanations for sexual reproduction,[32] but it has one further implication, which Fisher ignored: The interaction is of the type that produces multiple peaks within the fitness surface. And that opens the way not only for mutations of large effect, but for Wright's model of evolution.

2.2

The empirically minded biologist will at this point ask what the "fitness" is that is in a state of perpetually frustrated increase, and what it would mean if the organism did succeed in overcoming its Tantalus torture and reach the maximum point. A hint is provided in another exchange of letters between Leonard Darwin and Fisher:

I agree as to there being an ideal organism, developed from a lion, which would probably be unlike any existing animal; this, I presume, in an unchanging environment. In other cases, with simple organisms, the real and ideal might be much alike.... The ideal lion can be no further evolved by [natural selection]. What then is to set evolution again working? It can only be a change of environment. If organisms often reached the ideal stage, changes of environment would be of great importance. Organisms living in the sea ought then to be much less evolved than organisms living on land. This is not markedly the case.... Hence I think changes of environment are probably not of supreme importance.[33]

It is tempting to think that the ideal toward which the organism is evolving is a kind of *Überlöwe* roaring between the strokes of midnight in *Zarathustra*, and that the concept of the maximum point of fitness owes something to Fisher's interest in eugenics.[34] Certainly the "fitness" itself had a eugenic meaning. Fisher defines it as the rate of population growth, "a well-defined statistical attribute of the population,"[35] then hints that in the real world its increase would be expressed as an increase in the standing crop (the equilibrium population size or perhaps biomass),[36] but declares in apparent contradiction that it is "qualitatively different for every different organism."[37] The reader is left to infer that it is an abstract mathematical concept, a bit like entropy, derived from the rate of population growth, but not equal to it, and able to predict evolutionary results, or ecological changes, like an increase in the standing crop. It is, in short, "competitive ability," for this process of increasing "fitness' is disrupted by the fertility patterns in advanced societies—those people who are most "fitted" by their "moral" character to advance the cause of civilization being socially promoted and hence induced to have smaller families—leading to a loss of average "fitness" by the entire society, and the eventual decline of the civilization. In the "barbarian" societies to which civilizations succumb, there is no such decline, for in them the tangible reward for social worth is greater fertility.[38]

Fisher's interests in eugenics were in halting this decline, rather than in pushing the human race toward *Übermenschheit*: "... The Darwinian process of natural selection is yet capable of acting in ways which generally speaking are not progressive, so that we may, in a sense, regard mankind, unless it rises to the task of helping itself and guiding its own evolution, as being at the mercy of non-moral forces which might mould or hammer it into most undesirable shapes. I think ... that we must regard the human race as now becoming responsible for the guidance of the evolutionary process acting upon itself."[39] Although he favored a limited form of negative eugenics, discouraging reproduction by the bearers of dominant disabling traits,[40] his own "bootstrap" theory of sexual selection[41] showed him the bizarre results, like the tail of the peacock, that might arise from well-intentioned but erroneous positive eugenics.

Considering the history of the United Kingdom or even Europe after 1914, his fear of the decline of "civilization," if not his diagnosis of a genetic cause, had some rational foundation. But it is, I hope, not necessary to point out that a pessimistic conservative, even an adherent of Gobineau, is a long way from a Nietzschean: Fisher, although he could write of the "high perfection of existing adaptation,"[42] seems to have adhered fairly strongly to the view that the point of maximum fitness could not be attained: "... we can conceive, though *we need not expect to find*, biological populations in which the genetic variance is absolutely zero, and in which fitness does not increase."[43] But Leonard Darwin's letter hints at another problem, that approaching the high perfection of the maximum point leads to the organism getting "stuck" near the peak, and hence to evolutionary stagnation. An earlier letter of Fisher's reveals that he had been toying with an idea which might produce the solution:

... I feel that the situation of the species waiting for the lucky mutation to occur may be quite an unreal one. I am inclined to the idea that the main work of evolution lies in the discovery by trial of perhaps rare combinations of its existing variants which work better than the commoner combinations. A slight increase in the number of individuals bearing such a favourable combination will then set up selection in favour of all the genes in the combination, with marked evolutionary results. Many of these genes would have been previously rare mutant types (not necessarily rare mutations) unfavourable to survival.[44]

Here again, there are multiple peaks in the fitness surface; the rare combinations that work better than the common combinations must be occupying a "higher" peak. But as Fisher has stated it, the theory will not work; there is nothing to hold together the "better" combinations. Fisher was keenly aware that for any selective process to operate successfully, the variation that selection works on must not be dissipated, diluted, or blended; that fact was central to his thinking as he propounded, in the first chapter of the *Genetical Theory*, the thesis that Mendelian inheritance, which conserved genetic variation, would allow natural selection to work, where Darwin's blending theory of inheritance had failed. It was something of which he was proud: "... The assertion ... that particulate inheritance, so far from being antagonistic to Darwin's main theory, actually removed the principal difficulty with which it was encumbered ... was entirely new when I put it forward in 1930."[45]

But any theory that depends, not on Mendelian particulate genes, but on novel combinations of those genes, runs into the difficulty that these combinations will be dissipated by recombination and sexual reproduction in each generation. Presumably this is why Fisher abandoned the theory. What is needed is a system that will first create the advantageous new combination, and then keep it from being swamped by the rest of the population. And it is that system that Wright was discovering.[46]

3   Sewall Wright's Synthesis

3.1

In an infinitely large population the frequencies of the alternative states of the same gene (known to geneticists for purposes of precision or pedantry as "alleles," but usefully described in common speech by the generic term "genes") will be transmitted precisely to the next generation. In a finite population the frequencies of the genes will change in a random, stochastic walk, which has now been described with some precision, and many of whose properties were already known to Fisher and to Wright. As the action of selection is to alter the gene frequencies steadily in particular directions (usually just to increase or decrease them, but sometimes to hold them constant or to cause them to cycle), it would seem that this random "drift" of the gene frequencies resulting from the generation-by-generation sampling of a finite number of genes to set up the next generation could have one effect only: to spoil, as far as it is able, the adaptations produced by natural selection. At the most, it would be a "fluctuation" of neutral effect, producing random changes in the frequencies of genes that altered the functioning of the organism to such a small extent that they were of neutral effect on its survival or reproduction.

But just as mutation, which is a random disordering of the structure of the genetic material, acts as a supplier of the raw material from which adaptive improvements can be selected, so the random disordering of the genetic structure of a population might supply also the configurations from which further adaptive change could be produced. In strict analogy with the interaction of mutation and selection, what would be required would be the production of very many random disorderings of the genetic structure, followed by a selective process in which the less adaptive configurations are eliminated, and the more adaptive ones preserved. (If one does not like the use of "adaptive" in this context, one simply substitutes "some" for "more adaptive" and others for "less adaptive.") What Wright proposed was that the greatest opportunities for evolutionary advance would occur when the selection of differentially adapted *populations* was added to the conventionally understood process of selection, the differential survival of *individual organisms*.

The conditions under which the process would occur are quite restrictive.[47] The species must be divided into a number of isolated populations, preferably of various sizes, within which gene frequencies will drift at random, and consequently in varying directions. As a result there will now be an array of populations with different genetic structures. Two forms of selection are then possible: either some populations survive and some die out or (more usually) the more successful populations undergo demographic explosions, which simply swamp the less successful populations with emigrants.

3.2

The radical break that Wright saw between his theory and Fisher's was over the creation of evolutionary novelty; Fisher's theory, said Wright, accounted for "treadmill" evolution. Organisms would become better at doing what they were already doing. In the words of a modern journalist, it accounts for why there are better giraffes, but not for why there are giraffes in the first place.[48] How does an organism "invent" a new solution for the problems posed by the environment? If challenged, anyone wishing to use Fisher's model to explain longer-term evolution simply says that as populations inhabit diverse environments that themselves change with time, there is really no problem; the novelty is produced by the pressure exerted at some time by some particular local environment. The landscape oscillates like the surface of the sea; sooner or later, every valley shall be exalted, and every mountain and hill made low. The organism escapes from the peak because the peak itself vanishes.[49] Given a giraffid animal, it only requires some environment somewhere to select an isolated population for increased neck length (and there is nothing in the formulation that prohibits this from happening comparatively rapidly) for one to arrive at the promised end. The Fisherian theory is gradualist not in the sense of proposing a constant steady rate of change, but in proposing that there is, instead of the sudden creation of novelty, rather a continuous development of new forms, with each intermediate stage being better adapted than the one which went before it. "For these two theories [Lamarckism and Darwinian selectionism] evolution *is* progressive adaptation and consists in nothing else. The production of differences recognizable by systematists is a secondary by-product, produced incidentally in the process of becoming better adapted."[50]

True novelty requires some process that removes the organism from its current adaptive zone and places it in a new one. Mutationism, which Fisher was working hard to discredit, had proposed one mechanism: a new adaptive form might arise *de novo*, by alteration of the genetic material. Most such "monsters," as Goldschmidt later nicknamed them, would be maladaptive, but from them the small percentage of adaptive novelties would be picked by selection. Wright appears to have accepted the general discrediting of this mutationist point of view (only Goldschmidt was to keep it alive during the heyday of the Modern Synthesis), but proposed instead that random drift could provide the necessary creative scrambling of the genes. Unlike Fisher, who assumed for most purposes that evolution was adequately described by postulating one single point of maximum fitness, or adaptive peak, for an organism, Wright imagined that genes interacted with one another in sufficiently complex ways that several different genetic compositions of the organism could achieve adaptively satisfactory results.[51] To take a simple example, it might be a good survival strategy for a water baby to be either pink and small or large and blue (but not blue and small or large and pink). As the process of selection that Fisher described would allow the population to move closer only to one of the fitness maxima or adaptive peaks, it was effectively trapped with that particular solution to its environmental problems, unless the population became very small, and the amplitude of the stochastic change of gene frequencies consequently large. If that happened, the population might change its composition, in the face of selection, away from the current adaptive peak and, if it crossed one of the surrounding adaptive valleys or saddle ridges, could find itself approaching a neighboring peak with a new adaptive solution.

While it is possible within this model to propose that all that now has to happen is that the population size increases, leaving us with a genetically transformed population with an evolutionary novelty, Wright thought this unlikely; a species represented by a single population of such small size would be more likely to become extinct than to find a novel solution. He therefore suggested that the species which would achieve the most rapid—and creative—evolution would be those that were widespread and numerous, but divided into more or less isolated populations of individually restricted size. Those which came to occupy the best adaptive peak would then become the most successful populations, and would be selected in some way to replace all the other populations of the species.

3.3

Therefore in Wright's theory, evolutionary advance came about as much by breaking with the past as by improving on it. In Fisher's scheme, advance also occurred above the level of individual selection; the process of evolution was self-reflexive, and after individual adaptations had been selected the whole evolving system improved its genetic structure and its ability to adapt; genes changed their dominance relations, and, perhaps more important, their linkage relations, as a result of their interactions; advantageous genes became more advantageous, and detrimental genes more detrimental. Fisher went as far as he could go, while maintaining compatibility with work on mutation (particularly that of H. J. Muller[52]) in restoring the Victorian view that natural selection did not merely preserve, but created and moulded; it moulded even the genetic material itself. And the supply of mutational variation was so finely divided that in no way could it be seen as providing even fragments of adaptations ready-made. Selection had been restored to its supreme, Darwinist, position.

The flexibility required for evolutionary advance was achieved not by shifting into new adaptive territory, but by using sexual reproduction and genetic recombination to maintain the most flexible state of the genome, to allow an efficient passage of the great flux of new mutations that kept the population perpetually adapted to its changing environment.

4   Why Did Fisher Think Wright Wrong?

4.1

Wright's theory looks very different from Fisher's in one important respect: for Wright, stochastic changes speeded up evolution and widened the possibilities open to it. For Fisher, as we shall see, the random loss of variation from the smaller populations—the chief actors in Wright's drama—put a brake or regulator on the speed of evolutionary change; it is the large populations that have the greatest evolutionary potential.

Fisher not only disagreed with Wright's theory, but founded, with E. B. Ford in Oxford,[53] a whole English tradition of disagreeing with it. Wright had seen random drift as an escape from evolutionary determinism.[54] To oppose such a stochastic model of evolution seems a very odd position indeed for Fisher, who had pioneered much of modern statistical science. Fisher was indeed not in any way a determinist. As a general proposition he declared himself convinced of the indeterminacy of the future (for otherwise it would be predicted as easily as the past is remembered) and the fundamentally statistical nature of causation, so that the future was genuinely being created, in the sense that only one out of a large number of statistically possible futures was actually realized. It was at the point of interaction between the freely willing and choosing organism and its environment, in which the choice of action influenced the chances of life, death, and reproduction—and hence drove the process of natural selection—that this genuinely indeterminate future was created.[55] Natural selection, said Fisher, differed "from nearly all causal laws in requiring no rigid determinism whatever."[56]

Yet what had initially been a cordial relationship, if not a one-hundred-percent accord, between Fisher and Wright[57] declined into an almost bitter disagreement during the 1930s, mostly over the creative role of random drift. I shall consider six hypotheses that might explain Fisher's antagonism to Wright's theory:

1. that it did not fit with the logic of Fisher's overall model;

2. that Fisher considered it an objectively incorrect description of the real world (a useful and often neglected hypothesis in the history of science);

3. that Fisher thought the mathematical formulations inadequate;

4. that Fisher had some emotional antagonism to it;

5. that it did not fit in with some aspect of Fisher's general philosophy (but obviously something other than a philosophy of rigid determinism);

6. that Fisher and Ford had interests in further models that were in conflict with Wright's.

4.2

Wright's and Fisher's models were certainly not founded on different premises or interests. Fisher had already exchanged letters with Leonard Darwin in which the problem of evolution by selection getting "stuck" had been raised, and he had apparently been attempting to solve the problem with his theory, which he rightly rejected—in the form in which he had it—about evolution discovering rare beneficial combinations. Further, if one follows Fisher's theory as he finally presented it in the *Genetical Theory* through to its logical conclusions, one finds all the necessary premises of Wright's theory. Fisher accepts that there are interactions between genes of the type that produce multiple peaks, and that these interactions are common enough to sustain sexual reproduction and recombination as the counterforce preventing the genome from "congealing."[58] He further proposes that the important genes in evolution are those with individually small effects on individual fitness; in fact, the smaller the effect of the gene, the larger, according to Fisher's argument, will its importance in evolutionary advance become. But these are just the genes that will be most likely to undergo significant stochastic changes in frequency. Although a follower of Fisher's argument might want to reject Wright's (or for that matter Fisher's) theory on the grounds that it did not provide an accurate or useful description of reality, they certainly could not accept the one and reject the other with any consistency of logic.

Why then did Fisher reject Wright's formulation? "... I am willing to be convinced," he wrote to Wright, "not of the importance of subdivision into relatively isolated colonies, which I should agree to at once, but that I have overlooked here a major factor in adaptive modification, which is what at present I am not convinced of." [59] While it is unfortunately true for most of us that there is nothing more annoying than the elegant solution that one has just missed, Fisher's doubts were probably well founded in objective arguments. We have seen how he had already rejected his own suggestion of selection "discovering" new, adaptive combinations, presumably on the valid grounds that there was nothing, if these occurred in a large population, that could prevent them from being "blended" by mating with the rest of the population. Wright's model was in principle open to the same objection: A very low level of immigration—we now know that it need be no more than one successfully breeding individual in every other generation[60]—from the rest of the species range, can put an effective stop to the drifting of a gene in a small "isolated" population. Fisher thought that populations were seldom so very isolated, and that if they were, they were probably on their way either to extinction or to becoming new species: "I am not sure that I agree with you as to the magnitude of the population number n. To reduce it to the number in a district requires that there shall be *no* diffusions [i.e. immigration followed by successful breeding] even over the number of generations considered. For the relevant purpose I believe n must usually be the total population on the planet, enumerated at sexual maturity, and at the minimum of the annual or other periodic fluctuation." [61] Local natural selection, on the other hand, will only be swamped in this way if it is very weak indeed, or if immigration is much stronger: "I do think that differential selective action in different stations [ecological niches?] or regions may be exceedingly important, even if there is a steady diffusion of germ plasm between them." [62] In conclusion: "... Isolation, whether geographical or physiological, while of immense importance to the problem of fission, is not a primary factor in adaptive modification, save in the subordinate sense that fission is a necessary condition for divergent adaptation. Sewall Wright, however, at present thinks otherwise, and there are very few men who have a better right to form their own opinion." [63]

Fisher, one can see from this and other letters, had in the late 1920s and early 1930s a great respect for Wright's theoretical acumen. The incomprehension of knaves and fools might merely irritate him, but rejection of his theories by Wright could well have been more deeply wounding. And Wright was extremely skeptical about the evolution of dominance,[64] a theory to which, to judge from its importance to members of the Oxford school, Fisher was deeply attached, and which required populations so large that very slight coefficients of selection would be effective over great periods of time. It was this theory, rather more than Fisher's general model, that was threatened by random genetic drift.

4.3

In later years, Fisher's respect for Wright's mathematics waned. In at least two of his papers Fisher's criticism topples over into irrationality.[65] From conversations with former students of Fisher's, I gather that his two most trenchant criticisms were of Wright's treating the mean fitness as if it were a potential function, analogous to a magnetic field, a possibility that Fisher said he had considered, but rejected, when formulating the Fundamental Theorem (a criticism made in his

1941 paper on fitness and spelled out clearly in a later letter to Kimura[66]), and of a non sequitur in the way Wright had then related this potential function to the frequencies of the genotypes in order to develop his picture of a multiple-peaked surface. This criticism also is clearly enunciated in another of Fisher's published letters.[67]

Wright at first conflated a multidimensional *grid*, in which the vertices represented each of the possible genotypes that could be adopted by a member of the species, with a multidimensional *graph* in which any point on the continuous scales represented a particular composition of the population. A third dimension then represents the average fitness of the population. Only if all populations in the system are completely homozygous are these two models equivalent. The best way of making the difference clear is to draw the two-dimensional versions of each (figure 1). Fisher believed that while the peaks and valleys appeared to be clearly separated on the grid, in the multidimensional version of the graph, they would not be: the more dimensions there were, the less the probability that any high point was surrounded by lower ground in all *n* dimensions. Most apparent peaks would in fact be cols or saddle ridges in some dimension or other. This amounted to the conclusion that the highest overall peak in the landscape probably was attainable by gradual progress from ridge to ridge, under natural selection, and that all Wright's theory proposed was that populations could take shortcuts across the valleys.

But these criticisms made only a late appearance. Indeed the second can hardly have been in Fisher's mind in 1930. unless we have a fine case of motes and sunbeams, as he made the same type of error himself. His argument about large and small mutations and their effects on fitness and his Fundamental Theorem repeatedly fail to make clear whether the "organism" Fisher is discussing is a single individual or a whole population or even species.[68] It is fair to think that the opacity of presentation reveals a confusion of thought. Fisher's derivation of the Fundamental Theorem also involves exactly the same confusion as does Wright's model between a grid of possible genotypes and a continuous scale of genotype (or gene) frequencies.

Thus one can represent the fitness of an individual organism on one axis of a graph and use points on the other axis to represent the genotypes. If the fitness of each possible genotype is plotted on the graph, one can then draw a regression line (Fisher called it the "average effect" of substituting the gene in question). This is analogous to Wright's grid of genotypes. One can also plot the average fitness of a whole population of organisms against the frequencies of the genes in the population and obtain a curve—or in the multidimensional case, a surface—whose slope at any one point is given simply by its first derivative. This is the equivalent of Wright's graph of populations plotted against their gene frequencies, with a surface representing their fitness (figure 1). It is not difficult to show that Fisher's first proof of the Fundamental Theorem[69] depends on the assumption that the average effect is mathematically identical with this derivative—in other words, on the conflation of the discontinuous grid with the continuous scale of gene frequency. In fact, the derivative, under certain side conditions, is equal to *twice* the

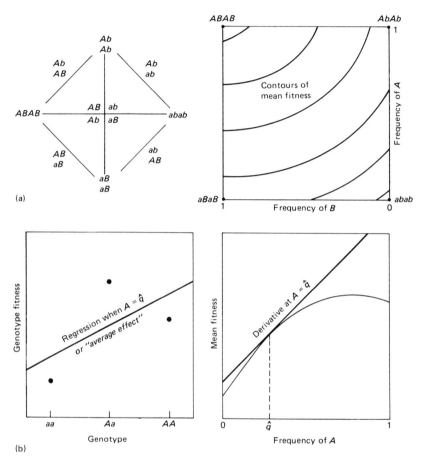

Figure 1
Both Wright and Fisher had two conceptions of the genetic structure of populations, the
one a grid of genotypes, the other a continuous field of genotype frequency. (a) Wright's
conceptions, based on his own figures in various publications (top): the genotype grid
(left) and the adaptive topography, in which fitness is plotted as contour lines representing
a third dimension. (b) A pictographic representation of Fisher's two conceptions (bottom):
first the grid of genotypes, each with its own fitness, from which one can calculate the
regression of fitness on genotype (left); second the field of gene frequency, against which
the mean fitness of the whole population may be plotted as a continuous curve (at any
given gene frequency, the derivative of this curve can be taken). In the first edition of
The Genetical Theory of Natural Selection, Fisher assumed that the regression and the
derivative were equal, or perhaps even the same; contrariwise, in a random mating
population, one is twice the other, a fact that he incorporated into the proof of the
Fundamental Theorem in the second (1958) edition.

average effect, and this was taken into account in Fisher's later version of the theorem.[70] He never discussed fully the side conditions.

5 The Problem of Nonadaptive Differences

5.1

Another tradition in evolutionary biology that aided this polarization of opinion was the old debate among systematists about the adaptive significance of the differences between races and species, and even between higher taxonomic categories.[71] Were these differences produced by natural selection? If they were not, then random genetic drift was an excellent candidate for the process that would account for the truly significant part of evolution—the origination of organic diversity—and natural selection provided one accepted that it occurred, would have the role only of adapting species to their environments. Darwin's central equation of adaptation and diversity under the one blanket cause of natural selection would turn out to be wrong. The question, having lain dormant since the later days of the Modern Synthesis, has again become important in the debate over the punctuational theory of evolution.

Although it is without doubt a minor work in every sense, *The Variation of Animals in Nature* by Robson and Richards[72] may fairly be taken as representing some of the thinking on this subject in the 1930s; both Provine and Gould[73] regard it as "highly influential." Much of the time the authors are so shy of forming an opinion that I am reminded of the grandfather's aphorism in *A Kid for Two Farthings*: "An open mind is like a sieve: it can't hold anything." But minor authors often reflect the general current of muddled opinion among the majority of workers in a field much better than the confident and incisive writings—right or wrong—of the great masters. Sheppard recommended the book to me when an undergraduate as showing the position that the Oxford school had been at pains to refute. The following quotes seem to summarize Robson's and Richards' views fairly:

... we do not believe that Natural Selection can be disregarded as a possible factor in evolution. Nevertheless, there is so little positive evidence in its favour, so much that appears to tell against it, and so much that is as yet inconclusive, that we have no right to assign to it the main causative rôle. . . . [74]

The theory of Natural Selection ... postulates that the evolutionary process is unitary, and that not only are groups formed by the multiplication of single variants having survival value, but also that such divergences are amplified to produce adaptations. . . .

[But] there is a good deal of evidence that suggests that races and species arise independently of the survival value of their characters, unless we are prepared to make a very large appeal to ignorance.

We attach considerable importance to the facts assembled ... which suggest that the divergence of races and species is not influenced by selection. . . .

. . . [and] the very convincing suggestion that a great deal of specific and racial differentiation is due to isolation and chance survival.[75]

They favored also the idea that the organism itself might have some degree of autonomy, that it could not be molded by selection as a waterdrop might be molded to a complex surface, but was on the contrary like a stiff tennis ball, to be pressed into the surface only by a considerable force, and able to spring back to its own shape once the pressure has been released.[76]

5.2

The concept of species and race differences being nonadaptive was insufficiently analyzed in the literature at this time for its meaning to be fully clear. Strictly, it would mean that the character by which the races or species differ was not now, and had not at any time been, produced or maintained by natural selection. (In that event it would be a purely neutral character, and the difference would result solely from the stochastic fixation of neutral alleles, or the stochastic alteration in their frequencies, if either of the races or species still showed genetic variation for the character.) That is to say, supposing the difference to be in the length of the wings,

1. The establishment of the short-winged phenotype in one race, and the long-winged phenotype in the other race, was *not* produced because natural selection favored individuals with long wings over individuals with short wings in the one, and conversely in the other race.

2. If a short-winged mutation were now to occur in the long-winged race, or a long-winged within the short-winged race, these mutations would be selectively neutral and would not be removed or favored by selection.

Weaker interpretations are also possible. The difference could perhaps be called nonadaptive if *only* statement 1 *or* statement 2, but not both, were true, and it might be considered valid if one or both of these statements were true of one race but not the other. The statement that the difference is nonadaptive might, on the other hand, mean that while the forms were originally established in the races by selection at the individual level, and are indeed maintained that way to this day (so that *neither* statement 1 nor 2 is true), this difference contributes nothing to the success or failure of the races as races, or of the species as species. A vaguer meaning with the same intent would be that the short wings could in no way be considered a superior adaptation to the environment of the short-winged race, compared with the adaptation produced by the long wings in the environment of that race: in other words that the two forms are indeed adapted to different environments, and that the adaptations are equally satisfactory. Yet another meaning would be that short and long wings are adaptations to the *same* environment, both equally good. Operationally this would mean something like "If a long-winged mutation occurs in the short-winged race, then this mutation will be neutral, and similarly for the long-winged race, but other wing lengths, certainly those that are longer than long and shorter than short, but perhaps also intermediates between long and short, will be at a selective disadvantage."

This surely does not exhaust the possible meanings of the statement "The difference in character X between races A and B is a nonadaptive difference," but even so it is a long enough and diverse enough list to suggest that any discussion or attempt to verify the statement empirically is likely to founder in a welter of mutually incomprehensible meanings.

As natural selection plays a key role in Wright's model, it clearly only explains "nonadaptive" differences in one of the weaker meanings. But given the lack of analysis, it was inevitable that he should be interpreted as implying the stronger meaning, and that a refutation of that should be seen as a refutation of Wright's whole theory. In fact, as Provine and Gould[77] have shown, Wright tended to conflate the two kinds of meaning himself.

5.3

Wright's theory was in evolutionary terms highly original, and perhaps a greater break with the evolutionary tradition than he realized. Both the Lamarckian and Darwinian models had postulated that organic diversity and organic adaptation had a common cause—in Lamarck, the direct adaptation of the evolving line to the needs of the individual, and in Darwin, natural selection. Even though the linkage between adaptation and diversity might seem to be less close in creationism, insofar as all things in the world must have a common ultimate cause in God, and no more proximal causal connection is proposed, yet the fact that God must have wisely designed the organisms to be fitted to the stations in life for which He intended them gives adaptation and diversity a closer link within the Divine thought than, say, between moral laws and planetary motions. One strong tradition in preevolutionary taxonomy had been that one should reflect that connection in the classification by grouping organisms according to those characteristics that were most directly important to their way of life and to their physiological functioning.

Ironically, the uncoupling of adaptation and diversity that was provided by Wright's theory, and in which there was truly a break with the evolutionary past, was the culmination of a movement in taxonomy that had drawn its strength precisely from the evolutionary theory; in the *Descent of Man*, Darwin had deprecated his own neglect of "useless structures" in the *Origin*.[78] For the attempt to erect an evolutionary classification seemed to demand that the characters used for defining the groups be those that were not adaptive,[79] and so ran no risk of producing resemblances that were purely the result of adaptive convergence; as Darwin had remarked, no one paid any attention, as far as classification went, to the similarities between a whale and a fish. Thus two streams of thought, both bequeathed from pre-Darwinian taxonomy by Darwin, one of which linked adaptation firmly to diversity and the other of which put them no less firmly asunder, came into collision in the persons of Fisher and Wright.

While the tradition that important differences were nonadaptive was therefore quite an old, even preevolutionary one, it had remained a kind of taxonomists' tradition, until Wright produced a dynamic and genetic explanation. For Wright's theory could provide the mechanism for the postulated nonadaptive changes between races and species. The simpler and more naive interpretation would be that the differences were in selectively neutral characters, established purely by random drift, and were therefore nonadaptive in the strict and general sense. A more subtle view, too difficult probably for many workers to grasp, would be that the differences were nonadaptive in a much more restrictive sense, having been established through Wright's combined interaction of natural selection and random drift. Robson and Richards themselves seem to have been insufficiently versed in the vertiginous intellectual developments of population genetics (and who can blame them, for there were no simple textbooks to help the nonmathematical biologist) to adopt Wright's theory, and the problem of the differences between species was not one with which Fisher directly concerned himself. Little in the *Genetical Theory*, he said, would change if the world had been created in 4004 BC.[80] But, increasingly, differences between populations would be attributed by their discoverers to the working of genetic drift, and opposition to Wright's views could be expected from any biologist schooled in the alternative, adaptationist tradition.

Now not only was Fisher clearly brought up in some way within the adaptationist branch of Darwinism (his interest in taxonomy, the stamp-collecting branch of biology, must have been small indeed), but the nonadaptive tradition had already been introduced into evolutionary theory before Wright, by no one other than the mutationists:

"... There are plenty of characters to which it is exceedingly difficult to ascribe any utility, and the ingenuity of supporters of [the adaptationist] view has often been severely taxed to account for their existence. On the more modern view this difficulty is avoided. The origin of a new variation is independent of natural selection, and provided that it is not directly harmful there is no reason why it should not persist. In this way we are released from the burden of discovering a utilitarian motive behind all the multitudinous characters of living organisms.... The function of natural selection is selection and not creation.[81]

Ford's *Mendelism and Evolution*,[82] an influential, lucid, and nonmathematical account of his views and Fisher's, had been written as a reply—hence the title—to the still popular "dangerous fallacy" of mutationism.[83]

In general, Fisher's reaction to Wright's model might well have been that of the famous bowl of petunias: "Oh no, not again!"[84] He had spent a decade or so building an intellectual *tour de force*, a comprehensive theory that would demolish mutationism. And the generation of nonadaptive differences was not the only role in which random drift was being asked to play the same character as had been demanded of mutation. Fisher's theory had sought to cut mutation down to its smallest possible size: feeding to selection a supply of variation that was so finely divided as to be almost a fluid. The creative role of selection was at its maximum, of mutation at the minimum. Selection did not simply preserve what had been created by mutation; it moulded it very much as the original Darwinists had supposed. Fisher saw his championship of Darwinism as a considerable personal achievement:

... genetics is exposed more indefensibly that you seem to admit to the criticism of being anti-Darwinian, ... in the ... damning sense of factiously attacking and trying to discredit the far-reaching and penetrating ideas on the *means* of organic evolution which Darwin had originated. It was not only Bateson and de Vries, but almost the whole sect of geneticists in the first quarter of this century, who discredited themselves in this way.... The idea of polygenic Mendelism was frowned upon by both the biometricians and the geneticists when I published the paper[85] you cite in 1918.[86]

Now in Wright's theory random drift was to be given the job of creating the new forms for selection to preserve. Fisher must surely have considered that he was being asked to fight his battle with Punnett over again.

6 Fisher's Alternative Theory

6.1

Fisher was keenly aware of the problems raised by any process that tended to destroy the variation on which selection would work; he was pleased not only with his refutation of mutationism but also with the first chapter of the *Genetical Theory*, in which he showed that the particulate inheritance of Mendelism solved what he saw as the still outstanding obstacle to Darwin's theory: the loss of inherited variation under Darwin's theory of blending inheritance.[87] Random genetic drift had presented him with another such challenge; it could not only scramble the adaptive results of selection (in fact, this possibility may not have occurred to Fisher until later), but, because it eventually leads, in a finite, closed population, to all genes being represented by just one allele, fixed at 100% frequency in the population—and hence to total genetic uniformity—it too seems to set a limit on evolution by draining away the genetic variation. Fisher, he later recalled,[88] had reacted very negatively to the book by the Hagedoorns[89] in which this "Hagedoorn effect" was first pointed out, and he had been at pains to set up a theory of evolution in which it could be taken into account. In doing so, he started long before he encountered him, on a journey to a very different destination from Wright's. He had concluded that the half-life of the genetic variance—what he called the "relaxation time"—was a little less than one and a half times the population number. (Actually Fisher originally solved the problem incorrectly, and came up with $2.8n$, and it was Wright, as Fisher fully acknowledged, who found the correct value of $1.4n$.) This, in Fisher's view, let Darwinism off the hook once more; the Hagedoorn effect did not deplete the available variation at any appreciable rate: "... This confirms the value of $2n$ generations for the time of relaxation, found by a quite independent method by Professor Wright. The variance will then be halved by random survival in $2n \log 2 = 1.4n$ generations. *The immense length of this period for most species* shows how trifling a part random survival must play in the balance of influences which determines the actual variability."[90] This of course is true only if n is large; Fisher assumed that it would be, because it would be the population number of the whole species: "As few specific groups contain less than 10,000 individuals between whom interbreeding takes place, the period required for the

action of the Hagedoorn effect ... is immense." [91] If he had supposed that n was
the number of individuals in a local population, he would have had to come to a
very different conclusion. As early as 1922 Fisher had committed himself to the
view that the whole population of a species was the unit of evolution. [92]

6.2

A statement of Darwin's, that "widely distributed, much-diffused and common
species vary most," seemed to support the Hagedoorn contention, for the loss of
variation is more rapid the smaller the population. In collaboration with Ford, and
then on his own, Fisher confirmed the truth of the statement for color variation in
thirty-five species of moth, and then for the size of birds' eggs. [93] Statistically, there
was a correlation between the abundance of a species and its degree of variation.
As Fisher's theoretical explanation [94] for this fact is far from transparent, it is best
to let Ford summarize it for us:

> ... A large population offers a greater opportunity than a small one for the
> occurrence of very rare advantageous mutations, and for the establishment of the
> genes to which these give rise. Secondly, it can be shown that mutations producing
> genes of nearly neutral selective value must be excessively uncommon. ...
> Fisher ... has therefore pointed out the number of these 'neutral genes' which
> any species can maintain will be nearly proportional to the logarithm of its
> population. That is to say, when very rare, their existence will favour but little
> the variance of the commoner [species] which, however, can keep more of such
> genes in reserve. Changes in the environment may from time to time cause the
> effects of any of them to become slightly advantageous. The gene in question
> will then spread, giving rise to increased variability, reaching a maximum when
> it and its allelomorph are present in the population in equal numbers. ... Any
> difference in variability, when ascribable to the population [number] of the
> species concerned, will be due to those genes actually engaged in bringing about
> evolutionary change. Consequently, such observations demonstrate the spread
> in natural conditions of genes having small advantageous effects. [95]

In other words, Fisher attributed the difference in variability between rare and
common species not to their reservoir of neutral genes, but to their having a
different number of advantageous genes taking part in the perpetual flux of new
mutations.

This was mainly due to the fact that in a larger population more advantageous
mutants will appear per generation than in a small one, but also to the fact that
mutations that are so weakly selected as to be effectively neutral will, because of
the Hagedoorn effect, exist in a much larger reservoir in a large population than
in a small population (their number being proportional to ln{population number}).
This seems to make Fisher into what is now known as a "neutralist" (subscribing
to the view that the genetic variability of populations is in the main due to a
reservoir of neutral genes), but Fisher has something more subtle in mind: "A very
striking result is that a mutation can only be regarded as effectively neutral if the
selective intensity multiplied by the population number is small, so that the zone
of effective neutrality is exceedingly narrow, and must be passed over, one way or

the other, quite quickly in the course of evolutionary change."[96] That is, neutral mutations do not remain neutral for very long; changes of the external or internal environment alter the fitness of their bearers, so that they will become disadvantageous, and tend to disappear from the population, or more to the point, advantageous, whereupon they will be taken up into the flux of beneficial variation passing through the population.[97] They may even suffer several reversals of fortune, passing through neutrality to being detrimental, and back again to being advantageous, before finally establishing themselves at 100% of the population.[98] It is this flux of mutations that is expressed as the detectable variation of the moths and the birds' eggs, and it is larger in the commoner species because it is fed from a larger reservoir.

Random drift has, like mutation, been relegated to its smallest possible role in evolution, and one that is markedly subservient to selection: population size acts as a regulator valve, controlling the supply of new mutations that are fed into the flux of genetic change; and it is this flux that produces evolutionary advance.[99] Thus even while Wright had been discovering that stochastic effects might be creative, Fisher had been discovering that both disordering processes, mutation and the Hagedoorn effect, could be sent back to the supply department, the one feeding the finely divided variation, the other regulating—though not in any beneficial manner—its flow into the selective machinery.

Fisher concluded that when it came to evolutionary advance, the abundant species would win; the less common species would adapt more slowly, and would consequently lose what we now call the "arms race" with the common ones. This would itself impose an evolutionary pattern at a higher level—those abundant species that split too readily into several less abundant ones would doom themselves to extinction. Subdivision *within* a species would not contribute to evolution in the way Wright supposed because it was seldom complete enough to allow random drift, although it would offer a way out of evolutionary stagnation through adaptation to local conditions;[100] so much the better if migration put the whole genetic repertoire of the species at the disposal of each local population. Subdivision *of* a species could be a recipe for extinction.

6.3

Fisher had therefore produced his own modified version of the Hagedoorn effect. While maintaining that the number of individuals in a species was in general too large for the existing genetic variation to be drained away in a length of time that was of evolutionary significance, he had shown that the random loss of new mutants, while they were still rare, was an important evolutionary effect. As most mutants were in fact lost within a few generations of their first appearance, population size, in altering that rate of loss, could affect the speed of evolution. It would do that in two ways. The more important, Fisher thought, was that in a numerous species more beneficial mutations (that is, an absolutely greater number, not a higher percentage) would occur in every generation, with a resultingly larger number surviving the period of risk, becoming common, and entering the flux of beneficial variation. The lesser effect concerned those mutations that happened to be neutral. The more numerous species would hold a larger reservoir of these, which would, however, be at individually low frequencies, and which would not contribute either to the flux of variation passing through the species or significantly to the variability of the species until the internal or external environment changed. If they then became beneficial, they would be taken into the flux, and contribute both to variation and to evolutionary progress. The larger reservoir held by the larger species would result in a larger number of them entering this flux at any one time.

The objection that can be raised is that, as Fisher himself showed, a neutral mutation can attain high frequency, even fixation at 100% of the population, purely by random walk. Why then should there not be neutral genes in the flux, contributing to the variation of the species? Fisher's answer is ingenious but simple. He showed[101] that the number of copies of such a gene in the population could not greatly exceed the number of generations that had elapsed since the mutation had occurred (assuming that it only arises once). In a common species, a common gene may be represented by millions of copies. It follows that the span of time available is not great enough, and that at least in a numerous species, if a gene is found to exist in this number of copies, it must either have been pushed to such a frequency by selection or, less probably, by recurrent mutation.[102]

Why is the span of time not great enough? It sometimes appears to be suggested that the species cannot have existed for so long,[103] but this is an argument appropriate only for a creationist or punctuationist, who might suppose that the species started life with a genetic *tabula rasa*. Fisher stated clearly that the environment could not stay constant for so long, and that no gene could exist for such a time without becoming subject to selection. Although the argument may well not be valid for certain molecular changes in DNA or even proteins—a concept that was not available to Fisher—it is still a valid argument against neutral gene evolution as a general case.[104]

It was for this reason that although Fisher correctly calculated the probability $(1/2n)$ that a neutral gene would reach 100% frequency and therefore had the essential mathematics at his disposal, he did not then go on to develop the neutral theory of evolution forty years ahead of its time.[105] His model, although it is the general theory from which Kimura's neutral theory[106] is developed as a special case, has been much neglected; it has begun to reappear only in attempt to refute Kimura.[107]

The point that Fisher seems not to have taken into account is that in a smaller population, a gene can be subject to a wider range of strengths of selection and still remain effectively neutral. Neutral genes will therefore remain neutral for longer in the less abundant species, and on that account there will be more of them! However, it seems likely that from this cause their numbers will decrease linearly with population number, whereas they increase logarithmically with population number from random loss effects. Fisher's general point still stands.

6.4

Fisher had a commitment, then, reaching back to around 1920, not long after his successful synthesis of Mendelian inheritance and biometry, and deeply enmeshed with his developing synthesis of Darwinism and Mendelism, to seeing the whole population of a species as the unit of evolution, to seeing the main evolutionary process as the flux of successful, advantageous genes that were not so much producing absolute improvements in the organism as allowing it to swim against the predominant flow of time's arrow and remain adapted in the face of the terrifying second law of thermodynamics: "To the traditionally religious man, the essential novelty introduced by the theory of . . . evolution . . . is that . . . creation was not all finished a long while ago, but is still in progress. . . . In the language of Genesis we are living in the sixth day, probably rather early in the morning. . . ."[108] Natural selection was the immanent working of the Creative Will through the medium of Natural Law: ". . . the effective causes of evolutionary progress lie in the day-to-day incidents in the innumerable lives of innumerable plants, animals and men. We need not think of the effective causes as . . . imposed upon the organic world by the external environment; but where the organism meets its environment, where it succeeds or fails in its endeavours, where its potentialities are tried out in practice and incarnated in real happenings, there it is that the doctrine of natural selection locates the creative process."[109] In a limited way, the free individual, at the very point of creation, has been reconciled with Iago's cruel God (Fisher had little use for the sicklier kind of romantic poetry[110]). There was grandeur in this view of life. Through its applications in explaining, and if properly understood reversing the tendency for the thermodynamic chaos of barbarian peoples to swamp the world's great civilizations, there was a human mission as well: ". . . the Divine Artist has not yet stood back from his work, and declared it to be 'very good'. Perhaps that can be only when God's very imperfect image has become more competent to manage the affairs of the planet of which he is in control."[111] Fisher may have hoped that what the world sees as his great contribution—putting together Mendel and Darwin—was the lesser achievement compared with his synthesis of biology and human history. That has been a hope nurtured by many great scientists. It was not a scheme that took kindly to incorporating something a little different.

But the model is not necessarily incompatible with Wright's, which would seem only to imply that if the successfully evolving widespread species were split into subpopulations in the appropriate way, then even further evolutionary possibilities would be open to it. However, there are always the objections, objective and possibly true, that if the subpopulations are truly isolated, they must each be regarded as a separately evolving unit, relatively deprived of genetic variation by Fisher's version of the Hagedoorn effect, and in any case adapting independently to their own local environments, and that just the correct balance between sufficient isolation for effective drift by selected genes and sufficient coherence for the subsequent takeover of the species by the "winning" population must rarely be achieved. Fisher would have been extremely skeptical of the alternative ending, in which the less successful populations simply become extinct, leaving the "winner" to succeed through group selection. It is of course possible, as would be argued by modern proponents of Wright's theory,[112] that although the conditions may rarely be met, when they are they may have results out of all proportion to their rarity:

a new adaptive breakthrough or the foundation of a new embryonic *Bauplan*. But this possibility seems not to have been within the terms of discourse between Fisher and Wright in the 1930s and 1940s. As it was, Fisher simply declared himself to be totally unconvinced that the conditions required for Wright's theory could hold: "I wish I could better understand your views on those points on which I differ from you, but on the points I have discussed with Lush, I see little chance that I shall ever do so."[113]

7   Ford's Alternative Theory

Ford developed his own theory of the role of small population sizes.[114] It was based on observations that he and his father had made on a colony of the marsh fritillary butterfly, which following a period of very small population sizes had undergone a population explosion, during which the patterns of the butterflies became extremely variable.

Both before and after the surge in population size, the population had presented a very uniform color pattern; moreover, the patterns "before" and "after" were different from each other. Ford, who, unlike Fisher and Wright, had a strong interest in field ecology, explained this by combining Fisher's rejected speculation about natural selection discovering new adaptive types with a postulate of Charles Elton to the effect that when a population was rapidly expanding, mortality and hence also natural selection were reduced nearly to zero: "When, after a period of great scarcity, a species is rapidly increasing in numbers, non-advantageous mutations tend to spread through the population. In the course of their spreading, they are likely to become incorporated with certain gene-complexes with which they give rise to characters having selection value. Thus periodical increases and decreases in numbers may result in more rapid evolution than stationary populations."[115]

The problem with this theory as it stands is, to repeat, that the new adaptive combinations, if they occur in the expanding population, will simply be swamped by recombination. The simplest, although not necessarily the correct, interpretation is the one originally conceived by Elton—that random drift has raised the frequencies of the relevant alleles. Rather little drift would occur while the population remained small, on account of the intense selective mortality. But random changes of gene frequency occurring in the generation immediately before the "population flush" will not be countered, as a result of the almost total relaxation of selection. It is during this period of expansion also that alleles rendered neutral in this way, and even those that are still slightly disadvantageous, have a lower than normal chance of stochastic loss, and so will remain in larger numbers in the population.[116] By the time selection is reimposed as the population size levels out, the genetic composition has found its way to the region of a new adaptive peak. Seen in this way, the theory is an attractive extension of Wright's, overcoming the difficulty of nonneutral genes drifting in the face of strong selection, and incorporating a universal, but rather ignored, fact of demography, the enormous instability of population numbers. That the model has always been presented by Ford as a counter to Wright's theory[117] speaks volumes for the state of Anglo-American relations.

Fisher seems to have remained wary of Ford's model, perhaps because he perceived how very close it came to conceding Wright's point. In writing to Ford about it, he found it opportune to present what he then saw as the flaws in Wright's theory.[118]

8 Ecological Genetics at Oxford

8.1

From this analysis, I believe it is correct to say that the crucial and testable difference between Wright's view of evolution and Fisher's is the extent to which the local populations of a species are genetically isolated from one another. But that question was to become submerged, as a result of the personal, ecological interests of Fisher's friend E. B. Ford and of the empirical findings of Ford's colleagues at Oxford, in a debate about the power of selection. As Ford put it, "... the recent recognition that advantageous qualities are frequently favoured or balanced in particular environments by far greater selection–pressures than had hitherto been envisaged ... [is] a discovery which could only be made by direct observation, yet it is one which profoundly influences evolutionary concepts."[119]

The tradition of "ecological genetics" founded by Ford at Oxford was one of experiment and observation in the field. Although owing much to Fisher, and showing considerable skepticism about genetic drift, the corpus of knowledge, like all good empirical science, built up a picture of evolution which adhered rigidly to no theory, and if it did not support Wright, failed also to go all the way with Fisher. Some members of the school were indeed rather limited in their respect for mere mathematical theory. An experimental approach is restricted to what can be achieved in practice, and this meant the study of visible characteristics in natural populations. What had been achieved can be seen by comparing the *Mendelism and Evolution* of 1931 with the *Ecological Genetics* of 1964: Fisher is honored in the latter work, which Ford tells us he planned after absorbing Fisher's theoretical work in 1928,[120] but the body of experimental data collected in the intervening years, and the conclusions from it, largely submerge Fisher's theoretical models. The concentration on "balanced polymorphism"—mechanisms that hold alternative alleles in populations indefinitely—sets up a picture of evolution very different from Fisher's perpetual flux.

The school's general conclusions about selection and drift can be summarized by quotations from its three most eminent senior practitioners:

Mathematical *speculations* are of great importance in demonstrating what can and what cannot happen under different circumstances, and, therefore, what type of data should be collected in the field. In the absence of data taken from wild populations, they cannot show what in fact happens in nature.[121]

[Genetic drift] is mathematically certain to occur, given the right conditions, and may even, as claimed, produce by a non-selective process particular combinations of genes which happen to be highly advantageous and can spread, once they have originated, by natural selection. On the other hand, selection coefficients actually determined are all (or almost all) much too high to allow drift to occur, even although they refer to apparently very trivial characters. And it may very well be doubted whether under the changing conditions normal in nature any gene can possess, except for a very short time, a selection coefficient near enough to neutrality for drift to become important in determining its distribution.[122]

Up to about 1940, it was assumed that the advantage possessed by genes spreading in natural populations rarely exceeded 1 per cent. As it is now known that it quite commonly exceeds 25 percent and is frequently far more..., it will be realized how restricted is the field in which random drift is of importance.[123]

Thus the Oxford school built up a strong tradition that random drift had no important role in evolution, which culminated in a vigorous dispute over Kimura's neutral theory.[124] To an outsider, this position might seem to be a very difficult one to sustain; how can one deny the importance of a process that is logically and mathematically inevitable (not merely "given the right conditions"), a necessary consequence of undisputed theorems? The position that they took was that natural selection was normally too strong for it. Stochastic change in gene frequency would always tend to be corrected by the deterministic effects of natural selection. Sheppard's undergraduate classes included an extension of the Dubinin and Romashchov experiment with black and white beads[125] to demonstrate, first, the effects of random drift acting alone and, then, its interaction with natural selection. We saw how easy it was, even with quite a small population size, to prevent the fixation of one allele or the other by imposing selection in favor of the heterozygote.

8.2

The empirical emphasis within the Oxford school was on the gathering of information about natural selection and other evolutionary forces as they operated in natural populations, an exercise sitting firmly within the tradition of natural history, or, if one likes, ecology, for it could not be done without an extensive knowledge of the organism and its way of life. Cain's repeated criticism of the proposal that the difference between two taxa is clearly of neutral selective value because it has no manifest function has been that such a statement reveals nothing but one's own ignorance of the life-style of the organism. Sheppard was less than wholehearted in his admiration for the work on *Drosophila pseudoobscura* by Dobzhansky and his colleagues, on account of the crucial gap in the account produced by our almost total ignorance of the natural history of the animal; the evolutionary changes were plain enough, and it had been possible to show that many of them must have been due to natural selection, but the form which that selection took had not been discovered.

The Oxford school's program for testing, or refuting, Wright's theory was therefore to examine evolutionary change in the context of a much more detailed understanding of natural history than had been customary. There was perhaps some confusion about which of Wright's theories was being tested: the theory that differences between races (or by implication populations) were neutral or the theory that random drift was causing adaptive shifts from one adaptive complex to another. The first theory is by far the easier to test. In fact, a full test of Wright's three-phase theory has never been performed. As we have seen, the citation of random drift as the likely explanation of the allegedly neutral taxonomic differences had caused the taxonomic theory and the theory of drift to become conflated. A refutation of the first (for instance, by showing that a substantial number of such taxonomic differences were not of merely neutral survival value) would not necessarily constitute a refutation of the more general theory of Wright.

But just the same, even if Wright's full theory was too elaborate to be tested with the means then, or now, available, the process must start with a change of gene frequency, in the direction opposite to that required by selection, as a result of restricted population size. An examination of the causes of changes in gene frequency would therefore be a practicable and valid way of testing Wright's theory, at least in part.

8.3

The Oxford school therefore embarked on a series of studies of natural selection, and of the change of gene frequency either in time or in space. The studies have been given lucid summaries by Ford,[126] and the empirical findings are not in serious doubt.

A single colony of the day-flying scarlet tiger moth was found to contain an aberrant genetic form which exhibited rather small differences from the normal number of spots on the wing. The homozygote, which hardly ever appeared in the colony as a result of the low frequency of the gene, was a moth of considerably different appearance, but the heterozygote, with a small amount of extra black pigment, was just the sort of variety that followers of the Robson and Richards school would be prone to dismiss as nonadaptive. Quite strong natural selection was in fact shown to act on the genes through selective mating and differences in mortality in the early stages, and as the population size was estimated annually, it was possible to show that the changes in gene frequency from year to year were too great to be accounted for by random drift.[127]

Better known were the studies of snail coloration initiated by Philip Sheppard and Arthur Cain. The apparently random and meaningless fluctuations in frequency of the different colors and stripe patterns on the shells of the highly variable *Cepaea nemoralis* were shown to be quite accurately related to the color and composition of the background on which the snails rested, those in woods tending to be pink and without stripes, those in grassy places yellow and striped. Song thrushes were shown to be producing the relevant natural selection by eating the snails that were most conspicuous in each habitat. Here a difference that had in fact often been cited as an obvious example of a neutral character subject to differentiation by random drift was shown to be under the control, perhaps the exclusive control, of natural selection.[128]

Best known was Kettlewell's investigation of industrial melanism.[129] Although this was not directed to the problem of drift, but rather to elucidating the forces which had produced what everyone was agreed was a simple adaptive change—the darkening of moths in industrial areas of England—the simple and clear demonstration that selection by birds, searching visually for the moths in their normal daytime resting places—fully exposed on the trunks of trees, and depending only on stillness and camouflage to survive until the following nightfall—was one that could not fail to impress every schoolchild with the efficacy of natural selection. What is more, the selective differential between the black and the pale forms was shown to be around 50%. With selection of that strength, populations would have to be minute, and remain so for considerable times, for random drift to have any evolutionary effect worth thinking about.

The empirical approach of the Oxford school to Wright's theory had been what we could call the "fairies in the garden" argument. A man who maintains that there are fairies in the garden may rightly be challenged to take an independent observer to Julian Huxley's garden to watch for them for a few evenings. If their non-appearance leads him to defend his proposition on the grounds that this garden is ecologically unsuited to them, but that they may be in J. B. S. Haldane's garden, and if similar watches in Haldane's, Ford's, and Fisher's gardens still fail to produce the little people, the general proposal is no more disproved than the infinite number of other fanciful hypotheses. Most neutral commentators would be inclined to think however that it could be for all practical purposes dismissed, pending a positive sighting. Likewise, repeated observations of rather strong selective effects, too strong to allow significant drift, or producing clear adaptive differences between local populations, left an onus of proof on the supporters of Wright's theory. The rest of the world tends to take the issue as provisionally answered in favor of selection and to go about cultivating its own, fairyless, garden.

9   Critique

9.1

But the empirical findings could not be interpreted as an unalloyed triumph for Fisher's model of evolution, whose central feature had been the predominance of genes with very small selective effects. While the Oxford school tended to write up their discovery of selection coefficients of up to 50% as vindicating Fisher's selectionism beyond his wildest dreams[130] (Fisher, so the story went, had simply been cautious in postulating selection coefficients of 1% or less, as nobody at that time had conceived that they could be stronger), it had in fact knocked a hole right through the middle of Fisher's portrait of Darwinism. And it would hardly do to admit that Fisher was in general right, and that the genes investigated in the field were merely the statistical exceptions to the general rule that selection coefficients were small (Fisher would no doubt have been willing to concede statistical exceptions), as that would immediately concede that the results were atypical of evolution in general and undo any claim that the findings had to represent it.

The Oxford school therefore developed, rather gradually and pragmatically, for none of them really favored grand theoretical schemes involving abstract quantities like "fitness," a view of evolution that was selectionist, but that was not really Fisher's, and that was ready to make a compromise with mutationism—the effects of genes that played important parts in evolution could be quite large. Natural selection was receiving its raw material in much bigger lumps than Fisher had imagined. Ford gradually changed his description of butterfly mimicry from the gradualist version of Fisher (even Ford's first descriptions of it are rather ambiguous) to something like the compromise version of E. B. Poulton,[131] in which both large and small mutations took part. The culmination was the adoption, by Clarke and Sheppard, of Wright's multipeak model to explain mimicry, with the shifts from peak to peak being produced, not by the random drifting of numerous mutations of individually small effect, but by the determinist selection of a mutation with large enough effects to take the organism clear across the adaptive valley.[132]

With the wisdom of hindsight, the Oxford school might be accused of having loaded the evidence by taking examples in which the genes produced such marked and distinct effects that it was virtually certain that strong natural selection would be discovered. This would be unfair, and unhistorical. With the exception of industrial melanism, the features investigated were exactly of the type that had widely been described as of obviously neutral survival value.

This was particularly true of human blood groups, regarded by geographers, for example, as the perfect guide to ethnic history precisely because they had no adaptive significance whatever.[133] With the demonstration of quite noticeable selection on the ABO system through susceptibility to infectious and constitutional disease, on top of the known selection on several other systems through hemolytic disease of the newborn, the selectionist view came to look very secure indeed. Even a "trivial" internal biochemical character turned out to be selected. But at this point, the Oxford school started to let their growing conviction that all selection coefficients were large go to their heads. Apart from these cases, natural selection on human blood groups has proved elusive, although a good many teams have tried to find it. Clear evidence from other studies on the genetics of human populations that gene frequencies of blood groups, biochemical markers, and even genes producing pathological conditions were subject to significantly large amounts of random drift[134] was simply ignored. Yet here, if melanism had been "Darwin's missing evidence," was Wright's missing evidence. The human population is, of course, a little exceptional, because of its genetically peculiar social structure in tribal periods and its tendency to send out small numbers of colonists in the period of European expansion, and so may suffer an exceptionally large amount of genetic drift. It is also exceptional in being the only population for which the pedigree and demographic history is well enough known for us to demonstrate random effects with any certainty. The fairy has at last materialized.

9.2

With the credibility of genetic drift restored, it becomes clear that the empirical work on selection had not settled the issue. Although the composition of populations of *Cepaea* does show a beautiful correlation with habitat, thus proving that some of the difference between populations is the result of selection, in no sense does this prove that the differences between populations within similar habitats are not due to random drift. In fact, Lamotte[135] produced rather good evidence that variance in the frequency of the color forms was correlated negatively with population size, and another study of populations of the snail in a marsh area where they had been all but wiped out by a flood made an excellent, and much disputed (and ignored), case that the differences between colonies in that area were produced purely by the random survival of the differently colored snails that founded them.[136]

The similar patchwork distribution of the various stripe patterns and colors on high chalk hills in England was probably produced by a similar kind of founder-drift, if not, as Wright[137] proposed, by the full operation of his shifting balance process. In the scarlet tiger, Wright[138] was able to show that the changes observed fitted very well to the theory that the gene under study was actually disadvantageous, and that Fisher and Ford had simply been watching its removal from the population under natural selection. The selectionist view would be that this might be true, but that the gene must first have increased in frequency under selection during a period when the environment temporarily changed. Wright points out that as this is the only population in which the gene has been observed, it is equally plausible that the frequency of the gene was raised purely by random drift during a population bottleneck, which might have occurred in the years just before the observations began.

It is possible to produce both selectionist and nonadaptation explanations for most of the observations that have been made (it being customary for selectionists to interpret what appear to be excellent cases of the randomizing effects of small population size by the similar randomizing effects of fluctuations in the environmental factors that produce natural selection), and selectionists will sometimes go to quite extreme lengths to support this view, even to the point of disputing the power of mutation. Thus the distribution of color blindness in human populations fits quite well with the assumption that natural selection ceases with the adoption of agriculture, leaving the frequency of the nonfunctional mutated color-vision gene to rise slowly but steadily as a result of mutation pressure.[139] Ford insists that this must be the result of selection *in favor* of color blindness in settled peoples, although the only tangential evidence he produces is that the color-blind can see insects that are camouflaged to a person with normal vision.[140] This might well be of use to a hunter-gatherer, but it is among such people that color blindness is *least* frequent.

9.3

In the context of what Gould[141] calls the "hardening" of the Synthesis, that is, a shift from pluralistic acceptance of random drift as a major force toward a hard-line selectionism—which if "pluralism" carries its usual connotations is presumably a bit like Reaganism—it can be seen that the Oxford school always took a "hard line." Ford was willing to accept that the characteristics of species and races could be nonadaptive (he attributed them to what were then known as correlative effects, or, as geneticists would put it, to the character being one of the multiple effects of a gene, some of which were subject to selection and some not),[142] but beyond this no concessions were made either to neutralism or to a significant role for random drift. If we view the debate not as one over the role of drift, but as a question of how large were the elements that were fed to the selective machinery for shaping into adaptations, then the Oxford school may be said to have "softened." They changed, almost imperceptibly and certainly without any dramatic announcements, the extreme gradualist thesis of Fisher into one that permitted much larger mutational steps, and, at least with regard to mimicry, ended by adopting multipeak models of the fitness surface.[143] Their disagreement with the latter-day mutationism of Goldschmidt finally boiled down not to a dispute over "large" mutations taking part in the evolution of mimicry—although there is always room for the largely fruitless question, "How large is large?"—but to a dispute over Goldschmidt's insistence that the adaptations produced by mutation were perfect *ab initio* and were never improved. "Softening" with respect to random drift, however, did not take place, and the excellent evidence for drift in human populations has continued to be studiously ignored. The first English population geneticist with unimpeachable credentials as an experimenter with natural populations to give proper weight to random drift was R. J. Berry, who was not raised in the Oxford tradition.[144]

9.4

Stochastic change in gene frequency then has clearly the role of producing maladaptive fluctuations in evolution, around whatever steady trends or dynamic equilibria are produced by natural selection. It is also extremely likely that it produces a steady rate of neutral evolutionary change in those parts of macromolecules where alterations have so little effect on the functioning of the organism that natural selection is effectively absent for very long periods of evolutionary time. Whether it plays the further role of creating new variants for selection to work on, either individually or through some kind of population selection, remains an open question.[145]

Had the Oxford school then failed to solve the problem, as Lewontin[146] claims? That depends on what one thinks the "problem" was. Certainly the empirical evidence of natural selection left undecided the role of random stochastic drift in increasing or decreasing the variation on which selection can work. But that had not been the only question for evolutionary geneticists when the synthesis was welded in the 1930s. Robson and Richards had indicated a glaring gap in the empirical evidence for the Darwinian theory; there was very little evidence indeed for the efficacy of natural selection in producing evolutionary change, and little satisfactory evidence even for its occurrence in natural populations: "The direct evidence for the occurrence of Natural Selection is very meagre and carries little conviction. . . . It is [they commented in an uncharacteristically clear statement] a very unsatisfactory state of affairs for biological science that a first-class theory should still dominate the field of inquiry though largely held on faith or rejected on account of prejudice." [147] His reminiscences [148] of the late 1920s in Oxford show that it was this lack of evidence that E. B. Ford perceived then, and it was this lack that he and his Oxford group had remedied abundantly. Whatever else remains to be thrashed out within the theory of evolution, there is no longer any question about the role of natural selection in producing adaptive change. Of that there is no possible, or even probable, doubt, whatever.

10 Moral

As an experimental scientist, I am biased toward the belief that one of the main motives behind a scientist's choice of theoretical models is his/her rational perception, right or wrong, of the success which the models have in describing the real world. Fisher clearly did reject Wright's central formulation because to Fisher it provided a description of the world that was unreal; populations of organisms were not normally both subdivided in such a way that substantial random drift could occur within the subunits, and yet at the same time sufficiently connected that a successful subunit could later take over the whole species. I have to admit this was a hunch; Fisher did not produce empirical evidence to back this view (he discusses human migration a bit in his letters, a possibly unfortunate choice in view of the strong evidence this has now provided for the workings of random drift); in fact, the empirical evidence one way or the other has yet to be found, and requires long-term observations of demographic structure and, in particular, of minute rates of successful migration that are almost impossible to perform. Fisher could have adopted a compromise position—that the small subunits would have to be full species in order to be sufficiently isolated, but that the stochastic events were of profound importance in the history of this new species. Such, if I understand it, is the view now adopted by Gould.

If we ask why Fisher chose to play his hunch in the one direction rather than the other, then we have to say that this did not result from any profound difference between his theoretical model and Wright's; the logical premises required to arrive at Wright's conclusions were all contained in some form or another in Fisher's model. And though Fisher came to denigrate Wright's approach, and to see the underlying philosophy as very different—Wright being concerned with continued evolutionary "progress" and Fisher with a Red Queen process in which change was unprogressive (*plus ça change, plus c'est la même chose*), in which it was not a species' level of adaptation that mattered, but its degree of adaptability in the face of a changing environment (the less numerous, less adaptable species heading to extinction)—this particular aspect of the schism did not become apparent until after Fisher had made his final declaration in 1931, that he would never understand Wright's theory.

I see Fisher as a man with a philosophy that integrated science, history, religion, and social action.[149] It admitted stochastic processes only as the essentially destructive and chaotic forces of the physical world, overridden in the living world by the organizing force of natural selection. Through it Creation continued, and as befits the Christian virtue of Hope, and the Christian doctrines of the value of sacrifice and suffering, and of Works, it was ultimately, and in a tough kind of way, beneficial.

As natural selection occurred at the interface of organism and environment, it was subject to will, and its course was not predestined or determinate. We had it in our free choice to aid the progress of evolution through applying our understanding of it in the techniques of eugenics. Fisher had found a midpoint between two threatening world views: meaningless chaos and rigid determinism.

As an intellectual *tour de force*, Fisher's theory could not but be impressive. But intellectual power is no more guarantee of correctness in a scientist than the necessary qualities of human leadership are warranty that a statesman is not about to take his people over the Gadarene plunge. When the allies of such an admirable leader in science begin to find things that are incompatible with the master's theory, they do not rush to announce the fact. The Oxford school, in its role as the experimental wing of Fisher's "party" in England, consisted of people, like Fisher, with a strong interest in natural selection, but perhaps for different reasons: a great liking for natural history in the case of Ford, and a strong commitment to disputing the "nonadaptive differences" school of taxonomy in the case of Cain. What they had in common was clearly membership in the English tradition which sees natural selection as supremely important, and which (Gould and I agree) is a continuation of, or counterpoise to, the national fascination with natural theology and the argument from design.[150]

When the Oxford school succeeded, so superbly, in demonstrating the power of natural selection, they chose to ignore the fact that in doing so they had cast grave doubt on the central gradualism of Fisher's theory—the view that only small mutations were of importance in evolution. If they had wanted to be nationalistic, the Oxford school could have declared themselves to have forged yet another Modern Synthesis, not simply between Darwinism and Mendel, but between two English traditions—Fisherism and the mutational theories of Punnett. Personal loyalties apart, I suspect it never crossed their minds; Punnett had already been dismissed. He was seen as a member of a dangerous and threatening tradition that sought to put down natural selection from its seat. And the wise world is always outside, waiting for the glimmer that will tell them the scientists have finally proved that "Darwin got it wrong," and that some other destiny shapes our ends.

As an experimental scientist, I find it comforting that empirical results have a habit of taking our understanding in directions that are markedly tangential to the "-isms" that start the experimentalist off on his research. Whatever their political, religious, and other predispositions—and the admiration for a revered teacher or friend is often stronger than all of these—scientists do have to confront the real world.[151]

Notes

1. M. Kimura, *The Neutral Theory of Molecular Evolution* (Cambridge: Cambridge University Press, 1983).

2. S. Wright, *Evolution and the Genetics of Populations*, Vol. III (Chicago: University of Chicago Press, 1977), Chapter 13.

3. It is not customary to append one's *curriculum vitae* to a paper, but in this case it seems germane to state that E. B. Ford, who was himself tutored by Julian Huxley, and who collaborated as friend and colleague with R. A. Fisher, acted as doctoral supervisor to P. M. Sheppard, who during his period as a graduate student collaborated with A. J. Cain on the study of *Cepaea* (see section 8 of this paper). I carried out my Honors Degree work in evolutionary genetics with Sheppard in Liverpool, and then became a doctoral student with Ford at Oxford. B. C. Clarke, a noted critic of M. Kimura's neutral theory, was a doctoral student of Cain.

4. For this kind of interpretation of Fisher as a "mechanistic materialist," see G. E. Allen, "The Several Faces of Darwinism: Materialism in Nineteenth and Twentieth Century Evolutionary Theory," in D. S. Bendall, ed., *Evolution from Molecules to Men* (Cambridge: Cambridge University Press, 1983), pp. 81–102.

5. "Credo in un Dio crudel che m'ha create/Simile a sè .... /Dalla vità d'un germe o d'un Atòmo/Vile son nato./Son scellerato/Perchè son uomo;/ E sento il fango originario in me." A. Boito, *Otello*, Act 2, Scene 2 (1887). "Fango originario" carries the dual meaning of "original sin" and "Urschleim."

6. J. R. G. Turner, "The Hypothesis That Explains Mimicry Explains Evolution: The Saltationist-Gradualist Schism," in M. Grene, ed., *Dimensions of Darwinism: Themes and Counterthemes in Twentieth Century Evolutionary Theory* (Cambridge: Cambridge University Press, 1983), Chapter 6.

7. Baudelaire, *La Géante*.

8. M. J. S. Hodge, "The Development of Darwin's General Biological Theorizing," in Bendall, ed., *Evolution*, pp. 43–62 (note 4).

9. A. R. Wallace, *Darwinism*, 2nd ed. (London: Macmillan, 1889), p. 99.

10. Wallace, *Darwinism* (note 9), pp. 62ff.

11. W. B. Provine, *The Origins of Theoretical Population Genetics* (Chicago: University of Chicago Press, 1971).

12. R. C. Punnett, *Mimicry in Butterflies* (Cambridge University Press, 1915). Norton's computations are mostly very accurate when checked against the best formulae now available (which are Haldane's), and the means Norton used to produce them are totally unknown. Norton's name is more widely known from its appearance in the dedication of *Eminent Victorians* than for this computational masterpiece.

13. R. C. Punnett, *Mendelism*, 3rd ed. (London: Macmillan, 1911), p. 135. The title of this book does not merely do homage to the discoverer of the laws of inheritance. It signals the setting up of "Mendelism" as an alternative evolutionary theory to "Darwinism."

14. J. B. S. Haldane, *The Causes of Evolution* (London: Longmans, Green, 1932), p. 110.

15. R. A. Fisher, "The Correlation between Relatives on the Supposition of Mendelian Heredity," *Transactions of the Royal Society of Edinburgh* 52 (1918), 399–433; reprinted in J. H. Bennett, ed., *Collected Papers of R. A. Fisher* (Adelaide: University of Adelaide Press, 1971–1974), No. 9.

16. R. C. Punnett, *Mendelism*, 5th ed. (London: Macmillan, 1919), Chapter XIV, "Intermediates."

17. R. C. Punnett, *Mendelism*, 7th ed. (London: Macmillan, 1927) Chapter XVI "Variation and Evolution." We should allow that the publishers may have limited the number of pages that could be reset.

18. R. B. Goldschmidt, *The Material Basis of Evolution* (New Haven: Yale University Press, 1940).

19. R. A. Fisher, "Darwinian Evolution [by] Mutations," *Eugenics Review*, 14 (1922), 31–34; reprinted in *Collected Papers* (note 15), No. 26.

20. R. A. Fisher, *The Genetical Theory of Natural Selection*, 1st ed. (Oxford: Clarendon Press, 1930).

21. 'Organism', an ambiguous word, is taken directly from Fisher. Is it an individual or a whole species? (See section 4 of this paper.)

22. Fisher, *Genetical Theory* (note 20), Chapter II, "The Fundamental Theorem of Natural Selection."

23. Jonathan Hodge has pointed out to me the strange and hitherto unnoticed fact that this theorem is named from the title of a section in a paper by Karl Pearson, "Mathematical Contributions to the Theory of Evolution. III. Regression, Heredity and Panmixia," *Philosophical Transactions of the Royal Society*, A187 (1896), 253–318; reprinted in E. S. Pearson, ed., *Karl Pearson's Early Statistical Papers* (Cambridge: Cambridge University Press, 1948). There is no similarity between Pearson's and Fisher's theorems.

24. Fisher, *Genetical Theory* (note 20), pp. 36ff.

25. Fisher, *Genetical Theory* (note 20), pp. 35–36.

26. J. H. Bennett, ed., *Natural Selection, Heredity and Eugenics, Including Selected Correspondence of R. A. Fisher with Leonard Darwin and Others* (Oxford: Oxford University Press, 1983).

27. R. Olby, "La théorie génétique de la sélection naturelle vue par un historien," *Revue de Synthèse*, 3rd Series (1981), 251–289.

28. Darwin to Fisher, 1925, *Natural Selection* (note 26), pp. 77–78.

29. Fisher, *Genetical Theory* (note 20), Chapter II, the section entitled "The Nature of Adaptation."

30. Fisher, *Genetical Theory* (note 20), Chapters III and VI, "The Evolution of Dominance" and "Mimicry"; see also Turner in Grene, ed., *Dimensions* (note 6).

31. Fisher, *Genetical Theory* (note 20), Chapter V, the section entitled "Equilibrium Involving Two Factors."

32. J. Maynard Smith, *The Evolution of Sex* (Cambridge: Cambridge University Press, 1978).

33. Darwin to Fisher, 1928, *Natural Selection* (note 26), p. 92.

34. B. Norton, "Fisher's Entrance into Evolutionary Science: The Role of Eugenics," in Grene, ed., *Dimensions* (note 6), Chapter 1.

35. Fisher, *Genetical Theory* (note 20), p. 37.

36. Fisher, *Genetical Theory* (note 20), p. 42.

37. Fisher, *Genetical Theory* (note 20), p. 37.

38. Fisher, *Genetical Theory* (note 20). Chapters VIII–XII; also R. A. Fisher, "Eugenics: Can It Solve the Problem of Decay of Civilisations?" *Eugenics Review*, 18 (1926), 128–136, reprinted in *Collected Papers* (note 15), No. 53. For a lucid summary of the latter paper, see Olby, "Theorie génétique" (note 27).

39. Fisher to P. de Hevesy, 1945, *Natural Selection* (note 26), p. 192.

40. See Fisher's paper of 1911 on "Heredity," printed in *Natural Selection* (note 26), pp. 51–58.

41. Fisher, *Genetical Theory* (note 20), Chapter VI, the section entitled "Sexual Selection."

42. *Natural Selection* (note 26), p. 279.

43. Fisher, *Genetical Theory* (note 20), p. 36, italics added.

44. Fisher to Darwin, 1928, *Natural Selection* (note 26), p. 88.

45. Fisher to K. Mather, 1942, *Natural Selection* (note 26), p. 236.

46. Haldane too was concerned with solving exactly the same problem. J. B. S. Haldane, "A Mathematical Theory of Natural and Artificial Selection. Part VII. Metastable Populations," *Proceedings of the Cambridge Philosophical Society*, 27 (1931), 137–142.

47. S. Wright, "Random Drift and the Shifting Balance Theory of Evolution," in K. Kojima, ed., *Mathematical Topics in Population Genetics* (New York: Springer, 1970), pp. 1–31.

48. R. Lewin, "Evolutionary Theory under Fire," *Science*, 210 (1980), 883–887.

49. E. B. Ford, *Ecological Genetics*, 1st ed. (London: Methuen, 1964), p. 33.

50. R. A. Fisher, "The Measurement of Selective Intensity," *Proceedings of the Royal Society of London*, B121 (1936), 58–62; reprinted in *Collected Papers* (note 15), No. 147.

51. As did Haldane, *Causes of Evolution* (note 14).

52. See also R. Olby, "Theorie génétique" (note 27).

53. Fisher and Ford first met in 1923. See J. Fisher Box, *R. A. Fisher: The Life of a Scientist* (New York: Wiley, 1978), pp. 180–181.

54. See S. Wright, "Biology and the Philosophy of Science," *The Monist*, 48 (1964), 265–290, and the references therein, particularly Wright's review of a book by Lloyd Morgan, in *Journal of Heredity*, 26 (1935), 369–373.

55. Fisher's chief writings on indeterminacy and free will are his unpublished under-graduate paper of 1911 on "Evolution and Society," later printed in *Natural Selection* (note 26), pp. 58–62; "Indeterminism and Natural Selection," *Philosophy of Science*, 1 (1934), 99–117, reprinted in *Collected Papers* (note 15), No. 121; and *Creative Aspects of Natural Law, The Eddington Memorial Lecture* (Cambridge: Cambridge University Press, 1950), reprinted in *Collected Papers* (note 15), No. 241; also letters to Leonard Darwin, 1938, *Natural Selection* (note 26), pp 159ff., and to C.S. Sherrington, 1947, *Natural Selection* (note 26), pp 261ff. These are lifelong beliefs, held from his undergraduate days, in harmony with his Anglican faith.

56. Fisher to C.S. Stock, 1936, *Natural Selection* (note 26), p. 264.

57. Correspondence between Fisher and Wright, *Natural Selection* (note 26), pp. 272ff. In the introduction to this collection, J. H. Bennett provides a rather full chronology of the letters and the various meetings between Fisher and Wright, pp. 40–48.

58. See section 2 of this paper for the meaning of this modern term.

59. Fisher to Wright, 1931, *Natural Selection* (note 26), p. 279.

60. See, for example, D. S. Falconer, *Introduction to Quantitative Genetics*, 1st ed. (Edinburgh: Oliver and Boyd, 1960), p. 77.

61. Fisher to Wright, 1929, *Natural Selection* (note 26), p. 273.

62. Fisher to Wright, 1931, *Natural Selection* (note 26), p. 279.

63. Fisher to A. B. D. Fortuyn, 1931, *Natural Selection* (note 26), p. 204.

64. S. Wright, "Fisher's Theory of Dominance," *American Naturalist*, 63 (1929), 274–279.

65. R. A. Fisher, "The Measurement of Selective Intensity" (note 50); "Average Excess and Average Effect of a Gene Substitution," *Annals of Eugenics*, 11 (1941), 53–63, reprinted in *Collected Papers* (note 15), No. 185. In the first of these papers, Fisher criticizes the concept of the adaptive landscape as being an attempt to postulate an external force, independent of natural selection on individuals but guiding the evolution of the population. The second paper repeats this criticism and attempts to demolish the equation that Wright used to describe the change of gene frequency in terms of the steepness of the surface from point to point as merely proving one of its own preconditions—see also Turner, "Changes in Mean Fitness under Natural Selection," in Kojima, ed., *Mathematical Topics* (note 47), pp. 32–78. Both criticisms were totally unfair to Wright.

66. Fisher, "Measurement of Selective Intensity" (note 50), and Fisher to M. Kimura, 1956, *Natural Selection* (note 26), p. 229; see also A. W. F. Edwards's review of Wright's *Evolution and the Genetics of Populations*, vol. 2, in *Heredity*, 26 (1971), 332–337.

67. Fisher to Ford, 1938, *Natural Selection* (note 26), pp. 201–202.

68. I have deliberately used the ambiguous term "organism" in recounting Fisher's theories here.

69. Fisher, *Genetical Theory* (note 20), pp. 34–35.

70. Fisher, "Average Excess" (note 65), and *The Genetical Theory of Natural Selection*, 2nd ed. (New York: Dover, 1958); see also J. R. G. Turner, "Changes in Mean Fitness" (note 65). The crucial point is that α is initially defined in *Genetical Theory*, 1st ed., as the derivative of mean fitness on gene frequency, although as used in the derivation of the formula for the variance it is the regression of phenotype on genotype. The final conflation is made on p. 54 of the 1941 paper, where α is redefined as the regression.

71. S. J. Gould, "The Hardening of the Modern Synthesis," in Grene, ed., *Dimensions* (note 6), Chapter 4.

72. G. C. Robson and O. W. Richards, *The Variation of Animals in Nature* (London: Longmans, Green, 1936). It should be noted that as a result of Robson's increasing mental illness, the book's publication was delayed from about 1933, and therefore reflects opinions rather earlier than those of its nominal date.

73. W. B. Provine, "The Development of Wright's Theory of Evolution: Systematics, Adaptation and Drift," in Grene, ed., *Dimensions* (note 6), Chapter 3; also Gould, "Hardening" (note 71).

74. Robson and Richards, *Variation* (note 72), p 316.

75. Robson and Richards, *Variation* (note 72), pp. 370ff.

76. Robson and Richards, *Variation* (note 72), p. 355.

77. Provine, "Development of Wright's Theory" (note 73); and Gould "Hardening" (note 71).

78. C. Darwin, *The Descent of Man and Selection in Relation to Sex*, Vol. I, 1st ed. (London: Murray, 1871), p. 152.

79. W. B. Provine, "Development of Wright's Theory" (note 6). "Cladistics" has finally, and mercifully, developed ways of producing evolutionary classifications without the need to make this distinction.

80. Fisher to J. S. Huxley, 1930, *Natural Selection* (note 26), p. 222.

81. Punnett, *Mendelism* (note 13), p. 132.

82. E. B. Ford, *Mendelism and Evolution*, 1st ed. (London: Methuen, 1931).

83. E. B. Ford, *Mendelism and Evolution*, 6th ed. (London: Methuen, 1957), p. vii.

84. D. Adams, *The Hitch-Hiker's Guide to the Galaxy* (London: Pan Books, 1979).

85. R. A. Fisher, "Correlation between Relatives" (note 15).

86. Fisher to K. Mather, 1942, *Natural Selection* (note 26), p. 236.

87. See quote from the same letter to Mather (as in preceding note) in section 2 of this paper.

88. Fisher to A. B. D. Fortuyn, 1931, *Natural Selection* (note 26), p. 204.

89. A. L. Hagedoorn and A. C. Hagedoorn, *The Relative Value of the Processes Causing Evolution* (The Hague: Martinus Nijhoff, 1921).

90. R. A. Fisher, "The Distribution of Gene Ratios for Rare Mutations," *Proceedings of the Royal Society of Edinburgh*, 50 (1930), 205–230, reprinted in *Collected Papers* (note 15), No. 86 (italics added).

91. R. A. Fisher, "On the Dominance Ratio," *Proceedings of the Royal Society of Edinburgh*, 42 (1922), 321–341, reprinted in *Collected Papers* (note 15), No. 24.

92. See also the letter to Wright (note 57) quoted in section 4.1 of this paper.

93. R. A. Fisher and E. B. Ford, "The Variability of Species in the Lepidoptera with Reference to Abundance and Sex," *Transactions of the Royal Entomological Society of London*, 76 (1928), 367–379, reprinted in *Collected Papers* (note 15), No. 72, preliminary announcement in 1926, *Collected Papers* (note 15), No. 52; see also R. A. Fisher, "The Relation between Variability and Abundance Shown by the Measurements of the Eggs of British Nesting Birds," *Proceedings of the Royal Society of London*, B122 (1937), 1–26, reprinted in *Collected Papers* (note 15), No. 153.

94. Fisher, *Genetical Theory* (note 20), Chapters IV–V, "Variation as Determined by Mutation and Selection"; also "Distribution of Gene Ratios" (note 90) and "Relation between Variability and Abundance" (note 93).

95. E. B. Ford, *Mendelism and Evolution*, 3rd ed. (London: Methuen, 1940), pp. 89–90.

96. Fisher to Wright, 1929, *Natural Selection* (note 26), p. 274.

97. Fisher, *Genetical Theory* (note 20), p. 95.

98. Fisher, "Distribution of Gene Ratios" (note 90), final paragraph of summary.

99. Olby, "Théorie génétique" (note 27), shows how essential to this model was the objective demonstration that mutation rates were in fact minute.

100. R. A. Fisher, "Population Genetics," *Proceedings of the Royal Society of London*, B141 (1953), 510–523, reprinted in *Collected Papers* (note 15), No. 252. See also Wallace's discussion of population size as a regulator of variation and hence of the evolution of domestic breeds in *Darwinism* (note 9), p. 98. See also Wallace's discussion of the evolution of natural populations under changed conditions, in which the most abundant species would be survivors on account of their greater ability to adapt (ibid., p. 116). Adaptation to local conditions is also much emphasized by Wallace. There can be little doubt that Fisher was greatly influenced by Wallace's thinking.

101. Fisher, "Distribution of Gene Ratios" (note 90); Fisher, *Genetical Theory* (note 20), p. 80.

102. Fisher, *Genetical Theory* (note 20), p. 80.

103. Fisher's first attempt at formulating this argument reads, "The time required for a single mutation to increase sufficiently in numbers till it affects, say, one-third of the genes of the species ... in the absence of selective advantage ... will be ... a number of generations of the same order as the number of individuals which there are in the species; for many species this would be at least a thousand times as long as the longest time allowable": "On Some Objections to Mimicry Theory; Statistical and Genetic," *Transactions of the Entomological Society of London*, 75 (1927), 269–2781, reprinted in *Collected Papers* (note 15), No. 59. Without further explanation, this *seems* to imply that species cannot exist for so many genera-

tions, and to ignore the fact that daughter species should inherit the genetic variation of their parents. Fisher clearly realized that the argument was either wrong or, more probably, poorly worded, and omitted this passage when the paper became a chapter of *Genetical Theory* (note 20).

104. R. C. Lewontin, *The Genetic Basis of Evolutionary Change* (New York: Columbia University Press, 1974).

105. Fisher, "Distribution of Gene Ratios" (note 90).

106. M. Kimura, "Evolutionary Rate at the Molecular Level," *Nature*, 217 (1968), 624–626.

107. E. B. Ford, *Genetics and Adaptation* (London: Edward Arnold, 1976), p. 11.

108. R. A. Fisher, "The Renaissance of Darwinism," *The Listener*, 37 (1947), 1001ff., reprinted in *Collected Papers* (note 15), No. 217.

109. Fisher, "Renaissance" (note 108).

110. Fisher to N. Barlow, 1958, *Natural Selection* (note 26), pp. 181–182.

111. Fisher, "Renaissance" (note 108).

112. S. J. Gould, "Is a New and General Theory of Evolution Emerging?" *Paleobiology*, 6 (1980), 119–130.

113. Fisher to Wright, 1931, *Natural Selection* (note 26), p. 280.

114. For example, Ford, *Mendelism and Evolution*, 1st ed. (note 82), pp. 75ff.; also *Ecological Genetics*, 1st ed. (note 49), Chapter 2, "Fluctuations in Numbers."

115. This is a summary of Fisher, Ford, and Elton concocted by J. R. Baker, 1931, *Natural Selection* (note 26), p. 281. See also the quote from Elton in Provine, "Development of Wright's Theory" (note 73).

116. Fisher, *Genetical Theory* (note 20), p. 82.

117. Ford, *Ecological Genetics* (note 49), pp. 36–37.

118. Fisher to Ford, 1938, *Natural Selection* (note 26), p. 201.

119. Ford, *Ecological Genetics* (note 49), pp. 9–10.

120. Ford, *Ecological Genetics* (note 49), p. xii.

121. P. M. Sheppard, *Natural Selection and Heredity*, 1st ed. (London: Hutchinson, 1958), p. 123, italics added. Mathematicians, he says, produce not "theories" but "speculations."

122. A. J. Cain, *Animal Species and their Evolution*, 2nd ed. (London: Hutchinson, 1963), p. 146.

123. Ford, *Ecological Genetics* (note 49), p. 32.

124. B. C. Clarke, "Darwinian Evolution of Proteins," *Science*, 168 (1970), 1009–1011.

125. See Beatty's chapter in this volume.

126. E. B. Ford, *Ecological Genetics*, 4th ed. (London: Chapman and Hall, 1975).

127. This work was conducted by Fisher, Ford, Sheppard, and others.

128. This work was conducted by A. J. Cain and P. M. Sheppard.

129. See, for example, H. B. D. Kettlewell, *The Evolution of Melanism* (Oxford: Clarendon Press, 1973).

130. For example, Ford, *Ecological Genetics* (note 126).

131. W. C. Kimler, "Mimicry: Views of Naturalists and Ecologists before the Modern Synthesis," in Grene, ed., *Dimensions* (note 6), Chapter 5.

132. C. A. Clarke and P. M. Sheppard, "Disruptive Selection and Its Effect on a Metrical Character in the Butterfly *Papilio dardanus*," *Evolution*, 16 (1962), 214–226.

133. H. J. Fleure, foreword to A. E. Mourant, *The Distribution of the Human Blood Groups* (Oxford: Blackwell, 1954).

134. L. L. Cavalli-Sforza and W. F. Bodmer, *The Genetics of Human Populations* (San Francisco: Freeman, 1971).

135. M. Lamotte, "Polymorphism of Natural Populations of *Cepaea nemoralis*," *Cold Spring Harbor Symposia on Quantitative Biology*, 24 (1959), 65–86.

136. C. B. Goodhart, "Variation in a Colony of the Snail *Cepaea nemoralis* (L.)," *Journal of Animal Ecology*, 31 (1962), 207–237.

137. Wright, "Random Drift" (note 47).

138. S. Wright, *Evolution and the Genetics of Populations*, Vol. IV (Chicago: University of Chicago Press, 1978), pp. 171ff.

139. R. H. Post, "Population Differences in Red and Green Color Vision Deficiency," *Eugenics Quarterly*, 9 (1962), 131–146.

140. Ford, *Ecological Genetics*, 1st ed. (note 119), pp. 124–125.

141. Gould, "Hardening" (note 71).

142. Ford, *Mendelism and Evolution*, 1st ed. (note 82), p. 78; see also Provine, "Development of Wright's Theory" (note 73).

143. Turner, "The Hypothesis That Explains Mimicry" (note 6).

144. For example, R. J. Berry, "Genetical Changes in Mice and Men," *Eugenics Review*, 59 (1967), 78–96.

145. Gould, "Hardening" (note 71).

146. Lewontin, *Genetic Basis* (note 104), pp. 232ff.; see also Lewontin's review of R. Creed, ed., *Ecological Genetics and Evolution: Essays in Honour of E. B. Ford*, in *Nature*, 236 (1972), 181–182.

147. Robson and Richards, *Variation* (note 72), pp. 310, 187.

148. E. B. Ford, "Ecological Genetics," in R. Harré, ed., *Scientific Thought 1900–1960* (Oxford: Clarendon Press, 1969), Chapter 8.

149. J. R. G. Turner, "Fisher's Evolutionary Faith and the Challenge of Mimicry," *Oxford Surveys in Evolutionary Biology*, 2 (1985), 159–196. In this paper, which forms one of a pair with the present chapter, I discuss Fisher's integration of science, religion, and eugenics, and his response to the other stochastic force of evolution, namely mutation.

150. Gould, "Hardening" (note 71); also Turner, "Changes in Mean Fitness" (note 65).

151. For complementary views of the subject of this chapter, see also W. B. Provine, "The R. A. Fisher–Sewall Wright Controversy and Its Influence on Evolutionary Biology," *Oxford Surveys of Evolutionary Biology*, 2 (1986), 197–219; also W. B. Provine, "Adaptation and Mechanisms of Evolution after Darwin: A Study of Persistent Controversies" in D. Kohn, ed., *The Darwinian Heritage* (Princeton: Princeton University Press, 1985), Chapter 28.

13 On the Prior Probability of the Existence of Life

Bernd-Olaf Küppers

The problem of the origin of life is analyzed from a probabilistic point of view. It can easily be shown that the prior probability of the spontaneous formation of a living system is near zero. The same is true even for the prior probability of the spontaneous formation of a single macromolecule carrying biological information or function.

In view of such low probabilities of nucleation, three explanations of the origin of life have been put forward, each laying a different emphasis upon the respective roles of chance and law in evolution: (1) the hypothesis of singular chance, (2) the hypothesis of vitalistic forces, and (3) the molecular-Darwinian approach. An analysis based on algorithmic information theory shows that for fundamental reasons the chance hypothesis cannot be proven and that for fundamental reasons the vitalistic hypothesis cannot be disproven. Only the Darwinian approach yields an explanation of the origin of life that is satisfactory within the framework of epistemological standards. The problem-solving capacity of the Darwinian approach is demonstrated with a game-theoretical model.

1 Introduction: The Molecular Roots of Living Systems

According to Darwin's theory of evolution, the degree to which an organism is adapted to its environment is a decisive criterion for its chance of survival. Consequently, the criterion of adaptation is reflected in the systematic and functional arrangement of structures on all levels of organization of living beings. Even on the level of biological macromolecules, highly organized and adaptive structures are encountered. This is seen with especial clarity in the example of hemoglobin.

Hemoglobin is a biological macromolecule built up from four subunits. Each subunit consists of a long protein chain made up of smaller units; these are the twenty naturally occurring amino acids. Each of the four protein chains is wrapped around a central, ring-shaped molecule, the heme group. When air is breathed into the lung, oxygen molecules fasten on to the heme groups and are transported with the hemoglobin from the lung to their destination in the tissue. Subsequently, the hemoglobin transports a waste product of metabolism, carbon dioxide, back to the lungs. The uptake and release of oxygen by hemoglobin are accompanied by a subtle (and reversible) change in the structure of the hemoglobin molecule; figuratively speaking, the hemoglobin works like a "molecular lung." Under given physiological conditions, the folding of the protein chains, and thus the biological function of the hemoglobin molecule, is determined completely by the order of the amino-acid residues in the chains. Replacement of one such unit may be enough to change the structure of the molecule and perhaps even to destroy its function.

The hemoglobin molecule is only one of countless examples of adaptation

I would like to thank John Beatty and Paul Woolley for their critical reading of the manuscript and for valuable discussions about the subject of this chapter.

among biological macromolecules. Even in a single bacterial cell, there are two to three thousand kinds of protein molecule, differing primarily in their amino-acid sequences, and consequently in their shapes and functions. Each of these is tailor-made to carry out a particular task, and most are indispensable for the organism.

The fact that a set of only twenty amino acids is sufficient to build up such a variety of structures and functions is made more plausible by a simple calculation. The smallest catalytically active protein molecules of the living cell, such as the electron-transporting enzyme cytrochrome c, consist of at least 100 amino-acid residues. A protein chain of this length has already

$$20^{100} \approx 10^{130}$$

alternative sequences. This makes it obvious that even the lowest stage of biological complexity opens up an almost unlimited multiplicity of possible structures. The macromolecular structures found in organisms are unique, in the sense that they represent a specific number of *optimized* sequences chosen out of an immense number of chemically equivalent alternatives.

It is plausible that the assembly and coordination of functional molecules within the cell should require a genetic program. Today we know that the information needed to construct a living organism is stored in a particular sort of molecule: the nucleic acids. Like the proteins, the nucleic acids are macromolecules—that is, they are built up out of smaller, repeating molecular units. These units, the nucleotides, function in the nucleic acid like the symbols of a written language.[1] The alphabet of the genetic language consists of four symbols, and these four symbols are generally denoted by the initials of their chemical names: for a desoxyribonucleic acid (DNA) these are A, T, G, and C. The linear sequence of the nucleotides in DNA encodes the entire genetic information of a living being, including the blueprints for the construction of all its proteins. For this purpose, three nucleotides are read together as a code word and represent a particular amino-acid residue. The following sequence of nucleotides is an excerpt from the genetic blueprint of the virus ϕX174:[2]

...CGTCCTTTACTTGTCATGCGCTCTAATCTCTGGGCA

TCTGGCTATGATGTTGATGGAACTGACCAAACGTCGTT

AGGCCAGTTTTCTGGTCGTGTTCAACAGACCTATAAAC

ATTCTGTGCCGCGTTTCTTTGTTCCTGAGCATGGCACT

ATGTTTACTCTTGCGCTTGTTCGTTTTCCGCCTACTGC

GACTAAAGAGATTCAGTACCTTAACGCTAAAGGTGCTT

TGACTTATACCGATATTGCTGGCGACCCTGTTTTGTAT

GGCAACTTGCCGCCGCGTGAAATTTCTATGAAGG...

In viruses and in bacteria the blueprints for all proteins are connected up in series in a single nucleic acid molecule. In higher organisms, the genetic information is divided among several DNA molecules. Nevertheless, for all organisms there is the same circular relationship between biological information and biological function; the genetic information encodes the blueprint of a complex molecular machinery whose principal task is the efficient reproduction of itself and the copying and transmission of its own blueprint from one generation to the next.[3]

The basis of this copying process is a physicochemical interaction that causes the basic units of a nucleic acid to join up in pairs according to the principle of *complementary base recognition*. This means that the nucleotides A and T form one kind of stable pair and the nucleotides G and C form another. Thus the first step in the duplication of DNA is the formation of a negative copy by the individual matching and subsequent joining up of complementary nucleotides along the whole length of the molecule to be copied. In a second copying step, the negative copy yields a positive one.[4] However, the Brownian motion of the molecules interferes with the copying process, so that there is always a certain probability that an error (a mutation) will appear. The genetic variability that results from such mutations is the source of changes on the phenotypic level and is thus the target of natural selection.

2 Life—an Irreducible Phenomenon?

The further that molecular biologists have penetrated into the secrets of the phenomenon of life, the more exactly they have been able to formulate the question

Life = Physics + Chemistry?

In other words: What properties are unique to living systems? Is there an irreducible life principle over and above mere matter?

This question is so difficult and so all-embracing that we shall restrict the following discussion to one aspect to it, and consider only the question of the *origin* of life. Even this we shall restrict to a general consideration of the transition from nonliving to living matter, and we shall not attempt to discuss higher-level issues, such as the origin of consciousness.

In the introduction, we discussed the fact that complex and perfectly adjusted interactions between biological macromolecules are the basis of all forms of life. With respect to the origin of life, the question formulated above can be narrowed down to the following: Do the laws of physics and chemistry cause biological macromolecules (1) to arise spontaneously from their basic units, and (2) to organize themselves into living systems?

The first part of this question belongs to the discipline known as prebiotic chemistry and can today be regarded as solved, at least in principle. The results of numerous experiments point to the conclusion that, under the conditions of the primeval earth, the two most important classes of biological macromolecules (proteins and nucleic acids) are able to form spontaneously in a noninstructed reaction from their basic units (amino acids and nucleotides, respectively), which in turn emerge as a result of normal chemical reactions.[5]

A "primordial soup," enriched in this way with complex macromolecules, certainly constitutes a *necessary* condition for the origin of living systems, but, as we shall now demonstrate, not a *sufficient* one. To do so we take up the second part of our question: Is it possible for biological macromolecules to organize themselves into living systems?

We start by adopting, for the time being, a naive scientific approach to the problem, and ask whether a living system could at some time have arisen by chance. This question we can answer on two levels, the phenotypic level of the molecular machinery of the living system, and the genotypic level of biological information. The molecular machinery consists basically of the cellular network of proteins, while the biological information is encoded in the detailed sequence of the nucleotides of the DNA of the organisms.

Let us first examine the phenotypic level. E. Wigner[6] has estimated the prior probability of the existence of a self-reproducing molecular machinery using an approach based on quantum theory. His calculation involves basically the determination of the structure of the quantum mechanical transformation matrix that describes the transition from a nonliving to a living state of matter. He reasons as follows. Assume that the state "life" is completely given in the quantum mechanical sense. Assume further that there is at least one state of the reservoir of nutrient that enables the organism to reproduce itself. The self-reproduction of a living system is then, formally, a specific interaction between the organism and the nutrient that leads to a material copy. If we represent the reproductive system of organism plus nutrient by a state vector in Hilbert space Ψ_A, then the process of reproduction is described by the transformation

$$\Psi_E = S\Psi_A$$

where S is an unitary transformation matrix and Ψ_E is the state vector of the system after the reproduction. We now replace Hilbert space by a finite-dimensional space. In particular, the state space of the organism shall be of dimension N and the state space of the metabolic products of dimension R, where N and R are of course very large numbers. On the assumption that S is a random matrix, Wigner was able to show that the number of equations required to describe the transformation is much greater than the number of components of the state vector that appear as unknowns in the equations. That is, there are N^2R transformation equations but only $(N + R + NR)$ unknowns. Since N^2R is very much greater than $(N + R + NR)$, it is arbitrarily improbable that the transformation equations can be satisfied by the unknowns.

Thus, as Wigner correctly argued, the laws of quantum mechanics lead to an arbitrarily low probability for the chance origin of a self-reproducing material system as a consequence of a gigantic fluctuation.[7]

One might object that a chance origin of a complex machinery was not necessary. Perhaps the appearance of a single information-carrying molecule was enough to program the assembly of a living system in the rich and complex primordial soup, just as in ontogenesis a complex organism develops from the fertilized egg cell.

If this is so, then the question of the prior probability of the origin of life must be formulated on the genotypic level: Is it possible for biological information to arise by pure chance, as a by-product of the spontaneous and random synthesis of a DNA molecule? To answer this question, we start by presupposing the most favourable case, in which the prebiotic scenario has already led to *potential* information carriers (DNA molecules) of sufficient length—for example, as long as the bacterial genome. Now we assume that each DNA molecule is the product of a purely random synthesis and that no chemical or other forces beyond pure chance determine its sequence. The probability of finding a nucleic acid molecule with a certain predefined sequence, such as that of a given gene, is then proportional to the number of combinatorially possible alternative sequences.

Even in the simple case of bacterial DNA (some 4 million nucleotides) the number of alternative sequences is almost inconceivable: $10^{2.4\ \text{million}}$! Thus the prior probability of the chance appearance of a bacterial genome is so low that neither the entire space nor the age of the universe could suffice to make the *random* synthesis of this information-carrying molecule even vaguely probable—to say nothing of the exact conditions on the primitive earth. Thus, to give a concrete example, the total mass of the universe is thought to be equal to that of some 10^{69} DNA molecules of the size of a bacterial genome. Therefore, even if the entire mass of the universe were made of DNA, as long as the sequences were random, we should not find a molecule among them with the sequence of a bacterial genome of today.

One might raise the objection that the above statistical calculation is based on the amount of information in the genome of a bacterium, while the historical origin of life may be assumed to have proceeded by way of simpler forms of life. An appropriate analysis of the corresponding expectation values shows, however, that not even an optimized enzyme molecule can reasonably be expected to arise in a random synthesis. Even the smallest catalytic protein molecules in the cell consist of at least 100 amino acids and thus possess more than 10^{130} possible sequences (see the introduction). Since there are no known physical or chemical principles that lead in a noninstructed synthesis to the preferred formation of any particular nucleotide sequence, the expected distribution of sequences in equilibrium will always be an arbitrary one, with the prior probability of occurrence for an information-carrying molecule around zero.[8] Thus traditional physics and chemistry leave the existence of living matter an unsolved riddle.

Even so, physicists and chemists have continued to produce hypotheses about the origin of life. J. Monod,[9] for instance, believed that the existence of life must be considered as the result of a *singular* random event, which certainly has never been and will never be repeated anywhere in the universe. N. Bohr,[10] on the other hand, had asserted many years earlier that life must be accepted as a fundamentally inexplicable fact of biology, just as classical physics views the quantum as a fundamentally inexplicable or irreducible element. W. Elsasser[11] and in moderate form also E. Wigner,[12] went one step further and postulated, in the sense of critical neovitalism, the existence of life-specific natural laws that direct, and thus sustain, the processes of living matter, but that are not reducible to physical regularity.

3   Limitations of Objective Knowledge in Biology

The epistemological question associated with the problem of the origin of life can be discussed with considerable precision if we formalize it somewhat. For this purpose let us take a closer look at the genetic blueprint of the primitive living being discussed in the introduction. Each of the letters A, T, G, and C represents a chemical building block of DNA and is thus a "letter" of the genetic language. The information contained in such a sequence of nucleotides can at once be translated into the language of information theory. Since the nucleic acids are built up from four classes of chemical building blocks, a binary representation (as used by computers) needs two digits to encode one nucleotide. We now define the code:

A = 00,

T = 11,

G = 01,

C = 10.

In this code, the excerpt from the hereditary information of the virus ϕX174 corresponds to the following sequence of binary digits (bits):

... 100111101011111100101111011110001101100110111011100001110111011010101011000

11101101011011001101001101111101001101010000101101001010000000100111100111

11000101101000011111111110110101111001110111111000001000010010101100110000

00100011101011011101010011001111111101111110111111010110100011000110010110

00101100110111111100101110111101100110111101111110011111111110100110101100

10110110010010110000000010001001111110000111001010111100001001101100000000101

11011011111101000101111001100101001001001111101101101011001001010101011011111

11110111001101011000001011110110100110100110011101000000111111110110011010100

000101....

Thus, in this formalism, the question of the origin of life is logically equivalent to the question of the origin of such a *defined* binary sequence.

    J. Monod[13] was convinced that from the macromolecular blueprints found in nature and their systematic comparison a general law could be deduced, which he called the "law of chance": "These structures are random in the sense that, even knowing the exact order of 199 residues in a protein containing 200 it would be impossible to formulate any rule, theoretical or empirical, enabling us to predict the nature of the one residue not yet identified by analysis."[14] Since the genetic code establishes an unambiguous correlation between the nucleotide sequence of the DNA and the sequence of amino-acid residues in the corresponding protein chain, Monod's statement about protein structure can be applied directly to the structure of DNA. Monod thus assumed that the obvious lack of regularity in the nucleotide pattern of biological information carriers reflects the random character of their origin.

Monod's hypothesis bears profound philosophical implications. Let us consider in more detail the concept of random sequences, in order better to analyze Monod's hypothesis. For this purpose, we can make use of some recent developments in algorithmic information theory, which allows the randomness of a series of symbols to be precisely defined and even to be measured. Since this new mathematical concept implies a fundamental limitation of objective knowledge in biology, we shall first give a brief outline of this concept, following closely a paper by G. J. Chaitin.[15]

Each of us has an intuitive idea of what we call a random sequence. Let us consider an example:

Sequence (a): 10101010101010101010101010101010101010101010101010.

Sequence (b): 10100011111001011001011100010110001010000001000111.

These two sequences of symbols have the same length (50 symbols) and both consist of the symbols "0" and "1" only. However, the sequence (a) follows a simple rule, according to which it can be extended indefinitely; the symbols "0" and "1" appear in strict alternation. The resulting pattern is clearly an *ordered* sequence. The sequence (b), on the other hand, does not seem to obey any rule (algorithm) that could be used to extend it. It is therefore reasonable to describe sequence (b) as a *random* sequence.

The classical method of generating a random sequence is to toss a coin repeatedly. It might be deduced from this that the manner in which a sequence of symbols originates determines whether or not it is random. This conclusion is, however, incorrect. If, for example, a coin is tossed 50 times and the result is noted, any one of the 2^{50} combinatorially possible sequences may emerge, since each sequence possesses the same prior probability. This is equally true of the two sequences of numbers of which we can see that one is ordered and the other is random.

We therefore clearly need a definition of the term "random sequence" that is independent of the origin of the sequence, but that accords with our intuitive ideas of ordered and disordered sequences. A new definition of randomness, developed independently by A. N. Kolmogorov and G. J. Chaitin,[16] does not take account of the origin of the sequence of symbols but considers only the characteristics of the sequence itself. To explain it one has to make use of some terms taken from information theory.

Consider again the two sequences of symbols above. The fixing of a single symbol demands one Yes/No decision, or one bit (binary digit) of information. If each sequence contains information amounting to N bits.

We now perform a thought experiment with the following information-theoretical problem. The inhabitants of a faraway planet, who use the same communication system as we do, are to receive a message containing the two sequences shown above and formulated in the most economical way possible. For sequence (a) we send out this instruction:

PRINT "10" ($N/2$) TIMES.

Clearly, the amount of information needed does not increase noticeably if N is increased even to very large values. The number of units of information stays roughly the same, however many units are contained in sequence (a). Thus sequence (a) is *compressible*.

Things are quite different with sequence (b). Here there is no redundancy in the sequence pattern, so we must transmit the entire sequence:

PRINT "101000111110010110010111000101100010100000001000111".

Therefore, in the second case, the quantity of information needed for the transmission instruction is proportional to the quantity of information in the sequence itself. The difference between sequences (a) and (b) is that the latter is not compressible, i.e., it cannot be expressed in a shorter form. In other words, sequence (a) can be expressed by an algorithm much shorter than the sequence itself, while sequence (b) cannot.

The degree of incompressibility of a sequence of symbols is indeed an adequate measure of the degree of its randomness. This fact is exploited in the following definition: *A sequence is defined as "random" when the shortest algorithm needed to generate it contains about as many information units as the sequence itself.*

The smallest number of bits sufficient to specify a given sequence is called the *complexity* of the sequence. If a given sequence is not random, its compressibility can in principle be demonstrated. For this purpose, it suffices to show that there is an algorithm whose coding is much shorter than the coding of the sequence itself. If, conversely, it has to be shown that a sequence is random, then it must be proved that there is *no* algorithm by which the sequence can be compressed. One of the most important discoveries within the realm of algorithmic information theory is the theorem that the randomness of a particular sequence can in partice never be proved. More precisely: In a formal system with n bits it is impossible to prove that a particular binary string is of complexity greater than $n + c$, where c is a constant that is independent of the particular system employed.[17]

The theorem of randomness is proved by formulating it as a halting problem for a universal Turing machine. A detailed treatment of the theorem and its philosophical implications for biology has been published elsewhere.[18] Here, only two of the main conclusions concerning the epistemological status of the chance hypothesis and the vitalistic hypothesis are summarized:

1. The hypothesis of Monod that the blueprint of a living organism is the product of random synthesis, because there exists no detectable sequence pattern, is not provable. It should be emphasized that this is an epistemological assessment; the theorem of randomness excludes not the existence of random sequences in general

but simply their demonstrability. One might object that the nonprovable character of the chance hypothesis is a feature that the chance hypothesis shares with all physical theories. However, Monod's statement does not only refer to an empirical fact. A deeper analysis of his arguments shows that the chance hypothesis effectively negates the causality principle, in that the role of chance in the process of Darwinian evolution is raised to the rank of an antiteleological law, for which Monod tried to give a positive proof along the lines indicated above.

2. All vitalistic hypotheses of the origin of biological information are irrefutable. This is easily understood within the realm of algorithmic information theory: Vitalistic hypotheses postulate the existence of an algorithm that embodies a specific biological law, according to which the blueprints are made up. Such an algorithm can only have the character of a law when it leads to sequences that are not random, that is, that are larger than the algorithm itself.[19] However, the existence of such compact algorithms cannot be disproved, since their nonexistence cannot be proved. On the other hand, no vitalistic hypothesis has yet led to the concrete formulation of such an algorithm. Vitalistic hypotheses are therefore only apparent solutions based on gaps in our actual knowledge of physics and chemistry.

Let us recapitulate our conclusions so far: *The chance hypothesis is fundamentally unprovable, while all vitalistic hypotheses are fundamentally irrefutable.* Thus the theorem of Chaitin, which is essentially an information-theoretical version of Gödel's theorem of the limitations of the axiomatic method, leads immediately to some important limitations of objective knowledge concerning the origin of life.

4   The Semantic Aspect of Biological Information

The statistical problems that arise in connection with the phenomenon of the origin of life appear to indicate that organisms are irreducible structures, whose scientific explanation transcends present-day physics and chemistry. This is the general tenor of two papers by M. Polanyi on the problem of reductionism in biology.[20] Polanyi's approach to this problem is of particular interest because it provides a paradigm of the epistemological difficulties that accompany the physicochemical interpretation of the origin of life.

In Polanyi's treatment, the so-called "control" of natural laws by system-specific boundary conditions has a particular importance. The meaning of this principle of control is best illustrated by reference to an analogy between machines and organisms. According to Polanyi, a machine is characterized by two levels: the level of its components, which are completely explicable in terms of the laws of physics and chemistry, and the higher level of the relations between the components, as those relations are dictated by the construction of the machine. The principle of construction, and thus the functions of the machine and its parts (e.g., the fact that the machine should serve some particular human need), follow certain technological criteria that are irreducible, that is, are not derivable from the laws of physics and chemistry. Equipped solely with the methods and concepts of physics and chemistry—Polanyi argues—one is not in a position either to explain or to describe a machine in its completeness, or even to identify it as a machine.

The conclusions reached for the machine are applied by Polanyi to living structures, insofar as these can be considered as complex biochemical machines. According to this view, living systems are also subservient to certain irreducible principles of construction that exist independently of the physicochemical laws determining the detailed behavior of the material component parts.

Polanyi has explained these ideas in detail by considering the role that DNA plays in the living organism. On the one hand, the particular nucleotide sequence in the genome of an organism encodes the information for the construction of the organism, as postulated by molecular-genetic determinism. On the other hand, however, the exact sequence of the DNA is not derivable from physicochemical laws; present-day understanding of the properties of nucleic acids indicates that all the combinatorially possible nucleotide patterns of a DNA are, from a chemical point of view, equivalent. Thus the laws of physics and chemistry cannot explain the origin of those particular patterns that carry biological information. Polanyi argues that the nucleotide sequence of a DNA molecule must therefore be considered as an *irreducible* boundary condition under which the laws of nonliving nature are constrained to operate in the living system.

Here we come up against the central point of the problem of the origin of biological information. Everything in biological structures that is governed by information is, according to the neo-Darwinian view of evolution, the result of a "material learning process," and has a certain *function* with regard to the preservation and development of the specific order of the living system. It is in this way that semantics become associated with genetic information. A theory of the origin of life must therefore necessarily include a theory of the origin of semantic (i.e., functional and not only structural) information. And precisely here lies the fundamental difficulty confronting every scientific theory of the origin of life. Physics and chemistry, at least, have traditionally ignored semantic phenomena. Again, Polanyi has expressed this aspect in the strongest way: "All objects conveying information are irreducible to the terms of physics and chemistry." [21]

This assertion provided the basic statement upon which Polanyi attempted to build up his thesis of the irreducibility of biological systems. The central question of the origin of life therefore emerges as the question of how far the semantic aspect of information can be formulated objectively and thus become an object of study in a mechanistically oriented science such as molecular biology. The central statement of Darwinian theory is that the generation of information follows the mechanism of evolution by natural selection.

Thus the next question is virtually forced upon us: Is the principle of natural selection also valid at the molecular, a priori nonliving level, and, if so, can it explain the origin of macromolecules carrying biological information?

Organismic biology has invariably denied this. [22] However, it has become clear, mainly through the work of S. Spiegelman and his coworkers, that natural selection is not a phenomenon restricted to autonomous living systems, but that it can also occur, under certain conditions, among biological macromolecules outside the living cell.[23] Furthermore, proceeding from the prerequisites of Spiegelman's experiment (the system must be open, self-reproductive, and prone to mutation), M. Eigen[24] was able to demonstrate that natural selection follows inevitably as a physically derivable extremum principle and that natural selection is just as inevitable a property of nonliving matter as is the acceleration of a body acted on by a force. We know that the nucleic acids fulfil all the material preconditions for *selective* self-organization. And indeed, since the pioneering work of Spiegelman, it has been shown in a number of experiments that selection in the Darwinian sense takes place among nucleic acid molecules in the test tube, if they are put under selection pressure, and that this leads to environmentally adapted RNA structures.[25]

5   The Origin of Biological Information—a Game-Theoretical Approach

It remains for us the examine the question of whether selection and evolution at the molecular level can solve the *statistical* problem described in section 2. A detailed formulation of the molecular-Darwinian theory can be found elsewhere.[26] I should like here to demonstrate the problem-solving capacity of the theory by appeal to a simple game-theoretical model.

In the following example, prompted by an idea of M. Eigen, the generation of biological information is to be simulated on a computer.[27] For reasons of perspicuity we make use of the similarity between the human and the genetic languages; a unit of genetic information will now be represented by a word in human language. This is not a serious restriction, since the information is in any case given to the computer in the form of a string of binary digits, and this can just as well represent some biological information (cf. section 3). We take a sequence of letters that represents a particular information content

NATURAL.SELECTION

and ask whether this sequence can be produced out of an unrelated initial sequence by an evolution mechanism in the Darwinian sense. We shall start with a random initial sequence, such as

ACVYIDPLOWRTSIOQA

We now define a semantic level on which the "evolving" sequence will be tested. This level is defined by the meaningful sequence

NATURAL.SELECTION

We need at least 5 symbols to represent a letter of the alphabet in a binary code. (With 5 binary digits, 2^5 "code words" can be produced, enough to encode 26 letters of the alphabet and some punctuation marks.) The 17 "letters" of the sequence NATURAL.SELECTION then represent a quantity of information amounding to 85 bits. A sequence of 85 bits has

$2^{85} \approx 10^{26}$

alternatives. The probability that the computer will produce a specific 85-bit sequence by generating zeros and ones at random is thus close to zero.

Instead of this we shall apply a Darwinian optimization principle. For this purpose we put the random, initial sequence into the computer, with the command that it reproduce this, copy the products of the reproduction, and so on. Thus we have introduced into the program the biological phenomenon of *self-reproduction*. We further program the computer so that the copying of single binary digits is not always exact, that is, sometimes a zero (or one) is replaced by a one (or zero), so that the progeny sequences occasionally contain wrong letters. The appearance of wrong letters is a purely random process, and thus it simulates the biological phenomenon of *mutation*. Since all sequences are copied, the mutants can also accumulate. However, for the amplification process we introduce a selective assessment: Every mutant sequence that agrees one bit better with the "meaningful" or reference sequence than its master copy will be allowed to reproduce more rapidly by a certain factor (the *differential advantage*).[28]

Thus at first the sequences are reproduced, with occasional errors, at a characteristic rate. We now allow the size of the overall population to grow up to 100 and then reduce it—by random choice—to 10 copies. The total population is thus on average constant. By the restriction of growth we apply a continual *selection pressure* to the system.[29] It can be shown that all sequences whose reproduction rates lie below the average reproduction rate of the distribution are squeezed out of the population and are thus excluded from further optimization. This raises the average reproduction rate, which rises as the system evolves and which finally reaches an asymptotic maximum value. At this point the distribution consists of the reference sequence and its stationary mutant distribution. In a similar manner, the "protosemantics" of biological information are laid down by a criterion of assessment, which is reflected by a sliding and self-adjusting threshold value.[30]

The computer output below shows three separate phases of evolution, namely, the compositions of the population after the 2nd, the 15th, and the 35th generations of reproduction (mutation rate, 1%; differential advantage, 2.7):

2nd generation	15th generation	35th generation
ACVYUFLLSSNSCEPOA	NITURAL.SULECSYO!	NATURAL.SELECTION
ACVYYFLLSSNTCIPOA	NITURAL.SULESTYO!	NATURAL.SELECTION
ACVYQFLLSSNTSIOOA	NITURAL.SULECTYO!	NATURAL.SELECTION
ACVYYFLLSSNTCGOOA	NITURAL.SYLECTYO!	NATURAL.SELFCTGON
ACTYUFLLSSNSCIOOA	!IVUZAL SULEETYO!	NATURAL.SELECTION
ACTYUFLLSSNSCIWOA	NITURAL.SULECTYO!	RATURAL.SELECTION
,CVYIELLSUNTSIKOA	NITURAL.SULECTYO!	NATURAL.SELECTION
ACVYQFLLSSNTCIOOQ,	NITURAL.SULCCSYO!	NATMRAL.SELEETIQN
ACVYIFLLWSNDCIKOA	!ITURAL.SULECDYC!	NATURAL.SCLUCTION
ACVYYFLLSSNTCGOOA	NGVMRAL.SULECTYO!	NATURAL.SELECSI.N

In the 35th generation, selection equilibrium has already been reached. It consists of (on average) 5 correct copies of the reference sequence and a stationary distribution of mutants arising from it.

We have thus, for this example, solved the statistical problem raised in section 2. Starting from a random sequence of symbols, we have arrived at a sequence that is only *one* of 10^{26} combinatorially possible alternatives, and whose prior probability was close to zero. To generate this sequence we needed neither a teleological principle nor a singular, random event. All that was necessary was a selection mechanism in the Darwinian sense. The effect of randomly caused hereditary changes was simulated in the computer by statistical variation of a given sequence of symbols. The selective assessment of mutants was based, as it is in biology, upon an advantage in rate of reproduction over against the other mutants present.

The simulation experiment shows that meaningful information can indeed arise from a meaningless initial sequence by way of random variation and selection. Since the appearance of mutants is, on the genetic level, completely indeterminate, the process of natural selection lays down a general gradient of evolution, but not the detailed path by which the local maximum will be reached.

If all combinatorially possible sequences of a molecular information carrier are denoted by coordinates in "information space," then the origin of biological information can be compared to a random walk in a many-dimensional mountain range, the profile of which is determined by the various reproduction rates. The only element of orientation is the requirement that the walk must lead uphill (neglecting fluctuations). This follows from the fact that the criterion of survival can be expressed by a sliding and self-adjusting threshold value, which excludes all paths leading downhill. But this topological analogy also confronts us with a weakness of our simulation experiment. In contrast to biological information, human language does not possess a semantically hierarchical structure. There are no independent "semi-meaningful" words. To this extent our computer experiment represents only a construction a posteriori of the generation of genetic information; that is, we started with a meaningful result (in our case the reference sequence) and showed, looking backward, that the statistical problem can in principle be solved. A construction a priori, however, does not seem possible. For if we could simulate the generation of genetic information under the genuine conditions of biological evolution—for example, without prior knowledge of the reference sequence or of the criteria for assessment of the mutants—then we would have solved a central problem of artificial intelligence. The computer would be able to generate information *de novo* in a process of self-organization, requiring nothing but an energy supply. Nevertheless, there have been successful attempts to improve this game-theoretical model by applying more realistic constraints.[31]

There is a further aspect to consider. Organisms are themselves a part of their environment, and the environment changes in response to the evolution of the organisms. Even a single organism provides the environment for its own genes, and it exerts an additional "internal" selection pressure on them. In contrast to our computer experiment, the final structures called for by the environment are not defined until evolution is well under way.

This implies a mutual dependence of "ultimate goal" and "present aim," which in turn makes it clear that Darwinian theory predicts a priori only the emergence of information in general, but not the detailed structure of this information. In consequence, the neo-Darwinistic view of the origin of life attempts not to reconstruct the historical course of this process, but simply to uncover its fundamental laws and principles that can be expressed in the language of physics.

Notes

1. The analogy between human and genetic language is extensive. It has been emphasized and analyzed in particular by V. A. Ratner, *Molekulargenetische Steuerungssysteme* (Stuttgart: 1977). Thus fundamental problems of the origin of biological information may be illustrated by examples from human language without loss of perspective.

A characteristic of every language is its syntax, that is, the hierarchically organized relationship of its symbols. Likewise, the molecular-genetic language possesses a syntactic dimension, since the symbols (nucleotides) constituting the genetic information do not function unrelatedly, but instead are arranged hierarchically into operational units. The nucleotides are the primary coding symbols of the molecular-genetic language. Three nucleotides represent a translation unit for one amino acid. About 100–1000 translational units (codons) make up a gene (or cistron) and encode the information for a single protein. Up to 15 cistrons make a transcriptional unit (scripton). Many hundred scriptons make up a reproducing unit (replicon). These are in turns subsumed into the meiotic unit (segregon) and the mitotic unit (genome). The hierarchically organized structure of genetic language reflects directly the hierarchical assembly of a living system. There are, however, limits to the analogy between human and genetic language. Thus, apart from certain regulatory phenomena, genetic language possesses no interrogatory characteristics; furthermore, the generative character of human language finds its genetic counterpart only within the context of the evolutionary codevelopment of all organisms.

2. F. Sanger et al., "The Nucleotide Sequence of Bacteriophage ϕX174," *J. Mol. Biol.*, 125 (1978), 225.

3. J. D. Watson, *Molecular Biology of the Gene* (New York: W. A. Benjamin, 1976).

4. RNA replication can also take place outside the cell and without the help of enzymic catalysis—(cf. R. Lohrmann et al.,: "Efficient Metal-Ion Catalyzed Template-Directed Oligonucleotide Synthesis," *Science*, 208 (1980), 1964.

5. S. Miller and L. E. Orgel, *The Origins of Life* (Englewood Cliffs, NJ: Prentice-Hall, 1973).

6. E. Wigner, "The Probability of the Existence of a Self-Reproducing Unit," in *The Logic of Personal Knowledge: Essays in Honor of Michael Polanyi* (London: Routledge and Kegan Paul, 1961).

7. Quite apart from the fact that the application of quantum mechanics to macroscopic processes is somewhat problematic, the premises of Wigner's conclusions must be examined more closely. Wigner's calculations are only valid if the transformation matrix **S** is a *random matrix*. In other words, his model presupposes that the reproduction process takes place by way of a noninstructed, that is, an "informationless," mechanism of association of appropriate material subunits. These premises do not apply to living systems. The self-reproduction of organisms is rather to be seen as a direct consequence of the physicochemical properties of the molecules of heredity, and is fully instructed. However, the "informationless" initial state assumed by Wigner most probably existed in the prebiotic phase of evolution, which molecular chaos reigned in a primordial "soup" fully devoid of functional order.

8. In the spontaneous synthesis of polypeptides under prebiotic reaction conditions, a certain selectivity among the amino acids incorporated has been observed. However, the effect is so weak that the central statements of this article remain true.

9. J. Monod, *Chance and Necessity* (London: Collins, 1972).

10. N. Bohr, "Licht und Leben," *Die Naturwissenschaften*, 21 (1933), 245.

11. W. Elsasser, *The Physical Foundation of Biology* (Oxford: Pergamon, 1958). W. Elsasser, *The Chief Abstractions of Biology* (Amsterdam: North-Holland, 1975).

12. E. Wigner, "The Probability" (note 6).

13. J. Monod, *Change and Necessity* (note 9).

14. J. Monod, *Chance and Necessity*, p. 95 (note 9).

15. G. J. Chaitin, "Randomness and Mathematical Proof," *Scientific American*, 232 (1975), 47.

16. G. Chaitin, "On the Length of Programs for Computing Finite Binary Sequences," *J. ACM* 13 (1966), 547; A. N. Kolmogorov, "Three Approaches to the Quantitative Definition of Information," *Problemi Peredachi Informatisii* 1 (1965), 3.

17. G. J. Chaitin, "Information-Theoretic Computational Complexity," *IEEE Trans. Info. Theor.*, IT-20 (1974), 10.

18. B.-O. Küppers, *Der Ursprung biologischer Information. Zur Naturphilosophie der Lebensentstehung* (Munich: Piper, 1986).

19. It is obvious that a set of empirical data only embodies a natural law when the (binary) symbol sequences that encode the empirical data are nonrandom, i.e., when there is an algorithm whose coding is significantly shorter than the coding of the empirical data and from which these data can be "reconstructed." In other words, the empirical data reveal the existence of a natural law when they display a regularity that makes them compressible. The algorithmic interpretation of that which is called a "natural law" was first introduced by R. J. Solomonov in "A Formal Theory of Inductive Inference," *Inform. Contr.*, 7 (1964), 1 (part I), 224 (part II).

20. M. Polanyi, "Life Transcending Physics and Chemistry," *Chemical and Engeneering News*, 45 (1967), 45; M. Polanyi, "Life's Irreducible Structure," *Science*, 160 (1968), 1308.

21. M. Polanyi, "Life Transcending Physics and Chemistry," p. 59 (note 19).

22. L. von Bertalanffy, "Biologie und Weltbild," in *Wohin führt die Biologie?*, ed. M. Lohmann (Munich: Deutscher Taschenbuch Verlag, 1977).

23. D. R. Mills, R. L. Peterson, and S. Spiegelman, "An Extracellular Darwinian Experiment with a Self-Duplicating Nucleic Acid Molecule," *Proc. Nat. Acad. Sci. USA*, 58 (1967), 217.

24. M. Eigen, "Self-Organization of Matter and the Evolution of Biological Macromolecules," *Die Naturwissenschaften*, 58 (1971), 465.

25. B.-O. Küppers, "Towards an Experimental Analysis of Molecular Self-Organization and Precellular Darwinian Evolution," *Die Naturwissenschaften*, 66 (1979), 228.

26. B.-O. Küppers, *Molecular Theory of Evolution* (Heidelberg: Springer-Verlag, 1983).

27. M. Eigen, "How Does Information Originate? Principles of Biological Self-Organization," in *Advances in Chemical Physics*, Vol. 38, ed. S. A. Rice, (New York: Wiley, 1980).

28. When we speak of a "reference sequence," this is not a retreat into a teleological form of argument but rather the price we must pay for the clarity of our example. On the genetic level there is of course *no* such reference sequence. However, the difference is irrelevant for the evolutionary origin of information, for which only the *differential* reproductive behavior within each distribution of biological information carriers is important.

29. The restriction of growth is not a necessary condition for selection in the Darwinian sense. On the other hand, if all species have identical reproduction rates, the restriction of growth is necessary to induce some kind of neutral selection, as has been described by M. Kimura, "The Neutral Theory of Molecular Evolution," *Scientific American*, 241 (1979), 93.

30. M. Eigen, "Self-Organization" (note 24).

31. W. Ebeling and R. Feistl, *Physik der Selbstorganisation und Evolution* (Berlin: Akademie-Verlag, 1982).

VI PHYSICS

The Probabilistic Revolution in Physics—an Overview
Lorenz Krüger

Probability entered physics at two levels: that of measurement and that of theory construction. In contrast to the social sciences, probability never entered physics at the levels of data description and inference in any significant way. At the measurement level it functions as the central conceptual element of the theory of observational errors, which was developed between about 1795 and 1840. At the theoretical level it became important in two essentially different ways and at two stages in the history of physics, seperated by about 75 years. The first stage, around 1870, is marked by the emergence of statistical mechanics as a new type of theory—indeed, as James Clerk Maxwell put it, as a "new kind of knowledge." The second stage is marked by Max Born's probabilistic interpretation of the quantum mechanical psi function in 1926.

These two stages are historically and conceptually related in complex ways that, in spite of many illuminating studies, are not yet fully understood and certainly cannot adequately be treated in this short overview. Nevertheless, they are almost universally considered to be distinct, the main point of the distinction being that, in statistical mechanics, probabilistic statements can meaningfully be applied only to sufficiently large aggregates of elementary systems, e.g., molecules or atoms, whereas in quantum mechanics such statements are typically applied to a single elementary system as well, though this application is not without problems. In the first case a probability statement is to be interpreted in terms of relative frequencies; these may be the relative numbers of elementary systems in a particular state picked at random from the entire aggregate or the relative time the whole aggregate spends in a given state within a sufficiently long reference time. In the second case a probability statement may also be understood to indicate the tendency or the propensity of a single elementary system to develop in a certain manner or to be found in a certain state if examined. Doubts concerning this interpretation of probability arise immediately, however, from the fact that such propensities can only quantitatively be assessed as relative frequencies in a large collective of similar systems. Nevertheless, they cannot, in typical cases, be interpreted as the outcome of the internal interaction between the parts of the collective or as the expression of its overall structure, whereas the latter interpretation is possible, indeed inevitable, in classical statistical mechanics. In a macroscopic sample of a radioactive substance, the decay of each single atom is independent of interatomic forces, and yet the entire sample obeys a common decay law. Since the existence of the law, according to quantum theory, is not due to the mutual correlation of the atoms, the application of probabilistic statements to single elementary systems cannot be avoided. On the other hand, if in classical statistical mechanics the state of thermal equilibrium is identified with the most probable state of a system consisting of many parts, it does not make sense to apply the probabilistic characterization to any one of the parts seperately.

Hence, under the assumption of two essentially different uses of probability in the construction of physical theories, we may distinguish three thresholds in the development of probabilistic ideas in physics: (1) the discovery of the law of errors at the end of the eighteenth century, (2) the emergence of classical statistical mechanics around 1870, and (3) the establishment of quantum mechanics around 1926.

The first threshold marks the beginning of the systematic evaluation of data in physics. The relevant phase of this development was short, since not much more than the Gaussian distribution of errors is more than occasionally used in physics. Further sophistication in this field, e.g., randomization of experimental data or refined description of data collectives as developed in mathematical statistics, was not needed, because the attempts of the physicists were always concentrated on the experimental isolation and control of a very few basic variables. These were, moreover, assumed to be related in ways that could be derived from theoretical assumptions rather than inductively by scrutinizing the data with the help of ever more sophisticated techniques of statistical inference (in contrast to the social sciences). Even the contemporary increase in data processing in big science and computerized experiments is generally managed by means of theoretical assumptions about potential disturbing factors rather than data analysis as such.

The second threshold marks the beginning of a type of theorizing that has become a fundamental component of physics ever since. This development has witnessed several transformations, extensions and improvements, e.g., in non-equilibrium thermodynamics and synergetics, but also in classical areas like ergodic theory. It is, moreover, relevant to the further development of quantum theory, as is borne out by the existence of the thermodynamics of quantum systems and the macroscopic and irreversible character of all measurement processes, those of quantum phenomena included.

The third threshold marks not only a beginning, namely, that of "normal research" (in the sense of Thomas Kuhn) in quantum physics, but also the end of an explorative phase of a novel use of probability in physics, for which the ground had been prepared since Max Planck began publishing on black-body radiation in 1900. The extended explorative phase of quantum theory was inevitably impregnated by the complex and conceptually difficult overlap of the two theoretical uses of probability distinguished above. The phases between the thresholds may be briefly and crudely characterized as follows.

The development of error theory in the physical sciences, especially stimulated by the needs of the astronomer and the land surveyor, but also important to the experimenter in general, began around the middle of the eighteenth century with attempts to justify the widely practiced maxim of using arithmetical averages of observed values as approximations of the true value of the magnitude in question. These justifications were based on the assumption of certain probability distributions of the observational errors. Since these assumptions were all arbitrary, Karl Friedrich Gauss, in 1797, reversed the direction of the argument and derived the normal (or "Gaussian") distribution of errors from a few plausible and merely qualitative principles. The method of least squares, which had been introduced independently in 1805 by Legendre and was first published in 1808, followed from Gauss's argument. By then the new method had already celebrated its first triumph: the prediction of the position of the small planet Ceres, which was discovered on January 1, 1801, by Piazzi, and was subsequently lost and finally rediscovered late in 1801 by the help of Gauss's calculations. The probabilistic aspect of error theory became much clearer through theorems proved by P. S. Laplace in the years 1809 to 1811 according to which a sum of many independent errors is always normally distributed. This result was applied by G. H. L. Hagen in 1837 and F. W. Bessel in 1838; they explained the normal distribution of errors by assuming that each single observable error is composed of many independent elementary errors, an idea that had been first suggested by Thomas Young in 1819. Later developments of error theory concentrated on refinements and tried to weaken the assumptions necessary to obtain normal distributions. One may say that already in the 1830s Hagen and Bessel, in viewing observable errors as superpositions of mutually independent elementary errors, used probability not just as a tool in the evaluation of measurements but also as an explanatory concept. This explanatory use, however, was still restricted to the domain of experimental methodology.

It was James Clerk Maxwell who, in 1860, showed that the velocities of the molecules in a gas are also distributed about a mean value, determined by the temperature of the gas, according to the law of errors. He based his derivation on the statistical independence of the changes in the three velocity components under collisions. Though his arguments started from abstract probabilistic assumptions rather than the dynamical analysis of motion, they led to new experimental phenomena. Among other things, Maxwell inferred that the viscosity of gases should be independent of the density, a prediction he found "very startling" but was able to confirm by experiments of his own a few years later. This was the first probabilistic explanation of basic physical states or processes; it thereby announced our second threshold, the beginning of a new type of physical theorizing.

Maxwell later obtained the velocity distribution in a more mechanical but essentially statistical treatment of the collisions of the molecules. Ludwig Boltzmann, in 1872, used the collision method in an ingenious way to derive a mechanical equivalent of the law of entropy. But since he was unable to explain how purely mechanical laws, according to which all motions can be reversed in time, could possibly produce an irreversible change, he was led, in 1877, to disregard the actual collisions once more and to determine the intrinsic probability of a given observable state of a system as proportional to the number of theoretically distinguishable microscopic realizations of this state. This method was adopted by Max Planck in radiation theory about two decades later; it lent itself to quantum theoretical use, because it disregards completely the actual path of the system through space.

In order to justify within the framework of classical mechanics the transition from the actual motion of the system to distributions of properties of its constituents at a given time, Maxwell, following a lead from Boltzmann, in 1878 introduced a peculiar assumption, later called "the ergodic hypothesis." It states that a physical system in equilibrium will assume all conceivable states that are compatible with the conservation of energy. Maxwell made clear, however, that the hypothesis could only be true by virtue of suitable random disturbances of the system from outside. Since about 1930 the probabilistic character of ergodic properties has become quite clear in what virtually has grown into a seperate branch of mathematics: ergodic theory.

Maxwell introduced still another novelty, the ensemble approach. Instead of studying distributions within the actual physical system at hand, one can equivalently consider an assembly of possible systems distributed over all states compatible with the observed magnitudes. In this case the probability may be interpreted purely epistemically as giving the chance of finding one of these possible systems to be the actual one. Willard Gibbs worked out this approach in 1902, and it has rivaled Boltzmann's collision approach ever since. The ergodic hypothesis can be conceived as a justification of the epistemic assumptions in terms of actual physical processes. As is described by von Plato in this part, this point of view attracted many scientists, among them Einstein. Its major merit is to give probability a *fundamentum in re* without admitting the reality of the merely possible or the merely potential.

The first distinct step in the prehistory of our third threshold, the discovery of quantum mechanical probability, may be seen in the eventual use, if not the justification, of inherent and irreducible probabilities characterizing single physical systems. Before it could be taken, some further discoveries had to be made. In order to explain certain phenomena of the interaction of light and matter, especially the photoelectric effect, Einstein, in 1905, suggested that light exists in discrete quanta, an idea that had been implicit in Planck's radiation law of 1900. In his atomic model of 1913 Niels Bohr connected this idea with the discrete line structure of atomic spectra; he explained the emission and absorption of light by atoms in terms of the transitions of single atoms from one discrete energy level to another. Einstein, in 1916, took the decisive step. He considered an aggregate of molecules or atoms that can only assume discrete energy states and react with radiation. He assumed the entire system of particles and light to be in a state of equilibrium as defined by classical statistical mechanics. In addition, however, he assigned inherent probabilities to a given state of a particle characterizing its tendency to move to another such state, either spontaneously or when stimulated by the surrounding radiation. From these assumptions he was able to derive Planck's law and Bohr's frequency condition for atomic spectra in a very simple way. Moreover, he could show that a continuous distribution of radiation, as opposed to particlelike light quanta, would give results incompatible with Planck's law and with experiment. He explicitly recognized that, in his new approach, the single elementary process was left to chance; but he enclosed this key term of the new quantum era in quotation marks ("Zufall"), and expressed his confidence in being on the right track nevertheless ("trotzdem").

During the first decades of the twentieth century physicists also came to regard other phenomena in probabilistic terms, e.g., Brownian motion and radioactive decay. Though it was by no means clear which, if any, of these phenomena would in the end have to be covered by irreducibly probabilistic theories, their existence added to the general tendency to break away from the deterministic world view that had widely prevailed during the nineteenth century. Some scientists and philosophers converted to a fundamentally indeterministic or probabilistic world view long before, at any rate independently of, quantum theory (e.g., the mathematicians Emile Borel and Richard von Mises, the physicists Franz Exner and Marian von Smoluchowski, and the philosophers Charles S. Peirce and Hans Reichenbach, to mention only some of the more prominent figures). This development, however, had many sources also outside physics, which therefore lie outside the scope of this overview. The reader may be referred to volume 1 of this work for further study.

   In a series of papers published from 1918 through 1922, Bohr showed that, given the probabilistic basis laid by Einstein, major features of atomic and molecular line spectra could be understood, e.g., intensities of lines and the polarization of radiation. Apart from basic assumptions concerning the discrete nature of matter and energy and the existence of inherent transition probabilities, Bohr used a number of principles familiar from nonquantum mechanics and thermodynamics, among them hypotheses of an ergodic type and Boltzmann's relationship between the microscopic probability of a physical state and its macroscopic entropy. Moreover, his work was based on the principle of correspondence, which says that, in the limit of small energy differences and radiation of long wavelengths, classical and quantum theories should give identical results. This stage of the theoretical development, therefore, did not allow for a clear separation between a quantum mechanical interpretation of probability and a classical one. The subtle ways in which Bohr distinguished but also united a statistical and an "individualistic" view of probability are analyzed in detail by Wise in volume 1 of this work. In general, physics around 1920 shows clearly the conceptual overlap of the two theoretical uses of probability distinguished above.

   A most remarkable effort of reconciling explicitly the inherent probabilities of quantum transitions of single systems on the one hand and the statistical idea of probability as referring to a system of many interacting particles considered as a whole on the other was made in 1924 by Bohr, Kramers, and Slater (by the last, however, not without misgivings). These authors assumed that, while all atoms were permanently coupled with each other via a virtual field of radiation, any single exchange of energy and momentum between the field and an atom occurred independently of any other but according to probabilities determined by the field. Only in this way, the authors thought, could two accepted but apparently incompatible features be reconciled: the continuous nature of light waves and the discontinuous character of their interaction with matter. Yet this solution of the problem involved the claim that a fundamental principle of current physics, the conservation of energy and momentum, would be valid only on the macroscopic average and must be violated in single elementary processes of emission and absorption of light. This last implication of the theory, however, was immediately disproved in an experiment by Bothe and Geiger, who were able to correlate quanta of x rays and electrons emerging from the same interaction.

   It followed that if there was a mechanism correlating locally separate elementary processes, then it would be of a completely different nature—or else one would have to put up with probabilities that entirely depend on the internal structure of single elementary systems. Both of these ideas appear to contain some part of the truth as well as a host of physical and philosophical problems, most of which are still a matter of research and debate today. In 1926 Max Born proposed to interpret Schrödinger's wave function, a basic mathematical device of the final quantum theory, as giving the probability for a physical system of showing certain properties when examined. He thereby opened up our current problem situation concerning probability in physics. Less than a year later Werner Heisenberg shed new light on what was at stake when he discovered the uncertainty relations, according to which, say, the position and the velocity of a particle cannot simultaneously be determined with arbitrary precision. At least one of the two values can only be characterized by a probability distribution. He thereby showed not only that all predictions of future states of a quantum mechanical system from given initial conditions will be merely probabilistic, but also that every description of such a state at any given time will necessarily involve probabilities, hence indeterminacy.

The central place probability has acquired in a fundamental theory of matter and light posed an old question anew, but made it more difficult than ever before to dismiss it: Is probability not also to be viewed as ontic, i.e., as a feature of reality, rather than exclusively as epistemic, i.e., as a feature characterizing our state of knowledge? Moreover, since 1926 probability has been connected, among other applications, with a magnitude of a wavelike nature, hence with interference phenomena that are at variance with deep-rooted and otherwise well-confirmed intuitions concerning the combination of probabilities of different and locally separable events. It seems, therefore, fit to conclude this volume with two contributions by Cartwright, contrasting a more realistic attitude with a pragmatic attitude toward the new entity, quantum mechanical probability. Whatever future insights will teach us, there can be no doubt that quantum physics has become one of the cornerstones of our present-day probabilism.

14 Probabilistic Physics the Classical Way

Jan von Plato

This chapter concentrates on the approach to probability in physics in which probability is identified as the limit of a time average of single physical systems. It extends from Boltzmann in 1868 to the early 1930s, with emphasis on developments after the turn of the century. The time average notion of probability is historically bound to systems of classical physics. A central theme of this presentation is how various researchers could reconcile the determinism of classical systems with a physical (in contrast to epistemological) notion of probability. The views of Einstein, von Mises, and various ergodic theorists are addressed in particular. A rather strong claim is made about the importance of an objective, time average notion of probability for the successes of Einstein's works in statistical physics.

Physics turned out quantal rather than classical, which seemed to create an impasse for the time average interpretation of probability in physics. The second half of this chapter is devoted to a study of how a purely probabilistic formulation of the ergodic theory evolved, one that is not tied to those specific features of classical theories that make us call them deterministic. This development parallels the 'probabilification' of physics in quantum theory, but is not significantly affected by it.

1 Introduction

It is a commonplace that classical physics is deterministic and modern physics indeterministic. However, well over one hundred years ago some classical physicists started thinking that perhaps their physics is not so deterministic after all. In the 1860s, Ludwig Boltzmann started his program of deriving the macroscopic behavior of matter, especially that in gaseous form, from a mechanical microscopic dynamics of molecules. Probabilistic calculations had been applied in gas theory already, around the middle of the last century, by Krönig and Clausius.[1] Another early champion of probabilistic methods in classical physics was James Clerk Maxwell. The well-known "Maxwell's demon" is a popular illustration of the emerging new statistical interpretation of the macroscopic behavior of matter; it was gradually accepted that the tendency of matter to evolve toward equalization of temperature differences, for example, is not a universally valid law of the causal sort.[2] Instead, it is a statistical property, i.e., a property that holds for the most part, but not necessarily in all cases.

The modality of something holding 'statistically' can be interpreted in several ways. One way is Aristotle's 'for the most part of time'. Another is to treat a (possibly hypothetical) class of 'all possible cases' as formed of entities having some kind of simultaneous existence. The qualification 'statistical' then naturally means that a property thus qualified holds for 'most cases'. We find both of these approaches illustrated in a quotation from Boltzmann where he compares Maxwell's and his own interpretations of the concept of probability: "There is a difference in the conceptions of Maxwell and Boltzmann in that the latter characterizes the probability of a state by the average time under which the system is in this state, whereas the former assumes an infinity of equal systems with all possible initial states."[3] In modern terminology, we might differentiate between the two views as follows. Suppose we have two bodies of different temperatures brought into contact with each other. The temperatures naturally tend to equalize, though not by any necessary law of nature, but rather only with a high probability. It is therefore a statistical property, and the two interpretations of probability give us two views on how to treat the physical example. According to Boltzmann, one considers a single system as it evolves in time according to the mechanical laws of motion of its constituent molecules. The probability of temperature equilibrium is high; i.e., the relative time the system is in this equilibrium is near one. Boltzmann advocated a *single-system approach*, characteristic of some interesting later developments in ergodic theory, to be discussed below. Maxwell, on the other hand, proposed what, since Gibbs, has been called an ensemble approach.[4] Here one postulates a probability distribution over suitable physical variables, such as the initial conditions of a mechanical system of a given kind (e.g., of a given total energy or of a given number of molecules). The relative 'amount' of systems of the ensemble that are in equilibrium is near to one. However, the ensemble cannot exist physically, even if its use in calculations leads to agreement with what is observed. One way to circumvent this conceptual difficulty is to think that the probability distribution of the single systems in the ensemble represents the degree of ignorance as to the real state of the actual system in front of our eyes.[5] Nevertheless, both Maxwell and Boltzmann certainly thought that their statistical methods would lead to physically meaningful results. The epistemological flavor of the ensemble approach is therefore somewhat different from that of the subjectivist school on probability that is familiar today. Probability in physics cannot be merely an indication of 'degree of ignorance' or 'degree of belief', for some such 'degrees' lead to physically wrong results, whereas others do not.

In this essay, I shall concentrate on the development of the single-system approach. For one thing, it is significant in giving an unambiguous physical interpretation of the difficult and complex notion of probability. The probability of a state (or a set of states) is identified as a limit of a time average that the system is in this state (set of states). This interpretation has its problems, but they were not present in most of the work I shall consider. The reason is that the mathematics involved in the physics did not grow to the level of sophistication needed to bring out these difficulties.[6]

There is a definite connection between the probability-as-time-average and probability-as-limit-of-relative-frequency views, clearly seen by the physicists whose work I discuss. If the frequentist view on probability is taken as such, that is, if every occurrence of the word "probability" is substituted by "limit of relative frequency," no conceptual contradiction follows even if the limit was that of a (in some sense) deterministic process. Likewise, the limits of time averages of classical statistical physics are uniquely determined by a motion of molecules of the classical 'deterministic' kind. There is no subjective element, of the kind indicated by the 'probability as degree of belief' view, involved in the values of probability here. But as in the frequentist interpretation, one would like to be able to apply the concept of probability to shorter-than-infinite time spans, for, as Keynes said, "In the long run we are all dead." Finite relative frequencies certainly differ from the supposedly existing limit. The same holds for finite time averages. Realizing that these latter deviations are observable under suitable circumstances turns out to be one of the most significant events in the history of probability in physics. Consider a single system, as is done under the program initiated by Boltzmann—let us say the gas in this room. Its pressure in a given location is not uniform in time. Pressure is determined by the average kinetic energy of molecules. The individual energies naturally vary, but the great number of molecules tends to equalize this variation. We have, with high probability, a uniform pressure at different times and locations. However, in a sufficiently small volume, the pressure varies observably. If the infinite time average is taken, this variation is lost. It is also lost if we take an infinity of systems, as in the ensemble approach. Boltzmann, and probably also Gibbs, thought that the fluctuations, which must exist on the molecular level, according to the statistical interpretation, could never be observed.[7] In this sense, molecules would behave differently from dice, where we can see relative frequencies fluctuate.

It was Albert Einstein who first established, through purely theoretical considerations, that fluctuations at the atomic level are, under suitable conditions, of a magnitude that could possibly be observed. The fluctuations he suggested were soon identified as a familiar phenomenon, namely, Brownian motion. The fluctuation theory also soon led, through extensive experimentation, to the universal acceptance of the atomic structure of matter. This certainly was a scientific revolution of considerable magnitude. Its initiator, whose work established a few other revolutions as well, was Einstein. In his earliest works, inspired by Boltzmann's statistical physics, he gradually shifted from an inarticulate ensemble approach to the single-system approach. This latter gives physically real fluctuations in time, instead of dispersion in an ensemble of systems having only a conceptual existence. When one reads Einstein's works of the relevant period, one finds that he placed great emphasis on the interpretation of probability as a limit of a time average. It comes therefore as somewhat surprising that these explicit views on probability seem not to have been discussed seriously before, even though their advocate is the greatest hero of contemporary physics. One reason for this negligence may lie in the conventional interpretation (or rather, misinterpretation) of Einstein's views on quantum theory. Common wisdom has it that he was a determinist according to whom "God does not play dice." We shall see that this should rather have been put as "God does not play quantum dice." Einstein certainly took the statistical character of quantum mechanics as a transitory stage in the development toward, as he said, a complete theory of atomic processes. But it would not be right to think that he regarded classical statistical physics as somehow unfinished if it was not reducible to the sort of situations that we encounter in the archetypally deterministic world of celestial motions. In short, I shall try to show that the determinist Einstein did not regard the statistical character of classical statistical physics as imposed on the theory through our ignorance—accidental from the point of view of physics—of true deterministic molecular processes.

A second line of thought, in addition to fluctuation considerations, is to give a certain definition of chance within classical physics, and thereby to make the shorter-than-infinite run meaningful. I refer to ideas about instability of classical motion as the origin of physical chance. The idea clearly occurs in Maxwell's famous lecture on free will.[8] Independently, the philosopher and (more famously so) physiologist Johannes von Kries also defined chance as instability with respect to slight changes in initial conditions, in systems that display the classical motion of macroscopic bodies. It took Poincaré to make the definition famous. The physicist Marian von Smoluchowski, well known for his theoretical and experimental contributions to fluctuation theory, argued especially forcefully for a definition based on instability of objective chance and objective physical probability.[9]

Einstein and von Smoluchowski began their work in statistical physics largely under the assumption of classical dynamics of molecular motions. But in both, we can see the gradual weakening of this assumption. The reason was that they could not account for new phenomena that could only be understood on the basis of quantum physics. Einstein's approach to quantum theory was along the lines of statistical mechanics.[10] In his efforts to formulate a satisfactory theory, he gave up the assumption of classical molecular dynamics, as will be seen. For von Smoluchowski, the phenomenon was radioactivity. He admitted that a mechanical explanation should be impossible.[11] But even if both of these scientists gave up the universal applicability of the classical approach, they equally forcefully maintained its statistical and probabilistic methods. A more radical departure was given by the purely probabilistic statistical physics of Richard von Mises, the well-known probability theorist. He claimed that classical laws of motion are *incompatible* with statistical physics, though he was not always absolutely committed to this view. On occasions, he said that the assumption of classical laws of motion for molecules is irrelevant, not applicable in any physically meaningful way, or that it is just idle. His argument for the incompatibility, which he presented in 1920, was not based on quantum theory. He suggested a genuinely probabilistic formulation of physical statistics along independent lines.

Late in 1931, the single-system approach experienced a major breakthrough, as John von Neumann and George D. Birkhoff established their *ergodic theorems* for classical systems fulfilling certain conditions. Ironically, this took place after the probabilistic (Born's interpretation of the wave function) and indeterministic (Heisenberg's relations) character of the new quantum mechanics was established. In the final three sections of this essay, I shall return to some of these developments, the reactions of the classical ergodic theorists to the new waves of thinking in physics (these being sometimes represented by one and the same person—von Neumann, for example), and the purely probabilistic formulation of the single-system approach as a theory of abstract dynamical systems, given by Alexander Khintchine and Eberhard Hopf.

2   Einstein's Views on Probability

2.1   Probability as Time Average

It seems to be a commonly held opinion that Einstein advocated a subjective notion of probability. As was mentioned above, the reason for this view appears to be the claim that classical physics somehow excludes objective probability. As an example of such interpretations, we may refer to a letter of Einstein's in the English edition of Sir Karl Popper's *The Logic of Scientific Discovery*.[12] With reference to earlier correspondence, Einstein writes, "I wish to say again that I do not believe that you are right in your thesis that it is impossible to derive statistical conclusions from a deterministic theory. Only think of classical statistical mechanics (gas theory, or the theory of Brownian movement). Example: a material point moves with constant velocity in a closed circle; I can calculate the probability of finding it at a given time within a given part of the periphery. What is essential is merely this: that I do not know the initial state, or that I do not know it precisely!"[13] In 1959, Popper suggests that Einstein was here thinking of a subjective interpretation of probability.[14] This suggestion is related to a more general claim according to which probabilities in classical systems are always subjective or ignorance based. In a popular view, Einstein was a determinist ("God does not play dice"), so that statistical or probabilistic notions would only appear as expressions of ignorance, or as a transient state toward a deterministic theory. This train of thought supposes that genuinely ('nonsubjectively') probabilistic theories and classically deterministic theories are incompatible opposites. In what is known as 'the propensity interpretation of probability', Popper further argues that deterministic systems (in the sense of classical mechanics) require a subjective interpretation of probabilities.[15] True to this view, he naturally wants to interpret Einstein as advocating a subjectivist view of probability. Some authorities, like Max Born, in an essay on Einstein's statistical theories,[16] say that they "have found no definite statement of his about the question 'What is probability'." With this in mind, it proves useful to find out what Einstein's views on probability were. Surprisingly, Einstein's works on statistical mechanics contain extensive discussions on the interpretation of the concept of probability.[17]

Einstein's background for his work on statistical mechanics was Boltzmann's molecular gas theory. From this comes Einstein's conception of probability.[18] Boltzmann had already in 1868 and 1871 identified probability as a *limit of a time average*. This interpretation is followed by many physicists up to our own day. It is related to the interpretation of probability as a limit of relative frequency. In the latter, the existence and uniqueness of limiting frequencies is based on probabilistic assumptions about the events concerned. In the time average conception, there are no random trials, but, instead, the dynamical motion of the system together with initial conditions 'samples' the state space, hopefully yielding limits of time averages. Their existence and uniqueness are (in the best of possible worlds) proved from the physical description of the system together with one special assumption.[19] An alternative view of probability is that of Maxwell and Gibbs. Here a probability distribution over an ensemble (a continuous set) of systems is postulated. According to Gibbs's followers, the ensemble distribution is not a physical one. It is rather justified by its 'usefulness' in giving the right results. The main problem with this approach is how the unphysical a priori probabilities can lead to any physically meaningful conclusions. Einstein's work of 1902 is similar to Gibbs's statistical mechanics, though he was unaware of Gibbs's work.[20] In his next paper, of 1903, assumptions are made that are equivalent to the ergodicity of the system.[21] First it is assumed that the system has a constant total energy, i.e., that it is stationary or isolated from any surroundings. Second, the existence of constants of motion other than energy is denied. After this it is assumed that there is, for a stationary system, a definite limit value (*bestimmter Grenzwert*) for the time spent in a given region of state space. It follows that probability becomes identified as the limit of a time average.

The claim is sometimes made that Einstein's statistical mechanics abstracts from the classical mechanics of motion, and uses only one mechanical condition, the conservation of volumes of regions of the state space. This conservation property, called Liouville's theorem, holds for isolated mechanical systems that have a fixed total energy. We see that in fact Einstein assumed more: the absence of constants of motion other than total energy.[22] The validity of this assumption depends on the dynamical law of motion of the system. Einstein's next assumption, the existence and uniqueness of limits of time averages, follows from the previous one (as we know from Birkhoff's ergodic theorem of 1931). By this assumption, Einstein in fact requires that the underlying mechanical motion be ergodic. The conservation property as well as ergodicity can both be formulated either as conditions on classical dynamical systems or as more general probabilistic conditions. Their status with respect to the question, to what extent Einstein freed his statistical mechanics from mechanical assumptions, is exactly the same.

A more explicit explanation of the notion of probability is given in the famous 1905 article on the photoelectric effect, or the light quantum hypothesis.[23] In §5 of that work, we read the following:

In the calculation of entropy by the methods of molecular theory, the word 'probability' is often used in a sense that is not in accordance with the definition of probability as it is given in the calculus of probability. Specifically, 'cases of equal probability' are often determined hypothetically also in cases for which the theoretical pictures used are sufficiently fixed to allow of a deduction instead of a hypothetical determination. I shall show in a separate work that one gets along completely, in considerations of thermal processes, with what are known as 'statistical probabilities'. By this, I hope to dispense with a logical difficulty that still hinders the execution of Boltzmann's principle. Here I shall give, however, only its general formulation, and its application to certain specific cases.

In case it is meaningful to speak of the probability of a state, and if, further, any growth of entropy can be taken as a passage to a more probable state, the entropy S_1 of a system is a function of the probability W_1 of its instantaneous state.

There now follows a derivation of the connection between entropy growth and probability. The result is this: $S - S_0 = (R/N) \ln W$, where S_0 is the entropy of the initial state, R is the gas constant, and N the number of molecules in a 'Grammolekül'. W is a 'statistical probability', which is, as in 1903, the limit of a time average.

The beginning of the quotation, on the meaning of probability, is not unambiguous. The traditional definition of probability is that it is the number of favorable cases divided by the number of all possible cases that are here 'equiprobable'. In Boltzmann's formula for entropy, probabilities were determined combinatorially, as the relative numbers of 'complexions'. Why is this 'equiprobability' derivation, contrary to appearance, not in accordance with the one given in the calculus of probability? Because, I think, Einstein meant that in the calculus of probability, probabilities are defined as limits of relative frequencies, i.e., as 'statistical probabilities'.[24] The 'deduction', he promises, would be a replacement of a hypothesis of equiprobable cases treated combinatorially, by a determination of limits of time averages. This is indicated by his view that the absence of constants of motion other than total energy makes these limits coincide with the distribution of states determined by the energy. He had in fact already once made such a replacement, in §7 of his paper of 1903. The state space is there divided into a finite number of regions of the same volume (microcanonical measure). To prove that the system spends asymptotically equal times in these regions, i.e., to prove that they are 'equiprobable', requires that there be only one (independent) constant of motion. Einstein's promised deduction therefore requires a proof of ergodicity, even if he certainly 'got along', as he put it, by the considerably less difficult task of computing microcanonical probabilities, taking ergodicity for granted as he did.

The "execution of Boltzmann's principle" is discussed twice later, first in a note[25] on the limits of validity of the second law of thermodynamics. According to the "molecular theory of heat," fluctuations of parameters are inevitable, whereas thermodynamics says these are constants.

The derivation of the fluctuation formula is treated in greater detail, when, in 1910, Einstein returns to the foundations of thermodynamics and statistical mechanics.[26] He formulates again what he calls Boltzmann's principle as $S = (R/N) \ln W +$ constant. He refers to the usual interpretation of the probability W, which is "the number of different possible ways (complexions) in which the state, incompletely defined through the observable parameters of a system in the sense of molecular theory, can be thought realized."[27] Einstein continues by saying that it can be questioned whether this interpretation of Boltzmann's principle can have any sense "without a complete molecular-mechanical or other theory representing completely the elementary processes." His next important step is this: Boltzmann's principle has a content independent of any theory of elementary processes, provided that one generalizes the result of molecular kinetics according to which irreversibility is only apparent. If there is no irreversibility in principle, and if energy is the only constant of motion of the isolated system, the system will make a 'fair sample' of the state space, which gives us statistical probabilities.[28] These are exactly limits of time averages.

In Einstein's review of "the present state of the radiation problem" of the year 1909, we again find an emphasis on statistical probability and Boltzmann's principle:

If one takes the point of view that the irreversibility of processes in nature is only an *apparent* one, and that an irreversible process consists of a transition to a more probable state, one must then give a definition of the probability W of a state. The only definition that can be considered, according to my opinion, would be the following:
    Let $A_1, A_2, \ldots, A_l$ be all the states an isolated system of a given energy can take, or, to be more precise, all the states we are able to distinguish in such a system, with given aids. According to the classical theory, the system takes after a definite time a certain of these states (e.g., A_l), and remains in it (thermodynamical equilibrium). But according to the statistical theory, the system takes, in an irregular sequence, all of the states $A_1, A_2, \ldots, A_l$ over and over again. [In a footnote, Einstein says that this is the only tenable view, as one can immediately see from the properties of Brownian motion.] If one observes the system for a very long time θ there will be a certain part t_v of this time such that the system is in state A_v for precisely this time t_v. t_v/θ will have a definite limiting value, which we call the probability W of the state A_v.[29]

Einstein now defines entropy by Boltzmann's principle, and notes that neither Boltzmann nor Planck has given a definition of W. They give a "purely formal" definition of W as a number of complexions. Complexions are logically unnecessary, since one can define the probabilities statistically. Einstein then makes the strong claim that the relation between entropy and probability holds only if one follows his definition. Planck ought to have postulated the equations $S = (R/N) \ln W$ and $W =$ number of complexions only "under the additional condition that the complexions are chosen in such a way that they are found equally probable on the basis of statistical considerations, within the theoretical picture chosen by him." Happily, Planck dispensed with this requirement, but "it would not be appropriate to forget that Planck's radiation formula is incompatible with the theoretical basis Mr. Planck started with."[30] In the ensuing discussion of the theoretical basis of Planck's formula, Einstein notes the following possibility.[31] He had applied Boltzmann's principle for the calculation of entropy from a "more or less complete theory of the quantity W." But the principle can be reversed, to yield statistical probabilities from observed values of entropy. A theory that gives values of probabilities differing from these should be discarded. We note here a remarkable turn in Einstein's thought. His 'statistical probabilities' were like Boltzmann's; the latter's original requirement of ergodic motions has the characteristic consequence that limits of time averages can be strictly and logically identified with, as we say after Gibbs, microcanonical averages. Boltzmann's was a program for *deducing* statistical laws from the underlying dynamics. Similarly, Einstein hoped to derive his statistical probabilities from a "complete theory," though not necessarily a mechanical one. Nowadays we know that such a derivation is possible for at least some classical dynamical systems. Here the assumption of absolute continuity, together with ergodicity, leads to a determinate way of assigning probabilities to a system. Giving up the idea of such a derivation, or an analogous one without an underlying mechanics, Einstein now suggests that we make a statistical inference of sorts, namely, discarding a theory that gives probabilities not consonant with observation. It should be emphasized, however, that he did not take empirically determined probabilities as a final state of matters, but insisted on a subsequent theoretical determination.

2.2 The Importance of Fluctuations

If a system is in thermal contact with surroundings of constant temperature, any temperature differences should vanish. However, this claim has, according to the 'molecular theory of heat', only an approximate validity. The temperature of the system in fact fluctuates around that of the surroundings. These energy fluctuations are treated in Einstein's third paper (1904) on statistical mechanics.[32] The fluctuations follow a probability distribution of the familiar exponential form, and, as Einstein emphasizes, the probabilities are different from zero for all possible energy values, even if they are very small outside the value of highest probability. As we have seen in the preceding subsection, Einstein required of statistical physical systems that they have no constants of motion other than total energy. For a single system, probability is then the same as the limit of a time average. Therefore, we may conclude, events with positive probability must eventually take place. The concept of an energy fluctuation refers to the development in time of a single system and attains, on the basis of the above, physical reality in combination with Einstein's interpretation of probability. Another consequence of the 'molecular-kinetic' theory of heat, in addition to the fluctuation of parameters, is that there is no difference in principle between molecules and macroscopically observable particles, such as one has in a colloidal solution, for example.[33] It follows that one can consider the single particle as a 'system' whose kinetic energy fluctuates as a consequence of collisions with molecules. The existence of fluctuations is almost a consequence of the molecular structure of matter, but before Einstein's work, it was not realized that the fluctuations might be observable in suitable situations.[34] The motion that the molecular-kinetic theory predicts would, if observed, have as a consequence that "classical thermodynamics can no longer be looked upon as applicable with precision to bodies even of dimensions distinguishable in a microscope: an exact determination of actual atomic dimensions is then possible."[35] Einstein himself made one such determination in 1905, and soon the matter was subject to extensive experimental studies.[36] The emerging science of 'molecular statistics' was the final blow to antiatomistic schools of thought in physical science. Only when atoms started decaying in physical laboratories was atomism itself finally accepted in the circles that had resisted it the most—and this on the basis of a classical theory in which the structure of atoms is left completely open. The theory is classical, but this does not preclude its being probabilistic. In Einstein's work on Brownian motion, interpretative remarks are scarce.[37] At times, he refers to the 'irregular' and 'fortuitous' character of molecular motion, and introduces probabilistic concepts and conditions on this basis. As examples, we may refer to the assumption of probabilistic independence of the motion of a particle from that of other particles,[38] independence of the motions of one particle at sufficiently separated intervals of time,[39] and the emergence of a uniform distribution of particles in a solution as a consequence of the motion of molecules that "will alter their positions in the most irregular manner thinkable."[40] For the present, we can sum up Einstein's study of fluctuation phenomena as follows: By accepting the molecular theory with its inherently statistical character, by referring with probability to the behavior of a single system in time, and by establishing the ensuing fluctuations as physically observable phenomena, Einstein laid down the theoretical basis for the experimental vindication of atomism. Within a few years, the new science of molecular statistics led to the final acceptance of atomism. (I think this acceptance of atomism should be called a probabilistic revolution, or, more precisely, a consequence of *that* probabilistic revolution that made fluctuations a physical reality. But again, those who think there is, in principle, no place for a physical notion of probability within classical theories could still contend that the old ignorance interpretation of probability is somewhere lurking behind the fluctuations.) I have wanted here to emphasize the importance of probability *cum* time average, in principle no different from its discrete brother, relative frequency, for this development.

2.3 Einstein and Quantum Dice

Einstein's well-known uneasiness with the "statistical quantum theory" is closely related to his views on the relationship between statistical and ordinary mechanics of the classical sort. In fact, one sees that the uneasiness was not one about theories being statistical, but rather about the kind of justification we are able to give to statistical theories and laws. In his "remarks" on the essays brought together in *Albert Einstein: Philosopher-Scientist*,[41] he gives as his view that the quantum theoretical description ought to be interpreted as a description of an ensemble of systems. The statistical predictions of quantum theory are unacceptable only if they are claimed to be a description of individual systems. Instead, Einstein hoped for a future theory that would give a complete description of the latter. The relation of the present statistical quantum theory to that future theory would "take an approximately analogous position to that of statistical mechanics within the framework of classical mechanics."[42] Further, "If it should be possible to move forward to a complete description, it is likely that the laws would represent relations among all the conceptual elements of this description which, *per se*, have nothing to do with statistics."[43] Einstein, like Boltzmann earlier, believed that the probabilistic laws of statistical mechanics arise from the properties of an underlying mechanical motion, a motion that in itself has nothing to do with statistics. The analogy that he suggests is that a future complete theory of quantum phenomena will lead to statistical laws in a similar way. Incidentally, there is a way of basing the probabilistic laws of typical games of chance, such as playing dice, on a classical, complete theory of unstable motions of macroscopic bodies.[44] It fulfills the criteria Einstein set for giving a foundation for statistical laws. Therefore, indeed, his saying about God and playing dice could be modified in the way indicated.

Probability and statistics are in no way unacceptable or undesirable, provided that their use is justified by a basic complete theory. In all other ways of introducing probability into physics, it has to be *postulated*. It forms therefore a hypothetical element of physical theory that Einstein did not accept as the final state of affairs.[45]

3   The General Ergodic Problem

3.1   The Ergodic Problem in Physics

For the subsequent discussion, it will be necessary to have a somewhat more detailed view of the ergodic problem than became apparent in connection with the works of Einstein. I shall therefore treat the ergodic problem as it originated in physics, its formulation as a genuinely probabilistic question in the theory of abstract dynamical systems, and some of the more important points of interpretation. Ergodicity can easily be seen as one of the absolutely central notions in the foundations of probability. Its physical background makes it an especially important concept in attempts to justify objective interpretations of probability.

In classical dynamics, one starts from the classical mechanics of motion and tries to show that it can lead to a formulation of classical statistical mechanics. This program was started by Ludwig Boltzmann. In the early 1930s, the ergodic problem assumed its present formulation, with the development of the notion of an *abstract dynamical system*. This framework is independent of the classical physics from which it historically evolved. It is clearly formulated by Alexander Khintchine in a publication from the year 1933. I believe that Richard von Mises's views on probability, his criticism of classical dynamics, and his formulation of a genuinely probabilistic scheme for the treatment of problems of statistical physics at least partly inspired this development. It seems that this line of influence has been largely forgotten.

Ergodic theory evolved in the shadow of another theory of incomparably greater physical significance, namely, quantum mechanics. In section 8 I shall try to state what little I have found out about the attitudes and reactions of the ergodic theorists to the—at the time, novel—probabilistic and indeterministic features and interpretations of the quantum theory.

One can have different views on the tasks of ergodic theory. In the physical tradition that gave rise to it, one often sees its function as follows: The aim of ergodic theory is to determine the statistical properties of motions of bodies. By 'statistical properties', one refers (it is hoped) to well-defined properties of certain functions defined over the state space of the systems considered. One intuitive concept of 'statistical property' is that it is any property that holds, for a given system, with probability one. For illustrative purposes, let us take a simple example from everyday life. Suppose that we have a fair coin, one with probabilities half and half, and independent consecutive tosses. It is *not*, according to the mode of speech chosen above, a statistical property of this 'system' that this or that toss might give a particular result. Mere variation or contingency is not the hallmark of 'statistical property'. Rather, we require that there be regularity in statistical data about the system's behavior. If there is such regularity, it obtains a concise expression in the limit of an infinite number of tosses. Under the assumptions we have made, the law of large numbers states that the relative frequencies of events converge toward the values of the probabilities of these events. This holds 'except for a set of measure zero', or 'with probability one'. It is not a necessity that relative frequencies converge in the way the law of large numbers states, or that they converge at all. I think this is part of the meaning of 'statistical property'. If you consider finite relative frequencies, you get a result that says that the relative frequencies are, with a high probability, near the probability values of the events. Here 'statistical' appears almost in its everyday sense, as a designation of events that usually take place, but not necessarily always. The limit of finite relative frequencies brings these intuitive considerations back to the notion of 'statistical property' mentioned above, namely, probability-one events.[46]

Let S be the state space of a physical system. One can think of the unit square as an example where the x coordinate gives positions and the y coordinate momenta. Points of the state space are called *microstates s*. Consider the properties of functions defined over the state or phase space S. For example, f might be defined by giving it the value 1 if the argument s is in the lower left quadrant of the state space and the value 0 otherwise. Next consider the laws of motion of the system. These are condensed into transformations T_t that give, for an initial state s, the state $T_t s$ after t units of time. We are already assuming here that the transformations T_t depend only on the interval of time t, but not on the time that the transformation begins to act on the states. In classical dynamics, this assumption amounts to having a Hamiltonian function over the system that does not change in time, or in other words, we are considering an isolated, conservative system of motion. The transformations T_t form, in the classical case, a group with one real-valued parameter t. This essentially allows one to go from the state s to $T_{t_1} s$ to $T_{t_2} T_{t_1} s$ by one single transformation $T_{t_1 + t_2} = T_{t_2} \cdot T_{t_1}$. Sometimes we say that such systems are stationary, anticipating the corresponding notion of abstract dynamics. With stationary systems, the physical characteristics of the motion remain the same in time.

Let us now look at the function f defined above. If we start from the initial state s, the finite time average of f over the interval (t_1, t_2) is given by the integral $1/(t_2 - t_1) \int_{t_1}^{t_2} f(T_t s)\, dt$. We can start from 'zero time' and take its limit as $t \to \infty$.

$$\hat{f}(s) = \lim_{t \to \infty} 1/t \int_0^t f(T_t s)\, dt.$$

We obtain the limit of time average of f along the trajectory that starts from the initial state s. Now, we proposed that a statistical property of the system of motion be a property of functions over state space that holds with probability one. In certain cases, it luckily happens that

1. Limits of time averages $\hat{f}(s)$ exist, i.e., the integral above converges, except for a set of states of measure zero.

2. $\hat{f}(s)$ is a constant over S, except for a set of measure zero.

In the first case, we have the statistical property that limits of time averages exist; in the second, that they are unique.

The ergodic problem can now be given the following formulation:

Find conditions under which the above statistical properties are met, and determine what the limits of time averages are.

The latter part of this task is the central one in ergodic theory. Our basic aim is to determine theoretically the 'long-term' statistical properties of systems, if they have any such. Obviously, one crucial question is how and why we give measure zero to some sets of initial conditions, and a positive measure to others. In all cases where positive results on the ergodic problem are obtainable, the only latitude left for the solution is in deciding which sets of states receive measure zero.

Ludwig Boltzmann tried to give answers to the basic questions above in the following way. If a single trajectory goes through all the points of the state space, then, taking in unstated terms the convergence problem for granted, the limits are unique since there is in fact only one trajectory possible for the system. As an aside, we may wonder how Boltzmann could live with this assumption. It makes all talk about initial conditions empty. The world becomes, in figurative terms, devoid, in the last resort, of any contingency (namely, the initial conditions). The next obvious step for Boltzmann is to realize that one could weigh the values of functions of interest by lumping them together, instead of using an infinite time for this. So he integrated f over the state space S to get what is now called a *phase average:* $\bar{f} = \int_S f(s)\, ds$. One can think of this as a uniform probability distribution given over S. To answer the example problem above, just integrate and you get the answer: $\hat{f} = \bar{f} = \frac{1}{4}$. The relative amount of time the system spends in the lower left quadrant is one-quarter in the limit, granted that the approach works.

In the early 1910s, it was realized that ergodic motions of the Boltzmannian kind are in fact mathematically impossible.[47] It is not possible to let the trajectory wind through the state space so that it will cover *all* points. Instead, it was suggested that ergodic motions be defined by requiring them to visit, intuitively speaking, any regions of positive geometric measure of the state space. In the famous review of the foundations of statistical mechanics by the Ehrenfests in 1911, these were called quasi-ergodic motions.[48] The word 'quasi-' was subsequently dropped. With the emergence of Lebesgue integration, one could allow of exceptions to quasi-ergodic motions, if these had measure zero. The situation at the time of the Ehrenfest article was not as clear as it could have been. The following vital distinction was not made explicit: A *single* trajectory is called ergodic if it visits any open set of positive measure. The ergodicity condition for a *system* therefore requires that almost all trajectories be ergodic. In the older literature, the distinction between ergodic trajectories and ergodic systems is often obscured. Boltzmann's original ergodic hypothesis naturally leaves no space for such a distinction. On the other hand, if the distinction is made, his argument for the equality of time and phase averages is lost. In P. and T. Ehrenfest, the distinction between ergodicity and quasi-ergodicity is first made with respect to a system with a single trajectory.[49] They note that Boltzmann's and Maxwell's argument for the uniqueness of time averages would not apply to a system accepting one single quasi-ergodic trajectory.[50] They add immediately that for a quasi-ergodic system, and for each of its surfaces of constant energy, one has a continuous number of different trajectories to distinguish.[51] This is based on the following.[52] There are as many differential equations between the canonical variables of a Hamiltonian system as there are degrees of freedom. Therefore, there are as many integrals of the system. For time-independent systems, the first integral is the total energy. If that is fixed, there are still others that can have any real values and hence a continuous number of solutions of the equations. It is a bit mysterious why the Ehrenfests did not use Einstein's formulation of the ergodicity condition. As we noted above, Einstein saw that the existence of further constants or integrals of motion other than total energy leads to a situation where the statistical behavior of a system depends on (i.e., varies with) its initial conditions.[53]

After this historical digression, let us briefly see what the two fundamental properties in ergodic theory are, and what are their characteristic consequences. The first is stationarity in the form of a Hamiltonian invariant in time. The basic consequence is the existence of limits of time averages, except for sets of measure zero. The second is ergodicity, which requires that ('almost all') trajectories visit any open sets of positive measure. The characteristic consequence is that limits of time averages are unique, again with the exception of a set of measure zero.

In the application of ergodic theory to classical dynamics, one of course tries to prove that these two conditions are met. This is an extremely hard task, and has been successful for only some rather simple systems, like the Boltzmann gas model.

3.2   Abstract Dynamical Systems

Abstract dynamical systems look like this: We have a space S, a δ-algebra $\mathscr{A}$ of subsets of S, a normalized measure P over the $A \in \mathscr{A}$, and a transformation $T: S \to S$ such that if $A \in \mathscr{A}$, then $T^{-1}A \in \mathscr{A}$ and $P(T^{-1}A) = P(A)$. The first condition is simply that T is a measurable transformation, so that measures (probabilities) of sets TA are well defined. The second condition on T is that T is a measure preserving transformation with respect to P. A specific case is offered by stationary processes. Think of T as a shift in time and you obtain probabilities that are invariant under time shifts, which is the idea of stationarity in probabilistic terms. We will call systems with the property $P(T^{-1}A) = P(A)$ stationary. Call sets such that $T^{-1}A = A$ invariant. (If T is invertible, i.e., if T^{-1} is a transformation, we may write $TA = A$.) If invariant sets are always of measure 0 or 1, call the system ergodic.[54] The basic results are these: Any set A is visited by almost all trajectories with a converging frequency. A trajectory is a sequence s, Ts, TTs, $TTTs$, ..., $T^n s$, ... So, if x_A is the indicator of the set A, we have (where a.e. or 'almost everywhere' has the obvious meaning)

$$\lim_{n \to \infty} 1/n \sum_{i=1}^{n} x_A(T^i s) \qquad \text{exists a.e.}$$

This follows from stationarity; in fact it is characteristic of stationary systems. Next, if the system is ergodic, the sets A are visited with the right limiting frequency:

$$\lim_{n \to \infty} 1/n \sum_{i=1}^{n} x_A(T^i s) = P(A) \qquad \text{a.e.}$$

This is the probabilistic formulation of the ergodic theorem.

One could easily write down analogues in continuous time for the ergodic theorems of abstract dynamical systems. They would be similar to those of the previous subsection for classical systems. One aspect of ergodic theorems merits comment, namely, the well-known laws of large numbers give us sufficient conditions for the uniqueness of limits of relative frequencies. These conditions are that the events concerned have a constant probability and be independent of each other. In this situation, one would naturally like to have a converse for the result, i.e., necessary and sufficient conditions for the kind of conclusion about the uniqueness of limits of relative frequencies that the law of large numbers allows us to draw. Ergodicity is exactly such a necessary and sufficient condition, and therefore more basic than its special case, that of independent events. A notion of frequentist probability that lies within the usual measure theoretic formulation of probability is captured exactly by the condition of ergodicity.[55]

The alternative to the ergodic theory's time average approach is via ensembles. One introduces a continuous infinity of systems together with a probability distribution over them. An example is the uniform phase space density discussed above. The basic question is to explain why one gets physically right answers with this approach. The ensemble is certainly not a physically existent entity. Ergodicity, on the other hand, justifies the use of a certain probability measure for the determination of limits of time averages. But this approach is problematic, and besides, ergodic results are extremely difficult to obtain. We can see this by thinking in frequentist terms: What guarantee have we for assuming that limits of relative frequencies bear any significant relation to finite relative frequencies, which are, after all, the only things we can observe? The answer comes also via consideration of frequencies. If the dispersion of the finite relative frequencies is small, the probability is high that a given relative frequency closely approximates the limit. Or, in other words, if the dispersion is small, finite time averages are near the limits of time averages for most of the time.

4   The Earliest Ergodic Theorems

As was stated, there were some mathematical refinements in concepts in the 1910s, but they did not play a major role in physical applications. It is sometimes said that ergodic theory played no role at all, but this is not the case. Above, we discussed the statistical physics of Einstein in which ergodicity was postulated through its characteristic consequence, the uniqueness of limits of time averages. It is then possible to determine the limits through the microcanonical measure. Anyone who postulated the uniqueness consequently postulated the characteristic consequence of ergodic behavior, as, for example, did Einstein. It could be done in different terms. Einstein says in 1902 that if there are further constants of motion besides total energy, the properties of the distribution of states are not determined solely by total energy, but depend on initial conditions as well. Avoiding this dependence is another way of doing the job that ergodicity does.

The first ergodic theorem was about the equidistribution of decimals of irrational numbers. It was established almost simultaneously by the astronomer Bohl, the set theorist Sierpinski, and Hermann Weyl, in 1909–1910. Weyl's paper "Über die Gleichverteilung von Zahlen mod. Eins" of 1916 became the standard reference.[56] However, a couple of years earlier, Weyl delivered a lecture on "une application de la théorie des nombres à la méchanique statistique et à la theorie des perturbations."[57] In number theoretic terms, we can put the result as follows. If an irrational number is given, then multiplied by 2, 3, 4, ... and only the decimal part taken, the result will be a sequence of numbers between 0 and 1 that is uniformly distributed. The sequence is naturally only denumerable, so by equidistribution one means that the sequence visits each interval with the right limiting frequency. There are exceptions, such as the rationals, which form a periodic sequence, but they have measure zero. At that time, Lebesgue integration was of course known to such a top mathematician as Weyl. Earlier, in 1884 Leopold Kronecker had shown that an equivalent formulation of the situation in terms of rotations of a circle with an irrational angle gives a dense set on the circumference. (This result was, as I have shown, already known in medieval times.[58]) An example closer to physics was published in 1913 by König and Szücs. They showed that in the general case, the linear motion of a point inside a square or cube is dense.[59]

Weyl was very well aware of the physical applications of his result, as the title of his 1914 talk indicates. He noticed that his result leads to probabilities that coincide with limits of time averages, and makes possible a determination of these. Further, the astronomical application merits our notice. Weyl very clearly saw that ergodicity leads to the convergence of time averages. In astronomical terms, it leads to the existence of true mean motions—a great insight indeed. However, I shall not dwell more on Weyl's work since there is not much space for it. Rather, I assume that by 1916 the ergodicity of rotations was well established and known by competent mathematicians.

If we look at ergodicity as a general probabilistic notion, Emile Borel is surely one of the original contributors to the field. He showed in 1909 the strong law of large numbers for independent events.[60] In probabilistic terms, this is the first ergodic theorem. That this is not a far-fetched connection will be apparent through study of Richard von Mises's views on the nature of the ergodic problem. He represents a line of development that at first might seem slightly orthogonal to subsequent, now better-known, works, such as those by von Neumann and Birkhoff.

5   The Purely Probabilistic Physical Statistics of Richard von Mises

5.1   Probability

Richard von Mises's views on probability, and his theory of *Kollektivs*, are rather well known. It has even led to some new developments during the last twenty years, having in the meanwhile been considered a curiosity. His frequentist theory of probability was influential in the 1920s and early 1930s. Probability is a certain property of infinite sequences, namely, a limit of relative frequency in these sequences. The sequences have to be irregular or 'regellos' as von Mises required. In his view probability theory is part of theoretical physics, comparable to classical mechanics or optics.[61] It studies a well-defined class of phenomena, namely, mass phenomena. Its axioms define its basic concepts, and it has a deductive structure. In mechanics, we derive, e.g., later velocities from given initial ones. Likewise in probability, we derive other probabilities from given ones. If we think that interpretations of probability ought to answer two questions—first, what is chance? or how does it come about?; and second, where do we get the probability numbers from?—we see that von Mises at least tries to give answers to these questions. He builds them as basic postulates into his theory. Chance is *Regellosigkeit*. For von Mises, it was an intuitive idea. For many later workers, it became a precise concept, through the use of the theory of algorithms. The probability numbers, also, are postulated through the requirement that relative frequencies must converge. In the present-day formulations of von Mises's theory by Kolmogorov and Martin-Löf, it is shown that the new definition of *Regellosigkeit* has as its consequence the convergence of relative frequencies.[62] This holds uniformly, since, as von Mises required, the concept of probability does not apply to all sequences. Sequences that in the measure theoretic formulation are qualified as exceptions of measure zero are excluded since they are not *Kollektivs*.

5.2   Critique of Classical Mechanics

We see that von Mises postulated an element of randomness in his theory, and the question naturally arises whether that randomness is compatible with the supposedly deterministic classical dynamics that was still alive and well. Von Mises gives an answer in his paper "On the Present Crisis of Mechanics."[63] The basic question is, Can Newtonian mechanics give an explanation of the motions of all bodies? He addresses relativity, hydrodynamics, and statistical mechanics. In his well-known book *Wahrscheinlichkeitsrechnung* of 1931, he says that in this earlier paper, he stated that it is not possible to have a satisfactory theory of physical phenomena without the use of probabilistic basic concepts.[64] His reason is the following. Undeniably, there exist genuinely statistical sequences in nature. These cannot have any algorithm. (What exactly he meant by this is not clear, but at the least it should imply that there is no method of computing the sequence, even in principle.) Classical mechanics, on the other hand, has an algorithm. Therefore the two are incompatible.[65] When one reads the paper, there is nevertheless some difficulty in locating an exact statement. A much later paper, a talk on causal and statistical laws in physics of 1929, does not make the strong claim.[66] Instead, a more sophisticated opinion is given. The basic question is again, Can Newtonian mechanics explain the motions of all bodies? Von Mises concentrates on explaining, which, for him, is to reduce the more complicated to the simpler.[67] For example, one might reduce complicated planetary motions to a simple gravitational law of force. But if we consider the motion of matter on the atomic scale, the laws of force that operate are inaccessibly complicated. Classical mechanics becomes idle ('leer-laufend'), even if it is not contradicted.[68] The hypothesis that trajectories follow classical orbits is a weak remedy, for von Mises concludes, "An assumption from which we cannot determine whether it is right or not is not scientific." [69] The methods of classical mechanics leave the problem of motion on the atomic scale completely untouched, whereas those of probability theory give a result that is in accordance with experience.

5.3   Exclusion of the Ergodic Hypothesis from Physical Statistics

Next we come to von Mises's treatment of the ergodic problem. He had by now concluded that classical mechanics was not a relevant framework. He was naturally aware of the earlier attempts of Einstein and others at founding statistical physics on classical dynamics.

In 1920, he published an article in two parts in the *Physikalische Zeitschrift* in which he replaced the ergodic hypothesis of classical dynamics with a probabilistic formulation.[70] The ergodic (quasi-ergodic) hypothesis is the assumption that a mechanical system goes through all states (comes arbitrarily close to all states) in its evolution. As a consequence of Liouville's theorem, this amounts to giving the same average kinetic energy to all degrees of freedom of the system (equipartition of energy). In the case of specific heats, this is contradicted, so that the ergodic hypothesis has to be given up in many cases. (Quantum effects are not mentioned by him in this context.) But von Mises sees the task of the ergodic hypothesis differently: Given a microcanonical ensemble, one ought to be able to conclude that the microcanonical probabilities give the relative times that each of the systems of the ensemble spends in different regions of the phase space. Von Mises admits that the ergodic hypothesis makes this inference possible. However, it is not at all *necessary* for the inference. He calls the classical ergodic (or quasi-ergodic) hypothesis an ergodic hypothesis in the strict sense, and offers anything that does the following task as an "ergodic hypothesis in the wide sense". namely, to allow an interpretation of a combinatorially determined density, within a constructed ensemble, as the relative time of the actual motion of a specific mechanical system.[71] However, even this formulation does not satisfy him. It dispenses with the original mechanical ergodic hypothesis, but the *conclusion* is still unacceptable. He makes the point that a theory can lead to definite statements about the actual course of single systems only through mechanical equations, but never from probabilistic considerations.

Von Mises gives Brownian motion as an example of an unstated use of the ergodic hypothesis. The probabilistic calculations of the theory give a virtual ensemble (uninterpreted probability density); however, one tests the theory without hesitation by time averages. Therefore this approach is based on the ergodic hypothesis in the wider sense: The 'time ensemble' is a true image of virtual and spatial ensembles. But von Mises is not an ensemble man. He says that with virtual ensembles "there would not be any physical content in our considerations, for physics is concerned with the prediction of phenomena occurring in time."[72]

Von Mises comments on the Boltzmannian approach, as represented by Einstein, as follows. Here one starts with a time ensemble, defined through the relation between entropy and probability (entropy proportional to logarithm of probability): "But a deep contradiction remains in physical statistics, one that has not been conquered yet, namely that one takes from a certain point of view the course of events as completely determined through the physical equations (the differential equations of motion of the system); yet one thinks one can make definite statements about this course of events from a completely different point of view."[73] This, apparently, is the first statement that classical dynamics and genuinely probabilistic phenomena are in contradiction with each other.

I shall next briefly outline von Mises's own approach, as it would take too long to describe it in detail. He was an empiricist and took the punctual aspects of mechanics (that is, the use of continuous quantities) as idealizations. In fact, measurements always have to be of finite accuracy only. For von Mises, this is a consequence of the atomic structure of matter. The assumption of exact microstates behind observable macrostates or coarse grained states is idle ('leerlaufend'). By accepting only macrostates as empirically meaningful, one sees that the determination of the evolution of states in time of classical dynamics becomes empirically meaningless.

Probability is introduced as a hypothetical element. It should come in a form that leads to probabilistic laws about the time development of the system. This makes it possible to do the job of the ergodic hypothesis in its classical formulation: the replacement of virtual ensembles (probabilistic assumptions) with predictions of behavior in time. The specific structure that von Mises gave his theory can be condensed into the following. Only a finite number of macrostates are physically observable states. It is assumed that the course of events forms a Markov chain: Given the present macrostate of the system, the transition probability to the next macrostate depends only on the present state, but not on the previous 'history' of the system. If the transition probabilities remain the same in time and if they fulfill certain additional conditions, there will be a unique limiting distribution for the macrostates of the system.

It will soon be seen that von Mises's probabilistic ideas and ideals were not without influence in the transition from ergodicity of dynamical motions to ergodicity as a purely probabilistic and very general concept. But before that, let us see what transpired in the meantime.

6 Ergodic Theory in 1931: The Results of Koopman, von Neumann, and Birkhoff

In March 23, 1931, Bernard Koopman, a '26 PhD from Harvard, had a paper on "Hamiltonian Systems and Transformations in Hilbert Space" communicated in the National Academy of Sciences.[74] The paper consisted essentially of the following remark. Take square integrable complex valued functions f, g, ... over the state space of a conservative Hamiltonian system. Define transformations U_t over these functions by equations of the form $U_t(f(s)) = f(T_t s)$. These transformations allow us to keep track of the values of f under the motions T_t of the state point s without explicitly looking at the motion. The square integrable functions form a Hilbert space and the operators U_t are unitary in this space, e.g., linear and norm preserving. This observation marked the beginning of the operator treatment of classical dynamics. Note that indicator functions of subsets of the state space, through which probabilities are identified, belong to the Hilbert space.

Johann (John) von Neumann probably had the best grasp of properties of Hilbert spaces and their unitary operators, this having been his specialty in the formulation of the mathematical basis of quantum mechanics. Von Neumann wrote in his main paper (1932) that "the possibility of applying Koopman's work to the proof of theorems like the ergodic theorem was suggested to me in a conversation with that author in the spring of 1930." [75] He also noted that André Weil had in 1931 suggested a similar application. As a matter of fact, the collection of correspondence of George Birkhoff at the Harvard archives contains a letter by Weil in which he says that he already knew all the results of Koopman's paper, as well as many of von Neumann's, in 1928. But since he could not establish ergodicity in any specific cases, he did not publish this work—an example of youthful rigor indeed!

Von Neumann makes an assumption of ergodic motions, formulated through the absence of constants of motion other than one constant.[76] He then shows what is known as the mean ergodic theorem. If you take a part of the phase space, the average (mean, in other words) of the times that the different motions spend in this region approaches the measure of it as the time interval grows to infinity. Von Neumann's paper contains other remarkable insights—for example, the first sketch of the ergodic decomposition theorem. He also discusses, toward the end of the paper, the empirical meaning of measure theory. His long paper "Zur Operatorenmethode in der klassischen Mechanik" is an extended version of the original article.[77]

On October 22, 1931, von Neumann told Koopman and Birkhoff about his result. This led rapidly to Birkhoff's ergodic theorem. It says that under ergodicity, time averages approach unique limits not only on the averge, but for each single trajectory. (There is naturally a set of exceptions of measure zero.) Birkhoff published his result[78] in the *Proceedings of the National Academy of Sciences*, as did von Neumann, but was able to get the paper out in December 1931, whereas von Neumann's only came out in 1932. In an amusing note on "Recent Contributions to the Ergodic Theory," Birkhoff and Koopman give a sort of minihistory of the developments.[79] There is no evidence in Birkhoff's correspondence of any reaction on the part of von Neumann, though the latter published a paper on "Physical Applications of the Ergodic Theorem" in which he tried to argue that the mean ergodic theorem is all one needs in such applications.[80]

In a review of Birkhoff's book, *Dynamical Systems*, Koopman remarks that unstable solutions, if taken singly, are not physically meaningful.[81] In late December 1931, Birkhoff gave an informal talk on "Probability and Physical Systems,"[82] where he says that, as Koopman suggested, single solutions are not meaningful since initial conditions cannot be determined precisely. Instead, one ought to get general characterizations of the class of possible motions. He refers to Poincaré as the first one to use (intuitively) this sort of consideration: Poincaré's recurrence theorem is a 'probability-one' result. Birkhoff also says that physicists had "on an intuitive basis vaguely formulated certain types of theorems, one of which in precise form is ... the ergodic theorem."[83] So, right from the start, probability played a role within the ergodic theorems; but it took two or three years to free the ergodic theorem from the underlying classical dynamics. Khintchine and Eberhard Hopf were responsible for this development.

7   Probabilistic Formulation of the Ergodic Theorem

Alexander Khintchine was one of the foremost experts in probability theory in his time. Nowadays he is best known for his law of the iterated logarithm, which gives bounds to the oscillation of finite relative frequencies for situations where laws of large numbers apply.[84] His first paper on ergodicity is "Zur Birkhoff's Lösung des Ergodenproblems," where he says that "Birkhoff's results are a really essential advance in the mathematical foundations of statistical mechanics."[85] Because of the importance of the central result, he wants to give a simpler proof that generalizes it in two directions—he shows that the convergence of time averages holds for any stationary motions (i.e., measure-preserving transformations), and, second, he considers any Lebesgue integrable functions of phase space instead of only indicators of regions. The first part is the essential one. With it, the ergodic theorem is shown as a result of abstract dynamical systems. Classical systems of course form a special case. The hard and difficult part, in this special case, is to show that the condition for the theorem holds, i.e., that 'almost all trajectories go almost everywhere'.[86] In a footnote added after the paper was submitted, he notes that Eberhard Hopf had given a very similar result.[87] The content of Birkhoff's theorem is that in the indecomposable case (just another expression for the measure theoretic criterion of ergodicity), limits of time averages are equal to phase averages.

In another paper of 1933, "Zur mathematischen Begründung der statistischen Mechanik," Khintchine gives a more general philosophical discussion. It is published in the journal *Zeitschrift für angewandte Mathematik und Mechanik*,[88] in a *Festschrift* number for Richard von Mises. I will come back to the introduction of the paper later. Let me now just note how Khintchine compares the approaches of Birkhoff and von Mises. In the latter's theory, the succession in time of observations forms a Markov chain, so that the probability distribution of the events in the sequence depends only on the previous observation, but not on earlier ones. In cases in which this assumption is acceptable, it leads to great simplification. Birkhoff's approach is more general in that the effect of all of previous history can play a role. This statement naturally presupposes that 'Birkhoff's approach' refers to the proper probabilistic generalization of the ergodic theorem of classical systems.

We see that for Khintchine, von Mises's theory was a special case of the purely probabilistic ergodic approach. In the former, a specific hypothesis is made about the probabilistic law for successive events. The latter only requires that we consider systems stationary in the probabilistic sense. It is not farfetched to assume that Khintchine's knowledge of von Mises's work was the route by which he achieved this probabilistic formulation.[89]

8   Ergodic Theory and Quantum Theory

In this last section, views on the relation between the probabilistic aspects of ergodic theory and quantum theory will be reviewed. I shall not discuss the application of ergodic theory to quantum theory proper, even if works on this subject appeared in the period under study.[90] It is remarkable how very little was published on the topic of this section. The probabilistic and indeterministic character of the new quantum mechanics immediately became well known. Soon, indeterminism of the quantum mechanical kind became intimately connected with the very idea of physical probability. The development of ergodic theory within classical dynamics disturbs this ideal picture. But it did not seem to have disturbed some of the creators of these two theories. Von Neumann is a good example, for he contributed decisively to the proper mathematical formulation of both. He notes that there is a striking analogy between the operator treatment of classical and quantum mechanics, and that one can show a continuous change from quantum to classical mechanics by letting Planck's constant approach zero. But there is an essential difference, namely, that for a system with a finite volume state space, the spectrum is discrete in quantum mechanics, "whereas it seems that in the classical problem, a pure line spectrum is the general case."[91] Maybe von Neumann wanted strict mathematical results, akin to his proof of the impossibility of hidden variables in quantum mechanics, and therefore did not comment further on the issue.

As an incidental remark, let me quote the opinion of B. Hostinsky. He was a Czech mathematician who was one of the first to treat a game of chance from the physical point of view. In 1917, he gave equations of motion for an idealized Buffon's needle, and showed what probability distribution these lead to.[92] A little later, he described the roulette wheel as a mechanical system and found sufficient conditions for deriving equiprobability of red and black from the physical description.[93] He did not react favorably to the new mechanics. In the first volume of the famous journal of the Vienna Circle, *Erkenntnis*, a discussion on foundations of probability is reported. Hostinsky says, "There has been too much talk about indeterminism in the last years. In my view, there is determinism if in a given case a dependence between cause and effect has been established. If that does not succeed, it is better to say nothing than to talk about indeterminism."[94] And in 1934, Dirk Struik wrote in a paper[95] on probability within ergodic theory that there is "no probability without causality," that is, probability is not well defined outside the dynamical framework of classical ergodic theory.

The idea of applying ergodic theory to the macroscopic domain, of, for example, mechanically described games of chance, was continued by Eberhard Hopf, especially in his paper "On Causality, Statistics and Probability."[96] This is one of the most important papers on physical probability.

Bernard Koopman wrote (February 2, 1932) to G. D. Birkhoff about his work on the spectrum of classical and quantal operators through which he tried to relate the corresponding two theories to each other. In his opinion, "The insignificance of the individual stream-line (trajectory), and the statistical interpretations, suggest that there is a 'principle of indetermination' in classical dynamics as well as in quantum theory."

As a résumé of the different views on the role of ergodic theory within the development of physics, the following four lines of thought can be discerned. There were scholars such as Hostinsky and Struik who did not think well of the new quantum theory. A second line was to try to let both theories live in peaceful coexistence and to push the analogies as far as possible.[97] Yet another approach was to apply the theory to macroscopic phenomena, as in the work of Hopf. A fourth way out was Khintchine's. He had transformed the ergodic theorem into a result of probability theory, and was not therefore committed to classical dynamics. In the paper in honor of Richard von Mises, Khintchine wrote as follows:

Science had waited many years for this [Birkhoff's] fundamental success. However, it seemed to several experts to have come too late into being, and, in fact, physical statistics has meanwhile taken completely new ways. The continuous extension of state variables and the deterministic assumptions of the classical theory have gone more and more into the background. In addition to their historical importance, these two features of the classical theory seem to be applicable only as approximations in certain specific cases.

A mathematician, nevertheless, is convinced that a beautiful creation of his science—a success for which he has wrestled for years—never comes too late into being. I intend to show here that the result found by Birkhoff reaches far beyond the form given by him. It proves to be meaningful also for the present position of physical statistics. Not before long, I was able to prove that the analytic conditions for Birkhoff's theorem are totally inessential for its justification. Here it will be shown that this theorem can be given a purely probabilistic formulation in which neither determinism nor the continuous character of states is mentioned.

Notes

1. See I. Schneider, "Rudolph Clausius' erster Beitrag zur Einführung wahrscheinlichkeits-theoretischer Methoden in der Physik nach 1856," *Archive for History of Exact Sciences*, 4 (1974/75), 237–261, and S. G. Brush, *The Kind of Motion We Call Heat*, 2 Vols. (Amsterdam: North-Holland, 1976), for analyses of these early contributions.

2. The historical development of early statistical physics is treated extensively in Brush, *Motion* (note 1). I shall discuss this history only fragmentarily, as a general background for the later developments I wish to treat in detail.

3. L. Boltzmann, *Wissenschaftliche Abhandlungen*, F. Hasenöhrl, ed., 3 Vols. (Leipzig: 1909), vol. II, p. 582. The article is a short communication of Maxwell's work, in the *Wiener Anzeiger*, which explains the third person used.

4. This is not to say that Maxwell could not at times have considered single systems, not to speak of Boltzmann; Gibbs in fact credits him for the very idea of representative ensembles, in the preface of his book *Elementary Principles in Statistical Mechanics* (New York: Dover, 1960), originally published in 1902. von Neumann, writing in 1929, presents Boltzmann as a kinetic theorist, and Gibbs as an advocate of what we here call the Boltzmannian approach to statistical mechanics. As will be seen below, this interpretation of Gibbs's position is not at all singular. von Neumann's relevant work is "Beweis des Ergodensatzes und des H-theorems in der neuen Mechanik," *Zeitschrift für Physik*, 57 (1929), 30–70, especially pp. 30–31. It is reprinted in von Neumann's *Collected Works*, 6 Vols., ed. A. H. Taub (London and New York: Oxford, 1961–1963), in Vol. 1.

5. See R. C. Tolman, *The Principles of Statistical Mechanics* (London: Oxford University Press, 1938) for a representative of this aprioristic interpretation of the ensemble approach.

6. Just to indicate what some of these problems are, let me refer to the strong law of large numbers of Emile Borel of the year 1909. The law states that under certain assumptions, probabilities coincide with limits of relative frequencies, except for a set of measure zero. Why a certain set of nonconverging sequences receives a measure zero is a question left untouched by these physicists, who simply identified probability as a limit of a time average.

7. Gibbs was rather careful about this. See p. 74 of Gibbs, *Principles* (note 4).

8. The lecture is printed in L. Campbell and W. Garnett, *The Life of James Clerk Maxwell* (London: Macmillan, 1882, reprint New York and London: 1969).

9. Space does not admit a discussion of these matters, for which see J. von Plato, "The Method of Arbitrary Functions," *The British Journal for the Philosophy of Science*, 34 (1983), 37–47, and "Instability and the Accumulation of Small Effects," in M. Heidelberger and L. Krüger (eds.), *Probability and Conceptual Change in Scientific Thought*, Report Wissenschaftsforschung Nr. 22 (Bielefeld: Universität Bielefeld, 1982).

10. See A. Pais, *"Subtle Is the Lord . . .": The Science and The Life of Albert Einstein* (Oxford: Oxford University Press, 1982), for a discussion of Einstein's work on quantum theory and its background in statistical mechanics.

11. M. von Smoluchowski, "Über den Begriff des Zufalls und den Ursprung der Wahrscheinlichkeitsgesetze in der Physik," *Die Naturwissenschaften*, 6 (1918), 253–263. What he exactly does is the following. He gives the probability law of radioactive decay, and a mechanical model that reproduces that law. He says that he naturally does not believe that radium atoms really would be of that kind, but wants to show the possibility of classical models of chance phenomena in physics.

12. K. Popper, *The Logic of Scientific Discovery* (London: Hutchinson, 1959; German original 1934). Abbreviated below as *LScD*.

13. The letter was written in 1935. For full text see *LScD*, pp. 459–460.

14. See the footnote added to the English edition, *LScD*, p. 208.

15. See p. 66 of his "The Propensity Interpretation of the Calculus of Probability, and the Quantum Theory," in S. Körner (ed.), *Observation and Interpretation: A Symposium of Physicists and Philosophers* (London: Butterworth, 1957), pp. 65–70.

16. See Max Born, "Einstein's Statistical Theories," in P. A. Schilpp (ed.), *Albert Einstein: Philosopher—Scientist*, (La Salle, IL: Open Court, 1949), p. 175.

17. This is not the case for his works on Brownian motion. However, one can see that Einstein there uses the time average concept of probability, on pp. 8, 23, and 25 of the English translation, *Investigations on the Theory of Brownian Movement* (New York: Dover, 1956). Later, (p. 25), an ensemble of systems is introduced, "for the sake of simpler mode of expression and presentation." Einstein's other papers on statistical physics are more explicit on the concept of probability, notably a work of 1910 on opalescence, with some seven pages relating to this concept. It is not mentioned in standard references such as M. Klein, "Thermodynamics in Einstein's Thought," *Science*, 157 (1967), 509–516, or J. Mehra, "Einstein and the Foundations of Statistical Mechanics," *Physica*, A79 (1979), 447–477. The best account of Einstein's statistical physics is in the recent book by Abraham Pais, *Einstein* (note 10).

18. The exact relation of Einstein to Boltzmann's several works is not known. In his early papers, Einstein referred only to Boltzmann's *Lectures on Gas Theory*. Time averages do not appear explicitly there. However, in the paragraph on '*Ergoden*', Boltzmann refers to the time development of single systems. Einstein had studied Poincaré's book *La Science et l'Hypothese* (Paris: 1902), where the latter discusses probability (Chapter XI). See Pais, *Einstein* (note 10), p. 133.

19. This is the assumption of absolute continuity, discussed below.

20. A. Einstein, "Kinetische Theorie des Wärmegleichgewichtes und des zweiten Hauptsatzes der Thermodynamik," *Annalen der Physik*, 9 (1902), 417–433. In a short remark in *Annalen der Physik*, 34 (1911), 175–176, Einstein says that had he known Gibbs's work, he would have limited his own papers to some remarks. On this basis, it has been said that Einstein later adopted a Gibbsian standpoint. But even if we do not know what his remarks would have been, we shall see that he did not change his views on statistical probability in favor of postulated a priori ensembles. Even Gibbs himself, in his *Elementary Principles in Statistical Mechanics* (note 4), p. 17, writes that one may take probabilities as time averages over single systems if the ensemble view appears too unphysical.

21. A. Einstein, "Eine Theorie der Grundlagen der Thermodynamik," *Annalen der Physik*, 11 (1903), 170–187. Some of the technical notions will be given below.

22. This is stated explicitly by him in 1902: Einstein, "Thermodynamik" (note 20), p. 419. The existence of a constant of motion independent of total energy E "has the necessary consequence that the distribution of states is not determined by E only, but must necessarily depend on the initial states of the systems." The plural 'systems' here refers to his use of the microcanonical ensemble. If total energy is the only constant of motion, "the distribution of states of our systems ... is produced by itself (*von selbst hergestellt*) from any initial values of state variables which satisfy the condition for the value of energy": Einstein, "Thermodynamik" (note 20), pp. 418–419. I think this is a rather clear formulation of the view that the distribution of states along individual trajectories (i.e., the time averages) gives the (microcanonical) distribution. A further discussion of the effect of the existence of constants of motion for the statistical laws of a system can be found in von Plato, "The Significance of the Ergodic Decomposition of Stationary Measures for the Interpretation of Probability," *Synthèse*, 53 (1982), 419–432.

23. A. Einstein, "Über einen die Erzeugung und Verwandlung des Lichtes betreffenden heuristischen Gesichtspunkt," *Annalen der Physik*, 17 (1905), 132–148.

24. Further confirmation for this interpretation of the passage can be found by comparing it with a passage from Poincaré, *Science* (note 18), pp. 218–219. Poincaré discusses a game of chance that, when observed for a long time, leads us to judge "that the events are distributed in conformity with the calculus of probability, and here we have what I call *objective probability*." The chapter on probability in Poincaré's book was one that Einstein had read, as was noted above.

25. A. Einstein, "Über die Gültigkeitsgrenze des Satzes vom thermodynamischen Gleichgewicht und über die Möglichkeit einer neuen Bestimmung der Elementarquanta," *Annalen der Physik*, 22 (1907), 569–572.

26. A. Einstein, "Theorie der Opaleszenz von homogenen Flüssigkeiten und Flüssigkeitsge-mischen in der Nähe des kritischen Zustandes," *Annalen der Physik*, 33 (1910), 1275–1297. The statement that Einstein never worked on the foundations of statistical mechanics after 1904 does not hold.

27. A Einstein, "Opeleszenz" (note 26), p. 1296. The step from 'incomplete definition' to a subjectivist notion of 'incomplete information' is short. Einstein's emphasis on a statistical notion of probability is a healthy contrast in this respect.

28. Here lies the conceptual advantage of ergodic theory as compared with the ensemble theory. At some stage, the latter has to *postulate* randomness, e.g., repeated randomization through a 'random choice' of initial conditions.

29. A. Einstein, "Zum gegenwärtigen Stand des Strahlungsproblems," *Physikalische Zeit-schrift*, 10 (1909), 185–193, especially p. 187, section 4.

30. Einstein, "Stand" (note 29), pp. 187–188.

31. Einstein, "Stand" (note 29), p. 188, section 6.

32. A. Einstein, "Zur allgemeinen molekularen Theorie der Wärme," *Annalen der Physik*, 14 (1904), 354–362.

33. A. Einstein, *Investigations on the Theory of the Brownian Movement*, (New York: Dover, 1956), p. 3. The German original is Einstein's famous first paper on Brownian motion, of the year 1905.

34. Gibbs, *Principles* (note 4), pp. 74–75, was rather cautious about this, treating several cases where one could *not* observe fluctuations without making a definitive claim that all cases are of this kind.

35. A. Einstein, *Brownian Movement* (note 33), p. 2.

36. See G. L. de Haas-Lorentz, *Die Brownsche Bewegung und einige verwandte Erscheinungen* (Braunschweig: Vieweg, 1913), and R. Fürth, *Schwankungserscheinungen in der Physik* (Braun-schweig: Vieweg, 1920), for early reviews of these studies.

37. Compare note 17.

38. A. Einstein, *Brownian Movement* (note 33), pp. 12, 15.

39. A. Einstein, *Brownian Movement* (note 33), pp. 13, 14.

40. A. Einstein, *Brownian Movement* (note 33), p. 76.

41. P. A. Schilpp (ed.), *Albert Einstein: Philosopher—Scientist* (La Salle, IL: Open Court, 1949), pp. 668, 671.

42. P. A. Schilpp (ed.), *Albert Einstein* (note 41), p. 672.

43. P. A. Schilpp (ed.), *Albert Einstein* (note 41), p. 673.

44. See von Plato, "The Method" (note 9), for reviews of this topic.

45. After completing this section, I received from John Stachel, Editor of the papers of Albert Einstein, copies of two works directly bearing on the interpretation of probability. The first, Einstein's discussion remarks (with Poincaré, Lorentz, and others) in the famous *Conseil Solvay* of the year 1911, is printed in German in *Abhandlungen der Deutschen Bunsen-Gesellschaft*, 7 (1914). Einstein insists on the necessity of a physical, time average concept of probability that renders Boltzmann's equation (principle) directly into a physical statement (pp. 356–357). Lorentz makes the remark that Einstein is not following the Gibbsian method, to which the latter answers, "It is characteristic of this approach that one uses the (time) probability of a *purely phenomenologically* defined state. This has the advantage that one needs no specific elementary theory (e.g., statistical mechanics) as a basis" (p. 357). The second work is a set of unpublished lecture notes of Einstein's course on statistical mechanics in the summer of 1913, taken by Walter Dällenbach. Einstein considers there a point moving

along a closed curve. The probability of finding the rotating point on a given part of the curve is defined as the limit of the time average the point visits that part. Next Einstein considers another approach via ensembles: An infinite number of points is sent to rotate the curve—they form a stationary flow, which Einstein calls a "Systemgesamtheit." Probability is defined in the second approach as the relative number of points on the part of the curve under consideration. (Obviously 'relative number' should be read 'density'.) Einstein shows that time and ensemble averages coincide in this case. Next he considers the combination of two uniform rotations. He notes, in more modern terminology, that irrational rotations lead to ergodic trajectories. One may conclude that Einstein's remark to Popper was not a casual one.

46. This concept of 'statistical property' is not entirely unproblematical. See P. Martin-Löf, "On the Notion of Randomness," in Kino, Vesley, and Myhill (eds.), *Intuitionism and Proof Theory* (Amsterdam: North-Holland, 1970).

47. See Brush, *Motion* (note 1), Chapter 10, for discussion and references.

48. P. and T. Ehrenfest, *Begriffliche Grundlagen der statistischen Auffassung in der Mechanik* (Leipzig: Teubner, 1911).

49. P. and T. Ehrenfest, *Grundlagen* (note 48); see their note 89a, pp. 31–32.

50. P. and T. Ehrenfest, *Grundlagen* (note 48), p. 32.

51. P. and T. Ehrenfest, *Grundlagen* (note 48); see their note 90, p. 32. Oddly enough, the definition of quasi-ergodicity—made only for a single trajectory—was not reconsidered at this stage. It went without saying that all of the continuous number of trajectories had to be quasi-ergodic. This goes on on the same page where they indicate that a point moving on a torus can trace a dense, aperiodic trajectory, or a periodic one! As late as 1929, von Neumann flamboyantly writes that under ergodicity, "*The* system point in phase space comes ... arbitrarily close to all points of its energy surface" (italics added), J. von Neumann, "Beweis" (note 4), pp. 35–36.

52. P. and T. Ehrenfest, *Grundlagen* (note 48), p. 25.

53. P. and T. Ehrenfest, *Grundlagen* (note 48), p. 22.

54. This definition first appears in 1928 in G. D. Birkhoff and P. Smith, "Structure Analysis of Surface Transformations," *Journal de Mathématiques*, 7 (1928), 343–379. It is due to Smith.

55. Space does not permit a further discussion, for which see J. von Plato, "The Significance of the Ergodic Decomposition of Stationary Measures for the Interpretation of Probability," *Synthèse*, 53 (1982), 419–432.

56. *Mathematische Annalen*, 57 (1916), 313–352.

57. *L'Enseignement Mathématique*, 16 (1914), 454–467.

58. J. von Plato, "Nicole Oresme and the Ergodicity of Rotations," *Acta Philosophica Fennica*, 32 (1981), 190–197.

59. D. König and A. Szücs, "Mouvement d'un point abandonné a l'interieur d'un cube," *Rendiconti del Circolo Matematico di Palermo*, 36 (1913), 79–90.

60. E. Borel, "Les probabilités dénombrables et leurs applications arithmétiques," *Rendiconti del Circolo Matematico di Palermo*, 27 (1909), 247–271. Borel's work is discussed in detail in J. Barone and A. Novikoff, "A History of the Axiomatic Formulation of Probability from Borel to Kolmogorov: Part I," *Archive for History of Exact Science*, 18 (1978), 123–190.

61. See, for example, the preface of R. von Mises, *Wahrscheinlichkeitsrechnung* (Leipzig and Vienna: Franz Deuticke, 1931).

62. P. Martin-Löf, "The Definition of Random Sequences," *Information and Control*, 9 (1966), 602–619.

63. R. von Mises, "Über die gegenwärtige Krise der Mechanik," *Zeitschrift für angewandte Mathematik und Mechanik*, 1 (1921), 425–431, especially section 3, pp. 429–431.

64. R. von Mises, *Wahrscheinlichkeitsrechnung* (note 61); see his note 56, p. 561.

65. This intuition was one of the principal starting points of the work of N. S. Krylov, now published in English by Princeton University Press, *Works on the Foundations of Statistical Physics* (Princeton: 1979).

66. R. von Mises, "Über kausale und statistische Gesetzmässigkeit in der Physik," *Die Naturwissenschaften*, 18 (1930), 145–153.

67. Von Mises, "Physik" (note 66), p. 146.

68. Von Mises, "Physik" (note 66), p. 147.

69. Von Mises, "Physik" (note 66), p. 148.

70. R. von Mises, "Ausschaltung der Ergodenhypothese aus der physikalischen Statistik," *Physikalische Zeitschrift*, 21 (1920), 225–232, 256–262.

71. Von Mises, "Ausschaltung" (note 70), p. 226.

72. Von Mises, "Ausschaltung" (note 70), p. 227.

73. Von Mises, "Ausschaltung" (note 70), p. 227.

74. B. Koopman, "Hamiltonian Systems and Transformations in Hilbert Space," *Proceedings of the National Academy of Sciences*, 17 (1931), 315–318.

75. J. von Neumann, "A Proof of the Quasi-Ergodic Hypothesis," *Proceedings of the National Academy of Sciences*, 18 (1932), 70–82, quoted from p. 71. Only one year before the suggestion, the ergodic problem had been "absolutely unsurmountable by the present standard of science"—von Neumann, "Beweis" (note 4), pp. 30–31.

76. A constant of motion is a function f of state such that its value remains the same in the motion: $f(s) = f(T_t s)$ for all t. There can be only one constant constant of motion; if there are others, their values differ from one trajectory to another.

77. J. von Neumann, "Zur Operatorenmethode in der Klassischen Mechanik," *Annals of Mathematics*, 33 (1932), 587–642, with additions on pp. 789–791.

78. G. D. Birkhoff, "Proof of the Ergodic Theorem," *Proceedings of the National Academy of Sciences*, 17 (1931), 656–660.

79. Birkhoff and Koopman, *Proceedings of the National Academy of Sciences*, 18 (1932), 279–282.

80. J. von Neumann, *Proceedings of the National Academy of Sciences*, 18 (1932), 263–266.

81. B. O. Koopman, "Birkhoff on Dynamical Systems," *Bulletin of the American Mathematical Soceity*, 26 (1930), 162–166, especially p. 165.

82. G. D. Birkhoff, "Probability and Physical Systems," *Bulletin* (note 81), 28 (1932), 361–379.

83. G. D. Birkhoff, "Probability" (note 82), p. 366.

84. A. Khintchine, "Über einen Satz der Wahrscheinlichkeitsrechnung," *Fundamenta Mathematicae*, 6 (1924), 9–20.

85. A. Khintchine, "Zu Birkhoffs Lösung des Ergodenproblems," *Mathematische Annalen*, 107 (1932), 485–488, quotation from p. 485.

86. The second part is not deep, since it only reduces the study of arbitrary integrable functions to that of indicators, a usual thing to do in mathematics, but it has an interesting consequence, namely: If we think of the discrete time case, we notice that limits of relative frequencies are obtained, by simple arithmetic and a limit, from indicators of regions of phase space. A statistical property of a discrete sequence of numbers is of course always a function defined over the (elements of the) sequence, like relative frequency, number of runs of length two,

and so on. It is just a discrete form of a phase function (function defined over phase space). Now, the reduction to indicators means, in terms of frequencies, that the study of any statistical properties of a sequence is reducible to the study of frequencies.

87. E. Hopf, "On the Time Average Theorem in Dynamics," *Proceedings of the National Academy of Sciences*, 18 (1932), 93–100. I have previously discussed Hopf's views on probability rather extensively, in my two papers referred to in note 9. The views of Khintchine, one of the founders of modern probability theory, have received almost no attention so far by philosophers or others.

88. Khintchine, "Zur mathematischen Begründung des statistischen Mechanik," *Zeitschrift für angewandte Mathematik und Mechanik*" 13 (1933), 101–103.

89. As was mentioned above, Khintchine's philosophical views on probability have received almost no attention so far. In particular, a paper in Russian on "Von Mises' Doctrine of Probability and the Principles of Physical Statistics" of 1929 is relevant for the present purposes. Unfortunately, a study of this and some other works of Khintchine has to be postponed for reasons of inaccessibility.

90. The work of von Neumann ("Beweis des Ergodensatzes," note 4) is one example.

91. J. von Neumann, *Collected Works*, Vol. 1 (note 4), p. 595; see his note 72.

92. B. Hostinsky, "Sur une nouvelle solution du problème d'aiguille," *Bulletin des sciences mathématiques*, 44 (1917), 126–136.

93. B. Hostinsky, "Sur la méthode des fonctions arbitraires dans le calcul des probabilités," *Acta Mathematica*, 49 (1924), 95–113. The basic ideas go back to Poincaré's *Calcul des probabilités*, 1896, 2nd ed. (Paris: Gauthier-Villars, 1912), especially p. 148. The conception of typical games of chance as unstable mechanical systems was suggested more philosophically also by Johannes von Kries in 1886. These early works are discussed in the two works of ours referred to in note 9.

94. *Erkenntnis*, Vol. 1 (1930), p. 285.

95. D. Struik, "On the Foundations of the Theory of Probabilities," *Philosophy of Science*, 1 (1934), 50–70.

96. E. Hopf, "On Causality, Statistics and Probability," *Journal of Mathematics and Physics*, 1 (1934), 51–102.

97. The analogy, as far as physical statistics is concerned, has not led to success. The ergodic theorems do not allow of any obvious application to quantum systems, the reason being the feature that von Neumann had mentioned, viz., that of a pure point spectrum. The inapplicability became well understood only in the 1950s.

15 Max Born and the Reality of Quantum Probabilities

Nancy Cartwright

Max Born's introduction of the probabilistic interpretation of quantum mechanics was closely tied to his belief in particles: After collision, a rebounding electron will be in a state as near to a particle state as quantum mechanics allows. But its mathematical representation will be a spherical wave. What the spherical wave must represent, then, is the probability *for the electron to rebound in various particlelike states.*

It is commonly claimed that Born gave these probabilities a merely epistemic interpretation. But an epistemic interpretation cannot account for particle diffraction, and Born believed in particle diffraction as firmly as in particles. He could consistently do so because he assumed not just that the particlelike states exist, but that the spherical-wave probabilities—which account for diffraction—exist as well. Far from giving them an epistemic interpretation, for Born probabilities were real enough to exist and evolve on their own: In quantum mechanics, he explains, "We free forces of their classical duty of determining directly the motion of particles and allow them instead to determine the probability of states."

This chapter is an exercise in consistency: an example of compensating mistakes. Max Born is responsible for the probabilistic interpretation of quantum mechanics. How did he understand the probabilities he introduced? I used to have a rather simple, straightforward answer to this question. But I now think that answer depended first on one misunderstanding and then on a second to support the first. I am going here to describe these two mistakes, because my original interpretation of Born is a common one. Yet it cannot be the right one.

Here are the two views of Born's about which I was originally mistaken: (a) I have for a while now been urging that measurement has little to do with reduction of the wave packet.[1] Rather in various situations the wave packet reduces spontaneously, of its own accord, without the need for any external observer. Decay situations and scattering are paradigms. When I first began to read Born it seemed to me that he had exactly that view. I began to look further for confirmation: Does Born really believe that reductions can occur on their own, or did he hold what is now the conventional view that we have to wait for a final measurement? In fact it was hard to find an answer; Born never discusses the measurement problem. I now think I see the reason for this; it is true, as I had at first assumed, that Born does not believe in reduction of the wave packet on measurement. But that is because he does not believe in reduction of the wave packet at all, and that because he does not believe in wave packets. So: *Born rejects reduction of the wave packet.* I shall come back and explain this after I have described the second mistake in my original interpretation of Born. (b) The second mistake became clear during a dispute (October 1982) with Norton Wise, after we had read together in the ZiF group a number of Born's papers. "What criterion of reality did Born suppose in the early years?" asked Wise. I replied that he had no philosophical criterion then and that he needed none. He believed in particles, but that involved no sophisticated philosophical view; as Born frequently remarks,[2] his colleague James Franck was observing them in the laboratory next door every day. Obviously Born assumed that if he saw them, they existed. But I do not think he elevated—or needed to elevate—this truism into a philosophical *theory* of existence.

Wise claimed that Born's ontology was not so simple. Agreed, particles are real; but for Born *probabilities are real as well.* "How," asked Wise, "do these two kinds of reality relate? Are probabilities real for Born in exactly the same sense as particles? If not, in what sense?" The clearest statement in defense of Wise's view comes from Born's British Association lecture of August 1926:

This fact justifies, even demands, the existence of particles, although this cannot, in some cases as we have said, be taken too literally. There are electromagnetic forces between these particles ...; they are, so far as we know, given by classical electrodynamics in terms of the positions of the particles (for example, a Coulomb attraction). But these forces do not, as they did classically, cause accelerations of the particles; they have no direct bearing on the motion of the particles. As intermediary there is the wave field: the forces determine the vibrations of a certain function ϕ that depends on the positions of all the particles (a function in configuration space), and determine them because the coefficients of the differential equation for ϕ involve the forces themselves.[3]

Even more srikingly, a few paragraphs later he says, "We free forces of their classical duty of determining directly the motion of particles and allow them instead to determine the probability of states."[4] "What kinds of things are probabilities," asked Wise, "that they can be acted on by forces?"

I replied that this talk of forces 'determining probabilities' is just a manner of speaking. Born believed that there are particles, and that these particles make transitions, and that these transitions, though indeterministic, are nevertheless describable by laws: probabilistic laws. Born's ontology was no more complicated than that of classical physics. He believed in particles and he believed in laws, only for Born these laws were probabilistic rather than universal. But, what I had not realized and what I am going to try to show is that this defense depended on the assumption that Born's transitions were reductions of the wave packet. But Born does not believe in reduction of the wave packet, and this fact forces us to take seriously Wise's proposal that probabilities must have a certain very peculiar kind of reality. For Born, *probabilities are real in a special way.*

First, I consider Born on superpositions. Max Born is responsible for our current probabilistic interpretation of the quantum state function, for which he won the Nobel Prize almost 30 years after he introduced it. The interpretation was introduced in two papers in 1926, one a short note[5] in which he describes the results of his researches, the second a longer development.[6] Both papers are on the quantum mechanics of collision processes. There are three important characteristics of these papers of interest to us: (1) Born uses the Schroedinger formulation of quantum mechanics, but he gives it a Heisenberg matrix interpretation. Schroedinger had only recently 'proved' the equivalence of wave mechanics and matrix mechanics, but Born was not able to figure out a direct matrix treatment for collision and scattering problems. He says, "Of the different forms of the theory only that of Schroedinger turned out to be suitable for this task, and for this reason I regard it as the deepest version of quantum theory." [7] Heisenberg and Pauli were apparently annoyed with Born for this point of view, for they were obsessed by the paradoxes surrounding wave-particle duality and felt that Schroedinger's formulation obscured the force of these paradoxes. (2) Born treats the colliding particles *as particles*, and in fact his assumption that they are particles is central to the argument in which is developed Born's well-known probabilistic interpretation. He tells us, "The idea forces itself upon us that just as before the collision, so likewise after it, if the electron is sufficiently far away and the coupling is small, a definite state of the atom and a definite, uniform rectilinear motion of the electron must be definable. It is therefore a matter of representing this asymptotic behavior of the coupled particles mathematically." [8]

Now comes the calculation: "The dispersed wave produced by the disturbance asymptotically approaches [a spherical wave]." [9] Mathematically, the spherical wave is an integral, or superposition, over plane waves in all directions, with an amplitude designated by $\phi(\alpha, \beta, \gamma)$, for the direction with cosines α, β, γ. Plane wave, Born has already told us, represent 'a definite uniform rectilinear motion of the electron'. So *mathematically* we have a superposition of rectilinear states; but *physically* we have corpuscles with definite rectilinear motions. This conjunction is very suggestive. [9] Says Born, "If one wishes to understand this result in corpuscular terms, then only one interpretation is possible: $\phi(\alpha, \beta, \gamma)$ [or, says the note added in proof, $|\phi|^2$] defines the probability that the electron coming from [a particular direction] will be projected into [the new] direction α, β, γ." [10] So the probabilistic interpretation comes from the assumption that the electron is traveling in a definite direction both before and after the collision, despite the fact that its Schroedinger representation is a superposition of all possible directions. (3) The treatment that Born gives in this paper and in his subsequent papers (also in his earlier papers) is a *perturbation theory account*, and in fact this is crucial to his view.

Returning to (1), there is no doubt that when Born introduces probabilities he intends them to be *transition* probabilities: probabilities for the quantum system to make a transition from one state to another. An alternative would be to let the probabilities represent distributions of values of dynamical quantities, as in classical statistical mechanics. But clearly this is not what Born intended. He repeatedly stresses the importance of Heisenberg's original idea—following discussions in his own group in Göttingen—that in the quantum theory observable effects depend on transitions between *two* stationary states, and the mathematical representation uses *two* indices in which both states play symmetrical roles. When Born describes his own interpretative work in his Nobel Prize lecture in 1954, he draws a rectangular array (figure 1) with energy terms of the stationary states written twice over, horizontally and vertically. What is in the squares? He says, "Heisenberg banished the picture of electron orbits with definite radii and periods of rotation, because these quantities are not observable; he demanded that the theory should be built up by means of quadratic arrays of the kind [shown here]. Instead of describing

	E_1	E_2	E_3	$\cdots$
E_1	11	12	13	
E_2	21	22	23	
$\cdot$				
$\cdot$				
$\cdot$				

Figure 1

the motion by giving a coordinate as a function of time $x(t)$, one ought to determine an array of transition probabilities χ_{mn}."[11]

So Born's probabilities are undoubtedly transition probabilities. But what states are the transitions between? I originally modeled his view on our own contemporary picture for reduction of the wave packet. The formal treatment produces a super-position to represent the quantum system. At the moment of measurement, the system makes a transition from this superposition to one of the eigenstates of the measured observable, which make up this superposition. I assumed Born's transi-tions were like that, only one did not have to wait for the ultimate moment of measurement. Somewhere a little way from the scattering center the outgoing spherical wave reduced, discontinuously and indeterministically, into a plane wave traveling in a given direction. But this is not in fact Born's view.

For Born, transitions are never from superpositions to eigenfunctions. Instead, they are always from *one* eigenfunction to another. For it is the eigenfunctions, and never their superpositions, that represent the *physical* states that a system can take on. Born says this in his British lecture:

Every state of the system corresponds to a particular characteristic solution, an *Eigenfunktion*, of the differential equation: for example, the normal state the function ϕ_1, the next ϕ_2, etc. For simplicity we assume that the system was originally in the normal state; after the occurrence of an elementary process the solution has been transformed into one of the form

$$\phi = c_1\phi_1 + c_2\phi_2 + c_3\phi_3 + \cdots,$$

which represents a superposition of a number of *eigenfunctions* with definite amplitudes $c_1, c_2, c_3, \ldots$, giving the probability that after the jump the system is in the 1, 2, 3, ... state. Thus c_1^2 is the probability that in spite of the perturbation the system remains in the normal state, c_2^2 the probability that it has jumped to the second, and so on.[12]

Born describes the superposition merely as a *solution*; it is not a representation of a real physical state. One more passage makes this perfectly clear, a passage where he offers an argument for his view:

Schroedinger claims ... that in the individual atoms there can be simultaneously "a number of eigenoscillations excited." ... [But] the sentence "an atom can be in more than one stationary state" is not only against the entire sense of Bohr's theory, but also speaks directly against the natural, still undoubted significance of the particular stationary states, which belong to points of the line spectrum. Consider in particular some ionization process, that is, a transition from a point of the discrete spectrum into a point of the continuous (term) spectrum, so that the proper trajectory in the end is given in its asymptotic course directly as the electron flying away in a straight line, the trace of which has been made certain through the Wilson cloud chamber method. It also will not do to speak of the simultaneous existence of more states if one does not want to give up the natural significance of Wilson cloud chamber streaks and related occurrences as the passing through of corpuscles.[13]

We see here that Born's view about superpositions is intimately connected with his knowledge that there are particles, and that we observe their trajectories. What sense could it make to take as physically real a superposition in which the electron is both contained in an orbit around the nucleus and is also "flying away in a straight line"? As Born says, it won't do to speak of the simultaneous existence of a number of states if one does not want to give up the natural significance of Wilson cloud chamber streaks.

But this point of view is not trouble free. We can see the problem briefly by looking at Max Jammer's discussion in *The Conceptual Development of Quantum Mechanics*:

Summarizing Born's *original* probabilistic interpretation of the ψ function we may say that $|\psi|^2 \, dT$ measures the probability density of finding the particle within the elementary volume dT, the particle being conceived in the classical sense as a point mass possessing at each instant both a definite position and a definite momentum.... In spite of all these successes Born's original probabilistic interpretation proved a dismal failure if applied to the explanation of diffraction phenomena such as the diffraction of electrons. In the double-slit experiment, for example, Born's original interpretation implied that the blackening on the recording screen behind the double slit, with both slits open, should be the superposition of the two individual blackenings obtained with only one slit opened in turn.... It becomes clear on mathematical analysis that the ψ-wave associated with each particle interferes with itself and the *mathematical* interference is manifested by the *physical* distribution of particles on the screen.

The ψ-function must therefore be something physically real and not merely a representation of our knowledge.[14]

This is rather a caricature of Born's views. Jammer is probably more interested here in using Born to make a philosophical point than he is in describing Born's interpretation exactly. First, as I remarked, Born's probabilities are probabilities for *transitions to eigenstates*, and not, as Jammer proposes, for the *distribution of values*. Which eigenstates? That is determined by how the Hamiltonian is split into stationary and perturbing potentials, a split that is characteristic of perturbation analysis.

Second, Born is very keen on Heisenberg's uncertainty relations. It is difficult to be certain exactly what his view was, but he is very careful never to talk about particles 'in the classical sense as point masses possessing at each instant both a definite position and a definite momentum'. In this connection it is important to notice that the eigenfunction to which Born's transitions are made are themselves quantum states for which Heisenberg's uncertainty relations hold.

Third, it would be surprising if Born had offered an account of the Schroedinger function that was trivially inconsistent with the double-slit experiment, since he was convinced very early that particles could be diffracted, and he is very proud of the fact that he, in conjunction with his colleague Franck, set their student Elsasser to perform the 'first quantitative experiments' in support of de Broglie's hypothesis about the wave behavior of electrons.[15]

Still, there is a problem to which Jammer points. Let me put it in a simplistic modern formulation: Consider an ensemble of Born atoms, coupled loosely to the electromagnetic field. Each is in some particular stationary state or other. The appropriate representation for the collection then would be a mixture of these stationary states, as in classical statistical mechanics. But the theory provides a superposition, and there is no doubt that Born agrees that the superposition is required to predict the correct probabilistic behavior in the future.

His view seems to have a lot in common with current modal views like Bas van Fraassen's,[16] or Simon Kochen's,[17] or even the Everett-Wheeler splitting universe interpretation:[18] There is state that is the actual physical state of the system (recall that this is not a classical state with well-defined values, but is itself a quantum state obeying the uncertainty relations); and there is a different state that describes the probabilistic behavior of the system. Jammer says, "The ψ-function must therefore be something physically real." But for Born himself, since he explicitly denies that the superpositions are real physical states, it must be the probabilities themselves that are real.

We can express the view this way: For Born particles not only have their physical quantum state, but they also have physically real propensities over and above this state. There are two important aspects of the quantum probabilities, as Born uses them, that justify calling them propensities and ascribing a real existence to them. (1) The first is the one just noted. Two systems in the same eigenstates may nevertheless be correctly represented by different Schroedinger superpositions. That means that systems can be in identical states and subject to identical forces

and yet have different probabilistic behavior. Hence the propensity is a real characteristic independent of whatever characteristic(s) the quantum state is supposed to
represent. Notice that this is quite different from the more pedestrian story of
indeterminism and statistical law that I had hoped to tell. Normally, a statistical
evolutionary law will fix the probability for an outcome, given the current state,
even though it does not fix the outcome itself. But here not even the probabilities
are fixed by law, given the actual state. (2) The Schroedinger function evolves in
time; hence so do the probabilities. Born rejects reduction of the wave packet;
consistently, he claims that quantum probabilities evolve completely deterministically. Mathematically this is clear from the structure of the Schroedinger equation, where a description of the forces and of the ψ-function at one time fixes the
solution for the ψ-function at any other. This again bears out the propensity
interpretation: The forces act directly on the *probabilities* at one time to give
probabilities at other times; they do not evolve the probabilities by acting on the
actual state of the system.

A number of essays in these two ZiF volumes are concerned with the question
of whether probabilities in various sciences are to be given an ontological, or merely
an epistemic, interpretation. On Jammer's view of Born, his probabilities in one
sense would be epistemic—more detailed knowledge of the true momentum and
position of each particle would be possible. But in another sense they would be
ontological, or part of the natural order—as in classical mechanics, the appropriate
probability distributions for a given kind of situation are fixed by the laws of nature.

The view that Jammer attributes to Born is usually called "the statistical interpretation" of quantum probabilities, or sometimes "the ignorance interpretation,"
and it is widely rejected, for reasons like those that Jammer describes. It is commonly believed that a more robust "ontological" interpretation is required, though,
surprisingly, a good number of working quantum physicists remain vague about
exactly what this interpretation is to be. What I just called "the pedestrian story
of indeterminism" is one very standard approach; in classical statistical mechanics
the probabilities can be called 'real' because they are dictated by the boundry
conditions in a systematic and lawlike way. Nevertheless, the underlying laws of
evolution are presumed to be deterministic. On the 'pedestrian view' of quantum
mechanics, the most fundamental laws of evolution are themselves probabilistic,
or at least this is so for the laws that govern situations where reductions of the wave
packet occur.

But Born, we have seen, does not believe in reduction of the wave packet; nor
does he really believe that nature's laws are probabilistic. Quantum probabilities
for him are 'ontological' in a far more direct way: Probabilities, like particles, are
a part of nature itself; they are objects to be operated on and evolved by nature's
laws. So we have come after all to Norton Wise's position. Probabilities must have
a special reality of their own for Born, over and above the reality of the particles
and their states. Born said that quantum mechanics frees forces from their classical
duty of determining directly the motion of particles and allows them instead to
determine the probability of states; and I think we have good grounds to think that
that is what he meant.[19]

Notes

1. Nancy Cartwright, *How the Laws of Physics Lie* (Oxford: Oxford University Press, 1983). Chapter 9.

2. Cf. Max Born, *Mein Leben* (München: Nymphenburger Verlagshandlung, 1975), Chapter 19.

3. Max Born, "The Physical Aspects of Quantum Mechanics," in *Physics in My Generation.* (New York: Springer-Verlag, 1969), pp. 9–10. Originally in *Nature*, 119 (1926), 354–357.

4. Born, "Physical Aspects," p. 10.

5. Max Born, "Zur Quantenmechanik der Stossvorgänge," *Zeitschrift für Physik*, 37 (1926), 863–867.

6. Max Born, "Quantenmechanik der Stossvorgänge," *Zeitschrift für Physik*, 38 (1926), 803–827; translated in G. Ludwig, *Wave Mechanics* (Oxford: Pergamon Press, 1968).

7. Born, "Zur Quantenmechanik der Stossvorgänge," p. 864.

8. Born, "Zur Quantenmechanik der Stossvorgänge," p. 864.

9. Born, "Zur Quantenmechanik der Stossvorgänge," p. 865.

10. Born, "Zur Quantenmechanik der Stossvorgänge," p. 865.

11. Max Born, "Statistical Interpretation of Quantum Mechanics," in Max Born, *Physics in My Generation* (New York: Springer-Verlag, 1969), p. 91; originally in *Science*, 22 (1955), 675–679.

12. Born, "Physical Aspects," p. 10.

13. Max Born, "Das Adiabatenprinzip in der Quantenmechanik," *Zeitschrift für Physik*, 40 (1926), 167–192, on pp. 169–170.

14. Max Jammer, *The Conceptual Development of Quantum Mechanics* (New York: McGraw-Hill, 1966), p. 44.

15. Max Born, "Leben" Chapter 19 (note 2).

16. Bas van Fraassen, in E. Beltrametti, *Current Issues in Quantum Logic* (New York: Plenum Press, 1981).

17. Simon Kochen, Lecture presented to the Philosophy of Science Association, annual meeting, October 1978, San Francisco, California.

18. Cf. Neil Graham, *The Many Worlds Interpretation of Quantum Mechanics* (Princeton: Princeton University Press, 1973).

19. For a more detailed discussion of how Born introduced transition probabilities, cf. L. Wessels, "What Was Born's Statistical Interpretation," in Peter Asquith and Ronald Giere, *PSA 1980*, Vol. 2 (East Lansing, Michigan: Philosophy of Science Assn., 1981), pp. 187–200.

16 Philosophical Problems of Quantum Theory: The Response of American Physicists

Nancy Cartwright

In Europe the physicists who originated and developed the quantum theory were perplexed and troubled by it. Their worries centered on three problems: wave particle duality; the unvisualizability of the theory; and the fundamental statistical character of its laws. These were seen as grave problems; in Copenhagen in particular, they led to radical answers. In the United States, by contrast, few of the young physicists who took up the theory had any serious philosophical anxieties about it. This did not mean that they failed to understand the theory, nor that they were too philosophically uninterested or naive to worry about it. Rather, their attitude was entirely rational given the philosophy of science that most of them shared, a philosophy akin, but not identical, to the well-known American doctrines of pragmatism and operationalism.

This philosophy stressed two things: (1) hypothese must be verified by experiment and not accepted merely because of their explanatory power; and (2) the models that physics uses are inevitably incomplete and incompatible, even in studying different aspects of the same phenomenon. Their basic attitude was that the task of physics is not to explain but to describe—and we are very lucky if we can do that well.

1 Introduction

In continental Europe the originators of the new quantum theory and the early workers in the field were profoundly troubled by the theory they were creating. They were thoroughly committed to its correctness; yet they did not understand it. They were especially troubled by the failure of causality and the theory's inability to trace the history of systems in space and time. The best-known attempt to resolve the philosophical perplexities of the new theory was the Copenhagen interpretation, developed primarily by Niels Bohr, Werner Heisenberg, and Wolfgang Pauli. Nowadays philosophers distinguish two variants of the Copenhagen view: the weak Copenhagen interpretation and the strong. The two have in common the claim that quantum systems do not have familiar properties like position and momentum except when they are measured. Electrons, for example, are not in any place in particular. This is the view just about as Pauli held it: Measurement elicits properties that would not have been there otherwise. Bohr and Heisenberg supplemented this story by a metaphysical account of why it is true—the well-known doctrines of complementarity and the wholeness of action. The object and subject form an indissoluble whole, and this somehow has as a consequence that reality by its nature is fundamentally dual. Different incompatible descriptions will always be necessary; they cannot be consistently combined, nor is either eliminable.[1] These are drastic solutions to meet what were seen as drastic problems. The problems themselves have a special origin, as Heisenberg admitted at the time:

The experiments of physics and their results can be described in the language of daily life. Thus if the physicist did not demand a theory to explain his results

and could be content, say, with a description of the lines appearing on a
photographic plate, everything would be simple and there would be no need of
an epistemological discussion. Difficulties arise only in the attempts to classify
and synthesize the results, to establish the relation of cause and effect between
them—in short to construct a theory.[2]

In the United States the Copenhagen interpretation had little influence.[3] This is
not surprising—Americans suffered little philosophical anxiety about the theory.
This paper has two aims: the first, historical, to document the lack of philosophical
perplexity; the second, philosophical, to argue that it was entirely rational. The
lack of philosophical concern was accompanied by a widespread philosophy of
science that had no place for the problems that troubled the Copenhagen school.
This philosophy assigned theories a quite different role from that supposed by
Heisenberg: Very roughly, "the task of theory is to describe and not to explain."
Hence there was, quite reasonably, "no need of an epistemological discussion."

2   The Lack of Philosophical Anxiety

Werner Heisenberg visited the United States in 1929. He came directly from intense
discussions in Europe about how to interpret quantum theory. In America he found
no such discussion. He wrote, "While Europeans were generally averse and often
overtly hostile to the abstract, non-representational aspects of the new atomic
theory, the wave-corpuscle duality and the purely statistical character of material
laws, most American physicists seemed prepared to accept the novel approach
without too many reservations."[4]

Heisenberg's observation is borne out by the historical records. The Americans
who took up the theory and worked with it were all young, new PhDs, and there
is no evidence that they had any serious reservations. E. U. Condon remarked in
a 1929 review of a German work on developments in wave machanics, "I think it
worthwhile to quote here ... from the section on the probability concept and the
principle of causality because this topic occupies such a central place in discussion
today."[5] If the principle of causality was a primary topic of discussion among
physicists in the United States, their reflections never reached print. From the
invention of quantum mechanics in 1925 through the next six years, right up till
von Neumann sets very different kinds of philosophical problems, there is almost
no published American work evidencing serious philosophical concern. This includes the standard American journals in physics and related fields[6] as well as
articles by Americans in foreign journals. There is even little concern in the more
popular scientific magazines.[7] As late as 1929, for example, a *Scientific American*
article still found "The Strangest Things in Physics"[8] in the special theory of
relativity and not in quantum mechanics.

Nor do titles of papers at the American Physical Society meetings indicate
philosophical concern either.[9] In particular, at its December meeting in 1928 the
American Physical Society held a symposium on quantum mechanics to explain
the new theory to its members. The papers are reprinted in the *Journal of the*

Franklin Institute, 1929. A number of the best young Americans in the field gave talks, including J. C. Slater, E. U. Condon, J. van Vleck, E. H. Kennard, E. C. Kemble, Norbert Wiener, and H. P. Robertson. The papers gave straightforward descriptions of the new theory at a nontechnical level. Many of them stressed the statistical nature of quantum mechanics; yet there was none of the philosophical uneasiness and perplexity that Heisenberg had left in Europe. Van Vleck, for instance, at the end of his article listed five difficulties immediately facing the new theory, but none concern its deeper understanding or its radical implications about the nature of reality. Rather, they are (1) why are Cartesian coordinates privileged to give the correct Hamiltonians?; (2) the structure of the nucleus; (3) relativity corrections; (4) spontaneous radiation; and (5) the causes underlying the Pauli exclusion principle.[10]

This is also how the physicists of the time remember it. Robert Oppenheimer often remarks how unphilosophical he was.[11] He remembers that a lot of 'bad' philosophy was going on at the time (Oppenheimer spent much of the period in Europe), but that he himself was philosophically unconcerned because he had a good feel about how to work with the quantum theory.[12] Linus Pauling recalls that he had no worries about the theory.[13] Van Vleck admits he had no interest in philosophy at all;[14] and Slater felt that the philosophy at the Como and Solvay Congresses was just unnecessary fuss, "an entirely European phenomenon."[15] Both van Vleck and Slater, though, are accused by I. I. Rabi of being "too slide rule"; yet even he thought all the philosophical problems about quantum mechanics had been solved by Max Born's statistical interpretation. Rabi attended the 1927 Solvay Congress, but about the need for philosophy he thought that it was "all over by then," and certainly that the elaborations that Bohr was giving were unnecessary.[16]

In 1928 Kennard also maintained that it was all over: "The foundations of quantum mechanics can fairly be regarded as completed since the publication of Heisenberg's [uncertainty principle] paper last July."[17] D. M. Dennison is another physicist who was in Europe during the early development of the Copenhagen interpretation. He attended Heisenberg's colloquia and found Heisenberg's uncertainty principle beautifully unifying. Still, he did not take part in any of the philosophical discussions.[18] Like Rabi, the physical chemist H. C. Urey also got very little from Bohr and did not see the point of complementarity.[19]

Carl Eckart was different. He helped translate Heisenberg's Chicago lectures, which made him realize that despite his hopes, the problems with which the Copenhagen interpretation dealt would not go away; throughout his life he returned again and again to his philosophical concerns over the quantum theory. He was eventually much influenced by the logical positivist Rudolf Carnap, and his concerns over quantum mechanics even led him to the study of symbolic logic.[20] Eckart seems to be one clear exception among young American physicists.

As one further evidence of this one can compare the correspondence at the time of a number of Americans, like Kemble,[21] van Vleck,[22] Dennison,[23] or Oppenheimer,[24] with the well-known Europeans of the time. The correspondence of Einstein, Bohr, Born, Pauli, and a good many others is full of questions, views, and concerns about the broader meaning of the quantum theory. By contrast there is essentially no philosophy at all in the letters of these Americans.

In the context of the probabilistic revolution the indifference to Heisenberg's third problem—the purely statistical character of quantum laws—is particularly surprising. Max Born introduced the statistical interpretation of the wave function in 1926, in discussing scattering: $|\phi(x)|^2 dx$ gives the probability for a particle to be scattered into the small region dx, around x. In the United States the interpretation did not appear widespread in published papers until 1928, but then it was generally used without mention. Oppenheimer used it without remark. This is no surprise, though, for Oppenheimer was still in Europe at the time, and the year before had translated for *Nature* Pasqual Jordan's classic statement of the statistical interpretation. Kennard did defend it,[25] and van Vleck and Slater became almost official spokesmen for the interpretation.[26] What is amazing is that there is virtually no discussion of the change in the world picture that comes with the statistical interpretation—the change from a deterministic world to an indeterministic one. Indeterminism becomes codified in quantum mechanics almost without remark.[27]

For example, the two first American texts were published in 1929 and in 1930, both widely used for many years. The first is *Quantum Mechanics* by E. U. Condon and Philip Morse (New York: McGraw-Hill, 1929).[28] The second is A. E. Ruarck and H. C. Urey's *Atoms, Molecules, and Quanta* (New York: McGraw-Hill, 1930). Neither is an unphilosophical book (contrary to the claim of the English physicist C. D. Darwin,[29] who probably felt that way because he was struggling at the time with communications from Bohr defending the far more metaphysical Copenhagen interpretation.[30]) Condon and Morse begin with a long discussion of P. W. Bridgman's operationalism and the related views of Felix Klein; and later there is an entire explanatory section entitled "Wave Interference and the Uncertainty Principle." Ruarck and Urey also intersperse interpretive discussions, and present a very clear statement of the statistical interpretation.[31] Yet in neither book does either 'determinism' or 'indeterminism' appear in the index, and neither concept is discussed in the texts. The most we see is the pragmatic advice given by Ruarck and Urey: "The reader who feels disappointed that the information sought in solving a dynamical problem on the quantum theory is statistical and that the course of the individual system is not followed by our equations, should console himself with the thought that we seldom need any information other than that which is given by the quantum theory."[32]

The one clear exception in print is Bridgman, who wrote about indeterminism for *Harper's Magazine* in 1929. He is one of the few commentators who dwelt at all on the concept of wholeness, so central to the Copenhagen view. Bridgman was upset by the quantum theory, though the tone in *Harper's* is calm. The attitude he expressed is similar to that which Paul Forman attributes to Weimar physicists,[33] that the world has become unruly and untamable by the human intellect: "Our conviction that nature is understandable and subject to law arose from the narrowness of our horizons."[34] Bridgman is a surprising case, because he, like a large number of others, argued that his own operationalist philosophy eliminated the fundamental perplexities of the theory. Still, his anxieties do not seem to have been shared by his younger colleagues.

In his famous book *The Nature of the Physical World* (New York: Macmillan, 1928), A. S. Eddington linked the statistical nature of quantum laws with the long-standing philosophical problem of free will. He claimed that quantum indeterminism at last showed how free will was possible in a material universe. A. H. Compton became preoccupied by the issue; but the younger quantum phsicists, who actually worked with the theory, seem to have been especially indifferent. Not surprisingly, it does not enter into the professional physics journals; but apart from the efforts of Compton it is not a central topic in the popular science literature up to 1932. For example, though the statistical interpretation was explained in successive issues of *Scientific American*, the free will question arose only to be dismissed. There was some discussion in *Science*,[35] but the few discussions there were very sober, and did not involve any of the young quantum physicists.

This might suggest that these physicists were entirely unphilosophical. That was decidedly not the case. They had philosophical concerns and philosophical insights, but these were on a less grand scale than the problem of free will or the worries of the Copenhagen group. It helps to draw a distinction within the philosophy of physics, albeit a loose one, between foundational issues and issues that are far more speculative and abstract, such as 'Is the universe chancey?" or "Is objective knowledge of an independently existing material world possible?" Foundational issues, by contrast, tend to be local, and though they are not themselves either mathematical or experimental, often mathematical or experimental results are immediately relevant to them.

Americans were very concerned with foundational issues. Kennard was to my mind the most astute. He sorted out Heisenberg's confusion between measurement error in a single interaction and statistical error, and published a new proof of the uncertainty relations using only the definition of uncertainty as "mean square deviation."[36] In another article he argued, "As here stated [Heisenberg's] principle refers primarily to the *statistical* situation determined by the experimental conditions and by our knowledge of things. In Heisenberg's original paper great emphasis was laid upon the errors of a simultaneous observation of both quantities, but the 'error' of an observation has no meaning unless we can give a meaning to the 'true' value from which the observed value differs, and the definition of the true value must already involve the characteristic indetermination."[37] (Van Vleck noted a similar paradox involved in talking of averages while simultaneously denying the reality of the values averaged.[38]) Kennard also was concerned about the concrete connections between quantum and classical mechanics, and in the paper just quoted he gave a proof for a system of many particles of Ehrenfest's theorem that the mean position of a wave packet will follow the classical trajectory in a uniform field. Ruarck shared this concern; he also produced a generalization of Ehrenfest's theorem.[39] As one further example of a foundational study, Kennard noted that the selection of characteristic solutions of Schroedinger's equation was 'arbitrary' and post hoc, to agree with experiment. To remove this arbitrariness, he derived a number of the mathematical constraints on the solutions from 'the physical foundations of the theory'.[40]

Besides Kennard a number of other Americans did serious work on the uncertainty relations. These were seen as the philosophic core of the new theory; hence it was necessary to understand exactly what the mathematical relations were, and what they meant. This is the one important interpretive area where Americans made original contributions. Kennard, we have seen, rejected the view that the uncertainties came from uncontrollable effects of the measuring apparatus, and proved the Heisenberg principle as a relation between the statistical deviations in position and momentum. Condon pointed out that there could be individual states in which two noncommuting observables could both be well defined;[41] and H. P. Robertson gave the very first proof of the Heisenberg results for generalized coordinates.[42] Ruarck was puzzled by Heisenberg's claims that the uncertainty relations are due to the limits of accuracy in physical measurements, and proposed first one account and then another for how this would work in various thought experiments, both in the *Proceedings of the National Academy of Science* and in talks at the American Physical Society.[43] In fact Ruarck was quite philosophically active at the time; he gave three talks in a row at the American Physical Society on foundational issues—December 1927, February 1928, and April 1928.

Ruarck's third talk[44] was on yet another foundational issue that concerned Americans: Could a single system be in a superposition of different states, say, a periodic state (like an electron circling the nucleus) and an aperiodic state (free motion of an emitted electron)?[45] A final example is an experimental question, with obvious philosophical point: How long, Americans wanted to know, is a light quanta?[46]

Oppenheimer offers a contrast to these foundational interests. His attitude to the new quantum theory seems to have been entirely free of any kind of philosophical concern—a surprising difference from his later worries about quantum electrodynamics.[47] This is reflected in his publications at the time. He did not, even in passing, take up issues of philosophical relevance.[48] Instead he focused on pure physics questions like the Ramsauer effect or the quantum theory of two bodies. But Oppenheimer was clearly in a minority. On the whole, the young American physicists were untroubled by Heisenberg's deep problems about duality, chance, and visualizability; but they had a keen concern about more local, more concrete questions of what the theory meant and how it should be understood.

3   Operationalism

There is one well-known American philosophy that straightforwardly teaches that physics should not be concerned with Heisenberg's three problems. This is the operationalism of P. W. Bridgman, initially laid out in his *Logic of Modern Physics* (1922).[49] According to Bridgman a new concept may be introduced into physics only when a direct experimental procedure—an operation—is specified for determining whether the concept applies. The meaning of the concept is entirely given by the particular operation that introduces it.

The philosophical thesis is coupled with the universally accepted belief, following Heisenberg's uncertainty paper of 1927 and his illustration with the gamma ray microscope experiment, that wave and particle aspects can never be experimentally observed at the same time (nor can position and momentum, or any other pair of conjugate variables).[50] Together these two doctrines wipe out worries about unvisualizability and duality. To ask what happens when observations cannot be made is a meaningless question, and physics must not concern itself with meaningless questions.

Bridgman himself, in his one extended discussion of quantum mechanics in *Harper's Magazine*, said, "Position and momentum as expressions of properties which an electron can have are meaningless." [51] The Condon and Morse text agrees that this is the way to think about wave particle duality; in a section on Bridgman they claim that the limitations in nature on comeasurability give "the meaning of the wave-particle duality as it is coming to be understood." [52]

Unvisualizability is equally easy to deal with: It is meaningless to ask where the electron is in its orbit or how it makes the transition from one orbit to another. In his review of Bridgman's book, Oppenheimer quotes Heisenberg: "'Die Bahn ensteht erst dadurch, dass wir beobachten.' [The orbit comes into existence first through the fact that we observe it.] That, "Oppenheimer concludes," is the operational definition of the electronic orbit." [53]

The failure of causality is hardly a more complicated problem. Heisenberg had argued that in the assertion "if we know the present we can predict the future, it is not the deduction but the premise which is false." [54] Bridgman followed Heisenberg: "The precise reason that the law of cause and effect fails can be paradoxically stated; it is not that the future is not determined in terms of a complete description of the present, but that in the nature of things the present cannot be completely described." [55] But Bridgman, given his operationalist views, does not draw conclusions about reality from this, nor does he think it a meaningful topic for further discussion: "The failure of the law of cause and effect has been explained by a number of German physicists, who have emphasized the conclusion that we are thus driven to recognize that the universe is governed by pure chance; this conclusion does not, I believe, mean quite what appears on the surface, but in any event we need not trouble ourselves with further implications of this statement." [56]

This operationalist view was very popular in the United States. Bridgman was widely cited; and almost every one of the young quantum physicists stressed the importance of sticking to what is physically observable, and praised Heisenberg for basing his matrix mechanics on that insight.[57] Since these volumes on the probabilistic revolution are devoted specifically to the reception of statistical thinking, it is worth considering more examples of the operationalist philosophy applied to this issue. Linus Pauling in his Messinger lectures used the same line of argument as Bridgman.[58] So too did both Kemble[59] and van Vleck. Van Vleck is a good example to pursue, for recall that he was one of the two self-styled spokesmen for the statistical point of view. In his symposium article in the *Journal of the Franklin Institute*, van Vleck claimed, "There is one fundamental physical lemma which epitomizes the spirit of the whole statistical outlook." This is "Heisenberg's Indeterminacy Principle: It is impossible to specify or measure accurately and simultaneously both a coordinate q and its conjugate momentum p." It is "the experiment itself" that "spoils" the possibility.[60] Thus van Vleck followed Heisenberg. But only a little way. For Heisenberg after his 1927 paper was drawn deeper and deeper into the Copenhagen doctrines that reality is dual and that the subject and the object form an indivisible whole.[61] Like Bridgman, and unlike Heisenberg himself and the other members of the Copenhagen school, van Vleck drew no conclusions from this about the fundamental nature of reality, but instead stayed close to the operational facts: "... the future of a particle is only statistically determined. The reason for the ambiguity is not necessarily the breakdown of the 'causality principle' but rather to be imputed to the fact that the initial conditions are inevitably indeterminate because of Heisenberg's principle."[62]

Van Vleck was undoubtedly personally influenced by Bridgman. Indeed, Bridgman's influence on a large number of the young American quantum theoreticians was direct and explicit. Edwin Kemble, who wrote the first American PhD in quantum mechanics with Bridgman as his advisor,[63] and then worked as Bridgman's colleague most of his life, was devoted to Bridgman and to his philosophic ideals. Both van Vleck and Slater studied quantum mechanics with Kemble. Slater did his doctoral work with Bridgman and van Vleck with Kemble. Oppenheimer was an undergraduate student of both at Harvard, and corresponded with both Kemble and Bridgman during his studies in Europe. Oppenheimer also wrote the very favorable *Physical Review* review of Bridgman's *Logic of Modern Physics* in 1928. Although Condon was never at Harvard, the entire philosophical perspective of the Condon and Morse textbook is based explicitly on Bridgman's operationalism. Linus Pauling too reports that he accepted Bridgman's view: What did not lead to experimental results was not significant. Hence he never worried about the more 'penetrating' questions of interpretation.[64]

4   A Wider Philosophy Than Operationalism

Although operationalism eliminates the penetrating philosophical questions of quantum mechanics, it does so coarsely and without sufficient care. The kinds of questions Bohr wanted to ask are dismissed as meaningless; hence they are debarred from physics. Their elimination depends on the Heisenberg doctrines about comeasurability; yet physics is not allowed to ask why there are limits in nature on comeasurability, or at least is not allowed to give the kinds of answers that appealed to the Copenhagen group. But without a deeper understanding of why comeasurability must fail, the indeterminacy relations can be taken only as generalizations by inductive enumeration, to be tried again and again in each new experimental setup. Condon and Morse say, of the uncertainty principle, "It is a generalization from physical experience and like all such is to be abandoned if a single exception to it can be found."[65] Ruarck and Urey, shortly after a reference to Bridgman and to Heisenberg's gamma ray experiment, similarly point out the empirical character of uncertainty: "Perhaps the reader will feel that if the dynamics of the collision were investigated in more detail we should be able to follow the value of [the momentum] as a function of the time. The point is that the consequences of such a calculation cannot be checked by any experiment which has been devised *up to the present time*."[66]

Operationalism itself, as a theory of meaningfulness, is a narrow doctrine that deals with quantum problems with little subtlety or insight. But it is only a small part of the story. For it is embedded at the time in a wider philosophy that denies to physics the possibility of the kind of deeper understanding that could give the desired insight. Different authors develop this view in different ways, and some do not develop it at all; hence any statement of it will necessarily be both general and vague. The underlying ideas, I think, can be summarized in two short theses: (1) The task of physics is to describe what it can as accurately as it can, at the same time striving for simplicity and economy of presentation. In particular, the task of physics is not to explain. (2) Physics should not postulate hypotheses, but should accept only what is experimentally verifiable.

The two doctrines complement each other. It is generally the nonexperimental hypotheses that provide 'deep' explanations; and when theories in physics change, it is usually the explanatory claims that get radically revised, and not those directly verified.[67]

These two theses contradict recently popular philosophical views. They presuppose both that not all inference in physics is hypotheticodeductive[68] and also that there is a genuine distinction between what is and what is not directly experimentally verifiable.[69] But it would be a mistake for the modern reader to conclude from this that the views held by American physicists in the 1920s are of merely historical interest and have nothing to contribute to current philosophical thinking; for the recent doctrines about theory ladenness and hypothetical inference are in no way unavoidable and are at this moment themselves under serious attack.[70]

Both of the theses that I claim underlie a good deal of the thinking of the American quantum physicists about the new theory have familiar near relatives. But there are illuminating differences. Although an empiricist philosophy, this is not the positivism of Mach or the later Vienna Circle, or of the British sense-data philosophers, for it does not confine itself to sensations as a source of knowledge. This lack of attention to sensations is one important way that the view differs from the American pragmatism of the time as well. The view is not genuinely operationalist in the narrow sense either; there is little tendency in practice to follow Bridgman seriously in his identification of the concept with the operation that verifies it.[71] Nor is the doctrine antirealist. Physics does not describe the *results* of experiments, but rather the objects and properties that can be studied by experiment. K. K. Darrow lists three characteristics of atoms—mass, charge, and magnetic moment—that had certainly been experimentally verified by 1926;[72] and most of the American quantum physicists, impressed by the work of Compton, felt exactly the same about photons.

The role of speculative properties is more problematic. They may be ruled out altogether by thesis (2). On the other hand, they may be necessary if one is to have a theory at all; and a theory may be necessary if one wants to make any kind of detailed predictions or to give simple models that duplicate features we know to be true of the phenomena. Darrow is explicit on this point: "One is tempted to ask:

May we not discard the speculative features of past atom-models, and rebuild a world out of these experimental atoms [i.e., atoms with only experimentally verified properties] alone? This is impossible; more properties than mass and magnetic moment and detachable charge must be attributed to atoms if we are to copy most phenomena by means of them." [73]

Here Darrow explicitly links theses (1) and (2). Physics describes what it can as best it can, and it very well may not be able to describe everything; for to do so simply generally requires the postulation of further features—features beyond the reach of experiment—to make the links efficiently. As we shall see in the last section, Darrow goes beyond this and finds a reason for the inability of physics to give complete descriptions in the conflict between the complexity of nature and the need for simplicity of presentation. Indeed, Darrow was detailed and explicit in the development of these views; the prolegomena to his 1926 text, *Introduction to Contemporary Physics*, is a fine piece of philosophy.

But even those who were not so detailed in their philosophical accounts gave clear indications that they expected physics descriptions as a matter of course to be radically incomplete. At the American Physical Society symposium on quantum mechanics, both Condon and the older British-educated physicist W. F. G. Swann took pains to lay out what the particular questions were that quantum mechanics could deal with. In 1926 Kemble characterized the new quantum physics as "fundamentally . . . a scheme for calculating the frequencies and intensities of spectral lines in which the details of the Bohr model of quantized orbits are deliberately omitted from consideration." After 1926 the scope of the theory enlarged considerably, but any ultimate limitation on scope should not have affected Kemble's views, for even in the original it was as good as one could want: "The new method of calculation gives answers in harmony with the experimental facts with a minimum of special hypotheses and that is about all that could ever be claimed for Maxwell's equations." [74]

Probably because the much extolled Bohr model, which promised to deal fairly well with spectral lines, was almost useless for chemical problems, the physical chemists were most clear on the necessity of limiting the domain of a theory. According to Linus Pauling the primary question in his mind about the quantum theory was, "Is quantum mechanics, wave mechanics, for example, sufficiently close to being correct so that if we solve the equation we'll get the right answers in relation to the properties of atoms and molecules?—Not even going beyond that to the nucleus, but just atoms and molecules." [75] Darrow's view was stronger. He thought that the questions that physics deals with are dictated by no natural boundaries, but rather are a happenstance collection selected "because of their practical utility . . . ; some because they are spectacular; some because they were unusual or seemed strange, or fascinated someone having the zeal and the means to study them, some because they are easy to produce and easy to measure." [76] Others are neglected for similar human reasons.

The same view was expressed by the older American chemist Irving Largmuir, who argued his philosophical views again and again throughout his published works. "I think," argued Langmuir in a 1929 address linking physics—including the new quantum physics—with chemistry, "in trying to estimate the reliability of any of our scientific knowledge we should keep in mind that the whole complexion of a science may be made to change by the psychology of the investigators which governs the choice of the subjects that are investigated." [77] Elsewhere he stated succinctly the limitation on completeness: "Any model is an abstraction formed for a definite purpose." [78]

5   The Wider Philosophy Applied to Heisenberg's Problems

I have argued that by and large the attitude of American quantum physicists to the deeper philosophic problems of the quantum theory was one of complacency. In particular, they accepted the statistical nature of the theory explicitly and enthusiastically, without reservation and without qualms. The same was true of wave-particle duality. The problem was recognized, and dismissed. This was an entirely rational attitude given the philosophic context described in the previous section. In science typically one uses first one model, then another, to provide the best descriptions possible of a variety of features that one wants to study. To expect a single explanatory account from which a consistent and complete set of descriptions will follow is to adopt an ideal both quixotic and misguided. That is not how science operates.

Kemble drew the connection between this philosophic view and his attitude to wave-particle duality clearly and unabashedly. He urged that we should adopt a 'dualistic view' about wave-particle duality: 'Light consists of *both* spreading waves and corpuscles'. He said that this may seem

... an evasion of the fundamental question, '*Why* does light act in some respects like an assemblage of corpuscles and in other respects like a spreading wave phenomenon?' We assert, however, that in the last analysis the function of theoretical physics is to describe rather than to explain. Science seeks to interpret the infinitely complex world of direct experience as the outcome of fundamentally simple laws. The reduction of complexity to simplicity is the goal, and when it is attained, we prove that order underlies chaos and leaves the question 'Why?' still essentially untouched. Hence, discarding the questions as ultimately unanswerable, we may address ourselves to the task of describing what we observed in the most compact manner possible. [79]

John Slater is an example of someone who did not spend many words on philosophy, but whose work reveals an attitude just like Kemble's. In "Light Quanta and Wave Mechanics" Slater discusses the Compton effect, where we see particlelike behavior:

If we know merely that the plane wave was being scattered by the atoms we should use a solution of the wave equation in which a plane wave struck the atoms, and spherically plane waves came off from all of them [i.e., here we would have just wave-like behavior]. ... But actually in the experiment our information is more precise than this ... we know when and with what direction it [the recoil electron] was sent off. We must set up our wave representing the scattered light making full use of the observations....[80]

We do not know exactly how to do so, said Slater, but when we have done so, we shall certainly get a particlelike representation for the light in this case, as we require. This illustrates, he claimed, "the universal property of statistics, that the probability of an occurrence depends conspicuously on what is assumed known to start with."[81]

But be wary of the sense of "to start with." To derive a particlelike representation for the light after collision, we must use specific information about the ejected electron not predictable beforehand. If we do not use this information, we derive a wavelike representation. Our derivation then can in no way mimic the causal situation that confronts the incoming photon: Will it behave like a wave or a particle going out? For at that point the photon cannot have the information that we use to produce the right representation. To adopt the solution that Slater proposes is, as with Kemble, to give up an explanatory ideal. The theory can describe the photon's particlelike behavior. But it cannot explain why it behaves as it does.

This attitude to wave-particle duality connects immediately with Heisenberg's third worry about the nonrepresentational aspects of quantum theory. The theory does not show what is happening. But the situation is worse than that; the theory does not seem to leave room for anything sensible to be happening. In Davisson–Germer-type diffraction experiments, the electrons behave just like particles when they interact with the recording screen. How did these particles behave at the diffraction grating? If they did pass through the grating as particles, they certainly moved in a very peculiar way, contrary to any of our views about particle trajectories and their causes.

This is neither surprising nor objectionable from the point of view described in section 4. Our real knowledge about trajectories and their causes is radically incomplete, and our impressions of what is reasonable or even possible are based on hypotheses from previous theories, theories that themselves answered only particular questions and described only particular phenomena. Now we have a new theory with new hypotheses, and some new claims to knowledge, well verified in experiment. The new theory treats a large variety of problems, some old and some new; and we are increasingly expanding its range of problems. Yet there are many kinds of problems it does not treat and many kinds of questions it does not answer. That is to be expected from the very nature of scientific theorizing.

Langmuir expressed this clearly. Physicists, he argued, build models: "This involves a process of replacing the natural world by a set of abstractions which we have become very skillful in choosing in such a way as to aid in classifying and understanding phenomena...."[82] In so doing, we are replacing reality by 'a simplified model, a human abstraction, which is so designed by us that it has some of the properties of the thing we wish to displace'. For Langmuir, the progress of physics depends largely upon two things: (i) precise, operational characterization of concepts; and (ii) the issue of matter here, "the development of models (mechanical or mathematical) which have properties analogous to those of the phenomena we have observed." [83]

Langmuir, we see, did not require that a theory give a mechanical model of the phenomena. It need not provide a picture of what is going on. A mathematical 'model' that expresses 'relations that correspond in some simple way to that which is observed between measurable physical quantities' will do. [84] In fact, he said, "One begins to believe that mathematical theory is a far better model of the atom than any of the mechanical models which are possible." [85] A picture of how the atom is constituted, where the electrons are located, for example, and how they move, is not necessary; an efficient mathematical formula may well summarize better a large number of facts, even if it inevitably misses others.

This is the same kind of attitude we have already seen many adopt in the discussions of wave-particle duality. Condon, for instance, in his *Science* review article "Recent Developments in Quantum Mechanics" remarked, "Instead of trying to go back of this duality, quantum mechanics is attempting to give a self-consistent mathematical development of quantum physics, which recognizes the duality as fundamental." [86]

The older W. F. G. Swann expressed a similar view about mathematics versus mechanical models in his contribution to the American Physical Society symposium of December 1928. [87] To illustrate the correct way to treat a quantum mechanical problem, he gave a detailed example: An electron beam strikes a target, producing x rays, which fall on a plate, in turn ejecting electrons, which are caught in a Faraday cylinder. The plate itself heats and some of the evaporated atoms condense in the x-ray rube. Quantum mechanics can predict what would happen say, at the plate, but the treatment that does so will not also describe what happens in between. To get the prediction at the plate, one must keep the mathematics throughout, following through the wave function with the Schroedinger equation. Experimental physicists will want to interpolate what is happening to the particles in between. That, Swann argued, would be a mistake. It is not the kind of information that the new theory can provide.

Swann, though, is not a pure case. He not only had the philosophical view that I described in the last section as a near relative of American pragmatism. He also urged a pragmatic attitude in the more common everyday sense of the word. Here, for contrast with the more genuinely philosophical views we have been considering, is advice from a lecture given when Swann retired as the chairman of the physics section of the AAA in December 1924. Critical insight, he maintained, should be confined: "A power of critical insight which will enable us to show that, in the last analysis, nothing is real and most things are meaningless is all very well in its way; but it will not always carry its owner very far, and may frequently lead him into pessimism. Perhaps the most helpful condition is a combination of critical insight with a none too delicate conscience when smelling out the truth." [88]

Let us return to Swann's electron beam experiment, and the more reflective philosophical ideals that bear on it. The quantum silence about what is going on except at selected points where observations are made and where calculations are directed as well as the descriptive incompleteness that is associated with the ad hoc piecing together of wave and particle pictures are both to be expected from thesis (1). But more, they are to be extolled, by thesis (2).[89] Often theories must fill in gaps with unverified details to produce solutions for the problems in their domain. Quantum theory has a marked advantage. For it proceeds, as Kemble noted, 'with a minimum of special hypotheses'.

In turn, the statistical nature of the theory's laws becomes a virtue as well, and not a point of crisis, as Europeans more commonly saw it. Recall that van Vleck endorsed and approved the fact that in quantum mechanics, "the future of a particle is only statistically determined." He attributed that to Heisenberg's uncertainty principle. Van Vleck continued,

> These considerations may be illustrated by the scattering of alpha particles. . . .
> [I]n quantum mechanics . . . because of Heisenberg's principle . . . we cannot say
> how close a given alpha particle comes to the scattering center. However, the
> quantum mechanics does furnish directly formulas for the "distribution in
> angle" of the scattered alpha particles, which is what is observed experimentally;
> whereas the classical theory furnishes such formulas only with the intermediary
> of the nonobservable distance of closest approach and with it the supplementary
> statistical assumption that all distances of approach are equally probable.[90]

Or consider Darrow. Wave mechanics, he urged, gives mere probabilities:

> It thus appears, at first sight, that wave mechanics has fallen short of rendering
> the service we might expect from it, since it does not describe the behavior of
> an individual atom. This feature of the theory, however, is not to be considered
> a fault, but rather an evidence of perfection. If wave mechanics is to furnish a
> faithful picture of nature, without excess and without defect, it should not make
> any predictions that cannot be checked by experiment. Accordingly, it should
> not deal with individual atoms, for no statement about the individual behavior
> of an atom can ever be subject to an experimental test.[91]

Heisenberg's third problem about the statistical nature of quantum mechanics is no more a genuine worry from the point of view of the operationalist-pragmatist oriented philosophy of section 4 than is duality or visualizability. We come thus to the promised conclusion: Americans in general had little anxiety about the metaphysical implications of the quantum theory; and their attitude was entirely rational given the operationalist–pragmatist-style philosophy that a good many of them shared.

6   A Philosophical Postscript

If the two theses of section 4 make it reasonable to dismiss a number of philosophical worries about quantum mechanics, what makes these theses themselves reasonable? The second is clearly an epistemological thesis that has its grounds in views about the intimate connection between empirical knowledge and experience. The first may have epistemological grounds as well. For just one example, consider

Otto Neurath, who argued views very similar to thesis (1) in the early Vienna Circle, at just about the same time I have been considering here. For Neurath different incompatible and mutually irreducible descriptions of the same spatiotemporal processes are inevitable given the variety of interests science serves. The reason lies in the nature of experience, which does not supply a single set of underlying concepts sufficient to treat all the features we want: "In science there are no 'depths'; there is surface everywhere: all experience comes from a complex network, which cannot always be surveyed and can often be grasped only in parts." [92]

Although this remark arises from epistemological considerations in Neurath, it points to a metaphysical ground as well. When the Copenhagen group looked to Nature to account for the incompleteness and incompatibility of the quantum descriptions, Nature came out looking very peculiar. But there is a far less dramatic account, suggested by Neurath's remark, for our inability to give simple unified theories: Nature itself is complex through and through. This is a view I am particularly interested in because I believe it.

Science wants simple and economical summaries of the numerous complicated facts in its domains. It would also like these summaries to include only true laws of nature, presumably the "underlying" laws responsible for the facts. But if Nature is complex through and through, the two goals will be in conflict, and unified theories that are both simple and true will be impossible.

I have argued for this view at length in *How the Laws of Physics Lie*. But it seems that it is not a new view at all—this same account was argued by some of the very physicists I have been discussing here. I do not know how widespread the view was; nor is there the kind of very detailed investigation of the exact connection with the dualities and incompletenesses of quantum theory that one would ideally want. Still, a number of authors explicitly attribute the incompleteness of physics in general, and the quantum problems in particular, to the deep-seated complexity of Nature. I close by citing three of these.

First, consider Swann: "In pondering over the reason for the simplicity of nature's laws, it is well to remember that simplicity is imparted largely because mankind, being unconsciously wise, has chosen to give names to those things that are simply related." [93] The second example is Langmuir, who repeatedly argued that in the construction of theories we begin with highly complex nature; then we discard the less important features and replace a number of others with abstractions. For example, in the kinetic theory of gases, "What we really do ... is to replace in our minds the actual gases we observe and which have many properties which we do not understand by a simplified model, a human abstraction, which is so designed by us that it has some of the properties of the things we want to displace." [94]

But as usual it was Darrow who was most philosophically complete and explicit: "Most physicists feel that the complicated phenomena of this world differ from the simple ones not in that they conform to different laws, but in that the same fundamental laws are in the latter case discernible and in the former cases untraceable.... To believe this is of course an act of faith." [95] It is clear from the rest of the text that Darrow takes this to be an unwarranted act of faith.

Notes

1. For a far more complete description of the Copenhagen interpretation, its development, and its spread, see J. L. Heilbron, "The Copenhagen Spirit in Quantum Physics and its Earliest Missionaries," manuscript (Office for the History of Science and Technology, University of California, Berkeley).

2. Werner Heisenberg, *The Quantum Theory* (Chicago: University of Chicago Press, 1930), p. 1.

3. Heilbron (see note 1) claims that the efforts of the Copenhagen group to spread the doctrine of complementarity and to apply it beyond physics were not very successful anywhere, and this included the United States. Even with respect to the interpretation as a way of solving conceptual problems in quantum mechanics, he argues as I do here, "American physicists ignored complementarity and came to grips with uncertainty without wincing" (p. 20).

4. Werner Heisenberg, "Atomic Physics and Pragmatism (1929)," *Physics and Beyond* (New York: Harper and Row, 1972), p. 94. Heisenberg's emphasis on simplicity in this article contrasts with views I shall describe in the last section of this paper.

5. E. U. Condon, review of *Einführung in die Wellenmechanik*, by F. Frenkel, *Physical Review*, 34 (1929), 1066.

6. Specifically, *Physical Review*, *Reviews of Modern Physics*, *Science*, *Proceedings of the National Academy of Science*, *Bulletin of the National Research Council*, *Journal of the Optical Society of America*, *Chemical Reviews*, and *Journal of the Franklin Institute*.

7. Such as *Scientific American*, *Scribner's Magazine*, and *Scientific Monthly*.

8. "The Strangest Things in Physics," *Scientific American*, 140 (1929), 498–500.

9. See the indexes to the *Physical Review*, 1925–1931.

10. J. H. van Vleck, "The Statistical Interpretation of Various Formulations of Quantum Mechanics," *Journal of the Franklin Institute*, 207 (1929), 475–498.

11. Cf. *Archives for the History of Quantum Physics* (inventoried in Thomas S. Kuhn, John L. Heilbron, Paul Forman, and Lani Allen, *Sources for the History of Quantum Physics*, Memoirs of the American Philosophical Society, Vol. 68 (Philadelphia: American Philosophical Society, 1967), housed in the History of Science Library, University of California, Berkeley), interview with Robert Oppenheimer; also *Letters and Recollections* (Cambridge, MA: Harvard University Press, 1981).

12. *Archives for the History of Quantum Physics*, Oppenheimer.

13. *Archives for the History of Quantum Physics*, interview with Linus Pauling, p. 13.

14. *Archives for the History of Quantum Physics*, interview with J. H. van Vleck, II.

15. *Archives for the History of Quantum Physics*, interview with John Slater, p. 22.

16. *Archives for the History of Quantum Physics*, interview with I. I. Rabi.

17. E. H. Kennard, "On the Quantum Mechanics of a System of Particles," *Physical Review*, 31 (1928), 876–890, p. 876.

18. *Archives for the History of Quantum Physics*, interview with D. M. Dennison, III.

19. *Archives for the History of Quantum Physics*, interview with H. C. Urey, II.

20. *Archives for the History of Quantum Physics*, interview with Carl Eckart.

21. *Sources for the History of Quantum Mechanics*, correspondence of E. Kemble.

22. *Sources for the History of Quantum Mechanics*, correspondence with J. van Vleck.

23. *Sources for the History of Quantum Mechanics*, correspondence with D. M. Dennison.

24. R. Oppenheimer, *Letters* (note 11).

25. Kennard, "System of Particles" (note 17).

26. Cf. their articles in the *Journal of the Franklin Institute*, 1929; also the van Vleck correspondence with Slater (November 1928, in the *Sources for the History of Quantum Physics*), where both agree to "hammer on" the subject in their American Physical Society symposium.

27. Where there are remarks, indeterminism is not seen as problematic. E. Kemble's "General Principles of Quantum Mechanics," p. 176 in the first issue of *Reviews of Modern Physics (1929; the first volume is called Physical Review Supplement)*, is typical.

28. This text is falsely described by Wesley Brittin and Hales Odabassi in an honorary volume for Condon [Brittin and Odabassi, *Topics in Modern Physics: A Tribute to Edward U. Condon* (London: Adam Hilger, 1971), p. viii] as 'the first English language book on quantum mechanics'—overlooking George Birtwhistle *The New Quantum Mechanics*, published by Cambridge University Press one year before, and H. T. Flint's *Wave Mechanics: Being One Aspect of the New Quantum Theory* (London: Methuen) of the same year.

29. C. D. Darwin, *Nature*, 125 (1930), 560.

30. J. H. Heilbron, "Copenhagen Spirit," p. 19 (note 1).

31. A. E. Ruarck and H. Urey, *Atoms* (note 31).

32. A. E. Ruarck and H. Urey, *Atoms*, p. 622 (note 31).

33. Paul Forman, "Weimar Culture, Causality, and Quantum Theory, 1918–1927," *Historical Studies in the Physical Sciences*, 3 (1971), 1–115.

34. P. W. Bridgman, "The New Vision of Science," *Harper's Magazine*, 158 (1929), 443–451.

35. Paul R. Heyl, "The Perspectives of Modern Physics," *Scientific American*, 145 (1931), 168–170; R. S. Lillie, "Physical Indeterminism and Vital Action," *Science*, 65 (1927), 139–145; Leigh Page, review of Eddington's *The Physical World*, *Science*, 69 (1929), 624–626.

36. E. H. Kennard, "Zur Quantenmechanik einfacher Bewegungstypen," *Zeitschrift für Physik*, 44 (1927), 326–358.

37. Kennard, "System of Particles," pp. 345–346 (note 17).

38. J. van Vleck, "Statistical Interpretation," p. 481 (note 10).

39. A. E. Ruarck, "The Zeeman Effect and Stark Effect of Hydrogen in Wave Mechanics," *Physical Review*, 31 (1928), 533–538.

40. E. H. Kennard, "The Conditions on Schroedinger's Ψ," *Nature*, 127 (1931), 892–893.

41. E. U. Condon, "Remarks on Uncertainly Principles," *Science*. 69 (1929), 573–574.

42. H. P. Robertson, "The Uncertainly Principle," *Physical Review*, 34 (1929), 163–164.

43. A. E. Ruarck, "The Limits of Accuracy in Physical Measurement," *Proceedings of the National Academy of Science*, 14 (1928), 322–328; "Uncertainty Relations in the Motion of Free Particles," *Physical Review*, 31 (1928), 311–312, 709.

44. A. E. Ruarck, "The Force Equation and Some Related Theorems in Wave Mechanics," *Physical Review*, 31 (1928), 1133A.

45. Cf. R. W. Gurney and E. U. Condon, " Quantum Mechanics and Radioactive Disintegration," *Physical Review*, 33 (1929), 127–141; R. Oppenheimer, "Three Notes on the Quantum Theory of Aperiodic Effects," *Physical Review*, 31 (1928), 66–81; A. Ruarck and H. Urey, *Atoms* (note 31); B. Cassen, "Shot Fluctuations in Wave Mechanics," *Physical Review*, 33 (1929), 270A.

46. G. Breit, "The Length of Light Quanta," *Nature*, 119 (1927), 280–281; E. O. Lawrence and J. W. Beams, "Light, What Is It?" *Scientific American*, 138 (1928), 301–304; G. E. M.

Jauncey and A. L. Hughes, "An Attempt to Detect Collisions Between Photons," *Physical Review*, 35 (1930), 1439A; Paul R. Heyl, "How Big Is a Quanta?" *Scientific American*, 144 (1931), 328–330.

47. Peter Galison, "The Discovery of the Muon and the Failed Revolution against Quantum Electrodynamics," *Centaurus*, 26 (1983), 262–316.

48. With the possible exception of the question of whether single systems are in superpositions, which plays a role in his discussion of periodicity in "Three Notes" (note 45).

49. P. W. Bridgman, *Logic of Modern Physics* (New York: Macmillan, 1922).

50. W. Heisenberg, "Über den anschaulichen Inhalt der quantentheoretischen Kinematik und Mechanik," *Zeitschrift für Physik*, 43 (1927), 172–198.

51. P. W. Bridgman, "New Vision," p. 341 (note 34).

52. E. U. Condon and P. Morse, *Quantum Mechanics*, p. 17.

53. R. Oppenheimer, review of P. W. Bridgman's *Logic of Modern Physics, Physical Review*, 31 (1928), 145–146.

54. Quoted in Fredrich Hund, *History of Quantum Theory* (New York: Barnes and Noble, 1974), p. 162.

55. P. W. Bridgman, "New Vision," pp. 342–343 (note 34).

56. P. W. Bridgman, "New Vision," p. 343 (note 34).

57. Besides those already cited, see E. Kemble, *General Principles*; E. Kennard, "System of Particles," p. 876; A Ruarck and H. Urey, *Atoms*, Chapter 18; I. Langmuir, "Modern Concepts in Physics and their Relation to Chemistry," *Science*, 70 (1929), 385–396; K. K. Darrow, "New Light on Old Problems," in his *Introduction to Contemporary Physics* (New York: van Nostrand, 1926); and J. van Vleck, "Statistical Interpretation," p. 477. (note 10). Van Vleck even thought Born was an operationalist (*Archives for the History of Quantum Physics*, interview II, note 11).

58. *Archives for the History of Quantum Physics*, interview with Linus Pauling, II (note 11).

59. E. Kemble, "General Principles," Section 9 (note 27).

60. J. van Vleck, "Statistical Interpretation," p. 476 (note 10).

61. J. Heilbron, "Copenhagen Spirit," sec. II, 1 (note 1).

62. J. van Vleck, "Statistical Interpretation," p. 491 (note 10).

63. K. Sopka, *Quantum Physics in America* (Salem, NY: Ayer, 1980).

64. *Archives for the History of Quantum Physics*, interview with Linus Pauling, II (note 11).

65. E. Condon and P. Morse, *Quantum Mechanics*, p. 21.

66. R. Ruarck and H. Urey, *Atoms*, p. 619, italics added (note 31).

67. Cf. I. Hacking. *Representing and Intervening* (Cambridge: Cambridge University Press, 1983).

68. Contrary, for example, to Bas van Fraassen in *The Scientific Image* (Oxford: Clarendon Press, 1980).

69. Contrary to the claims that all experiment is mortally 'theory laden'. Cf. Paul Feyerabend, "Problems of Empiricism," in R. Colodny, *Beyond the Edge of Certainty* (Englewood Cliffs, NJ: Prentice-Hall, 1965).

70. See I. Hacking, *Representing and Intervening*, or Dudley Shapere, *Reason and the Search for Knowledge* (Dordrecht: D. Reidel, 1984), for an attack on theory ladenness; and Clark Glymour, *Theory and Evidence* (Princeton: Princeton University Press, 1980), and Nancy

Cartwright, *How the Laws of Physics Lie* (Oxford: Oxford University Press, 1983), for alternatives to the method of hypothesis.

71. This distinction resembles that of Stanley Goldberg in "Being Operational vs. Operationism," manuscript (Hampshire College, Amherst, MA). The strong version, operationism, that Goldberg describes depends on Bridgman's theory of meaning and results in the consequence, for example, that each separate operation identifies a different concept. The weak version, being operational, is similar to thesis (2) here.

72. K. K. Darrow, *Contemporary Physics*, p. xxii (note 57).

73. K. K. Darrow, *Contemporary Physics*, p. xxiii (note 57).

74. E. Kemble, review of Max Born's *Problems of Atomic Dynamics, Physical Review*, 28 (1926), 423–424.

75. *Archives for the History of Quantum Physics*, interview with Linus Pauling, II-120.

76. K. K. Darrow, *Contemporary Physics*, p. xvi (note 57).

77. I. Langmuir, "Modern Concepts," p. 393 (note 57).

78. I. Langmuir, "Science as a Guide to Life," *General Electric Review*, 1934. Reprinted in Langmuir's *Collected Works*, Vol. 12, p. 297.

79. E. Kemble, "General Principles," p. 160 (note 27).

80. J. C. Slater, "Light Quanta and Wave Mechanics," *Physical Review*, 31 (1928), 895–899. p. 899.

81. J. C. Slater, "Light Quanta," p. 899 (note 80).

82. I. Langmuir, "Modern Concepts," p. 390 (note 57).

83. I. Langmuir, "Modern Concepts," p. 390 (note 57).

84. I. Langmuir, "Modern Concepts," p. 390 (note 57).

85. I. Langmuir, "Modern Concepts," p. 394 (note 57).

86. E. U. Condon, "Recent Developments in Quantum Mechanics," *Science*, 68 (1928), 193–195, p. 194.

87. W. F. G. Swann, "Wave Mechanics," *Journal of the Franklin Institute*, 207 (1929), 457–466.

88. W. F. G. Swann, "The Trend of Thought in Physics, II," *Science*, 61 (1925), 452–460, pp. 459–460.

89. See note 57.

90. J. van Vleck, "Statistical Interpretation," pp. 491–492 (note 10).

91. K. Darrow, *Contemporary Physics*, p. 319 (note 57).

92. Otto Neurath, "Empirical Sociology," *Empiricism and Sociology*, ed. Marie Neurath and Robert S. Cohen (Dordrecht: D. Reidel, 1973), p. 306.

93. W. F. G. Swann, "The Trend of Thought," p. 427 (note 88).

94. I. Langmuir, "Modern Concepts," p. 276 (note 57).

95. K. K. Darrow, *Contemporary Physics*, p. xix (note 57).

Name Index for Volumes 1 and 2

Page numbers in boldface refer to volume 2.

Subject Index for Volumes 1 and 2

Page numbers in boldface refer to volume 2.

causal, 121, 123, 126, 130, 134, **13**, **379**, **395–396**
of chance, 139, 140, 143, 147, **241**, **360**
of comparative judgment, **15**, **27–28**, **56**, **64–65**
correlation of, 76–77
deterministic, 64, 68, **14**, **49**
dynamical vs. statistical, 79–80
of entropy, 66, 81, 85n26, **320**
of error, 136, 171, 382, 383
hierarchy of, 76–79
of large numbers, 171, 174, 177–178, 181–182, 192, 196, 198, 342, 345, **153**, **159**, **160**, **393**, **394**, **399**, **402n6**
of mechanics, 413
of nature, 138, 140, 147, **415**
Newton's second, 8
nomothetic, **13**, **14**
numerical, **108**
plurality of, 78–79
probabilistic, 76, **410**, **415**
second, of thermodynamics, 80, 226, **313**, **318**, **337**, **385**
statistical, 95, 113, 351, 358, 359, 361, 365, 382–384, **125**, **415**, **418**, **420**, **430**
Law finding, **67**
Lawful order, 78
Lawlike relationships, uncovering, **172**
Lawlikeness, 124, 127
League of Nations, **180**, **194**
Least squares, 153–154n131, 268, 272–273, 280, **172**, **174–176**, **178**, **180**, **181**, **193**
method of, 192, 201, 202, 206–211, 268–269, 272–273, 277–279, 281
Leipzig, 123, 131, 133, 136, 137, 143, 379, 380, 389
Levellers, 27
Leyden jar, 12, 14, 20
Liberalism, 198–199, 377, 382, 383, 385, 389, 391, 395, 396, 398, 401, 407
Lie group, 223
Life, origin of, **357**, **359**, **360**, **363**, **364**, **368**
Life table, 327
Limit theorem, central, 192, 290
Limitation, 201
Linear operator models, **83**, **90**
Linnean Society of London, 35
Liouville's theorem, 229, **396**
Logic, 130–131, 281, **213**
inductive, **17**
Logic of facts, **211–216**, **218**
Logical positivism, 147
Lottery, 136, 140, 141, 159, 160, 179–181
Lübeck, 380

Macromolecules, biological, **355–357**, **360**, **365**
Magnitudes, internal scalable, **73**, **83**, **86–88**, **90**, **92**, **96**
Malthusianism, 343
Man, **278**, **282**
Mankind, **278**
Marginal approach in measuring economic variables, **153**, **156**, **159**, **165**
Marginal utility, 174–175, 179–180
Marriage, 329
Martingale, 167
Materialism, 119, 131, 133
Materialist, 121
Materialist explanation of evolution, **316**
Mathematics, 124–125
Matter, 148n16
Mean(s), 139–142, 169, 170, 320–321, 329–330, 334, **208**
successive, **208**, **209**
Mean deviation, 142
"Mean man." *See* Average man; *Homme moyen*
Meaning of words, 8, 9, 18–21
Measure(s)
indirect, **75**, **76**
institutionalized, **24**
Measurement, 137, 215, 224, 261–262, 265–266, 274, **13–15**, **17**, **23**, **26**, **29**, **56**, **64**, **78**, **160**, **171**, **172**, **178**, **179**, **192**, **193**
statistical, **147**, **148**, **158**, **159**, **163**
test of fitness and reliability of, **161**, **162**
Measurement error, 262, 272, **60**, **173–175**, **178**, **179**, **187**, **188**, **192–194**, **419**, **421**
Measurement instrument, 262, 263, 271, 277, **17**, **26**, **161**, **163**
Measurement theory, **23**, **58**
Mechanical philosophy, 29
Mechanics, 69, 80
celestial, 136
classical (dynamics), 146, **238**, **382–384**, **386**, **388**, **395**, **396**, **398**, **400**, **401**
laws of, 413
Newtonian, 8–10, 12, 60–63
rational, **141**, **142**, **144**
statistical, 39, 40, 53, 78, 79, 82, **384**, **387**, **389**, **415**
Mechanistic philosophy, 60
Mechanization, 274, 277, **12**, **17**
Median, 137, 142
Medicine, 37, 39
Melanism, industrial, **342**, **343**
Memory strength, **92–94**, **97n8**
Mendelian Revolution, 32